NUMERICAL METHODS OF CURVE FITTING

NUMERICAL METHODS
OF
CURVE FITTING

BY

P. G. GUEST

University of Sydney
Australia

CAMBRIDGE
AT THE UNIVERSITY PRESS
1961

CAMBRIDGE UNIVERSITY PRESS
Cambridge, New York, Melbourne, Madrid, Cape Town,
Singapore, São Paulo, Delhi, Mexico City

Cambridge University Press
The Edinburgh Building, Cambridge CB2 8RU, UK

Published in the United States of America by Cambridge University Press, New York

www.cambridge.org
Information on this title: www.cambridge.org/9781107646957

First published 1961
First paperback edition 2013

A catalogue record for this publication is available from the British Library

ISBN 978-1-107-64695-7 Paperback

CONTENTS

PREFACE

The aim of this book is to provide an introduction to the methods of treating series of observations. The field covered embraces portions of both statistics and numerical analysis, and one might adopt the sub-title 'The Combination of Observations', in the sense used by Brunt many years ago, to describe the contents. The book is intended primarily for students and graduates in physics, and the types of observation discussed will be those most commonly met with in routine work in the physical sciences. It is hoped that the book will be useful as a reference work for statisticians and biologists, since much of the material presented here does not find a place in statistical textbooks but is only available in the original literature.

Part I (Chapters 1 to 4) deals with observations of a single variable ('curve' of zero degree). Much of this material will, of course, be familiar to statisticians, and Part I is certainly not intended as a substitute for a good text on statistics, but rather as a rapid summary of those portions of statistics used in the reduction of routine physical measurements. Subjects such as analysis of variance, factorial design, etc., are deliberately omitted. The fitting of straight lines is dealt with in Part II (Chapters 5 and 6). Some of the results derived in this part are special cases of general results for polynomials of arbitrary degree, but the fact that the majority of 'curves' fitted are straight lines warrants the treating of the linear case separately. Part III (Chapters 7 to 12) deals with the fitting of polynomial curves and of special types of curve. In the final chapter a number of typical examples are worked out in detail. These examples are intended to serve as a guide for those who want to fit a curve without going into the underlying theory. In a book of reasonable size it is not possible to treat all the topics relating to curve fitting, but it is hoped that the major topics have been covered and that other work can be located with the aid of the bibliography.

There has been a tendency in recent years for books on numerical analysis to omit numerical examples illustrating the applications of the methods. In the present work an attempt has been made to obtain a better balance between theory and practice. Each method is illustrated not only with an example but also with a full calculating scheme, so that the beginner can proceed along well-tried paths. However, it is also intended that the book should cover the theoretical aspects of curve fitting, and full derivations of all formulae are given. A knowledge of the calculus is presumed, and this background should enable most of the derivations to be followed. Some use is made of matrix notation in establishing a few of the more complicated results, but only the very simplest matrix operations are required.

The calculating schemes are designed primarily for a desk calculating machine, although most of the calculations in the first two parts can be done without the aid of a machine. In some cases a number of different schemes are given, for it is not wise to be dogmatic about which scheme is 'best'; the choice of the best scheme depends very much on the computing facilities and on the particular problem. In Section 12.1 there is a guide to assist in the choice of the calculating scheme most suited to any one of the commonly occurring types of problem. With high-speed automatic computers the routines for curve fitting will be very similar to those given for desk machines. Two routines for use with high-speed computers are discussed in Section 7.2. However, each computer has its own staff and manual, and the advice and instructions given by these should certainly be followed.

This book was written in England and Canada while the author was on sabbatical leave from the University of Sydney. The author wishes to express his thanks to the librarians of the various universities he visited for the facilities placed at his disposal. He also wishes to acknowledge his indebtedness to his wife Elizabeth for her patience and care in the typing of the manuscript.

The author is indebted to Professor Messel and the Nuclear Research Foundation within the University of Sydney for their generous financial support.

PART I

SINGLE VARIABLES

CHAPTER 1

GENERAL THEORY FOR A SINGLE VARIABLE

In this chapter an account is given of the theoretical concepts which are required for the treatment of observations of a single variable. The discussion is confined to those parts of the theory which can be developed without the assumption of a particular form for the frequency distribution of the observations. The practical methods of estimation based on a small number of observations are discussed, and illustrated by examples.

1.1 PROBABILITY AND FREQUENCY

If a large number N of observations η are made of a quantity, and the number of these observations which have the value y is N_y, the probability that a particular observation will yield the value y is defined as

$$\Pr\{\eta = y\} = \Pr\{y\} = \operatorname*{Lt}_{N\to\infty} N_y/N. \tag{1}$$

The probability of obtaining a value y in a single observation is then by definition proportional to the frequency of occurrence of the value y in a long series of observations. Since $\sum_y N_y = N$,

$$\sum_y \Pr\{y\} = 1. \tag{2}$$

For two variables ξ and η, the probability that a pair of observations yields the values x and y is written $\Pr\{xy\}$. This probability may be evaluated by considering a large number N of pairs of observations. The number of pairs for which $\xi = x$ will be denoted by N_x. Of these N_x, the number for which η has also the value y will be denoted by $N_{y|x}$. Then

$$\Pr\{xy\} = \operatorname*{Lt}_{N\to\infty} \frac{N_{y|x}}{N} = \operatorname*{Lt}_{N\to\infty} \frac{N_x}{N}\frac{N_{y|x}}{N_x},$$

and so

$$\Pr\{xy\} = \Pr\{x\}\Pr\{y\,|\,x\}, \tag{3}$$

where $\Pr\{y\,|\,x\}$ is the probability that $\eta = y$, given that $\xi = x$. Equation (3) is referred to as the product rule for probabilities.

It will often be true that the value obtained for η does not depend on the value x of ξ. If x and y are independent, so that

$$\Pr\{y\,|\,x\} = \Pr\{y\},$$

then the product rule takes the simpler form

$$\Pr\{xy\} = \Pr\{x\}\Pr\{y\}. \tag{4}$$

Since (3) can also be put in the form

$$\Pr\{xy\} = \Pr\{y\}\Pr\{x \mid y\},$$

it follows that

$$\Pr\{x \mid y\} = \Pr\{x\}\Pr\{y \mid x\}/\Pr\{y\}, \tag{5}$$

or, for variations of x with y fixed,

$$\Pr\{x \mid y\} \propto \Pr\{x\}\Pr\{y \mid x\}. \tag{6}$$

The proportional relation (6) is referred to as Bayes' Theorem. The three terms are called the posterior probability (i.e. the probability after y is fixed), the prior probability (before y is fixed), and the likelihood, respectively.

1.1.1 *Notation*

The symbol $f(y)$ will be used for the probability, $\Pr\{\eta = y\}$, that the observation or measurement will yield the value y. Usually the possible values y are not discrete, but form a continuous set. $f(y)$ is then defined so that $f(y)\,dy$ is the probability that the observation lies in a range dy centred at y. Thus $f(y)$ is referred to as the probability density function. It is also called the frequency function, since the probability of an observation lying in a given range is by definition proportional to the frequency of occurrence of values in that range in a long series of observations.

For a discrete distribution, from (1.1,2),

$$\sum_y f(y) = 1, \tag{1}$$

and for a continuous distribution the sum becomes the integral

$$\int f(y)\,dy = 1. \tag{2}$$

The probability integral or the distribution function is the integral of the frequency function. It is often denoted by $F(y)$, but here, following the usage of the *Biometrika Tables for Statisticians*, the symbol

$$P(y) = \int_{-\infty}^{y} f(u)\,du \tag{3}$$

will be employed. The probability integral $P(y)$ gives the probability that an observed value will be less than or equal to y. The differential of $P(y)$ is

$$dP(y) = P(y+dy) - P(y) = f(y)\,dy. \tag{4}$$

The probability that an observed value is greater than or equal to y will be denoted by $Q(y)$. Thus

$$Q(y) = 1 - P(y) = \int_y^\infty f(u)\,du. \tag{5}$$

If x and y are independent, then from (1.1,4) the probability that x and y lie simultaneously in the ranges dx and dy about x and y is

$$f_1(x)f_2(y)\,dx\,dy.$$

If this probability is written $f(x,y)\,dx\,dy$, where $f(x,y)$ is the combined frequency function, then, when x and y are independent,

$$f(x,y) = f_1(x)f_2(y). \tag{6}$$

1.2 EXPECTATION AND VARIANCE

If a very large number of measurements y of a quantity are made, the fraction of the observations lying in the range dy about y is identical with the probability that a single observation lies in that range, both being equal to $f(y)\,dy$. The average Y of the measurements as the number of observations approaches infinity is given by

$$Y = \int yf(y)\,dy,$$

and is often referred to as the expectation of y, written $E(y)$. Thus

$$E(y) = Y = \int yf(y)\,dy. \tag{1}$$

Y is also referred to as the population mean, the 'population' being simply the aggregate of all possible observations y.

If y and z are two variables, not necessarily independent, and their combined probability distribution is described by the frequency function $f(y,z)$, so that $f(y,z)\,dy\,dz$ is the probability that the observations lie simultaneously in the range dy about y and dz about z, then the expectation of the sum of y and z is

$$\begin{aligned} E(y+z) &= \iint (y+z)f(y,z)\,dy\,dz \\ &= \int y \int f(y,z)\,dz\,dy + \int z \int f(y,z)\,dy\,dz. \end{aligned}$$

Now $f(y,z)\,dz$ gives the probability density for y when the second variable lies in the range dz about z, and so the integral of this quantity over z must be $f(y)$. Hence

$$E(y+z) = E(y) + E(z), \tag{2}$$

and the expectation of a sum is the sum of the individual expectations.

For the product, if the two quantities are independent,

$$E(yz) = \iint yzf(y,z)\,dy\,dz = \iint yzf_1(y)f_2(z)\,dy\,dz,$$

and so
$$E(yz) = E(y)\,E(z). \tag{3}$$

The expectation of the product of independent variables is the product of their expectations.

The variance of a quantity is defined as the average of the squares of the deviations from the population mean Y. In symbols

$$\operatorname{var} y = E(y-Y)^2.$$

The square root of the variance is called the standard deviation or the standard error, and is denoted by σ. Thus

$$\operatorname{var} y = \sigma^2 = E(y-Y)^2 = \int (y-Y)^2 f(y)\,dy. \tag{4}$$

The variance is clearly a measure of the spread of the observations about the population mean.

Since $E(y) = Y$,

$$E(y-Y)^2 = E(y^2)-2YE(y)+Y^2 = E(y^2)-Y^2, \tag{5}$$

a result often useful in calculations.

The variance of the sum of two quantities is

$$E(y+z-Y-Z)^2 = E(y-Y)^2+E(z-Z)^2+2E(y-Y)(z-Z).$$

The last term is the quantity defined as the covariance of y and z,

$$\operatorname{cov}(y,z) = E(y-Y)(z-Z). \tag{6}$$

Then
$$\operatorname{var}(y+z) = \operatorname{var} y+\operatorname{var} z+2\operatorname{cov}(y,z). \tag{7}$$

If y and z are independent, it follows from (3) that

$$E(y-Y)(z-Z) = \{E(y-Y)\}\{E(z-Z)\} = 0.$$

Thus when y and z are independent

$$\operatorname{cov}(y,z) = 0, \tag{8}$$

$$\operatorname{var}(y+z) = \operatorname{var} y+\operatorname{var} z. \tag{9}$$

The variance of the product yz can be obtained by expanding

$$E(yz-YZ)^2 = E\{Z(y-Y)+Y(z-Z)+(y-Y)(z-Z)\}^2.$$

If the quantities y and z are independent, then, using (3),

$$E(yz-YZ)^2 = Z^2E(y-Y)^2+Y^2E(z-Z)^2+E(y-Y)^2(z-Z)^2,$$

the other terms vanishing since $E(y-Y)=0=E(z-Z)$. Thus

$$\operatorname{var} yz = Z^2 \operatorname{var} y + Y^2 \operatorname{var} z + (\operatorname{var} y)(\operatorname{var} z). \tag{10}$$

Usually the last term is much smaller than the others, and the approximation

$$\operatorname{var} yz = Z^2 \operatorname{var} y + Y^2 \operatorname{var} z \tag{11}$$

may be used.

If $\phi(y, z, \ldots)$ is an arbitrary function, then

$$\phi(y,z,\ldots) = \phi(Y,Z,\ldots)+\frac{\partial\phi}{\partial Y}(y-Y)+\frac{\partial\phi}{\partial Z}(z-Z)+\ldots. \tag{12}$$

If the deviations of the observations y from the population means Y are small, the higher order terms in the expansion (12) may be neglected, and so, by (2),

$$E\{\phi(y,z,\ldots)\} = \phi(Y,Z,\ldots). \tag{13}$$

If the variables $y, z, \ldots$ are independent,

$$E\{\phi(y,z,\ldots)-\phi(Y,Z,\ldots)\}^2 = \left(\frac{\partial\phi}{\partial Y}\right)^2 E(y-Y)^2 + \left(\frac{\partial\phi}{\partial Z}\right)^2 E(z-Z)^2+\ldots,$$

or

$$\operatorname{var}\phi(y,z,\ldots) = \left(\frac{\partial\phi}{\partial Y}\right)^2 \operatorname{var} y + \left(\frac{\partial\phi}{\partial Z}\right)^2 \operatorname{var} z+\ldots. \tag{14a}$$

When $\phi(y, z, \ldots)$ is a product of powers $Cy^a z^b \ldots$, then from (14a)

$$\frac{\operatorname{var}\phi}{\phi^2} = \frac{a^2 \operatorname{var} y}{Y^2}+\frac{b^2 \operatorname{var} z}{Z^2}+\ldots, \tag{14b}$$

or

$$\left(\frac{\text{S.D.}\,\phi}{\phi}\right)^2 = \left(\frac{a\,\text{S.D.}\,y}{Y}\right)^2 + \left(\frac{b\,\text{S.D.}\,z}{Z}\right)^2+\ldots. \tag{14c}$$

1.3 TYPES OF OBSERVED QUANTITY

The quantities observed in practical cases would appear to fall into one of two classes. Firstly, there are those quantities which are constant in magnitude, and which will be referred to as 'controlled' quantities. Many physical quantities are of this type—for example, the mass of a body, the velocity of light, etc. Secondly, there are those quantities which are inherently variable, the value of the various members of the population being distributed according to some frequency function $f(y)$.

Such quantities will be referred to as 'uncontrolled' quantities. Many of the quantities occurring in the biological sciences are of this class. Typical examples are the heights of men, the milk yield of cows, etc.

For controlled quantities, the observed value, population mean and true value all coincide, and an error-free observation will yield the true value Y of the quantity as determined by the experiment. There may, of course, still be unallowed-for systematic errors which cause the value Y to differ from that given by other experiments. For uncontrolled quantities an observation will yield a value y whose expectation is the population mean Y.

Either type of quantity may also be subject to experimental errors of observation. These errors are regarded as random quantities, equally likely to be positive or negative for any particular observation, which are produced by slight transient and unaccounted changes in the experimental conditions and apparatus. Thus if y' is the error-free or corrected value corresponding to an observed value y, the error is

$$\delta = y - y', \tag{1}$$

and the assumption of randomness leads to

$$E(\delta) = 0, \quad E(y) = y'. \tag{2}$$

For controlled variables $y' = Y$, and

$$E(y) = E(y') = Y, \tag{3}$$

so the population mean is the true value, while

$$\operatorname{var} y = \operatorname{var} y' + \operatorname{var} \delta = \operatorname{var} \delta \tag{4}$$

and the variance is a measure of the experimental error δ.

For uncontrolled quantities,

$$E(y) = E(y') = Y = Y', \tag{5}$$

and the population mean is unaffected by observational errors. Also

$$\operatorname{var} y = \operatorname{var} y' + \operatorname{var} \delta, \tag{6}$$

and the variance of the observed quantities y is greater than that of the error-free quantities y'.

1.4 ESTIMATION

In most cases the frequency function $f(y)$ is not known in detail, and a very large number of observations would be required to determine it. But the number of observations available, often referred to as the sample, is usually comparatively small. Hence some method of estimating the true value or population mean is required when the number n of observations y_i is small.

An estimate $\hat{Y}$ is said to be unbiased if its expectation is the true value Y of the quantity. Now any linear function

$$\sum_{i=1}^{n} \lambda_i y_i$$

will provide an unbiased estimate of Y, since

$$E\left\{\sum_{i=1}^{n} \lambda_i y_i\right\} = \sum_{i=1}^{n} \lambda_i E(y_i) = Y \sum_{1}^{n} \lambda_i,$$

and so

$$E\{\Sigma\lambda_i y_i / \Sigma\lambda_i\} = Y. \tag{1}$$

Which estimate will be the best depends on what is adopted as the criterion of 'best'. The most common criterion is a least-squares one, in which the estimate $\hat{Y}$ is chosen so that

$$\Sigma v_i^2 = \Sigma(y_i - \hat{Y})^2 \tag{2}$$

should be a minimum. Then differentiation leads to

$$\Sigma(y_i - \hat{Y}) = 0,$$

or

$$\hat{Y} = \Sigma y_i / n = \bar{y}, \tag{3}$$

where $\bar{y}$ is the sample mean. Thus the estimate obtained from the least-squares postulate is just the sample mean. This is the estimate almost always adopted. A discussion of other postulates is given in § 1.6.

The standard deviation σ can be estimated in terms of the deviations from the mean, usually called the residuals. For the ith residual

$$v_i = y_i - \bar{y}, \tag{4}$$

$$\begin{aligned} E(v_i^2) &= E(y_i - n^{-1}\Sigma y_j)^2 \\ &= E\{(y_i - Y) - n^{-1}\Sigma(y_j - Y)\}^2 \\ &= E\{(n-1)\,n^{-1}(y_i - Y) - n^{-1}\sum_{j \neq i}(y_j - Y)\}^2. \end{aligned}$$

If the observations are independent, $E(y_i - Y)(y_j - Y) = 0$, and

$$E(v_i)^2 = (n-1)^2 n^{-2} \operatorname{var} y + n^{-2}(n-1) \operatorname{var} y,$$

or

$$E(v_i^2) = \frac{n-1}{n} \operatorname{var} y = \frac{n-1}{n} \sigma^2. \tag{5}$$

Hence the quantity s^2 defined by the equation

$$s^2 = \Sigma v_i^2 / (n-1) \tag{6}$$

will provide an unbiased estimate of the variance σ^2. Of course, s will also provide an estimate of the standard deviation σ. It is interesting to note that s is not an unbiased estimate of σ, though the bias is almost always negligible. This point is discussed in § 2.5.3.

Σv_i^2 may be calculated by evaluating the individual residuals. It may also be calculated from the formula

$$\Sigma v_i^2 = \Sigma(y_i - \bar{y})^2 = \Sigma y_i^2 - n\bar{y}^2. \tag{7}$$

In estimating the standard deviations of combinations of observed quantities, it is usually necessary to replace in (1.2,9), (1.2,11), and (1.2,14*a*), the population means and variances by their estimates $\bar{y}$ and s^2 respectively.

1.4.1 *The arithmetic mean*

The mean has been adopted as the best estimate of the population mean or true value Y. Since it is a linear function of the observations,

$$\text{var } \bar{y} = n^{-1} \text{var } y. \tag{1}$$

Thus, using (1.4,6),

$$s^2(\bar{y}) = \Sigma v_i^2/n(n-1) \tag{2}$$

will provide an unbiased estimate of the variance of the mean $\bar{y}$.

It is perhaps worth while to emphasize the distinction between $\sigma(y)$, the standard deviation of an observation, and $\sigma(\bar{y})$, the standard deviation of the mean. The standard deviation of an observation is a measure of the spread of the individual observations about the true value or population mean, and its magnitude does not decrease as the number of observations is increased. The standard deviation of the mean does decrease as n is increased, being in fact proportional to $n^{-\frac{1}{2}}$. The larger the number of observations, the less the expected deviation of $\bar{y}$ from the true value or population mean.

Hence, for a controlled variable subject to error, increasing the number of observations will increase the accuracy of the estimate. But there is usually a practical limit to the number of observations that it is profitable to make, since there will almost certainly be undetected systematic errors which are not reduced by increasing n.

1.4.2 *Example*

In this example the refractive index μ of a glass prism will be calculated from measurements of the angle A of the prism, and the angle of minimum deviation θ for a ray of light passing through the prism. In any text-book on optics it is shown that

$$\mu = \frac{\sin \frac{1}{2}(A+\theta)}{\sin \frac{1}{2}A}. \tag{1}$$

The first step, the calculation of the mean values of A and θ from the observed values, is set out in Table 1.4.2. The residuals v_i and the sum Σv_i^2 have been calculated directly. As a check on the arithmetic, Σv_i^2 is also calculated from (1.4,7). The estimates of A and θ are

$$A = 60^\circ\ 27{\cdot}0' \pm 1{\cdot}3', \quad \theta = 43^\circ\ 28{\cdot}2' \pm 2{\cdot}3'. \tag{2}$$

Hence the estimated refractive index is

$$\mu = \frac{\sin 51^\circ\ 57{\cdot}6'}{\sin 30^\circ\ 13{\cdot}5'} = 1{\cdot}5645. \tag{3}$$

TABLE 1.4.2

Calculation of mean angles for Example (1.4.2)

A_i	v_i	v_i^2	θ_i	v_i	v_i^2
60° 27′	0	0	43° 29′	+0·8	1
60° 31′	+4	16	43° 16′	−12·2	149
60° 24′	−3	9	43° 18′	−10·2	104
60° 28′	+1	1	43° 35′	+6·8	46
60° 32′	+5	25	43° 28′	−0·2	0
60° 33′	+6	36	43° 34′	+5·8	34
60° 25′	−2	4	43° 30′	+1·8	3
60° 20′	−7	49	43° 25′	−3·2	10
60° 24′	−3	9	43° 27′	−1·2	1
60° 26′	−1	1	43° 40′	+11·8	139
Mean 60° 27′	Σv_i^2	= 150	Mean 43° 28·2′	Σv_i^2	= 487
Check $\Sigma A_i^2 - 10\bar{A}^2 = 7440\text{–}7290$			Check $\Sigma\theta_i^2 - 10\bar{\theta}^2 = 8440\text{–}7952{\cdot}4$		
$s(A_i) = \{\Sigma v_i^2/(n-1)\}^{\frac{1}{2}} = 4{\cdot}1'$			$s(\theta_i) = \{\Sigma v_i^2/(n-1)\}^{\frac{1}{2}} = 7{\cdot}4'$		
$s(\bar{A}) = n^{-\frac{1}{2}}\, s(A_i) = 1{\cdot}3'$			$s(\bar{\theta}) = n^{-\frac{1}{2}}\, s(\theta_i) = 2{\cdot}3'$		

Now

$$\frac{\partial\mu}{\partial A} = \frac{\cos\frac{1}{2}(A+\theta)}{2\sin\frac{1}{2}A} - \frac{\sin\frac{1}{2}(A+\theta)\cos\frac{1}{2}A}{2\sin^2\frac{1}{2}A} = \tfrac{1}{2}\mu[\cot\tfrac{1}{2}(A+\theta) - \cot\tfrac{1}{2}A], \tag{4a}$$

$$\frac{\partial\mu}{\partial\theta} = \tfrac{1}{2}\mu\cot\tfrac{1}{2}(A+\theta), \tag{4b}$$

and

$$\operatorname{var}\mu = \left(\frac{\partial\mu}{\partial A}\right)^2 \operatorname{var} A + \left(\frac{\partial\mu}{\partial\theta}\right)^2 \operatorname{var}\theta. \tag{5}$$

Hence substituting for the values μ, A, θ, $\operatorname{var}\mu$, $\operatorname{var}\theta$, the estimates (2) and (3),

$$\frac{\partial\mu}{\partial A} = -0{\cdot}467\mu, \qquad \frac{\partial\mu}{\partial\theta} = 0{\cdot}391\mu, \tag{6}$$

and

$$\operatorname{var}\mu = 1{\cdot}5645^2[1{\cdot}3^2\times 0{\cdot}467^2 + 2{\cdot}3^2\times 0{\cdot}391^2]\left[\frac{\pi}{180\times 60}\right]^2,$$

or

$$\text{S.D.}\,\mu = 1{\cdot}5645\,[1{\cdot}18]^{\frac{1}{2}}\left[\frac{\pi}{180\times 60}\right] = 0{\cdot}00049. \tag{7}$$

The last term in square brackets converts the standard deviation of the angle from minutes to radians, since in the formulae (4) and (6) for the differential coefficients the angles must be in radians.

The final estimate for the refractive index is then

$$\mu = 1{\cdot}5645 \pm 0{\cdot}0005. \tag{8}$$

1.5 OBSERVATIONS OF DIFFERENT WEIGHT

In many cases the observations y_i are not all of equal precision. In general the values y_i may belong to different populations whose means are all equal to Y, but whose standard deviations σ_i are different. Then, since σ_i^2 is a measure of the expected deviation from the true value for the observation y_i, the obvious generalization of the least-squares principle is to minimize

$$\Sigma(y_i - \hat{Y})^2/\sigma_i^2.$$

Clearly this is equivalent to minimizing

$$\Sigma w_i v_i^2 = \Sigma w_i(y_i - \hat{Y})^2, \tag{1}$$

where

$$w_i = \sigma^2/\sigma_i^2 \tag{2}$$

and σ is a constant of proportionality.

The quantities w_i are called the weights of the observations. The relative ratios of the w_i are fixed by the σ_i, but their magnitudes may be adjusted to any convenient values by altering the constant σ. From (2), for a given set of magnitudes w_i, σ is equal to the standard deviation of an observation of unit weight.

The least-squares estimate of Y is given by

$$\Sigma w_i(y_i - \hat{Y}) = 0,$$

or

$$\hat{Y} = \bar{y} = \Sigma w_i y_i/\Sigma w_i. \tag{3}$$

$\bar{y}$ is called the weighted mean.

The variance of the weighted mean is, from (3),

$$\operatorname{var} \bar{y} = \Sigma\left(\frac{w_i}{\Sigma w_i}\right)^2 \sigma_i^2.$$

Thus

$$\sigma^2(\bar{y}) = \sigma^2/\Sigma w_i, \tag{4}$$

or

$$\frac{1}{\sigma^2(\bar{y})} = \Sigma \frac{1}{\sigma^2(y_i)}. \tag{5}$$

In choosing the weights, estimated standard deviations $s(y_i)$ will usually have to be employed instead of the population parameters $\sigma(y_i)$. Then the estimated standard deviation of the mean will be of the form

$$\{s^2(\bar{y})\}^{-1} = \Sigma\{s^2(y_i)\}^{-1}, \qquad (6a)$$

or

$$s^2(\bar{y}) = \sigma^2/\Sigma w_i, \qquad (6b)$$

where

$$w_i = \sigma^2/s^2(y_i). \qquad (6c)$$

These formulae for the standard deviation of the mean are not quite accurate, since the weights w_i are experimental values and so would vary somewhat if the experiment were repeated. Taking this into account, Meier (1953) derived the formula

$$s^2(\bar{y}) = \frac{\sigma^2}{\Sigma w_i}\left\{1+\frac{4}{(\Sigma w_i)^2}\Sigma w_i(\Sigma w_j - w_i)/n_i'\right\}, \qquad (7a)$$

where

$$n_i' = n_i - 1 - \frac{4(n-2)}{n-1}, \qquad (7b)$$

n_i being the number of observations used in calculating the value $s(y_i)$. The correcting factor in brackets will be small if the n_i are reasonably large.

1.5.1 *Example*

Two separate determinations of the refractive index of a prism yield the values

$$\mu_1 = 1{\cdot}5645 \pm 0{\cdot}0005, \qquad \mu_2 = 1{\cdot}5637 \pm 0{\cdot}0009.$$

The weights will be taken as proportional to s_i^{-2}. If for convenience σ^2 is taken as $25\times 81/10^8$,

$$w_1 = \sigma^2/s_1^2 = 81, \qquad w_2 = \sigma^2/s_2^2 = 25,$$

and the weighted mean will be

$$\bar{\mu} = \frac{w_1\mu_1 + w_2\mu_2}{w_1+w_2} = \frac{81\times 1{\cdot}5645 + 25\times 1{\cdot}5637}{106} = 1{\cdot}5643.$$

From (1.5,6*a*),

$$\{s^2(\bar{\mu})\}^{-1} = 10^8\left\{\frac{1}{81}+\frac{1}{25}\right\} = \frac{106}{81\times 25}\times 10^8,$$

or

$$s(\bar{\mu}) = 0{\cdot}0004.$$

Alternatively, from (1.5,6*b*),

$$s(\bar{\mu}) = \sigma/\sqrt{\Sigma w_i} = 10^{-4}\sqrt{(25\times 81)}/\sqrt{(25+81)} = 0{\cdot}0004.$$

Hence the estimate of μ obtained by combining the two separate determinations is

$$\mu = 1{\cdot}5643 \pm 0{\cdot}0004.$$

If the numbers n_i of separate observations from which the values μ_1 and μ_2 were calculated are 10 and 11 respectively, the correcting factor in (1.5,7) is

$$1+(4/106^2)\,\{81(106-81)/9 + 25(106-25)/10\} = 1{\cdot}152,$$

and so the standard deviation of $\bar{\mu}$ should be increased by 7%.

1.5.2 *Deviations from the weighted mean*

The variances can be estimated from the deviations

$$v_i = y_i - \bar{y}$$

of the observations from the weighted mean. For

$$E(v_i^2) = E\left\{(y_i - Y) - \frac{\Sigma w_j(y_j - Y)}{\Sigma w_j}\right\}^2$$

$$= E\left\{\left(1 - \frac{w_i}{\Sigma w_j}\right)(y_i - Y) - \sum_{k \neq i} \frac{w_k}{\Sigma w_i}(y_k - Y)\right\}^2$$

$$= \left(1 - \frac{w_i}{\Sigma w_j}\right)^2 \sigma_i^2 + \sum_{k \neq i} \frac{w_k^2}{(\Sigma w_j)^2} \sigma_k^2$$

$$= \left(\frac{1}{w_i} - \frac{2}{\Sigma w_j} + \Sigma \frac{w_k}{(\Sigma w_j)^2}\right) \sigma^2,$$

and so $$E(v_i^2) = \left(\frac{1}{w_i} - \frac{1}{\Sigma w_j}\right) \sigma^2. \tag{1}$$

Hence $$E(\Sigma w_i v_i^2) = (n-1)\,\sigma^2,$$

and so $$s^2 = \Sigma w_i v_i^2/(n-1) \tag{2}$$

will provide an unbiased estimate of the variance of an observation of unit weight.

The estimated standard deviation of the weighted mean can then be obtained by inserting s^2 in (1.5,4). Thus

$$s^2(\bar{y}) = \Sigma w_i v_i^2/(n-1)\,\Sigma w_i \tag{3}$$

will provide an estimate $s(\bar{y})$ of the standard deviation of the weighted mean.

A special case which sometimes arises is that in which the y_i are themselves means of n_i individual observations y_{ki}, the observations in all the groups having the same standard deviation σ. Then

$$\sigma_i^2 = \operatorname{var} y_i = n_i^{-1} \operatorname{var} y_{ki} = n_i^{-1} \sigma^2. \tag{4}$$

Hence the y_i should be weighted as n_i, the number of separate observations in the group whose mean value is y_i. Thus

$$\bar{y} = \Sigma n_i y_i/\Sigma n_i \tag{5}$$

and $$s^2(\bar{y}) = \Sigma n_i(y_i - \bar{y})^2/(n-1)\,\Sigma n_i. \tag{6}$$

Formulae (5) and (6) may be of advantage when the number of original observations is very large, and it is desired to reduce the arithmetical calculations. The value

$$\bar{y} = \sum_i n_i y_i / \Sigma n_i = \sum_i \sum_k y_{ki} / \Sigma n_i$$

is the same as the arithmetic mean of the original observations. But if accurate estimates of σ are required, it is very much better to use the residuals of the original observations y_{ki} from $\bar{y}$, the formula being

$$s^2(\bar{y}) = \sum_{i,k} (y_{ki} - \bar{y})^2 / (\Sigma n_i - 1)\, \Sigma n_i. \tag{7}$$

1.5.3 *Choice of appropriate standard deviation formula*

A question of practical importance is which of the two formulae (1.5,6) and (1.5.2,3) for the standard deviation of the weighted mean should be used in any given case. It is, however, difficult to give a rule which covers all cases, and the choice is often a matter for individual judgment.

If the weights w_i and the standard deviations $s(y_i)$ are known fairly accurately, then the estimate $\sigma/\sqrt{\Sigma w_i}$ using (1.5,6) will usually be the more accurate one. In fact, if errors in the w_i are negligible, the ratio of the variances of the two estimates is $(\Sigma n_i - 1)/(n-1)$, as in § 1.6.3. But if it is suspected that the observations are discordant—that they do not all correspond to the same population mean or 'true' value Y, but possess systematic errors—then the formula $\sigma/\sqrt{\Sigma w_i}$ will greatly exaggerate the accuracy of $\bar{y}$ as an estimate of the physical quantity being measured. In such cases (1.5.2,3), which uses the residuals from $\bar{y}$, is much more satisfactory. It might be objected that it is not reasonable to combine discordant observations, but such a procedure is often unavoidable when the best value of a physical constant is required from the separate values obtained by different observers using different methods. It is recommended that $s(\bar{y})$ be evaluated by both methods if it is suspected that the y_i might be discordant.

A test for the homogeneity of different sets of observations—that is, for equality of both Y and σ for all sets—is given in § 3.6.2. If this test can be applied, and if it shows that the hypothesis of homogeneity is reasonable, the observations may all be pooled together to give a single set.

1.5.3.1 *Test for concordance.* It is shown in § 2.5.2 that, when the values y_i all have the same population mean or 'true' value Y, $\Sigma w_i (y_i - \bar{y})^2/\sigma^2$ is distributed 'as χ^2' with $n-1$ degrees of freedom.

Hence the value of this quantity will provide a test for the concordance of the observations. Details of tests using χ^2 are given in later chapters, but it is sufficient to state here that if $\surd\chi^2 - \surd(n-1)$ is greater than 2 the observations are most probably discordant, while if this difference is less than 1 there is no reason to suspect discordance.

When the weights w_i are merely estimates based on $s^2(y_i)$, the quantity

$$\Sigma w_i(y_i - \bar{y})^2/\sigma^2 = \Sigma(y_i - \bar{y})^2/s^2(y_i)$$

is still distributed approximately as χ^2 provided the numbers n_i of individual observations used in calculating $s^2(y_i)$ are not too small.

1.5.4 *Example*

Table 1.5.4 gives values of the ionization potential V for the hydrogen molecule obtained by different observers (Worthing and Geffner, 1943, p. 198). The errors specified are the so-called probable errors, $r_i = 0{\cdot}67s_i$ (§ 2.2.3). Taking σ as $1/0{\cdot}67 = 1{\cdot}33$, and the weights as equal to $1/r_i^2$, the values w_i are obtained. The weighted mean is

$$\bar{V} = \Sigma w_i V_i/\Sigma w_i = 22061/1429 = 15{\cdot}44.$$

If the possibility of systematic errors is ignored, (1.5,6*b*) gives

$$s^2(\bar{V}) = \sigma^2/\Sigma w_i = 1{\cdot}33^2/1429 = 0{\cdot}001238, \quad (1)$$

and $s(\bar{V})$ is 0·035.

TABLE 1.5.4

Ionization potential V for H_2 (Example 1.5.4)

V (volts)	w_i	v_i	$w_i v_i^2$	$\chi^2 = w_i v_i^2/\sigma^2$
16·5 ± 0·5	4	+ 1·06	4·5	2·5
17·1 ± 0·2	25	+ 1·66	68·9	39·0
15·6 ± 0·1	100	+ 0·16	2·6	1·5
15·4 ± 0·1	100	− 0·04	0·2	0·1
15·6 ± 0·1	100	+ 0·16	2·6	1·5
15·37 ± 0·03	1100	− 0·07	5·4	3·1
Sums	1429	—	84·2	47·7

The residuals v_i from the weighted mean are listed in the third column of Table 1.5.4. The value $\Sigma w_i v_i^2$ is 84·2, and so from (1.5.2,3)

$$s^2(\bar{V}) = \Sigma w_i v_i^2/(n-1)\Sigma w_i = 84{\cdot}2/5 \times 1429 = 0{\cdot}0118. \quad (2)$$

Since this is considerably greater than the value (1), systematic errors would be suspected. The χ^2 test for concordance gives

$$\surd\chi^2 - \surd(n-1) = \surd(\Sigma w_i v_i^2/\sigma^2) - \surd(n-1) = 6{\cdot}91 - 2{\cdot}24 = 4{\cdot}67$$

and so it seems certain that systematic errors are present.

The major contributor to χ^2 is the second observation, and it would seem desirable to perform the calculations with this value omitted. For the remaining five observations the calculations are shown in Table 1.5.4*a*. The weighted mean is

$$\bar{V} = \Sigma w_i v_i / \Sigma w_i = 21633/1404 = 15{\cdot}41.$$

From (1.5,6*b*),

$$s^2(\bar{V}) = \sigma^2/\Sigma w_i = 1{\cdot}33^2/1404 = 0{\cdot}001260,$$

and $s(\bar{V})$ is 0·035. The residuals from the weighted mean are given in Table 1.5.4*a*, and from (1.5.2,3)

$$s^2(\bar{V}) = \Sigma w_i v_i^2/(n-1)\Sigma w_i = 13{\cdot}73/4 \times 1404 = 0{\cdot}00244,$$

and $s(\bar{V})$ is 0·049. The χ^2 test for concordance gives

$$\sqrt{(\Sigma w_i v_i^2/\sigma^2)} - \sqrt{(n-1)} = 2{\cdot}79 - 2 = 0{\cdot}79.$$

Since this is less than unity, discordance is not indicated, and while the values are perhaps not completely concordant, the major systematic discrepancy has been removed by the omission of the second value in Table 1.5.4. The adopted value of V would then be $15{\cdot}41 \pm 0{\cdot}05$.

TABLE 1.5.4*a*

Calculation of V when the second value in Table 1.5.4 *is omitted*

V (volts)	w_i	v_i	$w_i v_i^2$	$\chi^2 = w_i v_i^2/\sigma^2$
16·5 ± 0·5	4	+1·09	4·75	2·69
15·6 ± 0·1	100	+0·19	3·61	2·04
15·4 ± 0·1	100	−0·01	0	0
15·6 ± 0·1	100	+0·19	3·61	2·04
15·37 ± 0·03	1100	−0·04	1·76	0·99
Sums	1404	—	13·73	7·76

1.5.5 *The combining of discordant observations*

There are three possible procedures in combining discordant observations, corresponding to the three following choices for w_i:

(*a*) $w_i \propto s_i^{-2}$;

(*b*) $w_i = 1$;

(*c*) $w_i \propto 1/(s_i^2 + s_i'^2)$, s_i' an estimate of the likely systematic error.

Ideally, the weights should be determined by the total (systematic plus random) errors of the observed values. Procedure (*a*) assumes that the random error s_i is also a measure of the systematic error s_i'. It is certainly often reasonable to assume that the scatter of the observations as given by $s^2(y_i)$ will be a measure of the accuracy of the method, and, by inference, of the likely systematic error. However, this will by no means be true in all cases, and the computor is clearly at liberty to vary the weights to correspond to what he believes to be the most likely systematic errors, as in procedure (*c*). His task is often made more difficult by uncertainty as to what the observer's final quoted error really represents. Some workers quote an error corresponding to $s(y_i)$, determined from the scatter of their observations, while others make a more or less arbitrary allowance for possible systematic effects in determining the final error. It is certainly the duty of the observer in his report to state clearly how he arrived at his quoted error.

Procedure (*b*) ignores the random errors, and adopts the simple arithmetic mean. This corresponds to the assumption of a single s' for all observed values such that $s' \gg s_i$.

1.5.5.1 *Example*

The three procedures described in Section 1.5.5 will now be applied to the six observations of Table 1.5.4. Procedure (*a*) has already been applied in § 1.5.4; it leads to the estimate $15{\cdot}44 \pm 0{\cdot}11$ volts. Procedure (*b*) gives

$$\bar{V} = \Sigma V_i/n = 15{\cdot}93, \quad \Sigma(V_i - \bar{V})^2 = 2{\cdot}5061,$$

$$s(\bar{V}) = \{\Sigma(V_i - \bar{V})^2/n(n-1)\}^{\frac{1}{2}} = 0{\cdot}289.$$

There are obviously many different forms of procedure (*c*). In one form, suggested by Cochran (1954*a*), s' is assumed the same for all observations, and it is calculated from the estimated total variance $\Sigma(y_i - \bar{y})^2/(n-1)$ by means of the equation

$$s'^2 + \Sigma s_i^2/n = \Sigma(y_i - \bar{y})^2/(n-1).$$

In the present example

$$s'^2 = 2{\cdot}5061/5 - 1{\cdot}33^2 \times 0{\cdot}3209/6 = 0{\cdot}4066,$$

and

$$w_i = \frac{1}{0{\cdot}4066 + 1{\cdot}33^2\, r_i^2}.$$

This leads to the following weights: 1·18, 2·09, 2·36, 2·36, 2·36, 2·45. Then $\bar{V}$ is 15·85, with an estimated standard deviation $1/\sqrt{\Sigma w_i} = 0{\cdot}280$. In this example the semi-weighted mean given by procedure (*c*) does not differ appreciably from the simple mean given by procedure (*b*). The assumption that the probable systematic errors are equal for each observed value is of course not very realistic in this case.

1.6 POSTULATES LEADING TO THE ARITHMETIC MEAN

It is shown in § 1.5 that the least-squares postulate leads to the weighted mean as the best estimate. It is possible to show that at least two other postulates lead to the same 'best' estimate.

1.6.1 *Minimum variance*

As shown in § 1.4, any linear estimate

$$\hat{Y} = \Sigma\lambda_i y_i/\Sigma\lambda_i \tag{1}$$

is an unbiased estimate of $\hat{Y}$. The variance of $\hat{Y}$ is

$$\text{var}\,\hat{Y} = \Sigma\frac{\lambda_i^2}{(\Sigma\lambda_i)^2}\sigma_i^2. \tag{2}$$

The minimum variance postulate states that the best estimate is that which leads to the least value of var $\hat{Y}$. Hence the λ_i are to satisfy

$$\frac{\partial}{\partial\lambda_j}\text{var}\,\hat{Y} = \frac{\partial}{\partial\lambda_j}(\Sigma\lambda_i^2\sigma_i^2)(\Sigma\lambda_i)^{-2}$$

$$= 2\lambda_j\sigma_j^2(\Sigma\lambda_i)^{-2} - 2(\Sigma\lambda_i^2\sigma_i^2)(\Sigma\lambda_i)^{-3} = 0,$$

or

$$\lambda_j\sigma_j^2 = (\Sigma\lambda_i^2\sigma_i^2)/\Sigma\lambda_i.$$

Since the right-hand side is independent of j,

$$\lambda_j\sigma_j^2 = \lambda_k\sigma_k^2,$$

or

$$\lambda_j = \sigma^2/\sigma_j^2, \tag{3}$$

where σ is a constant. Hence λ_j is identical with w_j defined by (1.5,2), and the minimum variance estimate (1) with λ_i given by (3) is just the weighted mean.

1.6.2 *Maximum likelihood*

The probability of obtaining a set of values y_i, given $f_i(y_i)$, is proportional to

$$L = \prod_{i=1}^{n} f_i(y_i|\,Y). \tag{1}$$

L is called the likelihood function. It will, of course, be a function of Y. The principle of maximum likelihood states that the best estimate $\hat{Y}$ of Y is that which when substituted in (1) makes L a maximum. That is, $\hat{Y}$ is the estimate for which the probability of obtaining the values y_i which were actually observed is a maximum. Then

$$\frac{\partial L}{\partial\hat{Y}} = 0, \quad \text{or} \quad \frac{\partial}{\partial\hat{Y}}\log L = 0. \tag{2}$$

Clearly little can be done unless the form of $f(y)$ is known. If it is assumed that the deviations follow the so-called normal law,

$$f(y|\,Y) \propto \exp\{-(y-Y)^2/2\sigma^2\}, \tag{3}$$

then (2) gives
$$\frac{\partial}{\partial \hat{Y}}\left\{\sum_i (y_i - \hat{Y})^2/2\sigma_i^2\right\} = 0,$$
or
$$\Sigma(y_i - \hat{Y})/\sigma_i^2 = 0.$$
Hence the maximum likelihood estimate is
$$\hat{Y} = \Sigma w_i y_i/\Sigma w_i,$$
where
$$w_i \propto \sigma_i^{-2},$$
and so is just the weighted mean.

Of course, if the function $f(y)$ is not of the normal type, the maximum likelihood estimate will not be equal to the weighted mean. The form of the deviation law would usually have to be assumed on somewhat meagre grounds; in most cases it will have some symmetrical form not very different from (3). So it is probable that the least-squares estimate and the maximum likelihood estimate will be very nearly equal. Since the former estimate is unbiased and readily calculated, it will almost always be adequate. It is interesting to observe that the least-squares estimate is identical with the minimum variance estimate, whatever the form of $f(y)$.

1.6.3 *Efficiency*

The efficiency of any estimate $\hat{Y}$ may be defined as the ratio of the variance of the least-squares estimate $\bar{y}$ to that of $\hat{Y}$. In symbols,
$$\eta(\hat{Y}) = \frac{\operatorname{var}\bar{y}}{\operatorname{var}\hat{Y}} = \frac{\sigma^2(\bar{y})}{\sigma^2(\hat{Y})}. \tag{1}$$
Similarly, the relative efficiency of two estimates may be defined as the ratio of their variances.

As an example, suppose that there are $n-1$ observations of unit weight and one of weight w. For the simple arithmetic mean $\hat{y} = \Sigma y_i/n$,
$$\operatorname{var}\hat{y} = n^{-2}\Sigma \operatorname{var} y_i = n^{-2}\sigma^2\{n-1+w^{-1}\}.$$
For the weighted mean $\bar{y}$,
$$\operatorname{var}\bar{y} = \sigma^2/(n-1+w).$$
Hence
$$\eta(\hat{y}) = n^2(n-1+w)^{-1}(n-1+w^{-1})^{-1}$$
$$\doteqdot 1 - \frac{w+w^{-1}-2}{n}. \tag{2}$$

In such a case the simple arithmetic mean would be quite efficient if w does not differ too greatly from unity and n is not too small.

Again, since the variance of the estimated standard deviation $s(\bar{y})$ can be shown to be proportional to $(n-1)^{-1}$ (§ 2.5.3), the relative efficiency of the estimates (1.5.2,6) and (1.5.2,7) is

$$\eta = (n-1)/(\Sigma n_i - 1),$$

which will usually be quite small.

1.7 MOMENTS AND CUMULANTS

1.7.1 *Characteristic function and cumulative function*

The characteristic function $\phi(t)$ corresponding to the frequency function $f(y)$ is defined by the equation

$$\phi(t) = \int_{-\infty}^{\infty} \mathrm{e}^{ity} f(y)\, dy, \tag{1}$$

i being the symbol indicating the imaginary part of a complex number. $\phi(t)$ is referred to as the Fourier transform of $f(y)$. Equation (1) may also be written

$$\phi(t) = E(\mathrm{e}^{ity}). \tag{2}$$

The moments μ'_r of the distribution are defined by the equations

$$\mu'_r = \int y^r f(y)\, dy = E(y^r). \tag{3}$$

In particular,
$$\mu'_1 = Y. \tag{4}$$

Then, from (2),
$$\phi(t) = E\Sigma(ity)^r/r! = \Sigma\mu'_r(it)^r/r!. \tag{5}$$

$\phi(t)$ is a moment-generating function, in the sense that the coefficient of $(it)^r/r!$ in the expansion of $\phi(t)$ as a power series in t is the rth moment μ'_r.

The moments about the mean are defined by the equation

$$\mu_r = \int (y-Y)^r f(y)\, dy = E(y-Y)^r. \tag{6}$$

Thus
$$\mu_1 = 0, \quad \mu_2 = \sigma^2. \tag{7}$$

The function
$$\psi(t) = \log \phi(t) \tag{8}$$

is called the cumulative function. If this function is expanded in powers of it, in the form

$$\psi(t) = \Sigma\kappa_r(it)^r/r!, \tag{9}$$

the coefficient κ_r is called the cumulant of order r.

The expansion of $\psi(t)$ can be accomplished by writing (8) as

$$\psi(t) = \log\left[1+\sum_1^\infty \mu_r'(it)^r/r!\right]$$

and using the logarithmic series. Thus

$$\psi(t) = \{\Sigma\mu_r'(it)^r/r!\} - \tfrac{1}{2}\{\Sigma\mu_r'(it)^r/r!\}^2 + \tfrac{1}{3}\{\Sigma\mu_r'(it)^r/r!\}^3 - \dots, \quad (10)$$

and the cumulants can be expressed in terms of the moments by comparing powers of it in (9) and (10). The first few expressions are

$$\left.\begin{aligned}&\kappa_1 = \mu_1' = Y, \quad \kappa_2 = \mu_2' - \mu_1'^2, \quad \kappa_3 = \mu_3' - 3\mu_2'\mu_1' + 2\mu_1'^3,\\ &\kappa_4 = \mu_4' - 4\mu_3'\mu_1' - 3\mu_2'^2 + 12\mu_2'\mu_1'^2 - 6\mu_1'^4.\end{aligned}\right\} \quad (11)$$

If the origin is chosen at the mean, the first moment is zero, and so, in terms of moments about the mean,

$$\left.\begin{aligned}\kappa_1 &= 0, \quad \kappa_2 = \mu_2 = \sigma^2,\\ \kappa_3 &= \mu_3, \quad \kappa_4 = \mu_4 - 3\mu_2^2.\end{aligned}\right\} \quad (12)$$

The cumulants are of considerable importance in theoretical discussions. Except for κ_1, they are unchanged in magnitude when the origin of y is changed. If

$$y' = y + \eta,$$

then for y the characteristic function is

$$\phi(t) = E(e^{ity}),$$

while for y' the characteristic function is

$$\phi'(t) = E(e^{ity'}) = E(e^{it(y+\eta)}) = e^{it\eta}\phi(t).$$

Hence $$\psi'(t) = \log\phi'(t) = it\eta + \psi(t),$$

and so κ_1 is changed by η while the other cumulants are unchanged. For this reason the cumulants are often referred to as semi-invariants.

1.7.2 *The inverse Fourier transform*

If $\phi(t)$ is the characteristic function satisfying (1.7.1,1), it can be shown that the corresponding frequency function is given by the inverse transform

$$f(y) = \frac{1}{2\pi}\int_{-\infty}^{\infty} e^{-ity}\phi(t)\,dt. \quad (1)$$

Hence if the characteristic function is known the frequency function can be determined.

The integral (1) can be most easily established as the limit of the familiar Fourier series

$$\phi(t) = \Sigma A(r\omega)\cos r\omega t + \Sigma B(r\omega)\sin r\omega t.$$

In terms of complex exponentials,

$$\cos\alpha = \tfrac{1}{2}(e^{i\alpha}+e^{-i\alpha}), \quad \sin\alpha = \frac{1}{2i}(e^{i\alpha}-e^{-i\alpha}),$$

and so

$$\phi(t) = \sum_{r=-\infty}^{\infty} C(r\omega)\, e^{ir\omega t}, \tag{2}$$

where the coefficients giving the amplitude and phase of the rth harmonic are in general complex. $C(r\omega)$ can be obtained in integral form by noting that

$$\int_{-\pi/\omega}^{\pi/\omega} e^{i(r-q)\omega t}\,dt = \int_{-\pi/\omega}^{\pi/\omega} \{\cos(r-q)\,\omega t + i\sin(r-q)\,\omega t\}\,dt$$

$$= \left.\begin{matrix} 0, & r \neq q; \\ 2\pi/\omega, & r = q. \end{matrix}\right\}$$

On multiplying each side of (2) by $e^{-iq\omega t}$ and integrating,

$$\frac{1}{\omega}C(q\omega) = \frac{1}{2\pi}\int_{-\pi/\omega}^{\pi/\omega} \phi(t)\, e^{-iq\omega t}\,dt. \tag{3}$$

The substitutions $q\omega = x, f(x) = C(x)/\omega$, are now made in (2) and (3). Then $\omega = \Delta x$, the interval between neighbouring values of x. Hence (2) is

$$\phi(t) = \sum_{x=-\infty}^{\infty} f(x)\, e^{ixt}\,\Delta x. \tag{4}$$

Thus when $\phi(t)$ satisfies (4), $f(x)$ is given by (3), which is

$$f(x) = \frac{1}{2\pi}\int_{-\pi/\Delta x}^{\pi/\Delta x} e^{-ixt}\,\phi(t)\,dt. \tag{5}$$

On letting $\Delta x \to 0$, in the limit if

$$\phi(t) = \int_{-\infty}^{\infty} f(x)\, e^{ixt}\,dx,$$

then $f(x)$ is given by

$$f(x) = \frac{1}{2\pi}\int_{-\infty}^{\infty} \phi(t)\, e^{-ixt}\,dt.$$

Thus the inverse relation (1) is proved.

1.7.3 *Linear sum of independent variables*

If y is a linear function of n independent variables y_j,

$$y = \Sigma \lambda_j y_j, \tag{1}$$

then the characteristic function for the variable y is

$$\phi(t) = E(\mathrm{e}^{ity}) = E \prod_j \mathrm{e}^{i\lambda_j y_j t}.$$

Since for independent quantities the expectation of a product is the product of the expectations,

$$\phi(t) = \prod_j \phi_j(\lambda_j t), \tag{2}$$

where $\phi_j(t)$ is the characteristic function for the variable y_j. The characteristic function of a sum is the product of the characteristic functions of the terms.

From (2),

$$\psi(t) = \sum_j \psi_j(\lambda_j t), \tag{3}$$

and the cumulative function of a sum is the sum of the cumulative functions of the terms. From the definition (1.7.1,9) of the cumulants,

$$\kappa_r = \sum_j \lambda_j^r \kappa_{j,r}, \tag{4}$$

where $\kappa_{j,r}$ is the rth cumulant of the variable y_j. In particular, if $y' = \lambda y$, the characteristic function for y' is

$$\phi'(t) = \phi(\lambda t), \tag{5}$$

and

$$\kappa_r' = \lambda^r \kappa_r. \tag{6}$$

1.7.4 *Central limit theorem*

If the variables in a linear sum of the form (1.7.3,1) are converted to standard measure by choosing the origins at the mean values and changing the scales so that $\mu_2 = \sigma^2 = \kappa_2$ is unity, then the sum becomes

$$y = \sum_{j=1}^{n} \lambda_j' y_j, \tag{1}$$

where

$$\lambda_j' = (\sigma_j/\sigma)\lambda_j. \tag{2}$$

Since the variables are now in standard measure, (1.7.3,4) gives for the second cumulant

$$1 = \sum_1^n \lambda_j'^2. \tag{3}$$

If no one of the λ_j' is very much greater than all the others, it follows that

$$\lambda_j' \sim n^{-\frac{1}{2}}. \tag{4}$$

For the third cumulant,

$$\kappa_3 = \sum_1^n \lambda_j'^3 \kappa_{3j} \sim n^{-\frac{1}{2}} \kappa_{3j},$$

and in general,
$$\kappa_k = \sum_1^n \lambda_j'^k \kappa_{kj} \sim n^{-\frac{1}{2}k+1} \kappa_{kj}. \tag{5}$$

If n is large the higher cumulants κ_k will be small, and so

$$\phi(t) \to e^{-\frac{1}{2}t^2}. \tag{6}$$

The frequency function whose characteristic function is $\exp -\frac{1}{2}t^2$ can be found from the integral (1.7.2,1),

$$f(y) = \frac{1}{2\pi} \int_{-\infty}^{\infty} e^{-ity} e^{-\frac{1}{2}t^2} dt = \frac{1}{2\pi} e^{-\frac{1}{2}y^2} \int_{-\infty}^{\infty} e^{-\frac{1}{2}(t+iy)^2} dt$$
$$= \frac{1}{2\pi} e^{-\frac{1}{2}y^2} \int_{-\infty+iy}^{\infty+iy} e^{-\frac{1}{2}t^2} dt.$$

From the theory of contour integration this integral is identical with

$$\int_{-\infty}^{\infty} e^{-\frac{1}{2}t^2} dt,$$

which is shown in § 2.1 to have the value $\sqrt{(2\pi)}$. Hence in standard measure the frequency function for a linear sum approximates to the form

$$f(y) = \frac{1}{\sqrt{(2\pi)}} e^{-\frac{1}{2}y^2} \tag{7}$$

which is known as the normal frequency function. The theorem that $f(y)$ approximates to the form (7) is referred to as the central limit theorem.

Clearly the larger the number n of variables y_j the more nearly will the frequency function approach the normal form. The approximation (7) is closer if the frequency functions $f_j(y_j)$ are symmetrical, since then κ_3 will vanish and the first additional term in $\psi(t)$ will be of the order of κ_4, or n^{-1}. The form (7) will be a very good approximation to the true frequency function if the original variables are themselves distributed approximately normally, for then the κ_{kj} in (5) will be small.

The discussion is based on the assumption that $\lambda_j' \sim n^{-\frac{1}{2}}$. For the simple mean $\Sigma y_j/n$,

$$\lambda_j' = (\sigma_j/\sigma)/n = n^{-\frac{1}{2}},$$

and λ_j' is exactly equal to $n^{-\frac{1}{2}}$. For all other cases the assumption will be fairly satisfactory provided no one term swamps the others. It is obvious that if any one term is very much larger than all the others the distribution will be similar to the distribution of that term.

1.8 NOTES AND REFERENCES

(1.1) The simple frequency definitions of probability given here are adequate for the purposes of this book. There are many different ways of developing probability theory, each with its own proponents. An interesting account of the difficulties associated with the various approaches is given by Kendall (1949).

(1.3) This classification of types of observed quantity is influenced by Berkson (1950).

(1.4) s^2 can be shown to be the best unbiased estimate of σ^2, in the sense that its variance is least; see Hsu (1938), Halmos (1946), and Nagler (1950).

(1.5) An elementary discussion of the combination of estimates from different experiments is given by Cochran (1954*a*).

(1.7) A fuller treatment of characteristic functions and cumulants is given by Kendall (1948, Chs 3 and 4).

The Gram–Charlier series can be used to expand general distributions in terms of the standardized normal variate X (Cornish and Fisher, 1937; Blom, 1954).

The use of Pearson's distributions is described in books by Elderton (1938) and Brunt (1917, Ch. IX).

CHAPTER 2

THE NORMAL DISTRIBUTION

The central limit theorem gives the normal frequency distribution a very prominent place in statistical theory, for in those cases where the deviations $Y-y$ can be regarded as due to the net effect of a number of small disturbances the frequency distribution will approach the normal form. Thus the error δ in an observation is regarded as being brought about by a number of small disturbances, and δ is usually assumed to follow a normal distribution. The statistical tests to be discussed in the remainder of Part I all assume that $f(y)$ is of the normal type. In this section the properties of the normal distribution will be discussed.

2.1 THE GAMMA FUNCTIONS

Certain properties of gamma functions are required in discussing the normal distribution. The gamma function is defined by the equation

$$\Gamma(n) = \int_0^\infty x^{n-1}\,\mathrm{e}^{-x}\,dx, \tag{1}$$

or, substituting y^2 for x,

$$\Gamma(n) = 2\int_0^\infty y^{2n-1}\,\mathrm{e}^{-y^2}\,dy. \tag{2}$$

If (1) is integrated by parts,

$$\Gamma(n) = [-\,\mathrm{e}^{-x}x^{n-1}]_0^\infty + \int_0^\infty (n-1)\,x^{n-2}\,\mathrm{e}^{-x}\,dx,$$

or

$$\Gamma(n) = (n-1)\,\Gamma(n-1). \tag{3}$$

In particular, if n is integral,

$$\Gamma(n) = (n-1)!, \tag{4}$$

since $\Gamma(1)$ is unity. (1) may be regarded as the generalization of $(n-1)!$ when n is non-integral.

The value of $\Gamma(\frac{1}{2})$ is often required. It can be found by considering the double integral

$$\int_{-\infty}^\infty \mathrm{e}^{-x^2}\int_{-\infty}^\infty \mathrm{e}^{-y^2}\,dy\,dx = \iint_{-\infty}^\infty \mathrm{e}^{-(x^2+y^2)}\,dx\,dy. \tag{5}$$

Changing to angular coordinates $x = r\cos\theta$, $y = r\sin\theta$, the right-hand side of (5) is

$$\int_0^\infty \int_0^{2\pi} e^{-r^2} r\, dr\, d\theta = 2\pi \int_0^\infty e^{-r^2} \tfrac{1}{2} dr^2 = \pi,$$

while the left-hand side of (5) is

$$2\int_0^\infty e^{-x^2}\left[2\int_0^\infty e^{-y^2}\,dy\right]dx = \{\Gamma(\tfrac{1}{2})\}^2.$$

Thus
$$\Gamma(\tfrac{1}{2}) = \sqrt{\pi}. \tag{6}$$

2.1.1 *The normalizing factor*

The normal distribution is, for a standardized variable with zero mean and unit variance,

$$f(X) = C\,e^{-\frac{1}{2}X^2},$$

the form given in (1.7.4,7). For a non-standardized variable y, $X = (y-Y)/\sigma$, and so

$$f(y) = C\,e^{-(y-Y)^2/2\sigma^2}. \tag{1}$$

The value of the normalizing factor C is found from the condition that $\int f(y)\,dy$ is unity. Since

$$\int_{-\infty}^{\infty} e^{-(y-Y)^2/2\sigma^2}\,dy = 2\int_0^\infty e^{-z^2}(2\sigma^2)^{\frac{1}{2}}\,dz = \sqrt{(2\sigma^2)}\,\Gamma(\tfrac{1}{2}),$$

the normal distribution is

$$f(y) = \frac{1}{\sigma\sqrt{(2\pi)}}\,e^{-(y-Y)^2/2\sigma^2}. \tag{2}$$

In the standardized form, $\sigma(X)$ is unity, and

$$f(X) = \frac{1}{\sqrt{(2\pi)}}\,e^{-X^2/2}. \tag{3}$$

2.1.2 *Moments*

The moments about the mean are

$$\mu_r = \int_{-\infty}^{\infty} (y-Y)^r f(y)\,dy,$$

and, setting $z^2 = (y-Y)^2/2\sigma^2$,

$$\mu_r = \frac{1}{\sigma\sqrt{(2\pi)}}\,(2^{\frac{1}{2}}\sigma)^{r+1}\int_{-\infty}^{\infty} z^r e^{-z^2}\,dz.$$

So

$$\mu_r = (\sigma\sqrt{2})^r\,\Gamma\{\tfrac{1}{2}(r+1)\}/\sqrt{\pi}, \quad r \text{ even};$$
$$= 0, \quad r \text{ odd}.$$

Now

$$\Gamma\{\tfrac{1}{2}(r+1)\} = \tfrac{1}{2}(r-1)\,\tfrac{1}{2}(r-3)\ldots\tfrac{1}{2}\Gamma(\tfrac{1}{2}) = 2^{-r/2}\sqrt{\pi}(r-1)\,(r-3)\ldots 1,$$

and so $$\mu_r = \{(r-1)\,(r-3)\ldots 1\}\,\sigma^r, \quad r \text{ even}; \qquad (1a)$$

$$= 0, \quad r \text{ odd}. \qquad (1b)$$

In particular, $\mu_2 = \sigma^2$ and $\mu_4 = 3\sigma^4 = 3\mu_2^2$.

The absolute moments when r is odd may be defined as

$$E\,|\,(y-Y)^r\,| = \int_{-\infty}^{\infty} |\,(y-Y)^r\,|\,f(y)\,dy$$

$$= \frac{1}{\sigma\sqrt{(2\pi)}}\,(2^{\frac{1}{2}}\,\sigma)^{r+1}\int_{-\infty}^{\infty} |\,z^r\,|\,\mathrm{e}^{-z^2}\,dz,$$

or $$E\,|\,(y-Y)^r\,| = (\sigma\sqrt{2})^r\,\Gamma\{\tfrac{1}{2}(r+1)\}/\sqrt{\pi}. \qquad (2)$$

The most important of these is the first absolute moment,

$$E\,|\,y-Y\,| = \sigma\sqrt{(2/\pi)}. \qquad (3)$$

2.1.3 *Cumulants*

The characteristic function is, if the origin is chosen at the mean,

$$\phi(t) = \frac{1}{\sigma\sqrt{(2\pi)}}\int_{-\infty}^{\infty} \mathrm{e}^{ity}\,\mathrm{e}^{-y^2/2\sigma^2}\,dy$$

$$= \frac{1}{\sigma\sqrt{(2\pi)}}\,\mathrm{e}^{-\frac{1}{2}\sigma^2 t^2}\int_{-\infty}^{\infty} \mathrm{e}^{-\frac{1}{2}\{(y/\sigma)-\sigma it\}^2}\,dy$$

$$= \frac{1}{\sigma\sqrt{(2\pi)}}\,\mathrm{e}^{-\frac{1}{2}\sigma^2 t^2}\,\sigma\sqrt{2}\int_{-\infty}^{\infty} \mathrm{e}^{-z^2}\,dz,$$

and hence $$\phi(t) = \mathrm{e}^{-\frac{1}{2}\sigma^2 t^2}. \qquad (1)$$

The cumulative function is

$$\psi(t) = -\tfrac{1}{2}\sigma^2\,t^2$$

and the cumulants beyond the second all vanish.

Conversely, if all the cumulants beyond the second vanish, the distribution is a normal one, as was shown in § 1.7.4.

2.1.4 *Sum of normally distributed variables*

If the variables y_j are normally distributed, $\kappa_{j,r}$ vanishes for r greater than 2. Hence, from (1.7.3,4), the cumulants of a linear sum

$$y = \Sigma\lambda_j\,y_j \qquad (1)$$

are

$$\kappa_1 = \Sigma\lambda_j\,\kappa_{1,j}, \quad \text{i.e.} \quad Y = \Sigma\lambda_j\,Y_j; \qquad (2)$$

$$\kappa_2 = \Sigma\lambda_j^2\,\kappa_{2,j}, \quad \text{i.e.} \quad \sigma^2 = \Sigma\lambda_j^2\,\sigma_j^2; \qquad (3)$$

$$\kappa_r = \Sigma\lambda_j^r\,\kappa_{r,j} = 0, \quad r > 2. \qquad (4)$$

Hence $\Sigma\lambda_j y_j$ is distributed normally with mean $\Sigma\lambda_j Y_j$ and variance $\Sigma\lambda_j^2 \operatorname{var} y_j$. This result is referred to as the reproductive property of the normal law.

2.2 TABLES RELATING TO THE NORMAL CURVE

The frequency curves for different normal variables differ only in the location of the mean and in the magnitude of the variance. If a change is made to the variable

$$X = (y - Y)/\sigma \tag{1}$$

the curves will all coincide. The standardized curve is

$$f(X) = \frac{1}{\sqrt{(2\pi)}} e^{-\frac{1}{2}X^2}. \tag{2}$$

$f(X)$ has been tabulated by many authors. Table 1 of the *Biometrika Tables* gives $Z(X) \equiv f(X)$ for the range 0(0·01)6·00. The curve of $f(X)$ is bell-shaped with a maximum of magnitude $1/\sqrt{(2\pi)} = 0{\cdot}3989$ at $X = 0$. The second derivatives vanish at $X = \pm 1$; these are points of inflexion, corresponding to the regions of steepest slope.

The function

$$P(X) = \frac{1}{\sqrt{(2\pi)}} \int_{-\infty}^{X} e^{-\frac{1}{2}x^2} dx \tag{3}$$

gives the probability that an observed value does not exceed X. This function is also given in Table 1 of the *Biometrika Tables*. The corresponding function

$$Q(X) = \frac{1}{\sqrt{(2\pi)}} \int_{X}^{\infty} e^{-\frac{1}{2}x^2} dx = 1 - P(X) \tag{4}$$

gives the probability that an observed value exceeds X.

Inverse tables giving X as a function of Q are often more useful in statistical tests. Table 4 of the *Biometrika Tables* lists X for Q in the range 0(0·001)0·5. Table 2.8*a* is a very short table of this type.

In words, the value $X = 1{\cdot}645$ when $Q = 0{\cdot}05$ implies that on the average the value of X exceeds 1·645 for 5% of the observations. Since $P(-X) = Q(X)$, these tables also give the probability of obtaining a value less than $-X$.

2.2.1 *Testing of observed values*

To test whether an observed value y is a reasonable one, on the hypothesis that the observations are distributed according to a normal law with mean Y and standard error σ, the quantity

$$X = (y - Y)/\sigma$$

is calculated. The probability of obtaining a value at least as great as this (or as small as this, if X is negative) can then be estimated from Table 2.8*a*.

It should be noted that this gives the probability of a single observation exceeding X. It does *not* give the probability of the largest of a set of n observations exceeding X. This latter probability is given by

$$Q_n(X) = nQ(X)\{P(X)\}^{n-1}. \tag{1}$$

For (1) is the product of the probabilities that $(n-1)$ observations are less than X and one observation exceeds X; the factor n takes into account that it may be any one of the n observations which exceeds X. The function $Q_n(X)$ is given in the right-hand half of Table 24 in the *Biometrika Tables*.

2.2.2 *Double-tail tests*

The tests described in the previous section may be called single-tail tests, in the sense that they give the probability of an observation lying in a particular 'tail' of the $f(X):X$ graph. One can also determine the probability of the observation lying in *either* of the two tails. That is, the probability of obtaining a deviation greater than $|y-Y|$, irrespective of the sign. Since the normal curve is symmetrical,

$$\Pr\{|X| \geqslant X_0\} = 2\Pr\{X \geqslant X_0\}.$$

Hence Table 2.8*a* may also be used for double-tail tests, if the levels Q are replaced by $2Q$.

2.2.3 *Probable error*

The value X for which

$$\Pr\{|y-Y|/\sigma \geqslant X\} = 0{\cdot}5$$

is $X = 0{\cdot}6745$. The quantity

$$\rho = 0{\cdot}6745\sigma \tag{1}$$

is called the probable error of an observation. The probability that the absolute deviation from the population mean $|y-Y|$ exceeds ρ in magnitude is one-half.

The probable error is widely used in physics to specify the spread or error of the observations. The term is practically unknown in the biological sciences and in statistics itself, where the standard deviation (also called standard error) is usually given. The standard error retains its significance if the deviations do not follow a normal law, while the probable error does not.

In the physical sciences at least it is very important to make it clear whether the error given is the standard error or the probable error. It should also be remembered in assessing the accuracy of an estimate that the probability of the error exceeding the standard error is about 1/3. To get something corresponding to a likely limit of error, the standard error should be multiplied by 2 or 3. From the tables

$$\Pr\{|y-Y|>2\sigma\} = 1/20; \quad \Pr\{|y-Y|>3\sigma\} = 1/400.$$

2.3 BIVARIATE NORMAL DISTRIBUTION

If two variables x and y, with origins at the population means, are each normally distributed when considered separately,

$$f(x) = \frac{1}{\sigma_x\sqrt{(2\pi)}}\exp -x^2/2\sigma_x^2, \quad f(y) = \frac{1}{\sigma_y\sqrt{(2\pi)}}\exp -y^2/2\sigma_y^2, \tag{1}$$

the combined frequency function can contain an additional term $e^{-\alpha xy}$. This extra term is allowable because, in integrating over x to give the normal distribution for y, it can be removed by completing the square, while no other term can be removed in this way. The general form for the combined frequency function is then

$$f(x,y) = \frac{1}{2\pi\sigma_x\sigma_y(1-\rho^2)^{\frac{1}{2}}}\exp -\frac{1}{2(1-\rho^2)}\left\{\frac{x^2}{\sigma_x^2}-\frac{2\rho xy}{\sigma_x\sigma_y}+\frac{y^2}{\sigma_y^2}\right\}. \tag{2}$$

It can be checked by integrating over x that the normal form for y is obtained and vice versa. The constant ρ is called the correlation coefficient.

The characteristic function for the bivariate distribution is defined as

$$\phi(t_x, t_y) = \iint_{-\infty}^{\infty}\exp(it_x x + it_y y)f(x,y)\,dx\,dy, \tag{3}$$

where $f(x,y)$ is given by (2). If the substitutions

$$\xi = x - it_x\sigma_x^2 - it_y\rho\sigma_x\sigma_y, \quad \eta = y - it_y\sigma_y^2 - it_x\rho\sigma_x\sigma_y,$$

are made in (3), then

$$\phi(t_x,t_y) = \exp -\tfrac{1}{2}(t_x^2\sigma_x^2 + 2t_x t_y\rho\sigma_x\sigma_y + t_y^2\sigma_y^2)\frac{1}{2\pi\sigma_x\sigma_y(1-\rho^2)^{\frac{1}{2}}}$$

$$\times\iint_{-\infty}^{\infty}\exp\left\{-\frac{1}{2(1-\rho^2)}\left(\frac{\xi^2}{\sigma_x^2}-\frac{2\rho\xi\eta}{\sigma_x\sigma_y}+\frac{\eta^2}{\sigma_y^2}\right)\right\}d\xi\,d\eta,$$

where the product, being of the form $\iint f(\xi,\eta)\,d\xi\,d\eta$, is unity.

Thus the characteristic function is

$$\phi(t_x, t_y) = \exp -\tfrac{1}{2}(t_x^2\sigma_x^2 + 2t_x t_y \rho\sigma_x\sigma_y + t_y^2\sigma_y^2). \tag{4}$$

The characteristic function is a moment-generating function, the moment

$$\mu_{rs} = E(x^r y^s)$$

being the coefficient of $(it_x)^r (it_y)^s/r!\,s!$. Thus

$$\mu_{20} = E(x^2) = \sigma_x^2; \quad \mu_{02} = E(y^2) = \sigma_y^2; \quad \mu_{11} = E(xy) = \rho\sigma_x\sigma_y. \tag{5}$$

2.3.1 *Estimation of* ρ

In practice the true values (assumed zero in the above discussion) are not known, and the correlation coefficient must be estimated in terms of deviations from the means. Now

$$E(x_i-\bar{x})(y_i-\bar{y})$$
$$= E(x_i y_i) - Ex_i\Sigma y_j/n - Ey_i\Sigma x_j/n + E(\Sigma x_j/n)(\Sigma y_j/n).$$

If the observations are independent, so that

$$E(x_i x_j) = 0 = E(x_i y_j),$$

then this becomes

$$\rho\sigma_x\sigma_y\{1-n^{-1}-n^{-1}+n^{-1}\} = \{(n-1)/n\}\rho\sigma_x\sigma_y,$$

and so

$$E\sum_i (x_i-\bar{x})(y_i-\bar{y}) = (n-1)\rho\sigma_x\sigma_y. \tag{1}$$

Thus

$$\sum_i (x_i-\bar{x})(y_i-\bar{y})/(n-1) \tag{2}$$

will provide an unbiased estimate of $\rho\sigma_x\sigma_y$. Usually σ_x and σ_y are also unknown and must be estimated from

$$\{\Sigma(x_i-\bar{x})^2/(n-1)\}^{\frac{1}{2}} \quad \text{and} \quad \{\Sigma(y_i-\bar{y})^2/(n-1)\}^{\frac{1}{2}}.$$

Hence

$$r = \frac{\Sigma(x_i-\bar{x})(y_i-\bar{y})}{\{\Sigma(x_i-\bar{x})^2\,\Sigma(y_i-\bar{y})^2\}^{\frac{1}{2}}} \tag{3}$$

will be an estimate of the correlation coefficient ρ.

2.3.2 *Expectation of product of absolute values*

In discussing the efficiency of another estimate s_1 of σ in § 2.6.3, the formula

$$E|x|\,|y| = 2\pi^{-1}\sigma_x\sigma_y[|\rho|\sin^{-1}|\rho| + (1-\rho^2)^{\frac{1}{2}}] \tag{1}$$

will be required. A brief outline of the proof of this formula will now be given.

Since

$$E|x|\,|y| = \iint_{-\infty}^{\infty} |x|\,|y| f(x,y)\,dx\,dy,$$

where $f(x, y)$ is given by (2.3,2), the substitution of angular co-ordinates defined by

$$x = \sigma_x z \cos\theta, \quad y = \sigma_y z \sin\theta$$

leads to

$$E|x|\,|y| = \frac{1}{4\pi}\sigma_x\sigma_y(1-\rho^2)^{-\frac{1}{2}}\int_0^{2\pi}\int_0^{\infty} z^2|\sin 2\theta|$$
$$\times\{\exp -z^2(1-\rho\sin 2\theta)/2(1-\rho^2)\}\,z\,dz\,d\theta.$$

Putting $$R^2 = z^2(1-\rho\sin 2\theta)/2(1-\rho^2),$$

the integral with respect to the R coordinate is $\Gamma(2) = 1$, and so

$$E|x|\,|y| = \frac{1}{2\pi}\sigma_x\sigma_y(1-\rho^2)^{\frac{3}{2}}\int_0^{2\pi}\frac{|\sin 2\theta|}{(1-\rho\sin 2\theta)^2}\,d\theta.$$

Replacing 2θ by ϕ,

$$E|x|\,|y| = \frac{1}{\pi}\sigma_x\sigma_y(1-\rho^2)^{\frac{3}{2}}\left[\int_0^{\pi/2}\frac{\sin\phi\,d\phi}{(1-\rho\sin\phi)^2}+\int_0^{\pi/2}\frac{\sin\phi\,d\phi}{(1+\rho\sin\phi)^2}\right]$$
$$= \frac{2}{\pi}\sigma_x\sigma_y(1-\rho^2)^{\frac{3}{2}}\int_0^{\pi/2}\frac{(1+\rho^2\sin^2\phi)\sin\phi\,d\phi}{(1-\rho^2\sin^2\phi)^2}.$$

This can be reduced by setting

$$|\rho|\cos\phi = (1-\rho^2)^{\frac{1}{2}}\tan\omega.$$

The value ω_0 of ω corresponding to $\phi = 0$ is $\sin^{-1}|\rho|$, and

$$E|x|\,|y| = \frac{2}{\pi}\sigma_x\sigma_y(1-\rho^2)^{\frac{3}{2}}\int_0^{\omega_0}\frac{\{1+\rho^2-(1-\rho^2)\tan^2\omega\}\times\{(1-\rho^2)^{\frac{1}{2}}\sec^2\omega\,d\omega\}}{|\rho|(1-\rho^2)^2(1+\tan^2\omega)^2}$$
$$= \frac{2}{\pi}\sigma_x\sigma_y|\rho^{-1}|\int_0^{\omega_0}\{(1+\rho^2)\cos^2\omega-(1-\rho^2)\sin^2\omega\}\,d\omega$$
$$= \frac{2}{\pi}\sigma_x\sigma_y|\rho^{-1}|[\rho^2\omega+\tfrac{1}{2}\sin 2\omega]_0^{\omega_0}$$
$$= \frac{2}{\pi}\sigma_x\sigma_y[|\rho|\sin^{-1}|\rho|+(1-\rho^2)^{\frac{1}{2}}].$$

This completes the proof of (1).

2.4 THE χ^2 DISTRIBUTION

If the normal law holds, the probability of obtaining a value in the range dy about y is of the form

$$C\{\exp -(y-Y)^2/2\sigma^2\}\,dy.$$

The probability distributions of the sums of the squares, for n observations, of the deviations from the true value, $\Sigma(y_i - Y)^2$, and from the mean, $\Sigma(y_i - \bar{y})^2$, will be derived in § 2.5. These will be shown to be of the χ^2 type, and in this section the derivation and properties of the χ^2 distribution will be discussed.

If z_i are ν quantities distributed normally about zero with standard deviation unity, the probability that a particular set of values will lie in the ranges dz_i about z_i is

$$dP(z_i) = \{C \exp - \tfrac{1}{2}\Sigma z_i^2\} \Pi \, dz_i. \tag{1}$$

The evaluation of the probability distribution of Σz_i^2 is done most simply by geometric methods. With a particular set of ν values z_i is associated a point in a ν-dimensional space whose coordinates are z_i. The term in brackets in (1) can be interpreted as a probability-density, since when multiplied by the volume element $\Pi \, dz_i$ it gives the probability that the point corresponding to a particular set of values lies in that element.

Clearly the probability-density is symmetrical about the origin, and a change to spherical coordinates is simply made. Denoting the radial coordinate by χ, Σz_i^2 is just the square of the distance of the point from the origin, and so equals χ^2. As regards the volume element, the volume of a hypersphere (i.e. the generalization of a sphere in ν-dimensional space) of radius χ will be proportional to χ^ν, and so the volume of a shell bounded by two hyperspheres will be proportional to $d\chi^\nu$, or to $\chi^{\nu-1} d\chi$. Hence in spherical coordinates (1) will be of the form

$$dP(\chi, \theta_k) = C(\exp - \tfrac{1}{2}\chi^2) \chi^{\nu-1} d\chi f(\theta_k) \Pi \, d\theta_k,$$

where the θ_k are the $\nu - 1$ angular coordinates. The constant C here will be different from that in (1). For simplicity the same symbol will be used for the constant of proportionality throughout the discussion. Since in the integrand the radial coordinate χ is independent of the angular coordinates θ_k, these may be integrated out to give the probability distribution

$$dP(\chi) = C(\exp - \tfrac{1}{2}\chi^2) \chi^{\nu-1} d\chi. \tag{2}$$

The range of χ is from 0 to ∞. From (2.1,2),

$$\int_0^\infty \exp(-\tfrac{1}{2}\chi^2) \chi^{\nu-1} d\chi = 2^{\frac{1}{2}\nu-1} \Gamma(\tfrac{1}{2}\nu),$$

and so the constant C in (2) is the inverse of this. Thus

$$dP(\chi) = \frac{1}{2^{\frac{1}{2}\nu-1} \Gamma(\frac{1}{2}\nu)} e^{-\frac{1}{2}\chi^2} \chi^{\nu-1} d\chi. \tag{3}$$

A variable whose distribution is of the form (3) is said to be distributed as χ^2 with ν degrees of freedom. The abbreviation d.f. or D.F. is often used for degrees of freedom.

It will be noted that (3) actually gives the distribution of χ, but the distribution is always spoken of as a χ^2 distribution. The actual form of the distribution of χ^2 is

$$dP(\chi^2) = \frac{1}{2^{\frac{1}{2}\nu}\,\Gamma(\frac{1}{2}\nu)}\,e^{-\frac{1}{2}\chi^2}\,\chi^{\nu-2}\,d\chi^2. \tag{4}$$

2.4.1 *Properties of the χ^2 distribution*

The characteristic function, from (2.4,4), is

$$\phi(t) = \frac{1}{2^{\frac{1}{2}\nu}\,\Gamma(\frac{1}{2}\nu)}\int_0^\infty e^{it\chi^2}\,e^{-\frac{1}{2}\chi^2}\,\chi^{\nu-2}\,d\chi^2$$

$$= (1-2it)^{-\frac{1}{2}\nu}\,\frac{1}{2^{\frac{1}{2}\nu}\,\Gamma(\frac{1}{2}\nu)}\int_0^\infty e^{-\frac{1}{2}z^2}\,z^{\nu-2}\,dz^2,$$

or

$$\phi(t) = (1-2it)^{-\frac{1}{2}\nu}. \tag{1}$$

Thus the cumulative function is

$$\psi(t) = \log\phi(t) = \nu(it) + 2\nu\frac{(it)^2}{2!} + 8\nu\frac{(it)^3}{3!} + \ldots, \tag{2}$$

and hence

$$E(\chi^2) = \nu, \quad \operatorname{var}\chi^2 = 2\nu. \tag{3}$$

The distribution may be shown to be approximately normal for large ν. For the characteristic function of $\chi^2/(2\nu)^{\frac{1}{2}}$ is, from (1.7.3,5),

$$\phi(t) = \left(1 - \frac{2it}{(2\nu)^{\frac{1}{2}}}\right)^{-\frac{1}{2}\nu},$$

and the cumulants then are

$$\kappa_2' = 1,\ \kappa_3' = \sqrt{(8/\nu)},\ \kappa_4' = 12/\nu, \text{ etc.}$$

Hence for large ν the distribution of $\chi^2/(2\nu)^{\frac{1}{2}}$, and so of χ^2, approaches the normal form. However, this approach is rather slow, and for small ν the distribution is decidedly unsymmetrical.

The probability of obtaining a value greater than a given value χ^2 is

$$Q(\chi^2) = \int_{\chi^2}^\infty dP(\chi^2).$$

Tables are available giving the values of χ^2 corresponding to different significance levels of Q. Table 8 of *Biometrika Tables* gives values of χ^2 for ν 1(1)30(10)100 and selected values of Q from 0·995 to 0·001. Similar tables are given in Fisher and Yates, and in the *Handbook of Chemistry and Physics*. Because of the asymmetry, tests using χ^2 tables are always single-tail tests.

2.4.2 *Expectation of* χ

The expectation of χ^2 is ν. The expectation of χ is very close to $\sqrt{\nu}$, but it is slightly different from this value. It can be evaluated from

$$E(\chi) = \int \chi dP(\chi) = \frac{1}{2^{\frac{1}{2}\nu-1}\,\Gamma(\frac{1}{2}\nu)} \int_0^\infty e^{-\frac{1}{2}\chi^2} \chi^\nu \, d\chi.$$

Thus $$E(\chi) = 2^{\frac{1}{2}}\,\Gamma\{\tfrac{1}{2}(\nu+1)\}/\Gamma(\tfrac{1}{2}\nu). \tag{1}$$

An approximate expression for the ratio of the gamma functions can be obtained in the following way. Suppose that ν is odd, so that $2q = \nu + 1$ (the discussion when ν is even follows similar lines). Then

$$\frac{\Gamma\{\frac{1}{2}(\nu+1)\}}{\Gamma(\frac{1}{2}\nu)} = \frac{\Gamma(q)}{\Gamma(q-\frac{1}{2})} = \frac{(q-1)!}{(q-\frac{3}{2})(q-\frac{5}{2})\ldots\frac{1}{2}\sqrt{\pi}}$$
$$= \frac{(q-1)!\,(q-1)!\,2^{2q-2}}{(2q-2)!\sqrt{\pi}}.$$

The factorials may be expanded by Stirling's approximation (Whittaker and Watson, 1940, p. 251),

$$n! \doteqdot (2\pi n)^{\frac{1}{2}}\,(n/e)^n\,(1 + 1/12n + \ldots).$$

Then

$$\frac{\Gamma\{\frac{1}{2}(\nu+1)\}}{\Gamma(\frac{1}{2}\nu)} \doteqdot 2^{-\frac{1}{2}}(2q-2)^{\frac{1}{2}}\{1+\tfrac{1}{12}(q-1)^{-1}\}^2\{1+\tfrac{1}{12}(2q-2)^{-1}\}^{-1}$$
$$\doteqdot 2^{-\frac{1}{2}}(\nu-1)^{\frac{1}{2}}\{1+\tfrac{1}{6}(\nu-1)^{-1}\}^2\{1+\tfrac{1}{12}(\nu-1)^{-1}\}^{-1}$$
$$\doteqdot 2^{-\frac{1}{2}}\,\nu^{\frac{1}{2}}\{1-\tfrac{1}{4}\nu^{-1}\}.$$

The expression (1) reduces to

$$E(\chi) = \nu^{\frac{1}{2}}(1-\tfrac{1}{4}\nu^{-1}), \tag{2}$$

the bias, or difference between $E(\chi)$ and $\nu^{\frac{1}{2}}$, being $\frac{1}{4}\nu^{-\frac{1}{2}}$.

The variance of χ is given approximately by formula (1.2,14*a*). Thus

$$\operatorname{var}\chi^2 \doteqdot \left\{\frac{\partial}{\partial\chi}\chi^2\right\}^2 \operatorname{var}\chi \doteqdot (2\chi)^2 \operatorname{var}\chi,$$

where the differential is evaluated for the value of χ corresponding to $E(\chi)$. Substituting the values ν for χ^2 and 2ν for $\operatorname{var}\chi^2$,

$$\operatorname{var}\chi \doteqdot \tfrac{1}{2}. \tag{3}$$

The ratio of the bias in (2) to the standard deviation is thus $\frac{1}{2}(2\nu)^{-\frac{1}{2}}$, and so the bias is usually neglected. Tables of $E(\chi)/\nu^{\frac{1}{2}}$ and of S.D. χ are given in the *Biometrika Tables* (Table 35, columns 2, 3, and 4).

It follows from (2) and (3) that

$$\chi' = \chi - \sqrt{\nu} \tag{4}$$

is approximately normally distributed about zero with standard deviation $1/\sqrt{2}$. The value of χ' for a given probability P varies only slightly with ν, and so χ' can be very easily tabulated. Table 2.8*b* is a table of the values χ' corresponding to various significance levels Q. This table can be used to test whether an observed value of χ^2 is reasonable or not. It is much more compact than a χ^2 table, but square roots have to be evaluated. Often it will be sufficient to remember that $\sqrt{2}(\sqrt{\chi^2} - \sqrt{\nu})$ is distributed roughly as X, the normal variate.

2.4.3 *The χ^2 distribution with one degree of freedom*

For the standardized normal distribution

$$dP(X) = f(X)\,dX,$$

and as the distribution is symmetrical the distribution of the absolute value is

$$dP(|X|) = 2f(|X|)\,dX = \frac{2}{\sqrt{(2\pi)}}\,e^{-\frac{1}{2}X^2}\,dX. \tag{1}$$

Comparison of this with (2.4,3) shows that $|X|$ is distributed as χ with 1 d.f., and X^2 is distributed as χ^2 with 1 d.f.

2.4.4 *Addition of χ^2 values*

If χ_1^2 and χ_2^2 are each distributed as χ^2 with ν_1 and ν_2 d.f. respectively,

$$dP(\chi_1, \chi_2) = C\,e^{-\frac{1}{2}\chi_1^2}\,e^{-\frac{1}{2}\chi_2^2}\,\chi_1^{\nu_1-1}\,\chi_2^{\nu_2-1}\,d\chi_1\,d\chi_2. \tag{1}$$

Transferring to angular coordinates χ, θ, defined by

$$\chi_1 = \chi\cos\theta, \quad \chi_2 = \chi\sin\theta, \quad \chi^2 = \chi_1^2 + \chi_2^2,$$

(1) becomes

$$dP(\chi, \theta) = C\,e^{-\frac{1}{2}\chi^2}\,\chi^{\nu_1+\nu_2-1}\,d\chi\,\sin^{\nu_2-1}\theta\,\cos^{\nu_1-1}\theta\,d\theta.$$

On integrating over θ,

$$dP(\chi) = C\,e^{-\frac{1}{2}\chi^2}\,\chi^{\nu_1+\nu_2-1}\,d\chi, \tag{2}$$

and the sum $\chi^2 = \chi_1^2 + \chi_2^2$ is distributed as χ^2 with $\nu_1 + \nu_2$ d.f.

2.5 THE DEVIATIONS FROM THE TRUE VALUE

For the set of n observations y_i of weights w_i, the combined probability distribution is

$$dP(y_i) = C\exp\{-\Sigma(y_i - Y)^2/2\sigma_i^2\}\,\Pi\,dy_i. \tag{1}$$

If the transformation

$$z_i = (y_i - Y)/\sigma_i \tag{2}$$

is introduced, it is seen that (1) becomes of the form (2.4,1). Hence

$$\Sigma(y_i - Y)^2/\sigma_i^2 = \Sigma w_i(y_i - Y)^2/\sigma^2$$

is distributed as χ^2 with n degrees of freedom.

However, the true value Y is usually unknown, and the residuals or deviations from the mean are of greater interest.

2.5.1 *Rotation of coordinate axes*

To determine the distribution of the residuals, it is necessary to investigate the transformations of variables corresponding in geometry to pure rotations of the coordinate axes. Consider a linear transformation of the form

$$z_i = \Sigma\lambda_{ij} y_j, \tag{1}$$

with the inverse transformation

$$y_i = \Sigma\mu_{ij} z_j. \tag{2}$$

Then

$$z_i = \sum_j \lambda_{ij} \sum_k \mu_{jk} z_k = \sum_k \left(\sum_j \lambda_{ij} \mu_{jk}\right) z_k,$$

and so

$$\sum_j \lambda_{ij} \mu_{jk} = \delta_{ik}, \tag{3}$$

where δ_{ik} is the symbol known as the Kronecker delta, taking the value unity when the indices i and k are equal and the value zero when they are different.

If the y_i are orthogonal, what conditions are imposed on the λ_{ij} for the z_i to be orthogonal also; that is, for the change from the y axes to the z axes to be equivalent to a rotation of the axes in space? Clearly the distance of a point from the origin is to be Σy_i^2 in the y coordinates, and Σz_i^2 in the z coordinates, and so

$$\sum_i y_i^2 = \sum_{j,k} \sum_i \lambda_{ij} \lambda_{ik} z_j z_k = \sum_i z_i^2 = \sum_{j,k} \sum_i \mu_{ij} \mu_{ik} y_j y_k. \tag{4}$$

Hence the conditions imposed on the λ_{ij} are

$$\sum_i \lambda_{ij} \lambda_{ik} = \delta_{jk}. \tag{5}$$

Also

$$\sum_i \mu_{ij} \mu_{ik} = \delta_{jk}, \tag{6}$$

and, on comparing (3) and (6),

$$\mu_{ij} = \lambda_{ji}. \tag{7a}$$

Thus the conditions imposed on the λ_{ij} can be put in the alternative form

$$\sum_i \lambda_{ji} \lambda_{ki} = \delta_{jk}. \tag{7b}$$

The conditions (5) can be obtained very easily by use of matrix notation. For (1) is

$$\mathbf{z} = \boldsymbol{\lambda}\mathbf{y},$$

and hence (4) is $$\mathbf{z}^T\mathbf{z} = \mathbf{y}^T\boldsymbol{\lambda}^T\boldsymbol{\lambda}\mathbf{y} = \mathbf{y}^T\mathbf{y},$$

and so $$\boldsymbol{\lambda}^T\boldsymbol{\lambda} = \mathbf{I} = \boldsymbol{\lambda}\boldsymbol{\lambda}^T,$$

which is the matrix form of (5). $\boldsymbol{\lambda}$ is referred to as an orthogonal matrix.

2.5.2 *The residuals*

Since $\Sigma w_i(y_i - \bar{y}) = 0$, where $\bar{y}$ is the weighted mean, it follows that the expression

$$\Sigma(y_i - Y)^2/2\sigma_i^2 = \Sigma w_i(y_i - \bar{y} + \bar{y} - Y)^2/2\sigma^2$$

can be put in the form

$$\Sigma(y_i - Y)^2/2\sigma_i^2 = \Sigma w_i v_i^2/2\sigma^2 + (\Sigma w_i)(\bar{y} - Y)^2/2\sigma^2. \tag{1}$$

If the scale of the coordinates in (2.5,1) is changed by the transformation

$$y_i^0 = y_i/\sigma_i = w_i^{\frac{1}{2}} y_i/\sigma,$$

$\Sigma(y_i - Y)^2/2\sigma_i^2$ becomes $\frac{1}{2}\Sigma(y_i^0 - Y_i^0)^2$, and the probability distribution of the y_i^0 is

$$dP(y_i^0) = C\{\exp - \tfrac{1}{2}\Sigma(y_i^0 - Y_i^0)^2\} \Pi\, dy_i^0. \tag{2}$$

The axes are now rotated by the orthogonal transformation

$$\left.\begin{aligned} z_1 = \Sigma w_i^{\frac{1}{2}} y_i^0/(\Sigma w_i)^{\frac{1}{2}} = (\Sigma w_i)^{\frac{1}{2}}\bar{y}/\sigma; \\ z_2, z_3, \ldots, z_n \text{ perpendicular to } z_1 \text{ and to one another.} \end{aligned}\right\} \tag{3}$$

Now the point corresponding to the true value is distant

$$\Sigma Z_i^2 = \Sigma Y_i^{02} = \Sigma w_i Y^2/\sigma^2$$

from the origin. But, from (3),

$$Z_1^2 = \Sigma w_i Y^2/\sigma^2,$$

and so $Z_2, Z_3, \ldots, Z_n$ all vanish, and

$$z_1 - Z_1 = (\Sigma w_i)^{\frac{1}{2}}(\bar{y} - Y)/\sigma.$$

Hence $$\Sigma(y_i - Y)^2/\sigma_i^2 = \Sigma(y_i^0 - Y_i^0)^2 = (z_1 - Z_1)^2 + \sum_2^n z_i^2, \tag{4a}$$

and, from (1), $$\sum_2^n z_i^2 = \Sigma w_i v_i^2/\sigma^2. \tag{4b}$$

Then (2) becomes

$$dP(\bar{y}, z_i) = C[\{\exp - \tfrac{1}{2}\Sigma w_i(\bar{y} - Y)^2/\sigma^2\}\, d\bar{y}]\left[\left\{\exp - \tfrac{1}{2}\sum_{i=2}^n z_i^2\right\}\Pi\, dz_i\right]. \tag{5}$$

The first term in square brackets shows that $\bar{y}$ is distributed normally about Y with variance $\sigma^2/(\Sigma w_i)$. The second term shows that $\Sigma w_i v_i^2/\sigma^2$ is distributed as χ^2 with $n-1$ degrees of freedom. The distributions of $\bar{y}$ and $\Sigma w_i v_i^2/\sigma^2$ are independent of one another.

2.5.3 *The estimated variance*

Since $\Sigma w_i v_i^2/\sigma^2$ is distributed as χ^2 with $\nu = n-1$ degrees of freedom

$$E(\Sigma w_i v_i^2/\sigma^2) = \nu,$$

and

$$\text{var}\,(\Sigma w_i v_i^2/\sigma^2) = 2\nu.$$

Hence

$$s^2 = \Sigma w_i v_i^2/\nu \tag{1}$$

will provide an unbiased estimate of σ^2, with variance given by

$$\text{var}\,s^2 = 2\sigma^4/\nu. \tag{2}$$

Similarly, s will provide a biased estimate of σ,

$$E(s) \doteqdot \sigma(1-\tfrac{1}{4}\nu^{-1}), \tag{3}$$

with standard deviation

$$\sigma(s) \doteqdot \sigma/(2\nu)^{\frac{1}{2}}. \tag{4}$$

Since the ratio of the bias term to the standard deviation is $(8\nu)^{-\frac{1}{2}}$, the bias in s is almost always neglected. In evaluating the accuracy of an estimate of σ, the estimated value s will usually have to be substituted for the unknown σ in (2) and (4).

It is almost never necessary to retain more than two significant figures for the standard deviation, since even with 50 observations the percentage standard deviation for the estimate s is 10%.

2.5.4 *Testing of estimated standard deviations*

From (2.5.3,1), $\nu s^2/\sigma^2$ is distributed as χ^2 with ν degrees of freedom. Hence the tables of $Q(\chi^2)$ can be used to test whether a value s obtained in a particular experiment for which σ is known is reasonable or exceptional.

For the measurement of prism angle (§ 1.4.2), many previous experiments might have shown that the standard deviation of an observation σ is 5′. For the observed value $s = 4{\cdot}1'$ (Table 1.4.2), $\Sigma v_i^2 = 150$, $n = 10$. Then χ^2 is $150/25 = 6$, with 9 d.f. From the tables, $Q(\chi^2) \doteqdot 0{\cdot}75$. Alternatively, $\chi' = \sqrt{6}-3 = -0{\cdot}55$, and from Table 2.8*b*, $Q(\chi') \doteqdot 0{\cdot}75$. That is, a value greater than the estimated standard deviation of 4·1′ would be expected in three-quarters of similar experiments, and an error less than 4·1′ in one-quarter. The value obtained in the particular experiment is neither exceptionally high nor exceptionally low, and there is no reason for supposing that the scatter is different from that normally present.

2.6 OTHER ESTIMATES OF STANDARD DEVIATION

There are two other formulae which have been widely used for the estimation of the standard deviation. Both assume that the distribution is of the normal form. The first formula employs the average of the deviations from the mean, irrespective of sign. In the second the range of the observations, that is, the difference between the largest and smallest values, is used.

2.6.1 *Properties of the residuals*

The residuals

$$v_j = y_j - \bar{y} = n^{-1}\left\{(n-1)\,y_j - \sum_{k \neq j} y_k\right\}$$

are linear functions of the y_k, and so from § 2.1.4 they are normally distributed about zero with variance

$$\sigma^2(v) = n^{-2}\{(n-1)^2 + (n-1)\}\,\sigma^2 = \{(n-1)/n\}\,\sigma^2. \tag{1}$$

The residuals are not independent but are correlated. Now

$$E(v_j\,v_k) = n^{-2}\,E\left\{(n-1)\,y_j - \sum_{q \neq j} y_q\right\}\left\{(n-1)\,y_k - \sum_{q \neq k} y_q\right\}$$

$$= n^{-2}\{-2(n-1) + (n-2)\}\,\sigma^2,$$

and so

$$E(v_j\,v_k) = -n^{-1}\,\sigma^2. \tag{2}$$

The coefficient of correlation is given by

$$\rho = E(v_j\,v_k)/\sigma(v_j)\,\sigma(v_k),$$

and from (1) and (2)

$$\rho = -1/(n-1). \tag{3}$$

2.6.2 *The estimate s_1*

From (2.1.2,3),

$$E\,|\,v_j\,| = \left\{\frac{2}{\pi}\,\sigma^2(v_j)\right\}^{\frac{1}{2}} = \left\{\frac{2}{\pi}\,\frac{n-1}{n}\right\}^{\frac{1}{2}}\sigma,$$

and so

$$s_1 = \left(\frac{\pi}{2}\right)^{\frac{1}{2}} \frac{\Sigma\,|\,v_j\,|}{\{n(n-1)\}^{\frac{1}{2}}} \tag{1}$$

will give an unbiased estimate of the standard deviation of an observation σ. This is sometimes known as Peters' formula. The numerical factor $\sqrt{(\pi/2)}$ is 1·253. The approximate formula

$$s_1 = \frac{1{\cdot}253\Sigma\,|\,v_j\,|}{n-\frac{1}{2}} \tag{2}$$

is more convenient for numerical calculations.

2.6.3 *Efficiency of* s_1

The variance of s_1 is given by

$$\operatorname{var} s_1 = E(s_1 - \sigma)^2 = E(s_1^2) - \sigma^2. \tag{1}$$

Now
$$E\{\Sigma |v_j|\}^2 = nE(v_j^2) + n(n-1)\, E|v_j|\,|v_k|, \tag{2}$$

and from (2.6.1,1), (2.6.1,3), and (2.3.2,1), this becomes

$$E\{\Sigma |v_j|\}^2 = (n-1)\,\sigma^2 + (2/\pi)\,(n-1)\,\sigma^2[\{(n-1)^2 - 1\}^{\frac{1}{2}} + \sin^{-1}\{(n-1)^{-1}\}].$$

Hence

$$\operatorname{var} s_1 = n^{-1}\,\sigma^2[\pi/2 + \{(n-1)^2 - 1\}^{\frac{1}{2}} + \sin^{-1}\{(n-1)^{-1}\} - n]. \tag{3}$$

A more useful form is obtained by expanding (3) in inverse powers of $(n-1)$. Thus

$$\sin^{-1}\{(n-1)^{-1}\} = (n-1)^{-1} + 0(n^{-3}),$$

$$\{(n-1)^2 - 1\}^{\frac{1}{2}} = (n-1) - \tfrac{1}{2}(n-1)^{-1} - 0(n^{-3}),$$

and
$$\operatorname{var} s_1 = n^{-1}\,\sigma^2\{(\pi/2) + (n-1) + \tfrac{1}{2}(n-1)^{-1} - n + 0(n^{-3})\}$$

$$= \frac{\sigma^2}{2(n-1)}\left\{(\pi - 2) - \frac{\pi - 3}{n} + 0(n^{-3})\right\}.$$

Hence to a very good approximation

$$\operatorname{var} s_1 = \frac{\pi - 2}{2(n-1)}\,\sigma^2. \tag{4}$$

Now, from (2.5.3,4),

$$\operatorname{var} s = \frac{\sigma^2}{2(n-1)},$$

and so the efficiency of the estimate s_1 is

$$\eta(s_1) = \frac{\operatorname{var} s}{\operatorname{var} s_1} = \frac{1}{\pi - 2} = 0{\cdot}876.$$

2.6.4 *Use of the estimate* s_1

The efficiency of s_1 will almost always be satisfactory, and this estimate has been very popular in the past. Its main advantage over the estimate s is that the residuals v do not have to be squared. However, with a modern calculating machine this squaring can be accomplished very rapidly. Since s_1 is only an estimate of σ if the distribution follows the normal law, while s is an estimate whatever the form of the law, the general use of s_1 is not recommended.

2.6.5 *The mean range*

The range w is defined as the difference between the largest and smallest values in the set of n observations. The mean range $E(w)$, the average value of the range for a large number of sets of observations, will be the difference between the mean largest and the mean smallest values.

The distribution function for the largest values will be $P^n(y)$, the probability that n observations are all less than y. Hence the probability that the largest observation lies in the range $y, y+\Delta y$, is

$$\frac{d}{dy}\{P^n(y)\}\,\Delta y.$$

The distribution function for the smallest values is $1-Q^n(y)$. For the probability that all the values exceed y is $Q^n(y)$, and so the probability that all the values do not exceed y, or that the smallest value is less than y, is $1-Q^n(y)$. Hence this is the distribution function for the smallest values, and the frequency function is

$$\frac{d}{dy}\{1-Q^n(y)\}.$$

The mean range is then given by

$$E(w) = \int_{-\infty}^{\infty} y\frac{d}{dy}\{P^n(y)\}\,dy - \int_{-\infty}^{\infty} y\frac{d}{dy}\{1-Q^n(y)\}\,dy. \qquad (1)$$

On integrating this by parts, integrals of the form

$$[yP^n(y)]_{-\infty}^{\infty} \quad \text{and} \quad [y\{1-Q^n(y)\}]_{-\infty}^{\infty}$$

will vanish, since they are of the order of $y\,e^{-\alpha y^2}$, which approaches zero as y approaches infinity. So (1) becomes

$$E(w) = \int_{-\infty}^{\infty} \{1-Q^n(y)-P^n(y)\}\,dy. \qquad (2)$$

If y is replaced by the standardized variable X on the right-hand side, (2) becomes

$$E(w/\sigma) = \int_{-\infty}^{\infty} \{1-Q^n(X)-P^n(X)\}\,dX, \qquad (3)$$

and the integral can be evaluated by numerical integration from the tables of $P(X)$ and $Q(X)$. The value of the integral is usually denoted by d_n. Then

$$E(w/\sigma) = d_n,$$

and so

$$s_R = w/d_n$$

will provide an unbiased estimate of the standard deviation σ.

The variance of s_R can also be found from similar integrals, but since the integrals have to be evaluated numerically, there is little point in writing out the explicit expressions. Table 2.8c lists the values of d_n and the corresponding efficiencies $\eta(s_R)$ for $n = 2(1)10$. The efficiency drops off quite rapidly for higher values of n.

2.6.6 *Grouping of observations when n is large*

If n is large, a more accurate estimate of σ is obtained by dividing the observations into N sets of ν observations, finding the range w_j in each set, and hence the mean range $\Sigma w_j/N$. Then the estimated standard deviation obtained from the ranges of the groups is

$$s_{GR} = \Sigma w_j/Nd_\nu. \qquad (1)$$

The variance of this estimate is

$$\operatorname{var} s_{GR} = N^{-1} \operatorname{var} s_R(\nu),$$

while for the estimate s,

$$\operatorname{var} s = \frac{\sigma^2}{2(n-1)} \doteqdot \frac{\sigma^2}{2N\nu}.$$

Hence the efficiency of the grouped estimate is

$$\eta(s_{GR}) = \frac{1}{2\nu}\left\{\frac{\sigma}{\text{S.D.}\, s_R(\nu)}\right\}^2.$$

Using the values for the last term tabulated in Table 2.8c, the efficiencies for $\nu = 2(1)10$ listed in Table 2.6.6 are obtained. It is seen that the efficiency is about 0·75 in the range $\nu = 6$ to 10.

TABLE 2.6.6

Efficiencies when the standard deviation is estimated from mean range in groups of ν values

ν	$\eta\{s_{GR}(\nu)\}$	ν	$\eta\{s_{GR}(\nu)\}$
2	0·437	7	0·753
3	0·605	8	0·754
4	0·686	9	0·751
5	0·723	10	0·745
6	0·742		

2.6.7 *Use of the range estimate s_R*

The range estimate is very useful as a rapid check on a more accurate estimate. Also, it is widely used in industrial testing and sampling procedures because it can be calculated so easily. But its validity depends on the deviations following a normal law, and its efficiency is rather low.

2.6.8 *Example*

The standard deviations of the observations A_i and θ_i in Example 1.4.2 are calculated by the three methods in Table 2.6.8. In each case the agreement between the different estimates is good, the differences being less than the estimated standard deviation of s.

TABLE 2.6.8

Estimation of σ by different methods

Angle of Prism A			
$\Sigma v^2 = 150$	$s = \sqrt{(150/9)}$	$= 4{\cdot}1$	S.D. $s \doteqdot s/\sqrt{18} = 1{\cdot}0$
$\Sigma \mid v \mid = 32$	$s_1 = 1{\cdot}253 \times 32/9{\cdot}5$	$= 4{\cdot}2$	
$w = 13$	$s_R = 13/3{\cdot}078$	$= 4{\cdot}2$	
Angle of Minimum Deviation θ			
$\Sigma v^2 = 487$	$s = \sqrt{(487/9)}$	$= 7{\cdot}4$	S.D. $s \doteqdot s/\sqrt{18} = 1{\cdot}7$
$\Sigma \mid v \mid = 54$	$s_1 = 1{\cdot}253 \times 54/9{\cdot}5$	$= 7{\cdot}1$	
$w = 24$	$s_R = 24/3{\cdot}078$	$= 7{\cdot}8$	

2.7 NOTES AND REFERENCES

(2.1) That the sum of two normal variables x and y follows a normal law can be established directly without the use of cumulants by a change of variables

$$u = x+y, \qquad v = \alpha x - \beta y,$$

in the combined frequency function, followed by integration over v.

(2.4) Approximations to the χ^2 distribution have been discussed by Blom (1954). It is suggested that the test using the quantity χ' might be useful in undergraduate teaching, since it does not require extensive tables.

(2.6) The efficiency of s_1 was derived by Helmert (1876), and was rediscovered by Fisher (1920). The present treatment is based on a paper by Guest (1951).

The distribution of s_1 is treated by Cadwell (1954).

Pearson (1950) gives some notes on the use of range. The simple derivation of (2.6.5,2) given here is due to Cox (1954).

2.8 TABLES

TABLE 2.8*a*

The normal distribution

$Q = \Pr\{(y-Y)/\sigma \geqslant X\}$, $y > Y$, single-tail test.
$Q = \Pr\{(Y-y)/\sigma \geqslant X\}$, $y < Y$, single-tail test.
$2Q = \Pr\{|Y-y|/\sigma \geqslant X\}$, double-tail test.

Q	$2Q$	X	Q	$2Q$	X
0·4	0·8	0·25	0·01	0·02	2·33
0·25	0·5	0·67	0·005	0·01	2·58
0·1	0·2	1·28	0·0025	0·005	2·81
0·05	0·1	1·64	0·001	0·002	3·09
0·025	0·05	1·96	0·0005	0·001	3·29

TABLE 2.8*b*

Values of $\chi' = \sqrt{\chi^2} - \sqrt{\nu}$ *at various significance levels* Q

ν \ Q	0·995	0·990	0·975	0·950	0·900	0·750	0·500
1	−0·99	−0·99	−0·97	−0·94	−0·87	−0·68	−0·33
2	−1·31	−1·27	−1·19	−1·09	−0·96	−0·66	−0·24
3	−1·46	−1·39	−1·27	−1·14	−0·97	−0·63	−0·19
5	−1·59	−1·49	−1·32	−1·17	−0·97	−0·60	−0·15
10	−1·69	−1·56	−1·36	−1·18	−0·96	−0·57	−0·11
20	−1·75	−1·60	−1·38	−1·18	−0·94	−0·54	−0·07
40	−1·77	−1·62	−1·38	−1·18	−0·93	−0·52	−0·05
80	−1·79	−1·63	−1·38	−1·17	−0·93	−0·51	−0·04
∞	−1·82	−1·64	−1·39	−1·16	−0·91	−0·48	0

ν \ Q	0·250	0·100	0·050	0·025	0·010	0·005	0·001
1	0·15	0·64	0·96	1·24	1·58	1·81	2·29
2	0·25	0·73	1·03	1·30	1·62	1·84	2·30
3	0·29	0·77	1·06	1·33	1·64	1·85	2·30
5	0·34	0·80	1·09	1·35	1·65	1·86	2·29
10	0·38	0·84	1·12	1·36	1·66	1·86	2·28
20	0·41	0·86	1·13	1·37	1·66	1·85	2·26
40	0·43	0·87	1·14	1·38	1·66	1·85	2·24
80	0·44	0·88	1·15	1·38	1·65	1·84	2·23
∞	0·48	0·91	1·16	1·39	1·64	1·82	2·19

TABLE 2.8*c*

Factors d_n *and efficiencies* η *in the estimation of standard deviation from the range*

n	d_n	S.D. s_R/σ	η	n	d_n	S.D. s_R/σ	η
2	1·128	0·756	0·64	7	2·704	0·308	0·84
3	1·693	0·525	0·78	8	2·847	0·288	0·83
4	2·059	0·427	0·83	9	2·970	0·272	0·82
5	2·326	0·372	0·84	10	3·078	0·259	0·80
6	2·534	0·335	0·84	20	3·735	0·195	0·68

CHAPTER 3

SOME STATISTICAL TESTS

3.1 DISTRIBUTIONS OF F AND t

3.1.1 *Beta functions*

The beta function $B(p,q)$ is defined by the equation

$$B(p,q) = \int_0^1 x^{p-1}(1-x)^{q-1}\,dx. \tag{1a}$$

If $\cos^2\theta$ is substituted for x, this is equivalent to the equation

$$B(p,q) = 2\int_0^{\pi/2} \cos^{2p-1}\theta \sin^{2q-1}\theta\,d\theta. \tag{1b}$$

The beta functions are related to the gamma functions by means of the equation

$$B(p,q) = \frac{\Gamma(p)\,\Gamma(q)}{\Gamma(p+q)}. \tag{2}$$

The proof of this equation is as follows. From (2.1,2),

$$\Gamma(p)\,\Gamma(q) = 4\int_0^\infty x^{2p-1}\,e^{-x^2}\,dx \int_0^\infty y^{2q-1}\,e^{-y^2}\,dy.$$

On changing to polar coordinates $x = r\cos\theta$, $y = r\sin\theta$, this becomes

$$\Gamma(p)\,\Gamma(q) = 4\int_0^\infty e^{-r^2}\,r^{2p+2q-1}\,dr \int_0^{\pi/2} \cos^{2p-1}\theta \sin^{2q-1}\theta\,d\theta$$

$$= \Gamma(p+q)\,B(p,q),$$

and so (2) is established.

From (1a) and (1b) it is obvious that

$$B(p,q) = B(q,p). \tag{3}$$

The quantity

$$B_\alpha(p,q) = 2\int_0^\alpha \cos^{2p-1}\theta \sin^{2q-1}\theta\,d\theta \tag{4}$$

is called the incomplete beta function. Tables of this function have been prepared by K. Pearson (1934). Table 16 of the *Biometrika Tables* is an inverse table, giving values of α corresponding to selected values of B.

3.1.2 *Distribution of F*

If χ_1 and χ_2 are two independent variables distributed as χ^2, then

$$dP(\chi_1, \chi_2) = C\,e^{-\frac{1}{2}\chi_1^2}\,e^{-\frac{1}{2}\chi_2^2}\,\chi_1^{\nu_1-1}\,\chi_2^{\nu_2-1}\,d\chi_1\,d\chi_2. \qquad (1)$$

The quantity F is defined as the ratio

$$\frac{\chi_1^2/\nu_1}{\chi_2^2/\nu_2}. \qquad (2)$$

Then $$\chi_1 = (F\nu_1/\nu_2)^{\frac{1}{2}}\,\chi_2.$$

To find the distribution of F, the variables in equation (1) are changed to F, χ_2. Now when χ_2 is constant,

$$d\chi_1 = \tfrac{1}{2}(\nu_1/\nu_2)^{\frac{1}{2}}\,F^{-\frac{1}{2}}\,\chi_2\,dF,$$

and so (1) becomes

$$dP(F, \chi_2) = C\,e^{-\frac{1}{2}\chi_2^2(1+F\nu_1/\nu_2)}\,\chi_2^{\nu_1+\nu_2-1}\,d\chi_2\,F^{\frac{1}{2}\nu_1-1}\,dF.$$

The variable χ_2 can be removed by the substitution

$$\chi^2 = \chi_2^2(1+F\nu_1/\nu_2).$$

Since χ^2 is distributed 'as χ^2' it can be integrated out, leaving

$$dP(F) = C(1+F\nu_1/\nu_2)^{-\frac{1}{2}(\nu_1+\nu_2)}\,F^{\frac{1}{2}\nu_1-1}\,dF. \qquad (3)$$

This gives the distribution of F. The constant may be evaluated by making the substitution

$$F\nu_1/\nu_2 = \tan^2\theta.$$

Then

$$\int dP(F) = 1 = C\int_0^{\pi/2} 2(\nu_2/\nu_1)^{\frac{1}{2}\nu_1}\cos^{\nu_2-1}\theta\,\sin^{\nu_1-1}\theta\,d\theta,$$

and $$C = (\nu_1/\nu_2)^{\frac{1}{2}\nu_1}/B(\tfrac{1}{2}\nu_2, \tfrac{1}{2}\nu_1). \qquad (4)$$

Equation (3) then becomes

$$dP(F) = \frac{(\nu_1/\nu_2)^{\frac{1}{2}\nu_1}}{B(\frac{1}{2}\nu_2, \frac{1}{2}\nu_1)}\,(1+F\nu_1/\nu_2)^{-\frac{1}{2}(\nu_1+\nu_2)}\,F^{\frac{1}{2}\nu_1-1}\,dF. \qquad (5)$$

The distribution function $P(F)$ can be expressed in terms of the incomplete beta function. Thus substitution of $F\nu_1/\nu_2 = \tan^2\alpha$ in (5) leads to

$$P(F) = \frac{B_\alpha(\frac{1}{2}\nu_2, \frac{1}{2}\nu_1)}{B(\frac{1}{2}\nu_2, \frac{1}{2}\nu_1)}. \qquad (6)$$

Hence numerical values of $P(F)$ can be found from tables of the incomplete beta function.

More convenient in practice are tables of F for given significance levels P. Table 18 of the *Biometrika Tables* gives the values of F

for seven values of $Q = 1-P$ from 0·25 to 0·001. Table V of Fisher and Yates' *Tables* gives the values of $e^{2z} = F$ for five values of Q in the same range. Similar tables are given in the *Handbook of Chemistry and Physics.*

The quantity

$$F' = \frac{\chi_1 \nu_2^{\frac{1}{2}} - \chi_2 \nu_1^{\frac{1}{2}}}{(\chi_1^2 + \chi_2^2)^{\frac{1}{2}}} = \frac{F^{\frac{1}{2}} - 1}{(F/\nu_2 + 1/\nu_1)^{\frac{1}{2}}} \tag{7}$$

is more complicated to evaluate than F, but varies much less with ν_1 and ν_2. Table 3.9*a* gives the values of F' for $Q = 0{\cdot}05$ and 0·01.

In all F tables the heading ν_1 refers to the estimate of higher χ^2/ν. That is, the subscript 1 denotes the variable for which

$$\chi_1^2/\nu_1 > \chi_2^2/\nu_2, \quad F > 1.$$

The levels are for single-tail tests, appropriate to testing whether χ_1^2/ν_1 is significantly greater than χ_2^2/ν_2. If it is required to test whether the two quantities are significantly different, irrespective of which is the larger, then a double-tail test is required and the values of F given in the tables correspond to significance levels $2Q$.

3.1.3 *Distribution of t*

If the variable χ_1^2 has $\nu_1 = 1$, so that $\chi_1 = |X|$, where X is the standardized normal variable, then the distribution of

$$F = X^2/(\chi^2/\nu)$$

is

$$dP(F) = \frac{1}{\nu^{\frac{1}{2}} B(\frac{1}{2}\nu, \frac{1}{2})} (1 + F/\nu)^{-\frac{1}{2}(\nu+1)} F^{-\frac{1}{2}} dF. \tag{1}$$

Hence the distribution of

$$t = \frac{X}{\chi/\nu^{\frac{1}{2}}} \tag{2}$$

is given by

$$dP(t) = \tfrac{1}{2} dP(|t|) = \tfrac{1}{2} dP(F^{\frac{1}{2}}),$$

or, using (1),

$$dP(t) = \frac{1}{\nu^{\frac{1}{2}} B(\frac{1}{2}\nu, \frac{1}{2})} (1 + t^2/\nu)^{-\frac{1}{2}(\nu+1)} dt. \tag{3a}$$

When the beta function is expanded this becomes

$$dP(t) = \frac{\Gamma(\frac{1}{2}\nu + \frac{1}{2})}{\Gamma(\frac{1}{2}\nu)\sqrt{\pi}\sqrt{\nu}} (1 + t^2/\nu)^{-\frac{1}{2}(\nu+1)} dt. \tag{3b}$$

The ratio t is often referred to as Student's ratio. The range of t is from $+\infty$ to $-\infty$. For large ν, $(1 + t^2/\nu)^{-\frac{1}{2}(\nu+1)}$ tends to $e^{-\frac{1}{2}t^2}$, as may be verified by writing out the binomial expansion. Thus the distribution of t approaches that of the standardized normal variable X. For small values of ν the probability of a large deviation is somewhat greater than in the normal case.

As with the normal curve, it is possible to use either a single-tail or double-tail test. Table 3.9*b* is a short table of values t for various significance levels $Q(t)$ and various values of ν. More extended tables are given in the *Biometrika Tables* (Table 12), and also by Fisher and Yates (Table III, double-tail test).

3.1.4 *t-test for linear function of the observed values*

If z is a linear function of the observations, Z the 'true' value of this function, and $s(z)$ the estimated standard deviation of z based on an estimate $s^2 = \Sigma w_i v_i^2/\nu$ of σ^2, then, provided the distributions of s and z are independent, the ratio

$$t = \frac{z-Z}{s(z)} \tag{1}$$

is distributed as t with ν d.f.

For if $$z = \Sigma\lambda_i y_i,$$

the estimate $s(z)$ is given by

$$s^2(z) = (\Sigma\lambda_i^2/w_i)\,s^2 = (s^2/\sigma^2)\,\sigma^2(z), \tag{2}$$

while $$\Sigma w_i v_i^2/\sigma^2 = \chi^2 = \nu s^2/\sigma^2. \tag{3}$$

Hence (1) can be written

$$t = \frac{(z-Z)/\sigma(z)}{s(z)/\sigma(z)} = \frac{X}{s/\sigma} = \frac{X}{\chi/\nu^{\frac{1}{2}}},$$

showing that the ratio (1) is distributed as t with ν d.f.

3.2 CHOICE OF SIGNIFICANCE LEVEL

In applying statistical tests to observed quantities, it is necessary to decide upon the significance level to be employed. If the discrepancy between the observed and expected values is not regarded as significant unless the probability of the occurrence of a discrepancy at least as 'bad' is less than Q, then Q is referred to as the significance level. The significance level effectively fixes the maximum acceptable discrepancy.

It is obvious that any discrepancy, however large, is possible even when the hypothesis about the magnitude of the quantity is true. Thus when Q is taken as 0·05 and the hypothetical value is correct, once in twenty observations (on the average) a larger discrepancy than that regarded as acceptable will occur, and the hypothesis will be wrongly rejected. The rejection of a true hypothesis is referred to as an error of the first kind. The chance of making an error of the first kind—of falsely deducing a deviation from the hypothetical conditions—is just the significance level Q.

Clearly the smaller the chosen value Q, the smaller will be the risk of making an error of the first kind. But the risk of missing a real deviation from the hypothetical conditions will be increased as Q is decreased. An error of this type is referred to as an error of the second kind. If Q is to be kept very small, the risk of making an error of the second kind can usually only be reduced by increasing the accuracy of the experimental value; for example, by taking further observations. Thus the standard deviation of the mean decreases as $n^{-\frac{1}{2}}$, and the spread $\bar{y}-Y$ for a given Q is correspondingly reduced. Smaller and smaller discrepancies between the postulated value Y and the observed value will then become detectable. This is illustrated by Example 3.3.1.

The choice of the significance level Q is then something of a compromise, depending mainly on the consequences of making a wrong decision and on the ease with which further measurements can be made. Values of 0·05 and 0·01 are typical in the statistical literature.

3.2.1 *Confidence intervals*

It is often possible to find confidence intervals for the population mean; that is, to find values Y_1 and Y_2 such that

$$\Pr\{Y_1 \geqslant Y \geqslant Y_2\} = \alpha, \tag{1}$$

where Y_1 and Y_2 are specific functions of the value y_0 obtained in the experiment. The meaning of (1) is that, if the experiment were repeated a large number of times, the statement $Y_1 \geqslant Y \geqslant Y_2$ would be true in a fraction α of the experiments. Y_1 and Y_2, being functions of y_0, would of course vary from one experiment to the next.

In particular, if the form of the distribution law depends only on the deviations $y-Y$ and not on the actual value of Y, so that

$$\Pr\{y_0 - Y \geqslant \eta_1\} = Q_1(\eta_1 + Y \mid Y), \tag{2}$$

then $$\Pr\{\eta_1 \leqslant y_0 - Y \leqslant \eta_2\} = Q_1 - Q_2 = \alpha,$$

or $$\Pr\{y_0 - \eta_1 \geqslant Y \geqslant y_0 - \eta_2\} = \alpha. \tag{3}$$

η_1 and η_2 are the values of $y-Y$ corresponding to the levels Q_1 and Q_2. Hence confidence limits for Y can be found.

For the normal law

$$X = (y - Y)/\sigma,$$

if X_1 and X_2 are the values corresponding to the levels Q_1 and Q_2, then η_1 and η_2 are σX_1 and σX_2, and so

$$\Pr\{y_0 - \sigma X_1 \geqslant Y \geqslant y_0 - \sigma X_2\} = \alpha. \tag{4}$$

There will be no unique confidence interval but an infinity of intervals for a given α, corresponding to all possible combinations of Q_1 and Q_2 whose difference is α. As an example, from Table 2.8*a*,

$$\Pr\{X \geqslant 1{\cdot}64\} = 0{\cdot}05, \quad \Pr\{X \geqslant -1{\cdot}64\} = 0{\cdot}95,$$

and so

$$y_0 + 1{\cdot}64\sigma \geqslant Y \geqslant y_0 - 1{\cdot}64\sigma \tag{5}$$

is a 90% confidence interval for Y. Similarly,

$$\Pr\{X \geqslant 1{\cdot}96\} = 0{\cdot}025, \quad \Pr\{X \geqslant -1{\cdot}44\} = 0{\cdot}925,$$

and

$$y_0 + 1{\cdot}44\sigma \geqslant Y \geqslant y_0 - 1{\cdot}96\sigma \tag{6}$$

is also a 90% confidence interval for Y. The confidence interval usually given is the smallest of all the intervals at the particular significance level. For the normal case the smallest interval is the one symmetrical about $X = 0$.

For the t distribution

$$t = (y - Y)/s(y),$$

the values of η_1 and η_2 are $t_1 s(y)$ and $t_2 s(y)$, where t_1 and t_2 are the values of t corresponding to the levels Q_1 and Q_2. Hence

$$y_0 - t_1 s(y) \geqslant Y \geqslant y_0 - t_2 s(y) \tag{7}$$

is a confidence interval for Y at significance level $Q_1 - Q_2$.

3.2.2 *Fiducial intervals*

If y_0 is the value given by the experiment, then it is possible to define for a given significance level α a quantity Y_1 such that, if the population mean were Y_1,

$$P(y_0 \mid Y_1) = \alpha.$$

Clearly, if α is close to unity, then it is very probable that the true value is less than Y_1. This leads to the consideration of a distribution of hypothetical values of Y, called the fiducial distribution. The fiducial distribution function $P_F(Y \mid y_0)$ is defined by the equation

$$P_F(Y \mid y_0) = 1 - P(y_0 \mid Y). \tag{1}$$

The associated frequency function will be

$$f_F(Y\,|\,y_0) = \frac{d}{dY}P_F(Y\,|\,y_0) = -\frac{d}{dY}P(y_0\,|\,Y)$$

$$= \frac{d}{dY}\int_{y_0}^{\infty} f(y_0\,|\,Y)\,dy,$$

and so
$$\frac{df_F(Y\,|\,y_0)}{dy_0} = -\frac{df(y_0\,|\,Y)}{dY}. \tag{2}$$

In cases where $f(y\,|\,Y)$ is a function of the deviations $y-Y$, as happens with the normal and t distributions, the two frequency functions are equal.

If Y_1 and Y_2 are the two values for which

$$P_F(Y_1\,|\,y_0) = 1-P(y_0\,|\,Y_1) = 1-\alpha_1 \tag{3a}$$

and
$$P_F(Y_2\,|\,y_0) = 1-P(y_0\,|\,Y_2) = 1-\alpha_2, \tag{3b}$$

then
$$P_F(Y_1\,|\,y_0)-P_F(Y_2\,|\,y_0) = \alpha_2-\alpha_1 = \alpha, \tag{3c}$$

and $Y_1 \geqslant Y \geqslant Y_2$ is said to be a fiducial interval for Y at the significance level α. Thus for the normal variable

$$P_F(y_0+1{\cdot}64\sigma\,|\,y_0) = 1-P(y_0\,|\,Y = y_0+1{\cdot}64\sigma) = 0{\cdot}95$$

and
$$P_F(y_0-1{\cdot}64\sigma\,|\,y_0) = 0{\cdot}05,$$

and so
$$y_0+1{\cdot}64\sigma \geqslant Y \geqslant y_0-1{\cdot}64\sigma$$

is a 90% fiducial interval for Y.

This interval coincides with the confidence interval (3.2.1,5). It will now be shown that a fiducial interval is always a confidence interval, in the sense that if the experiment were repeated a large number of times the population mean would lie within the fiducial interval in a fraction α of the experiments. For if Y is the true value, there are two values y_1 and y_2 such that

$$P(y_1\,|\,Y) = \alpha_1, \quad P(y_2\,|\,Y) = \alpha_2, \tag{4a}$$

while from (3)

$$P(y_0\,|\,Y_1) = \alpha_1, \quad P(y_0\,|\,Y_2) = \alpha_2. \tag{4b}$$

Thus, on comparing (4*a*) and (4*b*), $Y_1 \geqslant Y$ when $y_0 \geqslant y_1$ and $Y \geqslant Y_2$ when $y_2 \geqslant y_0$, and so $Y_1 \geqslant Y \geqslant Y_2$ when $y_2 \geqslant y_0 \geqslant y_1$. Now from (4*a*)

$$\Pr\{y_2 \geqslant y_0 \geqslant y_1\} = \alpha_2-\alpha_1 = \alpha,$$

and so
$$\Pr\{Y_2 \geqslant Y \geqslant Y_1\} = \alpha. \tag{5}$$

The reason for the introduction of fiducial probability is that by its use a fiducial frequency function can be determined for a quantity containing several variables by assuming that the ordinary laws of probability hold for fiducial probabilities. The standard example is the Behrens' test for the comparison of two means (Kendall, Vol. II, p. 91; Barnard, 1950).

3.3 TESTING THE MEAN

For the mean, the estimated standard deviation is

$$s(\bar{y}) = s/\sqrt{\Sigma w_i} = \{\Sigma w_i v_i^2/(n-1)\,\Sigma w_i\}^{\frac{1}{2}},$$

and the distributions of $\bar{y}$ and $s(\bar{y})$ are independent. So, by § 3.1.4,

$$t = \frac{\bar{y}-Y}{s(\bar{y})} \tag{1}$$

is distributed as t with $\nu = n-1$ d.f.

3.3.1 *Example*

In Example 1.4.2 the estimated prism angle is $60°\ 27{\cdot}0' \pm 1{\cdot}3'$, the standard deviation being based on 9 d.f. It is believed that the prism has been ground to an angle of $60°\ 25'$. Does the experimental value cast doubt on this hypothesis?

If Y is $60°\ 25'$, t is $2{\cdot}0/1{\cdot}3 = 1{\cdot}5$, with 9 d.f. Then from Table 3.9*b* $2Q$ is 0·17, and hence if the true value were $60°\ 25'$ a deviation (in either sense) at least as large as this would be obtained in 17% of such experiments. Hence there is little evidence that the true value differs from the supposed value $60°\ 25'$.

The question can be settled by taking a larger number of individual measurements. If the set is extended to 40 observations, the standard deviation of the mean, being proportional to $n^{-\frac{1}{2}}$, is approximately halved. If the estimate now obtained is $60°\ 26{\cdot}8' \pm 0{\cdot}7'$, then t is $1{\cdot}8/0{\cdot}7 = 2{\cdot}6$, with 39 d.f. The value $2Q$ is now only 0·01, and it is very unlikely that the population mean is $60°\ 25'$. It is very probable that there is something wrong, either with the original grinding procedure or with the apparatus for measuring the angle.

If the acceptance level had been arbitrarily set at 0·05, then with 10 observations the hypothesis that the true angle was $60°\ 25'$ would have been accepted, and an error of the second kind made. Increasing the number of observations to 40 increases the 'resolving power' of the experiment, and shows that the hypothesis is false.

3.3.2 *Confidence interval*

For the 90% confidence interval, based on the measurement $60°\ 27' \pm 1{\cdot}3'$, the values $t_1 = -1{\cdot}9$ and $t_2 = +1{\cdot}9$ (corresponding to $Q = 0{\cdot}95$ and $0{\cdot}05$ with 9 d.f.) may be used. Then the values ts are $\pm 1{\cdot}9 \times 1{\cdot}3 = \pm 2{\cdot}5$, and from (3.2.1,7) the confidence interval is

$$60°\ 24{\cdot}5' \leqslant Y \leqslant 60°\ 29{\cdot}5'.$$

The corresponding interval based on the measurement

$$60°\ 26{\cdot}8' \pm 0{\cdot}07' \quad \text{is} \quad 60°\ 25{\cdot}6' \leqslant Y \leqslant 60°\ 28{\cdot}0'.$$

3.4 COMPARISON OF TWO MEANS

Suppose that a set of n_1 observations leads to the mean $\bar{y}_1 \pm s(\bar{y}_1)$, where

$$s^2(\bar{y}_1) = \Sigma w_{1i} v_{1i}^2 / n_1 \nu_1.$$

If a second set of n_2 observations is made, leading to the mean $\bar{y}_2 \pm s(\bar{y}_2)$, the question arises as to whether the two means are in agreement. A t-test can be made to verify the hypothesis that the two sets of observations came from the same population with parameters $Y = Y_1 = Y_2$ and $\sigma = \sigma_1 = \sigma_2$.

If the hypothesis is true, $\bar{y}_1 - \bar{y}_2$ is normally distributed about zero with variance $\sigma^2(n_1^{-1} + n_2^{-1})$, and

$$X = \frac{\bar{y}_1 - \bar{y}_2}{\sigma(n_1^{-1} + n_2^{-1})^{\frac{1}{2}}}$$

is the standardized normal variate. Also $\Sigma w_{1i} v_{1i}^2/\sigma^2$ and $\Sigma w_{2i} v_{2i}^2/\sigma^2$ are each distributed as χ^2, independently of $\bar{y}_1$ and $\bar{y}_2$, and so

$$\chi^2 = (\Sigma w_{1i} v_{1i}^2 + \Sigma w_{2i} v_{2i}^2)/\sigma^2 = \sum_j \sum_i w_{ji} v_{ji}^2/\sigma^2$$

is distributed as χ^2 with $\nu_1 + \nu_2$ d.f. (§ 2.4.4). Hence

$$t = \frac{X}{\chi\nu^{-\frac{1}{2}}} = (\bar{y}_1 - \bar{y}_2) \left\{ \frac{(\nu_1 + \nu_2)\, n_1 n_2}{(n_1 + n_2) \sum_j \sum_i w_{ji}^2 v_{ji}^2} \right\}^{\frac{1}{2}} \tag{1}$$

is distributed as t with $\nu_1 + \nu_2$ d.f. In terms of $s(\bar{y}_1), s(\bar{y}_2)$, this is

$$t = (\bar{y}_1 - \bar{y}_2) \left\{ \frac{(\nu_1 + \nu_2)\, n_1 n_2}{(n_1 + n_2)\{n_1 \nu_1 s^2(\bar{y}_1) + n_2 \nu_2 s^2(\bar{y}_2)\}} \right\}^{\frac{1}{2}}. \tag{2}$$

3.4.1 *Example*

Consider the results of two different sets of measurements of prism angle:

1. $60° \; 27{\cdot}0' \pm 1{\cdot}3'$, $n = 10$, $\nu = 9$;
2. $60° \; 25{\cdot}2' \pm 1{\cdot}5'$, $n = 15$, $\nu = 14$.

Now $\Sigma w_{ji} v_{ji}^2 = n_1 \nu_1 s^2(\bar{y}_1) + n_2 \nu_2 s^2(\bar{y}_2) = 152 + 472 = 624,$

and so $t = 1{\cdot}8 \left\{ \frac{23 \times 150}{25 \times 624} \right\}^{\frac{1}{2}} = 0{\cdot}85.$

Since this is well above the 10% point, there is no evidence to suggest that the two sets of measurements are discordant.

3.5 RATIO OF STANDARD DEVIATIONS UNKNOWN

If the means to be compared are obtained by different experimental methods, the standard deviations σ_1 and σ_2 will in general be different in the two experiments. This will invalidate the previous discussion, which assumed a single value σ. If the ratio

$$\theta = \sigma_1^2/\sigma_2^2 \tag{1}$$

is known, then it is easy to show as in § 3.4 that

$$t = (\bar{y}_1 - \bar{y}_2)\left[\frac{(\nu_1+\nu_2)\,n_1\,n_2}{(n_1+\theta n_2)\{\theta^{-1}n_1\nu_1 s^2(\bar{y}_1)+n_2\nu_2 s^2(\bar{y}_2)\}}\right]^{\frac{1}{2}} \tag{2}$$

is distributed as t with $\nu_1+\nu_2$ d.f. However, the ratio θ is seldom known.

For cases where σ_1 and σ_2 cannot be assumed equal, Welch has introduced the variable

$$t_m = \frac{\bar{y}_1-\bar{y}_2}{s_m} = \frac{\bar{y}_1-\bar{y}_2}{\{s^2(\bar{y}_1)+s^2(\bar{y}_2)\}^{\frac{1}{2}}}$$

$$= \frac{\bar{y}_1-\bar{y}_2}{(\Sigma w_{1i}\,v_{1i}^2/n_1\nu_1+\Sigma w_{2i}\,v_{2i}^2/n_2\nu_2)^{\frac{1}{2}}}, \tag{3}$$

the denominator being simply the estimated standard deviation of the difference $\bar{y}_1-\bar{y}_2$. t_m will not be distributed exactly as t. However, it can be shown (§ 3.5.2) by comparing moments that t_m is distributed approximately as t with

$$\nu = \frac{(\theta n_1^{-1}+n_2^{-1})^2}{\theta^2 n_1^{-2}\nu_1^{-1}+n_2^{-2}\nu_2^{-1}} \tag{4}$$

degrees of freedom, θ being the ratio of the variances. Of course, this ratio is not known—if it were, the test using (2) could be employed. But it appears in practice that θ can vary within wide limits without affecting the value of t sufficiently to invalidate the test. The simplest choice for the value θ is the ratio of the estimated variances,

$$\theta = (\nu_1^{-1}\,\Sigma w_{1i}\,v_{1i}^2)/(\nu_2^{-1}\,\Sigma w_{2i}\,v_{2i}^2). \tag{5}$$

Then if s_{m1} and s_{m2} are the estimated standard deviations of the individual means, and $s_m = (s_{m1}^2+s_{m2}^2)^{\frac{1}{2}}$ the standard deviation of the difference, (4) becomes

$$\nu = \frac{s_m^4}{s_{m1}^4/\nu_1+s_{m2}^4/\nu_2}. \tag{6}$$

Accurate values for t_m at the 5% and 1% (single-tail) significance levels are given in Table 11 of the *Biometrika Tables* in terms of ν_1, ν_2, and the ratio s_{m1}^2/s_m^2.

3.5.1 *Example*

The following two measurements for the angle of a prism were obtained using different experimental methods:

1. 60° 17·3′ ± 2·4′, $n = 10$; 2. 60° 10·5′ ± 3·5′, $n = 20$.

From (3.5,3),
$$t_m = \frac{6{\cdot}8}{(5{\cdot}76+12{\cdot}25)^{\frac{1}{2}}} = 1{\cdot}60.$$

From (3.5,5) $$\nu = \frac{18\cdot01^2}{5\cdot76^2/9+12\cdot25^2/19} = 28\cdot0.$$

The value $t_m = 1\cdot60$ with 28 d.f. corresponds to about the 6% level for a single-tail test. So, while the means are not definitely discordant, there may still be some doubt about their agreement.

To use the *Biometrika Tables* it is necessary to calculate the ratio

$$\frac{\lambda_1 s_1^2}{\lambda_1 s_1^2+\lambda_2 s_2^2} = \frac{s_{m1}^2}{s_{m1}^2+s_{m2}^2} = \frac{5\cdot76}{18\cdot01} = 0\cdot32.$$

With $\nu_1 = 8, \nu_2 = 20$, ratio $0\cdot32$, $v \equiv t_m$ is $1\cdot70$ at the 5% confidence level. Hence the value $t_m = 1\cdot60$ is just above the 5% level, agreeing with the result obtained using the approximate distribution.

3.5.2 *Approximate distribution of* t_m

The value of t_m is

$$t_m = \frac{\bar{y}_1-\bar{y}_2}{\sigma_m}\frac{\sigma_m}{s_m}, \tag{1}$$

where σ_m and s_m are the standard deviation and estimated standard deviation of the difference $\bar{y}_1-\bar{y}_2$. If this is to be distributed approximately as t, it must be of the form $X\nu^{\frac{1}{2}}/\chi$, and so $\nu s_m^2/\sigma_m^2$ must be distributed approximately as χ^2 with ν d.f. But the actual value of this expression is

$$\chi_0^2 = \frac{\nu s_m^2}{\sigma_m^2} = \frac{\nu}{\sigma_m^2}\left[\frac{\sigma_1^2}{n_1\nu_1}\left(\frac{\Sigma w_{1i}v_{1i}^2}{\sigma_1^2}\right)+\frac{\sigma_2^2}{n_2\nu_2}\left(\frac{\Sigma w_{2i}v_{2i}^2}{\sigma_2^2}\right)\right], \tag{2}$$

where the terms in round brackets are distributed as χ^2 with ν_1 and ν_2 d.f. To find ν, the moments of each side can be equated. For the expectations, since $E(\chi^2)$ is ν,

$$\nu = \frac{\nu}{\sigma_m^2}\left[\frac{\sigma_1^2}{n_1}+\frac{\sigma_2^2}{n_2}\right] = \nu,$$

an identity. For the variances, since $\operatorname{var}\chi^2$ is 2ν,

$$2\nu = \left(\frac{\nu}{\sigma_m^2}\right)^2\left[\left(\frac{\sigma_1^2}{n_1\nu_1}\right)^2 2\nu_1+\left(\frac{\sigma_2^2}{n_2\nu_2}\right)^2 2\nu_2\right],$$

or
$$\nu = \frac{\sigma_m^4}{\sigma_1^4/n_1^2\nu_1+\sigma_2^4/n_2^2\nu_2}. \tag{3}$$

Thus if ν is given by (3), $\nu s_m^2/\sigma_m^2$ is distributed approximately as χ^2 with ν d.f. The first and second moments of $\nu s_m^2/\sigma_m^2$ and χ^2 coincide, though the higher moments may diverge. Hence t_m is distributed approximately as t with ν d.f.

3.5.3 *Behrens' test*

It is somewhat easier to find fiducial limits for $\bar{y}_1-\bar{y}_2$. The theory will not be discussed here (see Kendall, 1948, Vol. II,

p. 91). Fisher and Yates (1948) tabulate fiducial (double-tail) limits for

$$d = \frac{|\bar{y}_1 - \bar{y}_2|}{\sqrt{(s_{m1}^2 + s_{m2}^2)}} \tag{1}$$

in terms of ν_1, ν_2, and θ, where

$$\tan\theta = s_{m1}/s_{m2}. \tag{2}$$

The test using these quantities is known as Behrens' test.

3.5.4 *Example*

For the example of § 3.5.1,

$$\theta = \tan^{-1}\frac{24}{35} = 34^\circ, \quad \nu_1 = 9, \quad \nu_2 = 19.$$

From the tables, for $\nu_1 = 8, \nu_2 = 24, \theta = 30^\circ$ the variable d is 2·12 at the 5% fiducial level. Thus the significance level corresponding to the observed value

$$d = 6{\cdot}8/\sqrt{18{\cdot}01} = 1{\cdot}60$$

is certainly greater than 5% (double-tail test).

3.6 EXAMPLE OF THE USE OF THE F DISTRIBUTION TO COMPARE VARIANCES

One observer, in measuring an angle to the nearest minute, obtains a value $\Sigma v_1^2 = 27{\cdot}3$ based on 10 observations. A second observer obtains a value $\Sigma v_2^2 = 12{\cdot}1$ based on 15 observations. Is there sufficient evidence to indicate that the second observer is the more reliable?

Now $\Sigma v^2/\sigma^2$ is distributed as χ^2 with $\nu = n-1$ d.f. If it is assumed that the two observers are equally reliable, so that σ is the same in each case,

$$\begin{aligned} F &= (\chi_1^2/\nu_1)/(\chi_2^2/\nu_2) = \nu_2\,\Sigma v_1^2/\nu_1\,\Sigma v_2^2 \\ &= 14 \times 27{\cdot}3/9 \times 12{\cdot}1 = 3{\cdot}51, \end{aligned}$$

with $\nu_1 = 9$, $\nu_2 = 14$. From the tables, F is 3·21 at the 2·5% level and 4·03 at the 1% level. Hence on the hypothesis that the two observers are equally reliable such a high value of F would only have been found in 2% of such cases. So it is very probable that the second observer is more accurate than the first.

For the variable F' defined by (3.1.2,7),

$$\begin{aligned} F' &= \{\sqrt{(14 \times 27{\cdot}3)} - \sqrt{(9 \times 12{\cdot}1)}\}/\sqrt{(27{\cdot}3 + 12{\cdot}1)} \\ &= (19{\cdot}5 - 10{\cdot}4)/6{\cdot}3 = 1{\cdot}45. \end{aligned}$$

From Table 3.9a, the value of F' for $\nu_1 = \nu_2 = 12$ at the 5% level is 1·15 and at the 1% level 1·59. Hence the observed value of F' corresponds to the 2% level.

3.6.1 *F test for homogeneity*

The n observations y_i, with mean $\bar{y}$, may fall naturally into r separate groups, containing n_j observations, with means $\bar{y}_j$, where

$$n\bar{y} = \Sigma n_j \bar{y}_j. \tag{1}$$

For example, the groups may represent the readings taken on different days. It may be desired to test whether the whole set is homogeneous—whether all the observations can be regarded as having been taken under the same conditions. That is, whether all the y_i can be regarded as coming from a single normal population with mean Y and standard deviation σ.

If y_{ji} denotes an observation in the jth group,

$$\Sigma(y_i - Y)^2/\sigma^2 = \sum_j \sum_i \{(y_{ji} - \bar{y}_j) + (\bar{y}_j - \bar{y}) + (\bar{y} - Y)\}^2/\sigma^2.$$

On expanding the right-hand side, using the properties of the means,

$$\Sigma(y_i - Y)^2/\sigma^2 = \Sigma\Sigma v_{ji}^2/\sigma^2 + \Sigma n_j(\bar{y}_j - \bar{y})^2/\sigma^2 + n(\bar{y} - Y)^2/\sigma^2. \tag{2}$$

Now, from § 2.5.2, each term on the right is distributed as χ^2 independently of the others. The degrees of freedom of the terms of (2) are n, $\Sigma(n_j - 1) = n - r$, $r - 1$, and 1. Thus if the whole set is homogeneous, the ratio

$$\frac{\Sigma\Sigma v_{ji}^2/(n-r)}{\Sigma n_j(\bar{y}_j - \bar{y})^2/(r-1)} \tag{3}$$

will be distributed as F with $(n-r, r-1)$ degrees of freedom.

If the observations are of different weight, Σv_{ji}^2 is replaced in (3) by $\Sigma w_{ji} v_{ji}^2$ and n_j by $\sum_i w_{ji}$.

3.6.2 *Example*

The values y_{ji} in Table 3.6.2 represent readings of refractive index of air (referred to a convenient origin and scale) obtained on five different days. The observations are combined to give the means for each day $\bar{y}_j$ and the grand mean $\bar{y}$.

The residuals $v_{ji} = y_{ji} - \bar{y}_j$ have been separately evaluated in order to calculate Σv_{ji}^2. By a method similar to that used in establishing (3.6.1,2) it is easy to show that

$$\Sigma y_{ji}^2 = \Sigma v_{ji}^2 + n_j \bar{y}_j^2 \tag{1a}$$

and

$$\Sigma\Sigma y_{ji}^2 = \Sigma\Sigma v_{ji}^2 + \Sigma n_j (\bar{y}_j - \bar{y})^2 + n\bar{y}^2. \tag{1b}$$

These relations provide a very satisfactory check on the arithmetical calculations.

The F-test for homogeneity gives

$$F = \frac{488}{4}\Big/\frac{4064}{34} = 1{\cdot}03 \quad (\nu_1 = 4, \nu_2 = 34),$$

showing that there is no reason to suspect day-to-day changes in the instrument or in the composition of the air.

Statistical workers often rewrite (1*b*) as an analysis of variance table, in the form shown in Table 3.6.2*a*. From (3.6.1,2), the two middle mean squares are estimates of the variance σ^2. The other two mean squares would also be estimates of σ^2 if Y were zero.

TABLE 3.6.2

Observations of refractive index on different days (Example 3.6.2)

Day j	1	2	3	4	5	SUM
Observations y_{ji}	38 53 07 22 31 37 46 33 40 46	30 36 42 38 32 37 46 24	27 53 34 34 39 50	16 48 31 24 40 15 29 30 39	38 18 47 20 21 31	
Number n_j Σy_{ji} $\bar{y}_j$ $\bar{y}_j - \bar{y}$	10 353 35·30 +1·40	8 285 35·62 +1·72	6 237 39·50 +5·60	9 272 30·22 −3·68	6 175 29·17 −4·73	39 1322 $\bar{y}$ 33·90 $\Sigma n_j(\bar{y}_j - \bar{y})$ −0·14
Σy_{ji}^2	14037	10489	9871	9184	5779	49360
$n_j \bar{y}_j^2$ Σv_{ji}^2	12461 1577	10153 337	9362 511	8220 964	5104 675	45300 4064
				CHECK SUM		49364
				$\Sigma n_j (\bar{y}_j - \bar{y})^2$ $n\bar{y}^2$ CHECK SUM		488 44812 45300

TABLE 3.6.2*a*

Analysis of variance table for Example 3.6.2

Deviations		Sum of squares	d.f.	Estimated variance
Of grand mean from zero	$n\bar{y}^2$	44812	1	
Of each mean from grand mean	$\Sigma n_j (\bar{y}_j - \bar{y})^2$	488	4	122
Of observations from each mean	$\Sigma\Sigma v_{ji}^2$	4064	34	119
Of observations from zero	$\Sigma\Sigma y_{ji}^2$	49360	39	

3.7 THE REJECTION OF OUTLYING OBSERVATIONS

A number of tables have been devised giving the probability, assuming a normal frequency distribution, of obtaining large deviations from the true value and from the mean. These tables may be used to test whether there is something unusual about a particular observation, and hence whether it should be omitted from the series.

There is a very real danger of introducing bias by rejecting an outlying observation. The occurrence of such observations may in fact be an indication that the observations do not follow a normal law, and it is then obviously incorrect to apply tests based on the normal law. Most scientists would feel that it is wrong to reject an observation merely because it lies a long way from the mean. Certainly the decision to apply such a test should be made before the observations are taken, and not afterwards in an effort to improve the accuracy. The occurrence of outlying observations is an indication that the apparatus is not as well under control as had been hoped. Of course, it is perfectly legitimate to reject observations in which an obvious copying or reading error has been made.

The question of the rejection of outlying observations has been discussed by Jeffreys (pp. 188, 280, 287), Brunt (p. 129), and Wilson (p. 256), among others. It is only when the number of observations n is small that the question is of importance. When n is small the rejection of an observation may produce a large change in the estimate of mean and standard deviation, while when n is large the effect of any single observation is much less.

The tables designed for the testing of an outlying observation y_0 are listed below.

(*a*) Population mean and standard deviation known: *Biometrika Tables*, Table 24. Upper and lower percentage points of $|y_0 - Y|/\sigma$ at 10, 5, 2·5, 1, 0·5, and 0·1 per cent confidence levels; $n = 1(1)30$.

(*b*) Population standard deviation known: *Biometrika Tables*, Table 25. Percentage points of $(y_0 - \bar{y})/\sigma$; $n = 3(1)9$.

(*c*) Estimate of standard deviation s_ν based on ν d.f., independent of the set being tested: *Biometrika Tables*, Table 26. Five and one percentage points of $(y_0 - \bar{y})/s_\nu$; $n = 3(1)9$; ν from 10 to ∞ at irregular intervals. Due to Nair, and extended by him (Nair, 1952) to include 10, 2·5, 0·5, and 0·1 percentage points.

(*d*) Estimate of standard deviation s from set being tested: Grubbs (1950). 10, 5, 2·5, and 1 percentage points of $(y_0 - \bar{y})/s$; $n = 3(1)25$.

(*e*) Test based on range: Dixon (1950, 1951). Tables of

$$r_{ij} = \frac{x_{i+1} - x_1}{x_{n-j} - x_1},$$

where x_1 is the suspected observation.

An account of (*c*) and (*e*), with examples, is given by Proschan (1953).

3.8 NOTES AND REFERENCES

(3.1) Student was the pen-name of W. S. Gosset, a chemist who made substantial contributions to statistics. He discovered the distribution of t in 1908. 'Studentized' is an accepted adjective in statistical literature for describing a quantity whose distribution is independent of the scale parameter σ (Kendall, 1948, II, p. 80).

Lord (1947, 1950) discusses the use of range in place of standard deviation in a t-test. See also *Biometrika Tables*, Section 14.

Nekrassoff (1930) gives a nomogram for the t-test, while Crow (1945) gives a chart of χ^2 and t distributions.

(3.2) The theory of statistical estimation is only treated very briefly. The discussion of confidence and fiducial intervals follows Kendall (1948, II, Chs 19 and 20). Some writers do not distinguish clearly between the two intervals. The theory of estimation as developed by Neyman and Pearson requires the specification of alternative hypotheses to that under test, and is described in Ch. 26 of Kendall's book. A non-mathematical account is given by Wilson (1952, Ch. 8).

(3.5) A straightforward account of the combination of means with different variances is given by Cochran (1954*a*). References on Welch's method and related topics are: Aspin (1949), Welch (1937, 1947, 1951), Trickett and Welch (1954), Uttam Chand (1950), James (1951, 1954), and Meier (1953).

Bartlett (1937) and Hartley (1950) give tests for heterogeneity of variances. See also *Biometrika Tables*, Section 16.

(3.6) The effect of unequal group variances on the F-test for homogeneity of group means is discussed by Horsnell (1953).

3.9 TABLES

TABLE 3.9*a*

Values of $F' = (\chi_1 \nu_2^{\frac{1}{2}} - \chi_2 \nu_1^{\frac{1}{2}})/(\chi_1^2 + \chi_2^2)^{\frac{1}{2}}$ *for significance levels* $Q = 0{\cdot}05$ *(upper figure) and* $Q = 0{\cdot}01$ *(lower figure).* ν_1 *is d.f. of variable of greater* χ^2/ν. *Values of* Q *for single-tail test; for double-tail test levels are* $2Q = 0{\cdot}10$ *and* $0{\cdot}02$

ν_2 \ ν_1	1	2	3	6	12	24	∞
1	0·92 0·98	0·93 0·99	0·93 0·99	0·93 0·99	0·94 0·99	0·94 0·99	0·94 0·99
2	1·03 1·26	1·06 1·27	1·07 1·27	1·08 1·27	1·09 1·27	1·09 1·27	1·09 1·27
3	1·04 1·38	1·09 1·39	1·11 1·39	1·12 1·39	1·13 1·39	1·13 1·39	1·14 1·39
6	1.02 1·49	1·09 1·51	1·11 1·52	1·14 1·52	1·15 1·52	1·16 1·52	1·17 1·52
12	1·00 1·54	1·07 1·57	1·10 1·58	1·13 1·59	1·15 1·59	1·17 1·58	1·18 1·57
24	0·98 1·56	1·05 1·60	1·09 1·61	1·12 1·62	1·14 1·62	1·16 1·62	1·17 1·60
∞	0·96 1·57	1·04 1·62	1·06 1·64	1·10 1·65	1·12 1·65	1·14 1·66	1·16 1·64

TABLE 3.9*b*

Values of $t = X/\chi\nu^{-\frac{1}{2}} = (z - Z)/s(z)$ *for various significance levels* Q. $Q(t)$ *is the probability of obtaining a value greater than* t *(single-tail test).* $Q(t)$ *is also the probability of obtaining a value less than* $-t$. $2Q(t)$ *is the probability of obtaining a value greater in magnitude than* t*, irrespective of sign (double-tail test)*

Q $2Q$	0·1 0·2	0·05 0·1	0·025 0·05	0·01 0·02	0·005 0·01	0·0025 0·005	0·001 0·002	0·0005 0·001
ν								
1	3·1	6·3	12·7	31·8	63·7	127	318	637
2	1·9	2·9	4·3	7·0	9·9	14·1	22·3	31·6
3	1·6	2·4	3·2	4·5	5·8	7·5	10·2	12·9
5	1·5	2·0	2·6	3·4	4·0	4·8	5·9	6·9
8	1·4	1·9	2·3	2·9	3·4	3·8	4·5	5·0
12	1·4	1·8	2·2	2·7	3·1	3·4	3·9	4·3
24	1·3	1·7	2·1	2·5	2·8	3·1	3·5	3·7
∞	1·3	1·6	2·0	2·3	2·6	2·8	3·1	3·3

CHAPTER 4

DISCRETE DISTRIBUTIONS

The distributions to be discussed in this chapter are those appropriate to counting experiments, where the numbers of events falling into specified classes are determined. The values of the observed variables are then integral, and the distributions differ in certain respects from the continuous distributions discussed in earlier chapters.

4.1 THE BINOMIAL DISTRIBUTION

If the probability of an event occurring in a trial or experiment is denoted by p, then the probability of it not occurring is $q = 1-p$. When n trials are made, the probability of the event occurring in r of them is

$$f(r \mid n) = \frac{n!}{(n-r)!\,r!}\,p^r q^{n-r}. \tag{1}$$

For $p^r q^{n-r}$ gives the probability that a particular sub-set of the n trials should be successful—for example, $p^r q^{n-r}$ is the probability that the event occurs in the 2nd, 4th, 6th, ..., $2r$th trials and not in the others. The numerical factor $({}^nC_r)$ takes account of the fact that the order of the successful trials is immaterial. The frequency distribution (1) is called the binomial distribution.

The characteristic function can be defined in a similar way to the definition for the continuous distribution, the integral being replaced by a sum. Thus

$$\phi(t) = \sum_{r=0}^{n} \mathrm{e}^{irt} f(r) = \sum_{r=0}^{n} \mathrm{e}^{irt} \frac{n!}{(n-r)!\,r!}\,p^r q^{n-r},$$

and so

$$\phi(t) = (p\,\mathrm{e}^{it}+q)^n. \tag{2}$$

The cumulative function is

$$\psi(t) = \log \phi(t) = n \log\{1+p(\mathrm{e}^{it}-1)\}. \tag{3}$$

The cumulants are found by expanding this as a logarithmic series and collecting powers of $(it)^j/j!$. The first four cumulants are

$$\left.\begin{aligned} \kappa_1 &= np, & \kappa_2 &= npq, \\ \kappa_3 &= npq(1-2p), & \kappa_4 &= npq(1-6pq). \end{aligned}\right\} \tag{4}$$

The variance of r is npq. For the standardized variable

$$\mathscr{R} = \frac{r - np}{\sqrt{(npq)}}, \tag{5a}$$

the cumulants will be, from (1.7.3,6), the values (4) divided by $(npq)^{\frac{1}{2}j}$. Hence for the variable $\mathscr{R}$,

$$\kappa_2 = 1, \quad \kappa_3 = (npq)^{-\frac{1}{2}}(1-2p), \quad \kappa_4 = (npq)^{-1}(1-6pq), \tag{5b}$$

and, provided npq is not too small, the higher cumulants will be small.

The intervals $\Delta\mathscr{R}$ between neighbouring values of $\mathscr{R}$ are given by $\Delta r = 1$, and so

$$\Delta\mathscr{R} = (npq)^{-\frac{1}{2}}. \tag{6}$$

As n becomes larger, the discontinuous distribution of values $f(\mathscr{R})$ approaches more and more nearly to a continuous distribution whose frequency function is $f_c(\mathscr{R})$, where the probabilities of a value lying in the range $\Delta\mathscr{R}$ about $\mathscr{R}$ are equal in the two cases,

$$f_c(\mathscr{R})\,\Delta\mathscr{R} = f(\mathscr{R}). \tag{7}$$

Now the characteristic functions

$$\int f_c(\mathscr{R})\, e^{it\mathscr{R}}\, d\mathscr{R} \quad \text{and} \quad \Sigma f(\mathscr{R})\, e^{it\mathscr{R}}$$

will be very nearly equal for the two frequency functions, and so will be the cumulants. Hence, from (5b), $f_c(\mathscr{R})$ will be approximately normally distributed if npq is large,

$$f_c(\mathscr{R}) \to \frac{1}{\sqrt{(2\pi)}}\, e^{-\frac{1}{2}\mathscr{R}^2}.$$

Thus from (6) and (7),

$$f(\mathscr{R}) \to \frac{1}{\sqrt{(2\pi npq)}}\, e^{-\frac{1}{2}\mathscr{R}^2} \tag{8a}$$

and

$$f(r) \to \frac{1}{\sqrt{(2\pi npq)}}\, e^{-\frac{1}{2}(r-np)^2/npq}. \tag{8b}$$

Provided npq is not very small, the frequency function $f(r)$ will approximate closely to a continuous function of the normal form with mean np and variance npq.

4.2 THE TESTING OF HYPOTHESES BY THE χ^2 TEST

Suppose that the range of the observations is divided up into m groups, and the hypothesis predicts the numbers R_i of the observations which should fall into the ith group. The predicted number R_i equals Np_i, where N is the predicted total number and p_i the predicted fraction in the ith group. The distribution of the observed number r_i is of the normal form (4.1,8*b*),

$$f(r_i) \propto \exp[-\tfrac{1}{2}(r_i - R_i)^2/Np_i q_i], \tag{1}$$

and so $(r_i - R_i)^2/Np_i q_i$ will be distributed as χ^2 with 1 d.f. Hence

$$\chi^2 = \sum_{i=1}^{m} \frac{(r_i - R_i)^2}{R_i q_i} \tag{2}$$

will be distributed as χ^2 with m d.f. If m is fairly large, q will be quite close to unity, and in practical tests the form

$$\chi^2 = \sum_{i=1}^{m} \frac{(r_i - R_i)^2}{R_i} \tag{3}$$

is more usually employed.

When χ^2 has been calculated, the significance level $Q(\chi^2)$ can be found and a decision made as to whether the divergence of the r_i from the R_i is reasonable or not.

The form (1) is valid only if $Np_i q_i \doteqdot Np_i \doteqdot R_i$ is fairly large, since the higher order cumulants are negligible and the distribution of r_i approximates to the normal form only if this is so. The groups must be chosen so that none of the expected values R_i is very small. Five is often suggested as the minimum allowable value for R_i.

4.2.1 *Degrees of freedom*

Often the value N of the predicted total number of observations is not known, but must be estimated from the observed number n. Then the R_i are not known, and only the values

$$\hat{R}_i = np_i$$

are available. Since $\qquad \Sigma r_i = \Sigma \hat{R}_i = n,$

$$\Sigma(r_i - R_i)^2/R_i = \Sigma(r_i - \hat{R}_i)^2/R_i + \Sigma(n - N)^2 p_i^2/R_i.$$

It follows, as in § 2.5.2, that $\Sigma(r_i - \hat{R}_i)^2/R_i$ will be distributed as χ^2 with $(m-1)$ d.f. Hence

$$\sum_{i=1}^{m} (r_i - \hat{R}_i)^2/\hat{R}_i = \sum_{i=1}^{m} (r_i - np_i)^2/np_i \tag{1}$$

will be distributed approximately as χ^2 with $m-1$ d.f. One degree of freedom is lost because the values $\hat{R}_i$ are not independent, but $\Sigma \hat{R}_i = n$.

More generally, it may be assumed that each adjustable parameter (such as N) in the distribution law which must be estimated from the data reduces the degrees of freedom by unity. Thus

$$\sum_{i=1}^{m} (r_i - \hat{R}_i)^2/\hat{R}_i \tag{2}$$

may be assumed to be distributed approximately as χ^2 with $m-q$ d.f., if q parameters obtained from the data are used in calculating the predicted values $\hat{R}_i$.

Since the proof of this statement depends on the parameters being chosen to minimize

$$\chi^2 = \Sigma(r_i - \hat{R}_i)^2/R_i, \tag{3}$$

it is strictly only true when the parameters are in fact chosen in this way. A short outline of the proof will now be given.

Suppose the R_i are functions of q parameters B_j,

$$R_i = x_i(B_j).$$

Then if the estimate of B_j is b_j, the estimate of R_i is

$$\hat{R}_i = R_i + \sum_{j=1}^{q} (b_j - B_j)\, x_{ij},$$

where $$x_{ij} = \frac{\partial x_i}{\partial B_j}.$$

Thus $$\Sigma(r_i - R_i)^2/R_i = \Sigma\{r_i - \hat{R}_i + \Sigma x_{ij}(b_j - B_j)\}^2/R_i.$$

If the values b_j are chosen to minimize (3),

$$\Sigma(r_i - \hat{R}_i)\frac{\partial \hat{R}_i}{\partial b_j}\Big/ R_i = 0 = \Sigma(r_i - \hat{R}_i)\, x_{ij}/R_i \tag{4}$$

and

$$\Sigma(r_i - R_i)^2/R_i = \Sigma(r_i - \hat{R}_i)^2/R_i + \Sigma\{\Sigma(b_j - B_j)\, x_{ij}\}^2/R_i. \tag{5}$$

The two terms on the right-hand side of (5) may be shown to be distributed as χ^2, with $m-q$ d.f. and q d.f. respectively. The proof is on the same lines as that of §§ 2.5.2 and 8.2.1, and will not be repeated in full. If the substitution

$$\Sigma a_j T_{ij} = \Sigma b_j x_{ij}$$

is made, where the T_{ij} are orthogonal functions satisfying

$$\sum_i T_{ij} T_{ik}/R_i = 0,$$

then the second term in (5) becomes

$$\sum_i \frac{1}{R_i}\{\Sigma(a_j - A_j)\, T_{ij}\}^2 = \sum_j (a_j - A_j)^2 \sum_i T_{ij}^2/R_i,$$

and (5) is of the same form as (8.2.1,3).

4.2.2 *Example*

One thousand throws were made with a six-sided die. The number of times the various faces came uppermost are recorded in Table 4.2.2.

TABLE 4.2.2

Throws with a six-sided die (*Example* 4.2.2)

Face	No. of times r_i	χ^2
1	173	0·24
2	194	4·48
3	152	1·29
4	165	0·02
5	181	1·23
6	135	6·02
		13·28

On the hypothesis that the die is uniform, $p_i = 1/6$ and $\hat{R}_i$, the predicted number of times, is $1000/6 = 166\ 2/3$ for all six faces. Hence on this hypothesis the values $\chi^2 = (r_i - \hat{R}_i)^2/\hat{R}_i$ in the third column are obtained, and $\Sigma\chi^2 = 13{\cdot}28$. Since the total number n is used in estimating $\hat{R}_i$, the degrees of freedom are 5. The value

$$\chi' = \surd\chi^2 - \surd\nu = 3{\cdot}64 - 2{\cdot}23 = 1{\cdot}41,$$

and so from Table 2.8*b*

$$Q(\chi^2) = 0{\cdot}02.$$

This is, if the die were uniform, a value of χ^2 as high as this would be obtained once in fifty trials. Hence it is unlikely that the die is uniform.

Another example of the use of χ^2 in the testing of hypotheses is given in § 4.3.4.

4.3 THE POISSON DISTRIBUTION

4.3.1 *The counting of particles*

In many experiments in nuclear physics and cosmic-ray physics, it can be assumed that the average rate of arrival of the particles is a constant, denoted by the symbol μ. Then the probability of a particle arriving in a small time interval dt will be μdt. The form of the frequency function $f(r\,|\,t, \mu)$ giving the probability of r events occurring in time t will now be determined.

If dt is small, the probability of two events occurring in time dt will be negligible. Consider the probability of r events occurring in time $t+dt$. This can come about in one of two different ways; either:

(i) r events occurred in the interval $(0, t)$ and none in the interval $(t, t+dt)$; the probability of this is the product $f(r\,|\,t, \mu) \times (1-\mu dt)$; or

(ii) $r-1$ events occurred in the interval $(0, t)$ and one in the interval $(t, t+dt)$; the probability of this is the product

$$f(r-1 \mid t, \mu) \times \mu dt.$$

Since these two alternatives are mutually exclusive, the probability $f(r \mid t+dt, \mu)$ is the sum of the individual probabilities. Thus

$$f(r \mid t+dt, \mu) = f(r \mid t, \mu) + \mu dt\{f(r-1 \mid t, \mu) - f(r \mid t, \mu)\}.$$

Dividing by dt, and proceeding to the limit,

$$\frac{d}{dt} f(r \mid t, \mu) = \mu\{f(r-1 \mid t, \mu) - f(r \mid t, \mu)\}. \tag{1}$$

The solution of this equation is

$$f(r \mid t, \mu) = e^{-\mu t} \frac{(\mu t)^r}{r!}, \tag{2}$$

as may be verified by substituting (2) in (1).

Since the function depends on the product μt, and not on the individual values μ and t,

$$f(r \mid \mu t = \lambda) = e^{-\lambda} \frac{\lambda^r}{r!}. \tag{3}$$

A distribution whose frequency function is of the form (3) is called a Poisson distribution.

4.3.2 *Characteristic function*

The characteristic function for the Poisson distribution (written $\phi(z)$ instead of $\phi(t)$ to avoid confusion with the time variable) is

$$\phi(z) = \sum_{r=0}^{\infty} e^{irz} f(z \mid \lambda) = e^{-\lambda} \sum_{0}^{\infty} \frac{(\lambda e^{iz})^r}{r!},$$

and so
$$\phi(z) = \exp \lambda(e^{iz} - 1). \tag{1}$$

The cumulative function is

$$\psi(z) = \lambda(e^{iz} - 1), \tag{2}$$

and
$$\kappa_j = \lambda. \tag{3}$$

All the cumulants are equal to λ. In particular,

$$E(r) = \lambda, \quad \operatorname{var} r = \lambda. \tag{4}$$

In standard measure, the cumulants of $r/\lambda^{\frac{1}{2}}$ are $\kappa_j = \lambda^{-\frac{1}{2}j+1}$. Hence if λ is not too small, the higher cumulants will be unimportant and the distribution approaches the normal form, in the same way as does the binomial distribution.

4.3.3 *Estimation*

Since the expectation of r is λ, it follows that the observed number of arrivals r in time t will be an unbiased estimate of λ.

Hence $$\hat{\mu} = r/t \tag{1}$$

will be an unbiased estimate of the parameter μ. The variance of this estimate is given by

$$\text{var}\, \hat{\mu}t = \lambda,$$

and so $$s = r^{\frac{1}{2}}/t \tag{2}$$

will provide an estimate of the standard deviation of $\hat{\mu}$. Hence

$$\hat{\mu} = (r \pm r^{\frac{1}{2}})/t. \tag{3}$$

4.3.4 *Example*

The number of cosmic-ray particles detected by a coincidence telescope in an undergraduate laboratory experiment was recorded for a period of one hour (Guest and Simmons, 1953). In that time 87 particles were detected. Hence the estimate of μ, the number arriving per minute, is

$$\mu = (87 \pm 9{\cdot}3)/60 = 1{\cdot}45 \pm 0{\cdot}16.$$

The number of particles occurring in each minute interval was also recorded. Table 4.3.4 lists the number of minute intervals in which r particles were recorded.

TABLE 4.3.4

Number of minutes in which r particles were recorded

r	Observed no.	$f(r)$	Predicted no.	χ^2
0	13	0·235	14·1	0·09
1	22	0·340	20·4	0·13
2	14	0·247	14·8	0·04
3	7	0·119	7·1	0·00
4	4	0·043	2·6	0·04
>4	0	0·016	1·0	
			Sum	0·30

If the rate of arrival follows a Poisson distribution, t is unity and

$$f(r\,|\,\mu = \lambda) = e^{-\mu}\frac{\mu^r}{r!}.$$

Hence the predicted frequency function is

$$f(r\,|\,\mu = 1{\cdot}45) = e^{-1{\cdot}45}\frac{1{\cdot}45^r}{r!} = 1{\cdot}45^r/4{\cdot}263r!$$

and the predicted numbers in the groups are these values multiplied by 60. The predicted numbers are shown in Table 4.3.4.

To test the agreement between the observed and predicted numbers, χ^2 is calculated. The categories $n = 4, n > 4$ are grouped together, since the expectations in each are small. Since both the parameter μ and the expected total number are calculated from the data, two d.f. are lost. Hence χ^2 is 0·30 with 3 d.f. (χ' is $-1{\cdot}18$), and $Q(\chi^2)$, the probability of obtaining a higher value of χ^2, is 0·96.

The agreement between the observed and predicted frequencies is then much better than would normally be expected. The obvious conclusion would be that this set of readings has been specially selected as the example for the published paper because of the good agreement between the frequencies. But in fact this obvious conclusion happens to be untrue. The equipment was simply set up and the readings taken over a period of an hour. It was not discovered till later that the agreement was 'too good'. The temptation to obtain a further set of readings leading to a smaller χ^2 was resisted. The experiment has been repeated by a large number of students, who have obtained a wide range of values for $Q(\chi^2)$. The conclusion that there is something queer about the published set is in fact merely an error of the first kind.

4.4 THE POISSON DISTRIBUTION AS THE LIMITING FORM OF THE BINOMIAL DISTRIBUTION

The binomial distribution for rare events, with p very small and n sufficiently large so that np is finite, approximates to the Poisson form. For, using Stirling's approximation to the factorial,

$$\frac{n!}{(n-r)!} = \frac{(2\pi)^{\frac{1}{2}}\, n^{n+\frac{1}{2}}\, \mathrm{e}^{-n}}{(2\pi)^{\frac{1}{2}}\,(n-r)^{n-r+\frac{1}{2}}\,\mathrm{e}^{-n+r}} = n^r(1-r/n)^{-n+r-\frac{1}{2}}\,\mathrm{e}^{-r},$$

and so (4.1,1) becomes

$$\begin{aligned} f(r \mid n) &= \frac{1}{r!}\, n^r\left(1-\frac{r}{n}\right)^{-n+r-\frac{1}{2}} \mathrm{e}^{-r}\, p^r (1-p)^{n-r} \\ &= \frac{(np)^r}{r!}(1-p)^n\left[\left(1-\frac{r}{n}\right)^{-n}\mathrm{e}^{-r}\right]\left[\left(1-\frac{r}{n}\right)^{r-\frac{1}{2}}(1-p)^{-r}\right]. \end{aligned}$$

Now $$(1-p)^n = \left(1-\frac{pn}{n}\right)^n \doteqdot \mathrm{e}^{-pn}, \quad \left(1-\frac{r}{n}\right)^{-n} \doteqdot \mathrm{e}^r,$$

and since $p \sim r/n$ is small the last term in square brackets will be close to unity. Thus

$$f(r \mid n) \to \frac{(np)^r}{r!}\,\mathrm{e}^{-np},$$

and in the limit as $p \to 0$, $n \to \infty$, the binomial distribution approaches the Poisson form with $\lambda = np$.

4.5 SIGNIFICANCE LEVELS FOR THE POISSON DISTRIBUTION

The significance levels for the Poisson distribution can be found from the levels of a related χ^2 distribution. From (2.4,4), if $\nu = 2m$,

$$P(\chi^2 \mid m) = \frac{1}{2^m(m-1)!}\int_0^{\chi^2} e^{-\frac{1}{2}z^2}(z^2)^{m-1}\,dz^2.$$

Integrating this by parts,

$$P(\chi^2 \mid m) = \frac{1}{2^m(m-1)!}\left\{\left[\frac{1}{m}(z^2)^m e^{-\frac{1}{2}z^2}\right]_0^{\chi^2} + \frac{1}{2m}\int_0^{\chi^2} e^{-\frac{1}{2}z^2}(z^2)^m\,dz^2\right\},$$

or

$$P(\chi^2 \mid m) = \frac{(\frac{1}{2}\chi^2)^m}{m!}e^{-\frac{1}{2}\chi^2} + P(\chi^2 \mid m+1).$$

It follows that

$$P(\chi^2 \mid m) = \sum_{k=m}^{\infty}\frac{(\frac{1}{2}\chi^2)^k}{k!}e^{-\frac{1}{2}\chi^2}. \qquad (1)$$

For the Poisson distribution, the probability of obtaining a value greater than or equal to r,

$$Q(r \mid \lambda) = 1 - P(r-1 \mid \lambda), \qquad (2)$$

is from (4.3.1,3)

$$Q(r \mid \lambda) = \sum_{k=r}^{\infty}\frac{\lambda^k}{k!}e^{-\lambda}. \qquad (3)$$

Hence, comparing (1) and (3),

$$Q(r \mid \lambda) = P(\chi^2 = 2\lambda \mid \nu = 2r). \qquad (4)$$

It will be noted that, as the Poisson distribution is discrete, there is a finite probability of the value r occurring, and the relation $Q(y) = 1 - P(y)$ for the continuous distribution is replaced by the relation (2).

The probability of obtaining a value less than or equal to r is

$$P(r \mid \lambda) = 1 - Q(r+1 \mid \lambda) = 1 - P(\chi^2 = 2\lambda \mid \nu = 2r+2),$$

and so

$$P(r \mid \lambda) = Q(\chi^2 = 2\lambda \mid \nu = 2r+2). \qquad (5)$$

Hence the upper and lower limits for r at a given significance level can be found from the corresponding limits for the χ^2 distribution.

4.5.1 *Tables of Poisson limits*

Table 40 of the *Biometrika Tables* gives as functions of r the values λ_1 and λ_2 for which

$$P(r \mid \lambda_1) = \alpha, \quad Q(r \mid \lambda_2) = \alpha. \qquad (1)$$

The values λ_1 and λ_2 are given for five significance levels from 0·05 to 0·001, and for values of r 0(1)30(5)50.

The values λ_1 and λ_2 defined by (1) correspond to confidence limits for the parameter λ, as in § 3.2.2. For if r_1 and r_2 are the two values of r for which

$$P(r_1|\lambda) = \alpha, \quad Q(r_2|\lambda) = \alpha, \tag{2}$$

λ_1 is greater than λ when r is greater than r_1, and λ_2 is less than λ when r is less than r_2. So

$$\lambda_1 \geqslant \lambda \geqslant \lambda_2 \quad \text{when} \quad r_2 \geqslant r \geqslant r_1.$$

But, from (2), $r_2 \geqslant r \geqslant r_1$ in a fraction $1-2\alpha$ of a set of observations of r. Hence

$$\Pr\{\lambda_1 \geqslant \lambda \geqslant \lambda_2\} = 1-2\alpha. \tag{3}$$

The limits λ_1 and λ_2 can also be found from the square root approximation formulae

$$\lambda_1 = \{\surd(r+1)+X''\}^2, \tag{4a}$$

$$\lambda_2 = \{\surd r+X''\}^2, \tag{4b}$$

where the values X'' are given in Table 4.8*a* for selected values of r. X'' varies only slowly with r, and is normally distributed with magnitude one-half the standardized normal variate X when r is large.

4.5.2 *Example*

An observation with the cosmic-ray telescope yields 28 particles in 20 minutes. From the *Biometrika Tables*, $\lambda_1 = 40{\cdot}5$ and $\lambda_2 = 18{\cdot}6$ at the 5% (double-tail) level. The corresponding limits for $\mu = \lambda/t$ are 2·02 and 0·93.

For the square root approximation, from Table 4.8*a*, $X'' = \pm 0{\cdot}98$, and

$$\lambda_1 = (\surd 29+0{\cdot}98)^2 = 40{\cdot}5;$$

$$\lambda_2 = (\surd 28-0{\cdot}98)^2 = 18{\cdot}6.$$

4.5.3 *Effect of r being limited to integral values*

In the discussion of § 4.5.1, it was assumed that values r_1 and r_2 satisfying (4.5.1,2) could be found. Though this would be true for a continuous distribution, it is not true in counting experiments where the variable is limited to integral values. For such cases (4.5.1,2) is replaced by

$$P(r_1|\lambda) = \alpha_1, \quad Q(r_2|\lambda) = \alpha_2, \tag{1}$$

where r_2 is the smallest integer for which $\alpha_2 < \alpha$ and r_1 the largest integer for which $\alpha_1 < \alpha$. Then, if $r > r_1$, $P(r|\lambda) \geqslant \alpha$, while

$P(r\,|\,\lambda_1) = \alpha$, and so $\lambda_1 \geqslant \lambda$. Similarly, if $r < r_2$, $\lambda_2 \leqslant \lambda$. Hence

$$\lambda_1 \geqslant \lambda \geqslant \lambda_2 \quad \text{when} \quad r_2 > r > r_1. \tag{2}$$

But $$\Pr\{r > r_1\} = Q(r_1+1) = 1 - P(r_1) = 1 - \alpha_1$$

and $$\Pr\{r \geqslant r_2\} = Q(r_2) = \alpha_2,$$

and so from (2)

$$\Pr\{\lambda_1 \geqslant \lambda \geqslant \lambda_2\} = 1 - \alpha_1 - \alpha_2 > 1 - 2\alpha. \tag{3}$$

The true significance level will be greater than the nominal level $1-2\alpha$. However, the difference between the two levels will be small unless λ is small.

As an example, if λ has the value 15, the expressions

$$P(5\,|\,15) = 0{\cdot}002792, \quad P(6\,|\,15) = 0{\cdot}007631,$$
$$Q(26\,|\,15) = 0{\cdot}006185, \quad Q(27\,|\,15) = 0{\cdot}003312,$$

can be obtained by summing the individual terms of (4.3.1,3), which are listed in Table 39 of the *Biometrika Tables*. Now, from Table 40 for $1-2\alpha = 0{\cdot}99$, λ_1 is less than 15 when $r \leqslant 5$ and λ_2 is greater than 15 when $r \geqslant 27$. Hence

$$\Pr\{\lambda_1 \geqslant \lambda \geqslant \lambda_2\} = 1 - P(5\,|\,15) - Q(27\,|\,15) = 0{\cdot}993896,$$

and the true significance level is 0·994 rather than 0·99.

4.5.4 *The sum of two Poisson variables*

The characteristic function for the sum of two Poisson variables with parameters λ_1 and λ_2 is, from (1.7.3,2),

$$\begin{aligned}\phi(t) &= \phi_1(t)\,\phi_2(t)\\ &= \{\exp\lambda_1(\mathrm{e}^{it}-1)\}\{\exp\lambda_2(\mathrm{e}^{it}-1)\}\\ &= \exp\{(\lambda_1+\lambda_2)(\mathrm{e}^{it}-1)\}.\end{aligned}$$

The frequency function for the sum will be of the Poisson form,

$$f(r) = \mathrm{e}^{-\lambda}\frac{\lambda^r}{r!}, \tag{1}$$

with parameter $\lambda = \lambda_1 + \lambda_2$.

4.5.5 *Comparison of two estimates of a Poisson parameter*

On the hypothesis that two observed values r_1 and r_2 came from a population following a Poisson distribution with parameter λ,

$$f(r_1, r_2\,|\,\lambda) = (\mathrm{e}^{-\lambda}\lambda^{r_1}/r_1!)\,(\mathrm{e}^{-\lambda}\lambda^{r_2}/r_2!),$$

or $$f(r_1, r_2\,|\,\lambda) = [\mathrm{e}^{-2\lambda}(2\lambda)^r/r!]\,[r!/2^r r_1!\,r_2!], \tag{1}$$

where $r = r_1 + r_2$. The first term is $f(r\,|\,\lambda)$, from (4.5.4,1).

Now the values r_1 and r_2 will occur when the sum is $r = r_1 + r_2$, and one of the values is r_1. Hence, using the fundamental theorem for the product of probabilities, equation (1.1,3),

$$f(r_1, r_2 \mid \lambda) = f(r_1 + r_2 \mid \lambda) f(r_1 \mid r_1 + r_2, \lambda). \tag{2}$$

Thus, on comparing (1) and (2), $f(r_1 \mid r_1 + r_2)$ is the second term in (1) and

$$f(r_1 \mid r = r_1 + r_2) = \binom{r}{r_1} 2^{-r}. \tag{3}$$

So the probability, given the sum r, that a value for the smaller observation less than or equal to a particular value b will be obtained is

$$P(b \mid r) = \sum_{j=0}^{b} \binom{r}{j} 2^{-r}. \tag{4}$$

In the *Biometrika Tables*, Table 36 gives values of b corresponding to certain significance levels for values of r in the range 1(1)80. Hence a test can be made as to whether two values r_1 and r_2 are consistent.

The variance of the difference $r_2 - r_1$ will be 2λ, and, since $E(r_1 + r_2)$ is 2λ, $r_1 + r_2$ will be an estimate of the variance of the difference. It is found that the expression

$$\frac{r_2 - r_1 - 2}{(r_1 + r_2)^{\frac{1}{2}}} = \frac{r - 2(r_1 + 1)}{\sqrt{r}} \tag{5}$$

is distributed approximately as X, the standardized normal variate. Calculation shows that, when r exceeds 10, the values for r_1 obtained by equating (5) to various values of X are practically always the same as those given in the tables (occasionally, when the value given by (5) is near a half-integer, it will be rounded to the integer above or below that given by the tables).

It follows that the simple formula (5) provides an adequate test for the agreement between two values r_1 and r_2. The test will usually be a double-tail test, to determine whether the two values are significantly different. As in § 4.5.3, the fact that the distribution of r is discrete makes the true significance level somewhat greater than the nominal significance level for small values of λ. Table 36B of the *Biometrika Tables* shows the true levels for certain nominal levels as functions of λ.

4.5.5.1 *Example*

In Example 4.3.4, 87 particles were recorded in the hour. In another experiment 73 particles were recorded. Are these values discordant?

From (4.5.4,5)

$$X = (r_2 - r_1 - 2)/\sqrt{(r_1 + r_2)} = 12/\sqrt{160} = 0{\cdot}95.$$

From Table 2.8*a*, $\Pr\{|X| \geqslant 0{\cdot}95\} \doteqdot 0{\cdot}4$. So there is no evidence that the two values are discordant.

4.5.6 *F test for estimates of a Poisson parameter*

If r is fixed, it follows from (4.5,4) that $2\mu t$ is distributed as χ^2 with $2r$ d.f. Hence if the times t_1 and t_2 to count r_1 and r_2 particles are measured, the ratio

$$\frac{t_1/r_1}{t_2/r_2} = \frac{\hat{\mu}_2}{\hat{\mu}_1} \tag{1}$$

is distributed as F with $(2r_1, 2r_2)$ d.f., and the two estimates $\hat{\mu}_1$ and $\hat{\mu}_2$ can be tested for agreement. If the times t_1 and t_2 are fixed and the number of particles r_1 and r_2 arriving in these times are determined, it can be shown that the ratio (1) will still be distributed to a good approximation as F (Cox, 1953).

For the example of § 4.5.5.1,

$$F = 87/73 = 1{\cdot}19,$$

or
$$F' = \frac{\sqrt{(r_1 t_2)} - \sqrt{(r_2 t_1)}}{\sqrt{(t_1 + t_2)}} = \frac{\sqrt{87} - \sqrt{73}}{\sqrt{2}} = 0{\cdot}55.$$

4.6 COUNTING LOSSES

When the interval between two successive events is small the counting device may not record the second event. There is usually a finite interval following the recording of an event, variously referred to as the dead time, paralysis time, recovery time, or resolving time, during which the counter is insensitive, and any event occurring in that time will not be recorded. In such cases the number of events counted will be less than the number actually occurring, and it is necessary to apply some correction to allow for counting losses.

In theoretical discussions of counting losses two ideal types of counter are considered. In the first, referred to as a Type I counter by writers on probability theory (Feller, 1948), and as a Type II counter by some physicists (Korff, 1955), the dead time has a constant value τ. In the second type (referred to as Type II or Type I) the dead time has a constant value τ provided no new event occurs in τ. An event occurring during the dead time is not recorded, but it extends the dead time by an amount τ from the instant at which it occurred, and the counter only returns to a sensitive condition when there is an interval of length τ during which no new event occurs. Such a counter may be described as having an extended resolving time. A Geiger counter approximates to an ideal counter of constant (non-extended) resolving time, while a scintillation counter and a mechanical recorder both approximate to ideal counters with extended resolving times.

4.6.1 *Counters with fixed resolving times*

If μ is the rate of occurrence of the events and μ_0 the rate of counting, the counter is insensitive for a time τ for each event counted, and so over a long period t it is insensitive for a time $(\mu_0 t)\tau$. The number of events occurring in this insensitive time is $\mu(\mu_0 t\tau)$, and so, equating the number of events occurring to the number recorded plus the number missed,

$$\mu t = \mu_0 t + \mu\mu_0 t\tau,$$

or

$$\mu = \frac{\mu_0}{1-\mu_0\tau}, \quad \mu_0 = \frac{\mu}{1+\mu\tau}. \tag{1}$$

This gives the correction which should be applied to the observed counting rate to obtain the true rate.

4.6.2 *Counters with extended resolving times*

An event will be recorded in the small interval $t, t+dt$ if an event occurs in this interval, the probability of which is μdt, and if the counter is sensitive. The counter will be sensitive if no event has occurred in the interval τ preceding t, and the probability that no event has occurred in an interval τ is from (4.3.1,2) $e^{-\mu\tau}$. Hence the probability $\mu_0 dt$ of an event being recorded in the interval $t, t+dt$ is given by

$$\mu_0 dt = \mu dt\, e^{-\mu\tau},$$

and the recorded rate is

$$\mu_0 = \mu\, e^{-\mu\tau}. \tag{1}$$

The true rate μ can be found from the observed rate μ_0 by solving this equation.

If $\mu\tau$ is small, $e^{-\mu\tau}$ is approximately equal to $1/(1+\mu\tau)$, and the correction formulae (1) and (4.6.1,1) are very similar.

4.6.3 *Example*

The resolving time of a Geiger counter is $1{\cdot}6 \times 10^{-4}$ seconds. 91,254 counts were obtained over a period of 5 minutes from a gamma-ray source. What is the corrected counting rate?

The observed counting rate is

$$\mu_0 = 91{,}254/5 \times 60 = 304{\cdot}18 \text{ counts/sec.},$$

and so the corrected counting rate is

$$\mu = \mu_0/(1-\mu_0\tau) = 304{\cdot}18/0{\cdot}9513 = 319{\cdot}8 \text{ counts/sec.}$$

If the counter had had an extended resolving time, the corrected counting rate would have been given by

$$304{\cdot}18 = \mu \exp(-1{\cdot}6\times 10^{-4}\mu),$$

the solution of which is $\mu = 320{\cdot}2$ counts/sec.

Since the resolving time τ is seldom known very accurately, it is very desirable that $\mu\tau$ should be much less than unity if accurate rates are to be determined.

4.6.4 *Scaling circuits*

Mechanical recorders have a resolving time of the order of 0·1 second, and so are unsuitable for fast counting. To overcome this, it is usual to insert a scaling circuit between the input and the recorder so that only every Nth event actuates the mechanical recorder. If n_i is the number of events at the input,

$$n_i = nN+\nu, \quad \nu<N,$$

where n is the number shown on the recorder. Provision is usually made for observing from the state of the scaling circuit the number of additional events ν.

If N or more events occur in the dead-time τ of the mechanical recorder, counts will be lost. If the number of events lies between N and $2N$, one count will be lost, if the number lies between $2N$ and $3N$ two counts will be lost, and so on. In practically all cases the scaling factor is chosen so that this loss will be very small. The probability that at least N events occur in time τ is from (4.5,3)

$$Q(N\,|\,\mu\tau) = \sum_{k=N}^{\infty} \frac{(\mu\tau)^k}{k!} e^{-\mu\tau}, \tag{1}$$

and the probability that $2N$ or more events occur in time τ will be assumed negligible. Hence the probability that the recorder will lose one count during the interval τ is equal to $Q(N\,|\,\mu\tau)$, or $Q(N\,|\,\mu\tau)$ is the fraction of the counts lost.

It is customary to select a value of N such that this fraction is negligible rather than to attempt to apply corrections. The value of N for which $Q(N\,|\,\mu\tau)$ has the value α can be found from tables of the sum on the right-hand side of (1) (Molina, 1942). Formula (4.5.1,4*b*) can also be used, in the form

$$\mu\tau = (\sqrt{N}+X'')^2, \tag{2}$$

where the values X'' are those for the lower limit in Table 4.8*a*.

4.6.4.1 *Example*

The counting rate in Example 4.6.3 is about 300 per second. For what value of the scaling factor N will the counting loss be less than 0·1% if the mechanical recorder has a resolving time of 0·1 second?

Here $\mu\tau$ is 30, and from Table 4.8*a* X'' is about $-1{\cdot}5$ for $\alpha = 0{\cdot}001$. Hence (4.6.4,2) gives

$$N = (\sqrt{30}+1{\cdot}5)^2 = 49,$$

while, from Molina's Table II for $a = 30$, $Q(48 \mid 30) = 0{\cdot}001488$ and $Q(49 \mid 30) = 0{\cdot}000887$. A scaling factor greater than 48 is thus required.

4.7 NOTES AND REFERENCES

(4.2) An expository account of the χ^2 test is given by Cochran (1952); see also Cochran (1954*b*).

(4.4) The usual approach in statistics is to treat the Poisson distribution as the limit of the binomial. See, for example, Kendall (1948, I, Ch. 5).

(4.5) Surveys of testing and estimation methods are given by van Klinken and Prins (1954) and Walsh (1954). Approximations to the Poisson distribution are discussed by Blom (1954). Cox (1953) discusses approximate tests.

(4.6) Elementary treatments of counting losses are given by Lewis (1942), Bleuler and Goldsmith (1952), and Korff (1955). More extended discussions are given by Blackman and Michiels (1948), Feller (1948), and Elmore (1950).

4.8 TABLE

TABLE 4.8*a*

Values of X'' for the square root approximation to the Poisson distribution

α	0·001		0·005		0·01		0·025		0·05	
r	Lower	Upper	Lower	Upper	Lower	Upper	Lower	Upper	Lower	Upper
0	—	1·63	—	1·30	—	1·15	—	0·92	—	0·73
1	−0·97	1·62	−0·93	1·31	−0·90	1·16	−0·84	0·95	−0·77	0·76
2	−1·20	1·62	−1·09	1·31	−1·03	1·17	−0·92	0·95	−0·82	0·78
3	−1·30	1·61	−1·15	1·31	−1·07	1·17	−0·95	0·96	−0·83	0·78
5	−1·38	1·61	−1·20	1·31	−1·10	1·17	−0·96	0·97	−0·83	0·79
10	−1·44	1·60	−1·23	1·31	−1·13	1·17	−0·97	0·97	−0·83	0·80
20	−1·48	1·59	−1·26	1·31	−1·14	1·17	−0·98	0·98	−0·83	0·81
40	−1·50	1·58	−1·27	1·30	−1·15	1·17	−0·98	0·98	−0·83	0·81
∞	−1·55	1·55	−1·29	1·29	−1·16	1·16	−0·98	0·98	−0·82	0·82

PART II

REGRESSION THEORY AND THE STRAIGHT LINE

CHAPTER 5

REGRESSION CURVES AND FUNCTIONAL RELATIONSHIP

In this chapter an account is given of the basic theory of regression curves and of curves of functional relationship, as a prelude to the discussion of practical methods of curve-fitting in later chapters.

5.1 REGRESSION

The term 'regression' was originally introduced by Galton (1886), in an investigation of the relation between the heights of parents and the heights of their children. The observed heights are denoted by the symbols x_i and y_i respectively, and the range of the variable x is divided up into a number of small intervals $x_h \pm \frac{1}{2}dx$. If y_h denotes the mean of all the y_i in the interval centred on x_h, the points (x_h, y_h) were found to lie approximately on the straight line

$$y - \bar{y} = b_{1y}(x - \bar{x}), \tag{1}$$

where $\bar{x}$ and $\bar{y}$ are the means of all the observed x_i and y_i. The value obtained by Galton for the coefficient b_{1y} was about $\frac{2}{3}$. Hence if the height of the parent differs by an amount ξ from the mean $\bar{x}$ of the heights of all the parents, the height of the child will on the average differ by an amount $\frac{2}{3}\xi$ from the mean $\bar{y}$ of the heights of all the children. There is said to be a tendency in the next generation to return or regress towards the mean. Thus the name 'regression line' has attached itself to lines of the form (1), and is used generally, even though the two variables may be of a different nature and there can be no question of a regression in Galton's sense. Equation (1) is said to give the regression of y on x.

If the range of y is divided into small intervals $y_h \pm \frac{1}{2}dy$, and the means x_h of all the values x_i in the interval centred on y_h are plotted against y_h, the line obtained has a slope different from b_{1y}. The line is written

$$(x - \bar{x}) = b_{1x}(y - \bar{y}),$$

and is referred to as the line of regression of x on y.

5.1.1 *The general regression curve*

The distribution of the observations (x_i, y_i) will be given by the frequency function $f(x, y)$. This function can be split up in two ways:

$$f(x,y) = f_1(x)\,g_1(y\,|\,x), \quad f(x,y) = f_2(y)\,g_2(x\,|\,y). \tag{1}$$

$f_1(x)$ gives the probability of obtaining a value x, $g_1(y\,|\,x)$ gives the probability of obtaining a value y when the value x is specified. Both these functions are strictly probability-densities; when multiplied by dx and by $dx\,dy$ they give the probabilities.

The regression curve of y on x is defined to be the curve of the average value of y for a fixed x, as a function of the variable x. In terms of the frequency function the equation $U(x)$ of the regression curve is

$$U(x) = \bar{y}(x) = E(y\,|\,x) = \frac{\int y f(x,y)\,dy}{\int f(x,y)\,dy} = \frac{\int y g_1(y\,|\,x)\,dy}{\int g_1(y\,|\,x)\,dy}. \tag{2}$$

The regression curve may in the simplest cases be a straight line, but more generally it may be represented (at least approximately) as a polynomial

$$U_p(x) = \sum_{j=0}^{p} B_{pj}\,x^j \tag{3}$$

of degree p in x. In the determination of the regression of y on x, y is referred to as the dependent variable and x as the independent variable.

There will also be a regression curve of x on y,

$$U(y) = \bar{x}(y) = E(x\,|\,y) = \frac{\int x f(x,y)\,dx}{\int f(x,y)\,dx} = \frac{\int x g_2(x\,|\,y)\,dx}{\int g_2(x\,|\,y)\,dx}, \tag{4}$$

which will differ in form from the regression curve (2).

5.1.2 *Correlation ratio*

The correlation ratio of y on x, η_{yx}, is a measure of the ratio of the scatter of the observations from the regression curve to the scatter from the grand mean Y. It is defined by the equation

$$1 - \eta_{yx}^2 = E\{y(x) - E(y\,|\,x)\}^2 / E(y - Y)^2, \tag{1a}$$

where Y is the average value $E(y)$. If the scatter is independent of the x coordinate, (1a) can be written

$$1-\eta_{yx}^2 = (\operatorname{var} y \mid x)/(\operatorname{var} y). \tag{1b}$$

The form

$$\operatorname{var} y \mid x = (1-\eta_{yx}^2)\operatorname{var} y \tag{1c}$$

is often used in statistical literature. The variance on the left-hand side refers to deviations from the regression curve, the variance on the right to deviations from the grand mean.

5.2 TYPES OF VARIABLE

The variables which occur in practice seem to fall naturally into two distinct classes. Firstly, there are those variables which can be altered more or less at will by the experimenter, and which might be called 'controlled' variables. Such variables often occur in physical experiments. For example, in the measurement of the variation of electrical resistance with temperature, the values of temperature at which the measurements are made are under the control of the experimenter. Secondly, there are quantities which are inherently variable, where the values are outside the control of the observer. Such 'uncontrolled' variables are common in the biological sciences. The heights of parents and children in Galton's investigation (§ 5.1) are typical examples of uncontrolled variables.

The frequency function $f_1(x)$ associated with an uncontrolled variable is a fundamental characteristic of the quantity being measured. For controlled variables the frequency $f_1(x)$ of the occurrence of a particular value is at the discretion of the experimenter.

In point of fact, the procedure for the estimation of the regression curve $U_p(x)$ follows the same pattern whether the independent variable is controlled or uncontrolled. This comes about because (5.1.1,2) for the regression curve does not depend on the frequency function $f_1(x)$ but on the function $g_1(y \mid x)$.

Both types of variables may also be subject to experimental errors which cause the observed values x and y to differ from the true values x' and y'. The errors

$$\gamma = x - x', \quad \delta = y - y' \tag{1}$$

are attributed to the effects of small unaccounted changes in the experimental conditions. The basic assumption that will be made is that the errors γ and δ are random variables, equally likely to be positive or negative for any particular observation. Thus, whatever the values x' and y' may be,

$$E(\gamma\,|\,x',y',\delta) = 0, \quad E(\delta\,|\,x',y',\gamma) = 0; \tag{2a}$$

$$E(x\,|\,x',y',\delta) = x', \quad E(y\,|\,x',y',\gamma) = y'. \tag{2b}$$

For a variable which would be a controlled variable in the absence of errors, the presence of errors usually means that the observer can only set the value in the neighbourhood of a selected value; such a variable may be referred to as partially controlled. If the observed value x can actually be adjusted to any selected value (which will differ by an unknown amount from the true value), the variable x is again controlled.

When there are errors present, the experimental regression curves relating the observed variables x and y are usually different from the error-free or corrected regression curves relating the true variables x' and y'.

5.3 ESTIMATION OF THE EXPERIMENTAL REGRESSION CURVE

Equation (5.1.1,2) will give the regression of y on x if the frequency function is known. But the form of this function can only be determined from a very large number of observations, and in practice the number of pairs (x_i, y_i) is usually quite small. Hence some method of estimating the regression curve when only a small number of observations is available is required.

The usual procedure is to adopt the least-squares principle as the criterion for determining the best approximation

$$u_p(x) = \sum_{j=0}^{p} b_{pj}\,x^j \tag{1}$$

to the actual regression curve $U_p(x)$. On the least-squares principle the values b_{pj} are chosen so that

$$\sum_i v_i^2 = \sum_i \{y_i - u_p(x_i)\}^2 = \sum_i \left(y_i - \sum_j b_{pj}\,x_i^j\right)^2 \tag{2}$$

is a minimum. v_i is the distance measured in the y direction of the observed point (x_i, y_i) from the estimated regression curve. Differentiation of (2) with respect to the b_{pj} leads to the equations

$$\sum_i \left(y_i - \sum_j b_{pj}\,x_i^j\right) x_i^k = 0, \quad k = 0 \text{ to } p, \tag{3}$$

often called the normal equations.

Any coefficient b_{pj} obtained by solving (3) can be shown to be an unbiased estimate of B_{pj}, in the sense that if the experiment were repeated a very large number of times the mean of all the values b_{pj} obtained would tend to B_{pj}. For, from (5.1.1,2) and (5.1.1,3),

$$E(y_i \mid x_i) = U_p(x_i) = \Sigma B_{pj} x_i^j,$$

and so the expectation of the left-hand side of (3) is

$$\sum_j \{B_{pj} - E(b_{pj})\} \sum_i x_i^{j+k}.$$

This will only vanish for the $p+1$ value of the index k if the individual expressions in curly brackets vanish. Hence

$$E(b_{pj}) = B_{pj}, \tag{4}$$

i.e. b_{pj} is an unbiased estimate of B_{pj}.

5.3.1 *Postulates on which the estimation of regression curves may be based*

The least-squares principle may be accepted as a fundamental postulate, and this is perhaps the simplest procedure. Alternatively, the regression curve may be estimated on the basis of a maximum likelihood principle. This involves some assumption regarding the deviations of the observed y_i from the regression line values $\Sigma B_{pj} x_i^j$—that is, regarding the form of the frequency function $g_1(y \mid x)$.

If it is assumed that the deviations follow a normal law with standard deviation σ independent of x, then the probability of obtaining a value y_i is proportional to

$$\exp - \left(y_i - \sum_j B_{pj} x_i^j\right)^2 \Big/ 2\sigma^2. \tag{1}$$

Hence the probability of obtaining the observed set y_i is proportional to

$$\exp - \sum_i \left(y_i - \sum_j B_{pj} x_i^j\right)^2 \Big/ 2\sigma^2. \tag{2}$$

The values B_{pj} are unknown. The method of maximum likelihood states that the best estimates b_{pj} of the B_{pj} are those for which the probability of the occurrence of the y_i actually observed is a maximum. That is, the b_{pj} are chosen to minimize

$$\sum_i \left(y_i - \sum_j b_{pj} x_i^j\right)^2. \tag{3}$$

However, (3) is identical with (5.3,2), and so under these assumptions the maximum likelihood estimates are identical with the least-squares estimates.

For other forms of the deviation law the least-squares and maximum likelihood estimates will differ. But in most cases the deviation law will have some symmetrical form not very different from (1), and it is probable that the least-squares and maximum likelihood estimates will be very nearly equal. Since the former estimates are unbiased and readily calculated, they will almost always be adequate.

An alternative postulate is that the estimate b_{pj} be chosen so that its variance has the smallest possible value. It is interesting to observe that, whatever the form of the deviation law, the minimum variance postulate leads to estimates which are identical with the least-squares estimates. This is sometimes referred to as the Markoff theorem. The proof of this theorem will be postponed to § 8.3.

5.3.2 *Weights*

The least-squares form (5.3,2) will be appropriate when it can be assumed that the scatter of the observations about the regression curve is the same at all points—when the standard deviation σ is independent of x_i. If it is known that the scatter is different for different points on the curve, the value v_i^2 should be weighted by dividing by σ_i^2, the expectation of the square of the deviation. The least-squares principle then makes

$$\sum_i \left(y_i - \sum_j b_{pj} x_i^j\right)^2 \bigg/ \sigma_i^2 = \sum_i w_i \left(y_i - \sum_j b_{pj} x_i^j\right)^2 \bigg/ \sigma^2 \tag{1}$$

a minimum, with

$$w_i = \sigma^2/\sigma_i^2 \tag{2}$$

and σ a constant, the standard deviation of an observation of unit weight. Differentiation of (1) leads to the normal equations

$$\sum_i w_i \left(y_i - \sum_j b_{pj} x_i^j\right) x_i^k = 0. \tag{3}$$

It is clear that the maximum likelihood principle, assuming the deviation law (5.3.1,1), leads to the same equation, since the probability of obtaining the values y_i is proportional to

$$\exp - \sum_i (y_i - \Sigma b_{pj} x_i^j)^2/2\sigma_i^2 = \exp - \sum_i w_i (y_i - \Sigma b_{pj} x_i^j)^2/2\sigma^2. \tag{4}$$

It can be shown that the estimates obtained from any set of equations of the form

$$\sum_i \lambda_i \left(y_i - \sum_j \hat{b}_{pj} x_i^j\right) x_i^k = 0 \tag{5}$$

will be unbiased. For, as in § 5.3, the expectation of (5) gives

$$\sum_j \{B_{pj} - E(\hat{b}_{pj})\} \sum_i \lambda_i x_i^{j+k} = 0,$$

and so
$$E(\hat{b}_{pj}) = B_{pj}. \tag{6}$$

The weights $\lambda_i = w_i \propto \sigma_i^{-2}$ lead to the estimates of smallest standard deviation, by the Markoff theorem (§ 8.3). However, any other weights λ_i may be used if this seems desirable, the estimates so obtained being still unbiased but somewhat less accurate.

If the weights w_i cannot be taken as constant, their determination in any particular example may prove difficult. One special case of interest is that in which the error-free values x_i' and y_i' lie exactly on a smooth curve, so that x_i' and y_i' are connected by the functional relationship

$$y_i' = \Sigma B_{pj}' x_i'^j. \tag{7}$$

Then
$$\begin{aligned}\operatorname{var} y_i \mid x_i &= \operatorname{var}(y_i' + \delta_i) \mid x_i \\ &= \operatorname{var}\{\Sigma B_{pj}'(x_i - \gamma_i)^j + \delta_i\} \mid x_i.\end{aligned}$$

Thus if c_i' is the slope of the curve (7) at the point x_i, (1.2,14a) gives

$$w_i^{-1} \propto \operatorname{var} y_i \mid x_i = \operatorname{var} \delta_i \mid x_i + c_i'^2 \operatorname{var} \gamma_i \mid x_i. \tag{8}$$

Usually a rough estimate of the slope of the experimental regression curve will be a sufficiently good approximation to c_i'. It is clear that even if the standard deviations of the errors γ_i and δ_i are constant, the weights w_i will not be constant unless c_i' is constant—that is, unless the functional relationship is linear.

5.3.3 *Prediction*

Often the purpose of the curve connecting the variables x and y is to make predictions. That is, to estimate the most likely value of one variable, say y, when an observation of the second variable yields a value x_0. If this most likely value is taken to be $E(y \mid x_0)$, the prediction curve or calibration curve is simply the regression curve of y on x.

It should be emphasized that it is immaterial whether the observed values x and y contain experimental errors or not. The only requirement is that the measurement x_0 on which the prediction is being based should be made under the same experimental conditions as were the observations x_i, y_i, from which the regression curve was calculated. The predicted value y is the expected value an observation would yield under these experimental conditions. It is not necessarily an estimate of the error-free value y'.

5.4 THE ESTIMATION OF THE ERROR-FREE CURVE WHEN THE DEPENDENT VARIABLE IS SUBJECT TO ERROR

If y_i is subject to experimental error, while x_i is free from error,

$$y_i = y'_i + \delta_i, \quad x_i = x'_i, \tag{1}$$

there will be two regression curves giving the regression of y on x, the experimental curve

$$\bar{y}(x) = E(y \mid x) = \Sigma B_{pj} x^j, \tag{2}$$

and the error-free curve or corrected curve

$$\overline{y'}(x) = E(y' \mid x) = \Sigma B'_{pj} x^j. \tag{3}$$

But
$$E(y \mid x) = E(y' \mid x) + E(\delta \mid x) = E(y' \mid x),$$

as δ is a random variable, and so the two regression curves in fact coincide.

It follows that the estimated experimental regression curve can be used also to predict the value y' on the corrected or error-free regression curve corresponding to an observed value x_0. The effect of the error is merely to increase the standard deviation. For

$$\operatorname{var} y \mid x = E(y - \Sigma B_{pj} x^j)^2 = E(y' + \delta - \Sigma B_{pj} x^j)^2.$$

Since the variable δ is a random variable,

$$E(\delta \mid y', x) = 0.$$

Thus
$$E(y'\delta) = 0 = E(\delta x^j),$$

and so
$$\operatorname{var} y \mid x = \operatorname{var} y' \mid x + \operatorname{var} \delta \mid x, \tag{4a}$$

or
$$\sigma_y^2 = \sigma_{y'}^2 + \sigma_\delta^2. \tag{4b}$$

5.4.1 *Functional relationship*

One particular case of considerable importance is that in which the corrected points x' and y' lie exactly on a smooth curve, so that the error-free values are connected by the functional relationship

$$y' = \Sigma B'_{pj} x'^j. \tag{1}$$

The functional relationship can be regarded as a limiting form of regression curve, when the standard deviation $\sigma_{y'}$ of y' for fixed x' becomes zero—when there is a single value y' corresponding to the observed value x'.

When the independent variable is free from error, $x = x'$, the estimated experimental regression curve will also provide an estimate of the functional relationship between the error-free variables. The standard deviation σ_y will simply be that of the experimental error, σ_δ. This particular case corresponds to the classical curve-fitting problem, and most curves in physics will be of this class.

5.4.2 *Choice of independent variable in determining functional relationship*

The regression curve is, by definition, a single-valued function of the independent variable. Hence the functional curve and the regression curve can only coincide if the dependent variable is a single-valued function of the independent variable; otherwise the regression curve will be an average of the various branches of the functional curve. Thus if the regression curve is to be an estimate of the functional relationship, then the dependent variable must be a single-valued function of the independent variable, and this restriction may determine in a particular example which variable must be the dependent variable.

If the relationship is very roughly linear, with no maxima or minima within the range of observation, each variable is a single-valued function of the other. Hence the error-free regression curves of y' on x' and of x' on y' will both coincide with the functional relationship curve, and either variable may be chosen as the dependent variable in the determination of the functional relationship. But the estimation of the regression curve is quite complicated when the independent variable is subject to error, as will be seen in § 5.5. So if only one variable is subject to error, this variable should wherever possible be chosen as the dependent variable. To summarize:

(*a*) if the function has maxima or minima in the range of observations, it must be arranged that these are maxima and minima in the dependent variable y;

(*b*) if there are no maxima or minima, and only one variable is subject to error, this variable should be the dependent variable y.

5.5 THE INDEPENDENT VARIABLE SUBJECT TO ERROR

The experimental and corrected variables are now connected by the equations

$$x = x' + \gamma, \quad y = y' + \delta.$$

There are four possible regression curves relating one variable to the other. These are the experimental regression curve y on x, the corrected regression curve y' on x', and the two mixed curves y'

on x and y on x'. But if the error variable δ is a random variable

$$E(y\,|\,x) = E(y'\,|\,x), \quad E(y\,|\,x') = E(y'\,|\,x'). \tag{1}$$

Hence the mixed curves coincide with the experimental and the corrected curves respectively, and need not be considered separately. The experimental curve has been discussed already, so that there is only the corrected or error-free regression curve to be considered.

If a series of error-free observations x'_i, y'_i were available, then the least-squares normal equations would be, from (5.3.2,3),

$$\sum_i w_i\left(y'_i - \sum_j b'_{pj}\, x'^j_i\right) x'^k_i = 0. \tag{2}$$

But the observed values are x_i and y_i, and the error-free values x'_i and y'_i cannot be obtained by direct observation.

As a first approximation, the observed values x_i and y_i can be used for x'_i and y'_i in (2), giving

$$\Sigma w_i(y_i - \Sigma b'_{pj}\, x^j_i)\, x^k_i = 0. \tag{3}$$

The weights w_i should be inversely proportional to the variance, for fixed x'_i, of the difference $y_i - \Sigma B'_{pj} x^j_i$. By the usual rule, (1.2,14*a*),

$$\operatorname{var}(y_i - \Sigma B'_{pj} x^j_i) = \operatorname{var} y_i + \left\{\frac{\partial}{\partial x_i}(\Sigma B'_{pj} x^j_i)\right\}^2 \operatorname{var} x_i,$$

or

$$w_i \propto (\sigma^2_{yi'} + \sigma^2_{\delta i} + c'^2_i\, \sigma^2_{\gamma i})^{-1}, \tag{4}$$

where $\sigma^2_{yi'}$ is the variance of y'_i for fixed x'_i and c'_i is the slope of the curve at the point x_i. The solution (3) and (4) is that proposed by Deming in his book. Comparison with (5.3.2,5) and (5.3.2,6) shows that the coefficients in this first approximation are really estimates of the coefficients of the experimental regression curve.

In fact, though x_i is an estimate of x'_i, x^j_i is not an unbiased estimate of x'^j_i. For

$$x^j_i = (x'_i + \gamma_i)^j,$$

and so

$$E(x^j_i) = x'^j_i + \binom{j}{2} x'^{j-2}_i\, E(\gamma^2_i) + \dots. \tag{5}$$

Hence expressions for unbiased estimates of x'^j_i can be obtained if the moments (i.e. the expectations of the powers) of the errors γ_i are known. Substitution of these in (2) leads to a system of unbiased estimating equations for the b'_j.

The approximate values obtained from (3) will be very close to the actual values if the errors γ_i are small. Because the estimation of the corrected curve is so complex, the regression curve given by (3) is generally used in its place. In most of the subsequent chapters it will be assumed, unless the contrary is explicitly stated, that the error in the independent variable x can be neglected, the case where both variables are subject to error being only considered in §§ 6.5 and 11.2.

5.5.1 *Linear functional relationship*

If the relation between the error-free variables is linear,

$$y' = B'_0 + B'_1 x', \tag{1}$$

then the regression curve is given by

$$E(y \mid x) = E(y' + \delta \mid x)$$
$$= E(B'_0 + B'_1 x + \delta - B'_1 \gamma \mid x),$$

or

$$E(y \mid x) = B'_0 + B'_1 x - B'_1 E(\gamma \mid x). \tag{2}$$

The last term gives the difference between the functional line and the regression curve.

The frequency function $f(\gamma \mid x)$ will be proportional to the product of the error function $f_1(\gamma \mid x')$ and the frequency function $f_2(x')$, for the value $x' = x - \gamma$. If it is assumed that $f_1(\gamma \mid x')$ is independent of x',

$$f(\gamma \mid x) \propto f_1(\gamma) f_2(x - \gamma). \tag{3}$$

Hence $E(\gamma \mid x)$ will depend on the form of the frequency function $f_2(x')$. Only if $f_2(x')$ is symmetrical about the point x will the regression curve and the functional line coincide at this point.

If the distribution of values x' is of the normal form (with mean zero), then it can be shown that the regression curve is linear when $f(\gamma)$ is also normal. For then

$$f_1(\gamma) f_2(x - \gamma) \propto \mathrm{e}^{-\frac{1}{2}\gamma^2/\sigma^2} \mathrm{e}^{-\frac{1}{2}(x-\gamma)^2/\xi^2} = \mathrm{e}^{-\frac{1}{2}x^2/\xi^2} \mathrm{e}^{\frac{1}{2}b^2x^2} \mathrm{e}^{-\frac{1}{2}(a\gamma - bx)^2},$$

where

$$a^2 = \sigma^{-2} + \xi^{-2}, \quad ab = \xi^{-2}.$$

It follows that

$$E(\gamma \mid x) = \int \gamma f(\gamma \mid x)\, d\gamma \Big/ \int f(\gamma \mid x)\, d\gamma$$
$$= \int \gamma \{\exp - \tfrac{1}{2}(a\gamma - bx)^2\}\, d\gamma \Big/ \int \{\exp - \tfrac{1}{2}(a\gamma - bx)^2\}\, d\gamma$$
$$= \int a^{-1}(z + bx)\, \mathrm{e}^{-z^2}\, dz \Big/ \int \mathrm{e}^{-z^2}\, dz,$$

where

$$z = a\gamma - bx.$$

So

$$E(\gamma \mid x) = a^{-1} bx = (1 + \xi^2/\sigma^2)^{-1} x. \tag{4}$$

The regression curve is then a straight line of slope

$$B_1 = B_1'/(1+\sigma^2/\xi^2). \tag{5}$$

If the range of the observations is much greater than the standard error, the difference between the two slopes will be very small.

In many experiments $f_2(x')$ will be fairly flat over a region near the centre of the range, and will drop rather sharply at the ends of the range, as in Fig. 5.5.1*a*. This would correspond to an experiment where values of x' distributed more or less uniformly through a certain range were observed. Over the central region $f_2(x')$ is reasonably symmetrical, and, from (3), $E(\gamma \mid x)$ is very small. At the ends of the range $f_2(x')$ is decidedly unsymmetrical, and the extra term $E(\gamma \mid x)$ in (2) will be appreciable. The regression curve will not be a straight line but will be curved at each end, as in Fig. 5.5.1*b*.

5.5.2 *Predetermined variables*

An interesting case, first discussed by Berkson, occurs when the variable x is limited to certain definite values. Then for the value x it is reasonable to assume that $f(x')$ is symmetrical about x, and hence $E(\gamma \mid x) = 0$. For this special case the regression curve is a straight line coincident with the functional line.

The types of problem in which the values x_i may be predetermined are those in which the conditions may be adjusted so that the measuring instrument reads exactly x, but errors due to changes or uncertainties in calibration, or to difficulties in experimental procedure, cause the true value x' to differ from x.

For predetermined variables, where the true curve is of the second or third degree,

$$\begin{aligned} E(y \mid x) &= E\{\Sigma B_j'(x-\gamma)^j + \delta \mid x\} \\ &= \Sigma B_j' x^j + (B_2' + 3B_3' x)\, E(\gamma^2 \mid x), \end{aligned}$$

if $f(\gamma \mid x)$ is symmetrical. Hence if an estimate of $E(\gamma^2 \mid x)$ is available, the departure of the regression curve from the curve of functional relationship can be found.

5.5.3 *Other cases of coincidence*

The case discussed by Berkson is not the only one in which the regression line and the functional relationship line coincide. The two lines coincide in any experiment for which $E(\gamma \mid x) = 0$. Thus if it is decided to take observations over a definite range of x, then it is reasonable to assume that $E(\gamma \mid x)$ will vanish—that any observation is as likely to have a positive error as a negative one. On the other hand, if the range of x' is limited the two curves diverge at the ends, as was shown in § 5.5.1.

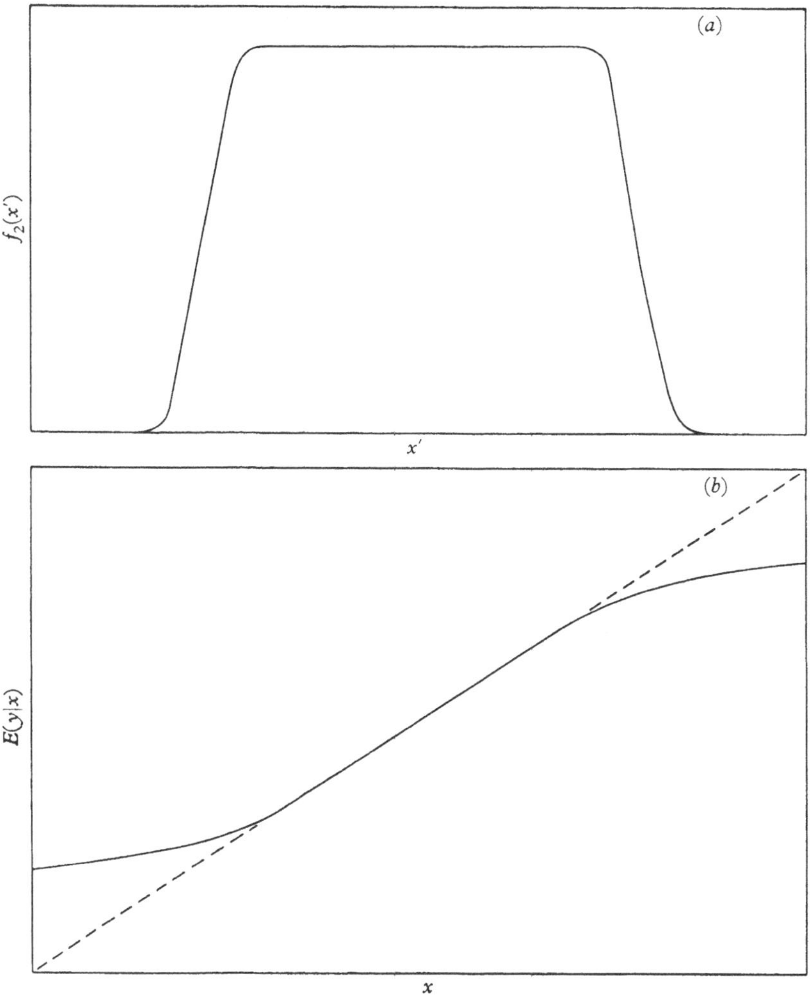

Figs. 5.5.1*a* and *b*. Graphs illustrating the curvature of the regression line.

5.6 NOTES AND REFERENCES

(5.5) General references to the cases where both variables are subject to error are: Kendall (1951), Lindley (1947, 1953), Berkson (1950), and Jessop (1952). A further discussion of the linear case is given in § 6.5.

Geary (1953) considers the fitting of a quadratic or cubic when x is a controlled variable subject to error.

CHAPTER 6

THE STRAIGHT LINE

For the straight line the least-squares principle leads to the normal equations whose solution is considered in § 6.1 and illustrated in Example 6.1.4. The calculating scheme of Table 6.1.4*a* is of general use, but in certain cases (especially when the observations are uniformly spaced) other schemes may be quicker. A guide to the choice of the calculating scheme is given in § 12.1.1.

The case where the independent variable x is subject to error is considered in § 6.5. However, the problem of estimating the slope of the line relating the corresponding error-free variables often cannot be solved exactly.

6.1 NORMAL EQUATIONS

When the regression curve or functional relationship curve is linear, the fitted curve is

$$u_1(x) = b_0 + b_1 x, \tag{1}$$

where the coefficients are chosen to minimize

$$\Sigma w_i\{y_i - u_1(x_i)\}^2.$$

They are given by the normal equations (5.3.2,3),

$$\Sigma w_i(y_i - b_0 - b_1 x_i) = 0, \tag{2a}$$

$$\Sigma w_i x_i(y_i - b_0 - b_1 x_i) = 0; \tag{2b}$$

or

$$b_0 \Sigma w_i + b_1 \Sigma w_i x_i = \Sigma w_i y_i, \tag{3a}$$

$$b_0 \Sigma w_i x_i + b_1 \Sigma w_i x_i^2 = \Sigma w_i x_i y_i. \tag{3b}$$

The solutions of these equations can be written down explicitly as follows:

$$b_1 = \{\Sigma w_i \Sigma w_i x_i y_i - \Sigma w_i x_i \Sigma w_i y_i\}/D, \tag{4a}$$

$$b_0 = \{-\Sigma w_i x_i \Sigma w_i x_i y_i + \Sigma w_i x_i^2 \Sigma w_i y_i\}/D, \tag{4b}$$

where

$$D = \Sigma w_i \Sigma w_i x_i^2 - (\Sigma w_i x_i)^2. \tag{4c}$$

If the observations are all of equal weight, w_i can be set equal to unity and Σw_i equal to n in all the formulae throughout the chapter. The equations (4a–c) simplify to

$$b_1 = \{n\Sigma x_i y_i - \Sigma x_i \Sigma y_i\}/D, \tag{5a}$$

$$b_0 = \{-\Sigma x_i \Sigma x_i y_i + \Sigma x_i^2 \Sigma y_i\}/D, \tag{5b}$$

$$D = n\Sigma x_i^2 - (\Sigma x_i)^2. \tag{5c}$$

The calculated values b_0 and b_1 can be checked by substitution in (3a, b).

6.1.1 *The origin of x at the mean*

A change in the origin of the x-coordinate will leave the slope of the least-squares line unaltered, but will change the constant term b_0. The system in which the origin of x is at the weighted mean leads to specially simple formulae for the coefficients and their variances. This coordinate system will be denoted by the symbol ξ, so that

$$\xi = x - \bar{x} = x - (\Sigma w_i x_i / \Sigma w_i) \tag{1}$$

and

$$\Sigma w_i \xi_i = 0. \tag{2}$$

The fitted curve in terms of the variable ξ will be written

$$u_1(\xi) = a_0 + a_1 \xi, \tag{3}$$

where a_0 and a_1 are the values of constant term and slope for the special variable ξ.

Equating the fitted values given by (6.1,1) and (3) at the point x,

$$b_0 + b_1 x = a_0 + a_1(x - \bar{x}).$$

Hence

$$b_1 = a_1, \tag{4a}$$

$$b_0 = a_0 - \bar{x}a_1 = a_0 - (\Sigma w_i x_i / \Sigma w_i)\, a_1. \tag{4b}$$

The slope and constant term for the variable ξ can be obtained by substituting ξ in (6.1,4a–c). Thus, using (2),

$$a_1 = \Sigma w_i \xi_i y_i / \Sigma w_i \xi_i^2, \tag{5a}$$

$$a_0 = \Sigma w_i y_i / \Sigma w_i, \tag{5b}$$

$$D = \Sigma w_i \Sigma w_i \xi_i^2. \tag{5c}$$

The value D, like the slope b_1, is invariant with respect to changes of origin. For, from (6.1,4c),

$$D = \Sigma w_i \left\{ \Sigma w_i x_i^2 - \left(\frac{\Sigma w_i x_i}{\Sigma w_i} \right)^2 \Sigma w_i \right\} = \Sigma w_i \Sigma w_i (x_i - \bar{x})^2 = \Sigma w_i \Sigma w_i \xi_i^2,$$

and so the value D does not depend on the origin.

If the origin of y is also chosen at the weighted mean, the constant term a_0 is zero. That is, the line passes through the point corresponding to the weighted means. In terms of the variable x, (5a) gives

$$b_1 = a_1 = \Sigma w_i(x_i-\bar{x})\,y_i/\Sigma w_i(x_i-\bar{x})^2$$
$$= \Sigma w_i(x_i-\bar{x})\,(y_i-\bar{y})/\Sigma w_i(x_i-\bar{x})^2. \qquad (6)$$

6.1.2 *Standard deviations of the estimates*

The weights w_i are inversely proportional to the variances, so that

$$\operatorname{var} y_i = \sigma^2/w_i. \qquad (1)$$

Since a_0 and a_1 are linear functions of the y_i,

$$\sigma^2(a_1) = \operatorname{var} a_1 = \Sigma w_i^2\,\xi_i^2 \operatorname{var} y_i/(\Sigma w_i\,\xi_i^2)^2 = \sigma^2/\Sigma w_i\,\xi_i^2, \qquad (2a)$$

$$\sigma^2(a_0) = \operatorname{var} a_0 = \Sigma w_i^2 \operatorname{var} y_i/(\Sigma w_i)^2 = \sigma^2/\Sigma w_i. \qquad (2b)$$

If the true values (or population means) corresponding to y_i are Y_i, then the true values of the coefficients are

$$A_0 = \Sigma w_i Y_i/\Sigma w_i, \quad A_1 = \Sigma w_i\,\xi_i Y_i/\Sigma w_i\,\xi_i^2. \qquad (3)$$

Hence

$$\operatorname{cov}(a_0, a_1) = E(a_0-A_0)\,(a_1-A_1)$$
$$= E\Sigma w_i(y_i-Y_i)\,\Sigma w_i\,\xi_i(y_i-Y_i)/\Sigma w_i\,\Sigma w_i\,\xi_i^2$$
$$= \Sigma w_i^2\,\xi_i \operatorname{var} y_i/\Sigma w_i\,\Sigma w_i\,\xi_i^2,$$

and so from (1) and (6.1.1,2)

$$\operatorname{cov}(a_0, a_1) = 0. \qquad (4)$$

Thus for the linear sum

$$z = \lambda_0 a_0 + \lambda_1 a_1, \qquad (5a)$$

$$\operatorname{var} z = \lambda_0^2 \operatorname{var} a_0 + \lambda_1^2 \operatorname{var} a_1. \qquad (5b)$$

For the fitted value, from (6.1.1,3),

$$\operatorname{var} u_1(\xi) = \operatorname{var} a_0 + \xi^2 \operatorname{var} a_1, \qquad (6a)$$

or

$$\sigma^2[u_1(\xi)] = \sigma^2\left\{1+\xi^2\frac{\Sigma w_i}{\Sigma w_i\,\xi_i^2}\right\}\Big/\Sigma w_i. \qquad (6b)$$

The standard deviations for the coefficients and fitted values in terms of the variable x can be found very rapidly from these formulae. From (2a), as $b_1 = a_1$,

$$\sigma^2(b_1) = \operatorname{var} b_1 = \sigma^2/\Sigma w_i(x_i-\bar{x})^2 = \sigma^2\,\Sigma w_i/D. \qquad (7a)$$

From (6.1.1,4*b*),

$$\sigma^2(b_0) = \operatorname{var} b_0 = \sigma^2\left\{1+\bar{x}^2\frac{(\Sigma w_i)^2}{D}\right\}\Big/\Sigma w_i$$

$$= \sigma^2\left\{1+\frac{(\Sigma w_i x_i)^2}{D}\right\}\Big/\Sigma w_i. \qquad (7b)$$

For the fitted value, from (6*b*),

$$\sigma^2[u_1(x)] = \sigma^2\left\{1+(x-\bar{x})^2\frac{\Sigma w_i}{\Sigma w_i(x_i-\bar{x})^2}\right\}\Big/\Sigma w_i$$

$$= \sigma^2\left\{1+(x-\bar{x})^2\frac{(\Sigma w_i)^2}{D}\right\}\Big/\Sigma w_i. \qquad (7c)$$

Equation (7*b*) corresponds to the special case of (7*c*) for which $x = 0$. The values b_0 and b_1 are not independent, but

$$\operatorname{cov}(b_0, b_1) = \operatorname{cov}(a_0-\bar{x}a_1, a_1),$$

or

$$\operatorname{cov}(b_0, b_1) = -\bar{x}\operatorname{var} a_1 = -\frac{\Sigma w_i x_i}{D}\sigma^2. \qquad (7d)$$

6.1.3 *Estimation of σ from the residuals*

Usually σ is not known, but must be estimated from the residuals

$$v_i = y_i - u_1(x_i) = y_i - u_1(\xi_i) = y_i - a_0 - a_1\xi_i. \qquad (1)$$

Now $E(v_i) = 0$, and, using (1.2,7),

$$E(v_i^2) = \operatorname{var} v_i = \operatorname{var} y_i + \operatorname{var} a_0 + \xi_i^2 \operatorname{var} a_i - 2\operatorname{cov}(y_i, a_0)$$
$$-2\xi_i \operatorname{cov}(y_i, a_1) + 2\xi_i \operatorname{cov}(a_0, a_1). \qquad (2a)$$

But

$$\operatorname{cov}(y_i, a_0) = E(y_i - Y_i)(a_0 - A_0)$$
$$= E(y_i - Y_i)\,\Sigma w_j(y_j - Y_j)/\Sigma w_j$$
$$= w_i \operatorname{var} y_i/\Sigma w_j = \sigma^2/\Sigma w_j,$$

or

$$\operatorname{cov}(y_i, a_0) = \operatorname{var} a_0. \qquad (2b)$$

Similarly,

$$\operatorname{cov}(y_i, a_1) = \xi_i \operatorname{var} a_1, \qquad (2c)$$

and so

$$E(v_i^2) = \operatorname{var} y_i - \operatorname{var} a_0 - \xi_i^2 \operatorname{var} a_1. \qquad (3)$$

Thus

$$E(w_i v_i^2) = \sigma^2\left\{1 - \frac{w_i}{\Sigma w_j} - \frac{w_i\xi_i^2}{\Sigma w_j \xi_j^2}\right\}$$

and

$$E\Sigma w_i v_i^2 = \sigma^2(n-2). \qquad (4)$$

Hence

$$s^2 = \frac{\Sigma w_i v_i^2}{n-2} \qquad (5)$$

will provide an unbiased estimate of the variance σ^2 of an observation of unit weight.

The quantity $\Sigma w_i v_i^2$ can be obtained without calculating the individual residuals v_i. For

$$\Sigma w_i v_i^2 = \Sigma w_i (y_i - a_0 - a_1 \xi_i)^2,$$

and

$$\Sigma w_i y_i = \Sigma w_i a_0, \quad \Sigma w_i y_i \xi_i = \Sigma w_i \xi_i^2 a_1, \quad \Sigma w_i \xi_i a_0 a_1 = 0.$$

Thus any of the alternative forms

$$\Sigma w_i v_i^2 = \Sigma w_i y_i^2 - a_0^2 \Sigma w_i - a_1^2 \Sigma w_i \xi_i^2, \tag{6a}$$

$$\Sigma w_i v_i^2 = \Sigma w_i y_i^2 - a_0 \Sigma w_i y_i - a_1 \Sigma w_i y_i \xi_i, \tag{6b}$$

$$\Sigma w_i v_i^2 = \Sigma w_i y_i^2 - (\Sigma w_i y_i)^2 / \Sigma w_i - (\Sigma w_i y_i \xi_i)^2 / \Sigma w_i \xi_i^2 \tag{6c}$$

may be used to calculate $\Sigma w_i v_i^2$. A convenient form in terms of the quantities occurring in (6.1,4*a*–*c*) is

$$\Sigma w_i v_i^2 = \Sigma w_i y_i^2 - (\Sigma w_i y_i)^2 / \Sigma w_i - b_1^2 D / \Sigma w_i. \tag{6d}$$

If the origin is chosen at the weighted mean of the values y_i, a_0 will vanish. Hence (6*a*) has the alternative form

$$\Sigma w_i v_i^2 = \Sigma w_i (y_i - \bar{y})^2 - b_1^2 \Sigma w_i (x_i - \bar{x})^2. \tag{6e}$$

6.1.4 *Example*

Table 6.1.4 shows the error y in a quartz clock, as deduced from transit-circle observations on the days x. In Table 6.1.4*a* the calculations for the fitted line and for the standard deviations are shown. The slope of the line, corresponding to the mean clock rate, is $84{\cdot}2 \pm 0{\cdot}6$ milliseconds per day.

In all such calculations, a very useful check on the arithmetic is obtained by evaluating the residuals v_i individually, and comparing the value of Σv_i^2 with that calculated by means of equations (6.1.3,6*a*–*e*). Part (*d*) of Table 6.1.4*a* shows the individual residuals. The sum of their squares is 0·0300, agreeing with the value 0·0301 obtained in Part (*c*). The sum of the residuals will be zero if no arithmetical slip has been made in calculating them. In addition, the signs of the residuals should be examined to see whether there are any systematic departures from the fitted line. The occurrence of a number of successive residuals of the same sign would indicate that the straight line does not fit the observations satisfactorily, and the proper curve is a polynomial of higher degree.

6.1.5 *Variation of standard deviation of fitted value with location of point*

Equation (6.1.2,7*c*) gives the standard deviation of the fitted value as

$$\sigma[u_1(x)] = \sigma(\Sigma w_i)^{-\frac{1}{2}} [1 + \{(\Sigma w_i)^2 / D\} (x - \bar{x})^2]^{\frac{1}{2}}. \tag{1}$$

Thus the standard deviation is a minimum at the mean value, and increases symmetrically on each side of the mean.

TABLE 6.1.4

Error y (in seconds) of a quartz clock as a function of time x (in days)

y	x	y	x
0·435	3	2·122	23
0·706	6	2·181	24
0·729	7	2·938	33
0·975	9	3·135	35
1·063	11	3·419	39
1·228	12	3·724	41
1·342	14	3·705	42
1·491	16	3·820	44
1·671	18	3·945	45
1·696	19	4·320	49

TABLE 6.1.4*a*

Calculations for Example 6.1.4

(*a*) Summations

Σy 44·645	Σxy 1457·543	Σy^2 130·322683
n 20	Σx 490	Σx^2 16324

(*b*) Coefficients b_0 and b_1

$D = n\Sigma x^2 - (\Sigma x)^2 = 86380$

$b_1 = (n\Sigma xy - \Sigma x \Sigma y)/D = 7274{\cdot}810/D = 0{\cdot}0842187$

$b_0 = (-\Sigma x \Sigma xy + \Sigma x^2\, \Sigma y)/D = 0{\cdot}168892$

Check $nb_0 + \Sigma x b_1 = 44{\cdot}645003 = \Sigma y.$

(*c*) Standard deviations

Σy^2	130·322683
$-(\Sigma y)^2/n$	99·658801
$-(b_1 D)^2/nD$	30·633746
$= \Sigma v^2$	0·030136; $s^2 = \Sigma v^2/(n-2)$ 0·001674; s 0·0409

$s(b_1) = s/\sqrt{(D/n)} = s/65{\cdot}7 = 0{\cdot}000623$

(*d*) Residuals

u	v	u	v
0·422	+0·013	2·106	+0·016
0·674	+0·032	2·190	−0·009
0·758	−0·029	2·948	−0·010
0·927	+0·048	3·117	+0·018
1·095	−0·032	3·453	−0·034
1·180	+0·048	3·622	+0·102
1·348	−0·006	3·706	−0·001
1·516	−0·025	3·875	−0·055
1·685	−0·014	3·959	−0·014
1·769	−0·073	4·296	+0·024

$\Sigma v^2 = 0{\cdot}0300$ $\Sigma v = -0{\cdot}001$

If the variable

$$k = (x-\bar{x})\,\Sigma w_i/\sqrt{(3D)} \tag{2}$$

is introduced,

$$\sigma[u_1(k)] = \rho_{10}(k)\,\sigma/\sqrt{\Sigma w_i}, \tag{3}$$

where

$$\rho_{10}^2(k) = 1+3k^2. \tag{4}$$

Hence the variation of standard deviation is described by the same function for all fitted straight lines. This function is tabulated in Table 6.7*a* for the range 0(0·05)0·3(0·1)3·0 of k. Equation (3) and Table 6.7*a* enable the standard deviation at a number of values of x to be calculated rapidly.

The factor 3 in (2) and (4) was introduced to agree with the treatment of the equally-spaced case to be given in § 6.3.1. Similarly, the suffixes 1 and 0 in $\rho_{10}(k)$ conform to the notation of § 8.4.5.

6.1.5.1 *Example*

Table 6.1.5 shows the calculation of the clock errors obtained from the straight line fitted in § 6.1.4, together with the standard deviations of these estimates calculated by (6.1.5,3), using the value s as an estimate of σ. Thus on the fiftieth day the estimate is $4{\cdot}380 \pm 0{\cdot}018$ seconds.

TABLE 6.1.5

Fitted values and standard deviations for clock errors (*Example* 6.1.4)

x	$u(x)\times 10^3$	k	$\rho_{10}(k)$	$s(u)\times 10^3$
0	169	−0·96	1·94	18
5	590	−0·76	1·65	15
10	1011	−0·57	1·40	13
15	1432	−0·37	1·19	11
20	1853	−0·17	1·04	10
25	2274	+0·02	1·00	9
30	2695	+0·22	1·07	10
35	3117	+0·42	1·24	11
40	3538	+0·61	1·45	13
45	3959	+0·81	1·72	16
50	4380	+1·00	2·00	18
55	4801	+1·20	2·31	21
60	5222	+1·40	2·62	24
65	5643	+1·59	2·93	27
70	6064	+1·79	3·25	30
75	6485	+1·99	3·59	33

6.1.6 *The straight line passing through the origin*

A special case which occurs rather frequently is that in which it is known on theoretical grounds that Y_i vanishes when $x_i = 0$; that is, the true line passes through the origin. Then B_0 vanishes and the equation of the regression line is

$$U_1(x) = B_1 x. \tag{1}$$

Hence the equation giving the least-squares estimate based on n observations x_i and y_i is

$$\frac{\partial}{\partial b_1} \Sigma w_i (y_i - b_1 x_i)^2 = 0,$$

and so
$$b_1 = \Sigma w_i x_i y_i / \Sigma w_i x_i^2. \tag{2}$$

Thus, proceeding as in the more general case,

$$\operatorname{var} b_1 = \sigma^2 / \Sigma w_i x_i^2, \tag{3}$$

$$\Sigma w_i v_i^2 = \Sigma w_i y_i^2 - b_1^2 \Sigma w_i x_i^2 = \Sigma w_i y_i^2 - (\Sigma w_i x_i y_i)^2 / \Sigma w_i x_i^2, \tag{4}$$

and
$$E(\Sigma w_i v_i^2) = (n-1)\sigma^2, \tag{5}$$

there being only one parameter b_1 which is estimated from the y_i.

Hence
$$s^2 = \Sigma w_i v_i^2 / (n-1) \tag{6}$$

will provide an estimate of the standard deviation σ.

For the fitted value,

$$u_1(x) = b_1 x, \tag{7}$$

and
$$\operatorname{var} u_1(x) = x^2 \operatorname{var} b_1. \tag{8}$$

6.1.6.1 *Example*

In measuring the elastic properties of a metal wire, a number of different masses were attached to the wire and the extension y measured for each mass. The measurements are given in Table 6.1.6, and the calculation of the extension per unit length in the lower part of this table. The value obtained is $(1{\cdot}530 \pm 0{\cdot}014) \times 10^{-2}$ inches/lb. wt.

TABLE 6.1.6

Extensions y in units of 10^{-2} inches for loads of x lb. wt.

y	x	y	x
1·9	1·35	7·7	5·20
2·9	1·86	9·5	6·13
3·7	2·42	11·7	7·82
5·5	3·36	14·6	9·43

Σyx 362·840 Σy^2 555·55

Σx^2 237·1223 $(\Sigma yx)^2/\Sigma x^2$ 555·211

b_1 1·530181 Σv^2 0·339 $s = \sqrt{\{\Sigma v^2/(n-1)\}}$ 0·220

$s(b_1) = s/\sqrt{\Sigma x^2}$ 0·0143

u	v	u	v
2·07	−0·17	7·96	−0·26
2·85	+0·05	9·38	+0·12
3·70	0	11·97	−0·27
5·14	+0·30	14·43	+0·17
Σv^2 0·345		Σv 0·00	

If the length L of the wire is measured as $127{\cdot}23 \pm 0{\cdot}05$ inches, and the radius R as $(1{\cdot}407 \pm 0{\cdot}010) \times 10^{-2}$ inches, the Young's modulus Y is

$$Y = \frac{1}{b_1}\frac{L}{\pi R^2} = 13{\cdot}37 \times 10^6 \text{ lb. wt./sq. in.}$$

The variance of Y will be given by the formula

$$\frac{\operatorname{var} Y}{Y^2} = \frac{\operatorname{var} b_1}{b_1^2} + \frac{4 \operatorname{var} R}{R^2} + \frac{\operatorname{var} L}{L^2}$$

$$= 0{\cdot}84 \times 10^{-4} + 2{\cdot}02 \times 10^{-4} + 15 \times 10^{-8} = 2{\cdot}86 \times 10^{-4}.$$

Hence the standard deviation of the estimated modulus is $0{\cdot}23 \times 10^6$.

6.1.7 *The bivariate normal distribution*

If x and y are uncontrolled variables, distributed normally about zero with variance σ_x and σ_y and correlation coefficient ρ,

$$f(x,y) = C \exp - \{x^2/\sigma_x^2 - 2\rho xy/\sigma_x \sigma_y + y^2/\sigma_y^2\}/2(1-\rho^2). \qquad (1)$$

If this is written in the form

$$f(x,y) = f_1(x)\, g_1(y \mid x),$$

$$g_1(y \mid x) = C \exp - (y - x\rho\sigma_y/\sigma_x)^2/2(1-\rho^2)\,\sigma_y^2. \qquad (2)$$

Hence, for a given value of x,

$$E(y \mid x) = (\rho\sigma_y/\sigma_x)\, x, \qquad (3a)$$

and $$\operatorname{var}(y \mid x) = (1-\rho^2)\,\sigma_y^2. \qquad (3b)$$

Thus the regression line of y on x passes through the origin and is of slope

$$B_{1y} = \rho\sigma_y/\sigma_x. \qquad (4a)$$

Similarly, the regression line of x on y,

$$x = B_{1x} y,$$

is of slope $$B_{1x} = \rho\sigma_x/\sigma_y, \qquad (4b)$$

and so $$B_{1y} B_{1x} = \rho^2. \qquad (4c)$$

If the estimates b_{1y} and b_{1x} of the slopes are obtained from the observations x_i and y_i, and if the origins of x and y are chosen at the means,

$$b_{1y} = \Sigma x_i y_i / \Sigma x_i^2$$

and $$b_{1x} = \Sigma x_i y_i / \Sigma y_i^2.$$

Hence the product

$$r^2 = b_{1y} b_{1x} = \frac{(\Sigma x_i y_i)^2}{\Sigma x_i^2 \Sigma y_i^2} \tag{5}$$

will provide an estimate of ρ, as was shown in § 2.3.1.

On comparing (3*b*) with (5.1.2,1*c*), it will be seen that the correlation ratio η and the correlation coefficient ρ are identical when the variables follow a bivariate normal distribution.

6.2 STATISTICAL TESTS BASED ON THE NORMAL LAW

If the deviations follow a normal law, then the probability of obtaining a value y_i is proportional to

$$\exp - (y_i - Y_i)^2/2\sigma_i^2 = \exp - w_i(y_i - Y_i)^2/2\sigma^2.$$

Hence, as in § 2.5,

$$\Sigma(y_i - Y_i)^2/\sigma_i^2 = \Sigma w_i(y_i - Y_i)^2/\sigma^2$$

is distributed as χ^2 with n d.f. Now

$$\begin{aligned}\Sigma w_i(y_i - Y_i)^2 &= \Sigma w_i(y_i - A_0 - A_1 \xi_i)^2 \\ &= \Sigma w_i \{(y_i - a_0 - a_1 \xi_i) + (a_0 - A_0) + (a_1 - A_1) \xi_i\}^2.\end{aligned} \tag{1}$$

From the normal equations,

$$\Sigma w_i(y_i - a_0 - a_1 \xi_i) = 0 = \Sigma w_i \xi_i(y_i - a_0 - a_1 \xi_i).$$

Also $\Sigma w_i \xi_i = 0$, and so the cross-products in the expansion of (1) disappear, and

$$\begin{aligned}\Sigma w_i(y_i - Y_i)^2/\sigma^2 = \Sigma w_i v_i^2/\sigma^2 + (a_0 - A_0)^2/(\sigma^2/\Sigma w_i) \\ + (a_1 - A_1)^2/(\sigma^2/\Sigma w_i \xi_i^2).\end{aligned} \tag{2}$$

By a rotation of coordinate axes, as in § 2.5.2, it can be shown that the three terms on the right-hand side are distributed as χ^2, with $n-2$, 1, and 1 d.f. respectively, and that the three distributions are independent. The proof is given for a polynomial of any degree in § 8.2.1, and will not be repeated in detail here.

Thus, as $\Sigma w_i v_i^2/\sigma^2$ is distributed as χ^2 with $n-2$ d.f.,

$$s^2 = \Sigma w_i v_i^2/(n-2)$$

will provide an unbiased estimate of σ^2. The standard deviation of s^2 is, from (2.5.3,2),

$$\text{S.D. } s^2 = \sigma^2/\sqrt{\{\tfrac{1}{2}(n-2)\}}, \tag{3}$$

and the standard deviation of s, from (2.5.3,4), is

$$\text{S.D. } s = \sigma/\sqrt{\{2(n-2)\}}. \tag{4}$$

The χ^2 tables can be used to test the significance of the departure of an observed value s from an expected value σ.

6.2.1 *Testing of slope*

From (6.2,2), a_1 is distributed normally about A_1 with variance $\sigma^2/\Sigma w_i \xi_i^2$, and the distributions of a_1 and s are independent. Hence, as in § 3.1.4,

$$t = \frac{a_1 - A_1}{s(a_1)} = \frac{b_1 - B_1}{s(b_1)} \tag{1}$$

is distributed as t with $\nu = n-2$ d.f. The quantity $s(a_1)$ is the estimated standard deviation of a_1, (6.1.2,2a) and (6.1.2,7a),

$$s(a_1) = s/(\Sigma w_i \xi_i^2)^{\frac{1}{2}} = \{\Sigma w_i v_i^2/(n-2)\, \Sigma w_i \xi_i^2\}^{\frac{1}{2}}. \tag{2}$$

Tables of $Q(t)$ can be used to test whether the deviation of a_1 from a hypothetical value A_1 is reasonable. In particular, setting $A_1 = 0$, the ratio

$$t = \frac{a_1}{s(a_1)} = \frac{b_1}{s(b_1)} \tag{3}$$

will test whether the slope of the line differs significantly from zero.

6.2.2 *Comparison of slopes and fitted values*

If two different determinations a_1' and a_1'' of slope are made, and if σ can be assumed to be the same in the two experiments, the ratio

$$t = \frac{a_1' - a_1''}{s(a_1' - a_1'')} \tag{1}$$

will be distributed as t with $\nu' + \nu'' = n' + n'' - 4$ d.f., by § 3.1.4. $s(a_1' - a_1'')$ is the estimated standard deviation of the difference, obtained from

$$s^2(a_1' - a_1'') = \{(1/\Sigma w_i' \xi_i'^2) + (1/\Sigma w_i'' \xi_i''^2)\}\, s^2, \tag{2}$$

where s^2 is the estimate of σ^2 given by the equation

$$s^2 = (\Sigma w_i' v_i'^2 + \Sigma w_i'' v_i''^2)/(n' + n'' - 4). \tag{3}$$

Explicitly, the ratio is

$$t = (a_1' - a_1'') \left\{\frac{n' + n'' - 4}{\Sigma\Sigma w_i v_i^2[(\Sigma w_i' \xi_i'^2)^{-1} + (\Sigma w_i'' \xi_i''^2)^{-1}]}\right\}^{\frac{1}{2}}, \tag{4}$$

or

$$t = (b_1' - b_1'') \left\{\frac{n' + n'' - 4}{\Sigma\Sigma w_i v_i^2[(\Sigma w_i'/D') + (\Sigma w_i''/D'')]}\right\}^{\frac{1}{2}}. \tag{5}$$

For the fitted values the ratio

$$t = \frac{u_1'(x) - u_1''(x)}{s} \left\{\frac{\sigma^2}{\sigma^2[u_1'(x)] + \sigma^2[u_1''(x)]}\right\}^{\frac{1}{2}} \tag{6}$$

can be used to test the divergence between the values given by two different determinations. The test of the coefficients b_0 corresponds to the special case $x = 0$ of (6).

If the two estimates are obtained by different experimental methods, so that σ is not the same in the two experiments, the ratio

$$\frac{b_1' - b_1''}{\{s^2(b_1') + s^2(b_1'')\}^{\frac{1}{2}}} \tag{7}$$

should be used, the test being either by Behrens' method (§ 3.5.3) or by Welch's method (§ 3.5). Similarly, for the two estimates of the fitted values the ratio

$$\frac{u_1'(x) - u_1''(x)}{\{s^2[u_1'(x)] + s^2[u_1''(x)]\}^{\frac{1}{2}}} \tag{8}$$

should be used.

6.2.3 *Example*

The clock errors in the 50-day period following that of Example 6.1.4 were also fitted by a straight line. Table 6.2.3 shows the comparison of the coefficients b_1 and b_0 for the two periods. For the slopes the value 1·18 for t corresponds to a value of Q of 0·24, while for the coefficients b_0 the value 2·03 for t corresponds to a value of Q of 0·05. Hence, although the mean clock rates are in reasonable agreement, the values at $x = 0$ are probably discordant and it seems very likely that the behaviour of the clock cannot be represented adequately by a single straight line covering the whole period.

TABLE 6.2.3

(*a*) *Comparison of slopes of fitted lines*

b_1'	0·08422	b_1''	0·08532	$b_1' - b_1''$	$1{\cdot}10 \times 10^{-3}$
n'	20	n''	25	$n' + n'' - 4$	41
$\Sigma v_i'^2$	0·0301	$\Sigma v_i''^2$	0·0585	Σv_i^2	0·0886
n'/D'	$2{\cdot}315 \times 10^{-4}$	n''/D''	$1{\cdot}727 \times 10^{-4}$	sum	$4{\cdot}042 \times 10^{-4}$

t (6.2.2,5) $1{\cdot}10 \times 10^{-3}\,(1{\cdot}145 \times 10^6)^{\frac{1}{2}} = 1{\cdot}18$

(*b*) *Comparison of values* $b_0 \equiv u_1(0)$

b_0'	0·1689	b_0''	0·0650	$b_0' - b_0''$	0·1039
D'	$8{\cdot}638 \times 10^4$	D''	$14{\cdot}475 \times 10^4$		
$(\Sigma x_i')^2$	$0{\cdot}2401 \times 10^6$	$(\Sigma x_i'')^2$	$3{\cdot}5721 \times 10^6$	$s^2 = \Sigma v_i^2/(n' + n'' - 4) = 0{\cdot}0886/41$	

(6.1.2,7c, $x = 0$) $\quad \sigma^2[u_1'(0)]/\sigma^2 = \{1 + (\Sigma x_i')^2/D'\}/n' = 0{\cdot}189$

$\sigma^2[u''(0)]/\sigma^2 = 1{\cdot}027 \quad$ sum 1·216

t (6.2.2,6, $x = 0$) $= 0{\cdot}1039(380{\cdot}6)^{\frac{1}{2}} = 2{\cdot}03$

6.2.4 *Tests for homogeneity*

When r different sets of observations y_{ji} are made, a straight line

$$u_{1j}(x) = a_{0j} + a_{1j}\,\xi_j \tag{1}$$

may be fitted to each separately. The distribution of the slopes a_{1j} will now be investigated, on the assumption that σ^2 and the true slope A_1 is the same in each set.

The variance of the slope a_{1j} is

$$\sigma^2(a_{1j}) = \sigma^2/\Sigma w_{ji}\,\xi_{ji}^2. \tag{2}$$

Hence the weighted mean of all the slopes will be

$$\bar{a}_1 = \Sigma W_j\, a_{1j}/\Sigma W_j, \tag{3}$$

where, from (2),

$$W_j = \sigma^2/\sigma^2(a_{1j}) = \sum_i w_{ji}\,\xi_{ji}^2. \tag{4}$$

The quantity $\Sigma W_j(a_{1j}-\bar{a}_1)^2/\sigma^2$, which is the weighted sum of the squares of the deviations of the slopes from their weighted mean, will be distributed as χ^2 with $r-1$ d.f. Also $\sum_j\sum_i w_{ji}\,v_{ji}^2/\sigma^2$ will be distributed, independently of the a_{1j}, as χ^2 with

$$\Sigma(n_j-2) = n-2r \text{ d.f.}$$

Hence on the postulate of homogeneity of slopes and standard deviations σ the ratio

$$F = \frac{\Sigma W_j(a_{1j}-\bar{a}_1)^2}{r-1}\Bigg/\frac{\Sigma\Sigma w_{ji}\,v_{ji}^2}{n-2r} \tag{5}$$

will be distributed as F with $(r-1, n-2r)$ d.f. This provides a test for the homogeneity of the slopes.

If the values of the slopes pass the test for homogeneity, $\bar{a}_1$ will provide an estimate of A_1, with variance given by the equation

$$\operatorname{var}\bar{a}_1 = \Sigma W_j^2 \operatorname{var} a_{1j}/(\Sigma W_j)^2 = (1/\Sigma W_j)\,\sigma^2. \tag{6a}$$

Since

$$s^2 = \Sigma w_{ji}\,v_{ji}^2/(n-2r)$$

is an unbiased estimate of σ^2, the standard deviation of $\bar{a}_1$ can be estimated from the equation

$$s^2(\bar{a}_1) = s^2/\Sigma W_j. \tag{6b}$$

The homogeneity of the values b_0 can be tested in a similar way. If the values b_0 and b_1 both pass the homogeneity test, the lines may be assumed to be all estimates of the same straight line.

It will be noted that the values a_0 would not be expected to be homogeneous in general, since a_{0j} is the fitted value at the point corresponding to the weighted mean $\bar{x}_j$ of the x_{ji} in the jth set, and the location of this point depends on the values x_{ji} and their weights w_{ji} in the particular set of observations.

6.2.4.1 *Example*

The constancy of the clock rate over the 50-day period of Table 6.1.4 can be tested by subdividing the observations into three groups and testing the homogeneity of the three values a_1. The calculations are carried out in Table 6.2.4. It is found that the spread of the slopes is somewhat less than would have been expected from the residuals, but the significance level corresponding to F is much greater than 5%, and so there is no reason to suspect anything untoward.

It will be noted that $\bar{b}_1$ is much less accurate than the estimate obtained in § 6.1.4. This is because the range of x and hence the value of $\Sigma(x-\bar{x})^2$ is very much greater for the complete set than for the subdivisions. In cases where the slopes and fitted values are in agreement, b_1 should be obtained from the complete set as in § 6.1.4.

TABLE 6.2.4

Test for uniformity of slopes in sub-sets

x	n	Σx	Σx^2	Σy	Σyx	Σy^2	D
3–14	7	62	636	6·478	64·636	6·608644	608
16–33	6	133	3135	12·099	284·262	25·783227	1121
35–49	7	295	12553	26·068	1108·645	97·930812	846

$b_1 D$	b_1	$(\Sigma y)^2/n$	$b_1^2 D/n$	Σv^2	$W = D/n$
50·816	0·08357895	5·994926	0·606735	0·006983	3648/42
96·405	0·08599911	24·397634	1·381791	0·003802	7847/42
70·455	0·08328014	97·077232	0·838215	0·015365	5076/42
					ΣW 16571/42

$$\bar{b}_1 = \Sigma W b_1/\Sigma W = 1402{\cdot}4610/16571 = 0{\cdot}08463346$$

$b_1-\bar{b}_1$	$(b_1-\bar{b}_1)^2$	
0·0010545	0·0000011120	$\Sigma W(b_1-\bar{b}_1)^2 = 0{\cdot}000666$
0·0013656	0·0000018649	
0·0013533	0·0000018314	$\Sigma v^2 = 0{\cdot}026150$

(6.2.4,5) $$F = \frac{26150}{14}\,\frac{2}{666}\,(\nu_1 = 14, \nu_2 = 2) = 5{\cdot}61$$

5% Significance level 19·4

S.D. of mean slope: $s^2 = \Sigma v^2/(n-2r) = 0{\cdot}001868$

From (6.2.4–6*b*)

$$s^2(\bar{b}_1) = 0{\cdot}001868 \times 42/16571 = 0{\cdot}000004735$$

$$s(\bar{b}_1) = 0{\cdot}00218$$

6.2.4.2 *Analysis of variance.* The sums of the squares of the deviations can be split up as shown in Table 6.2.4*a*. This scheme is an example of an analysis of variance table. Such tables are widely used in modern statistical practice. The agreement of the total sum of the squares with the value $\Sigma\Sigma w_{ji} y_{ji}^2$ provides a check on the arithmetical calculations.

TABLE 6.2.4*a*

Analysis of variance table (Example 6.2.4.1)

Deviations		d.f.	Sum of squares	d.f.
Of mean value in each set from zero	$\Sigma\{(\Sigma w_{ji} y_{ji})^2/\Sigma w_{ji}\}$	r	127·469792	3
Of mean slope from zero	$\Sigma W_j \bar{a}_1^2$	1	2·826075	1
Of each slope from mean slope	$\Sigma W_j(a_{1j}-\bar{a}_1)^2$	$r-1$	0·000666	2
Residuals	$\Sigma\Sigma w_{ji} v_{ji}^2$	$n-2r$	0·026150	14
Of observations from zero	$\Sigma\Sigma w_{ji} y_{ji}^2$	n	130·322683	20

6.3 EQUALLY-SPACED OBSERVATIONS OF EQUAL WEIGHT

When the observations are all of equal weight, and the interval $\Delta x = x_{i+1} - x_i$ between successive values of x_i is constant, the calculations can be considerably simplified. The variable x is replaced by the variable ϵ, where

$$x - \bar{x} = \epsilon\Delta x. \tag{1}$$

The spacing between successive values of ϵ at the points of observation is

$$\epsilon_{i+1} - \epsilon_i = (x_{i+1} - x_i)/\Delta x = 1,$$

and $\Sigma\epsilon_i = 0$. Hence the values ϵ_i are the integers or half-integers $-\tfrac{1}{2}(n-1), -\tfrac{1}{2}(n-3), \ldots, +\tfrac{1}{2}(n-1)$.

If the fitted curve in terms of the variable ϵ is written as

$$u_1(\epsilon) = a_0 + a_1\epsilon, \tag{2}$$

the coefficients a_0 and a_1 are, from (6.1.1,5),

$$a_1 = \Sigma\epsilon_i y_i/\Sigma\epsilon_i^2, \tag{3a}$$

$$a_0 = \Sigma y_i/n. \tag{3b}$$

When n is odd the values ϵ_i are $0, \pm 1, \pm 2$, etc., and the calculation of $\Sigma \epsilon_i y_i$ is simple. When n is even, the values ϵ_i are the half-integers $\pm \frac{1}{2}, \pm \frac{3}{2}$, etc., and it is usually more convenient to calculate $\Sigma(2\epsilon_i) y_i$, and a_1 from

$$a_1 = \tfrac{1}{2}\Sigma(2\epsilon_i) y_i / \Sigma \epsilon_i^2. \tag{3c}$$

The values $\Sigma \epsilon_i^2$ are listed in Table 6.7*b* for $n = 6(1)75$. The sum $\Sigma \epsilon_i^2$ is given by the formula

$$\Sigma \epsilon_i^2 = n(n^2-1)/12. \tag{3d}$$

For
$$\Sigma \epsilon_i^2 = \sum_{j=0}^{n-1} \{\tfrac{1}{2}(n-2j-1)\}^2,$$

and
$$\sum_0^r \tfrac{1}{4}(r-2j)^2 = \tfrac{1}{4}(r+1)r^2 - r\Sigma j + \Sigma j^2 = \tfrac{1}{12}r(r+1)(r+2).$$

On substituting $n-1$ for r, the expression (3*d*) is obtained.

6.3.1 *Standard deviations*

From (6.1.2,2*a*, *b*),

$$\sigma^2(a_1) = \operatorname{var} a_1 = \sigma^2/\Sigma \epsilon_i^2, \tag{1a}$$

$$\sigma^2(a_0) = \operatorname{var} a_0 = \sigma^2/n. \tag{1b}$$

Explicitly, formula (1*a*) is

$$\sigma^2(a_1) = 12\sigma^2/n(n^2-1) = 12\sigma^2/n^3(1-n^{-2}), \tag{2a}$$

and so
$$\sigma(a_1) = 3{\cdot}47 n^{-\frac{3}{2}}\sigma, \tag{2b}$$

neglecting the term n^{-2}.

For the fitted value,

$$\sigma^2[u_1(\epsilon)] = (\sigma^2/n)\left[1 + \epsilon^2 \frac{n}{\Sigma \epsilon_i^2}\right]. \tag{3}$$

The standard deviation of the fitted value can be put in the form (cf. 6.1.5,3)

$$\sigma[u_1(\epsilon)] = \sigma n^{-\frac{1}{2}} \rho_{10}(k), \tag{4a}$$

where, using (6.3,3*d*),

$$\rho_{10}(k) = (1+3k^2)^{\frac{1}{2}} \tag{4b}$$

and
$$k^2 = 4\epsilon^2/(n^2-1).$$

The approximate value

$$k = 2\epsilon/n \tag{4c}$$

will always be adequate. The function $\rho_{10}(k)$ is tabulated in Table 6.7*a*. The expression (4*a*) is useful when the standard deviations are required at a number of points. The region in which the observations lie—the region of interpolation—is between $k = +1$ and $k = -1$. The standard deviation at the extremities of the region of interpolation is twice that at the centre of the region. In the region of extrapolation beyond $|k| = 1$, the standard deviation increases steadily.

As in § 6.1.3, σ may be estimated by

$$s = \{\Sigma v_i^2/(n-2)\}^{\frac{1}{2}}, \tag{5a}$$

where Σv_i^2 may be calculated from any one of the formulae

$$\Sigma v_i^2 = \Sigma y_i^2 - na_0^2 - (\Sigma \epsilon_i^2)\, a_1^2, \tag{5b}$$

$$\Sigma v_i^2 = \Sigma y_i^2 - a_0 \Sigma y_i - a_1 \Sigma \epsilon_i y_i, \tag{5c}$$

$$\Sigma v_i^2 = \Sigma y_i^2 - (\Sigma y_i)^2/n - (\Sigma \epsilon_i y_i)^2/\Sigma \epsilon_i^2. \tag{5d}$$

6.3.2 *Return to the original variable*

Usually it will be necessary to obtain expressions for the coefficients and fitted values in terms of the original variable x. If the fitted curve is written

$$u_1(x) = b_0 + b_1 x, \tag{1}$$

the conversion formulae are as follows:

$$b_1 = a_1/\Delta x, \tag{2a}$$

$$b_0 = a_0 - b_1 \bar{x}; \tag{2b}$$

$$\sigma^2(b_1) = \sigma^2(a_1)/(\Delta x)^2, \tag{3a}$$

$$\sigma^2(b_0) = \sigma^2(a_0) + \bar{x}^2 \sigma^2(b_1); \tag{3b}$$

$$\sigma^2[u_1(x)] = \sigma^2(a_0) + (x-\bar{x})^2 \sigma^2(b_1), \tag{4}$$

$$\sigma[u_1(k)] = \sigma n^{-\frac{1}{2}} \rho_{10}(k), \tag{5a}$$

with

$$k = 2(x-\bar{x})/n\Delta x. \tag{5b}$$

6.3.2.1 *Dependence of variance of slope on range and on number of observations.* The variance of the estimate b_1 is

$$\operatorname{var} b_1 = \operatorname{var} y/(\Delta x)^2 \Sigma \epsilon_i^2.$$

If $\mathscr{R}(x)$ denotes the range of the variable x,

$$\mathscr{R}(x) = x_n - x_1 = (n-1)\,\Delta x, \tag{1}$$

$$\operatorname{var} b_1 = 12\sigma^2(n-1)^2/n(n^2-1)\,\mathscr{R}^2(x),$$

or

$$\sigma(b_1) = \{3{\cdot}47\sigma/\mathscr{R}(x)\sqrt{n}\}\{1-2/n\}, \tag{2}$$

and the standard deviation is inversely proportional to $\sqrt{n}$ and to the range of x.

If the variable x is not equally-spaced, it is shown in § 8.5 that the standard deviation of b_1 is equal to the expression (2) divided by a factor $f_1^{\frac{1}{2}}$ which usually lies between 0·8 and 1·3. This factor takes into account the variation in Σx^2 for a given value of $\mathscr{R}(x)$ due to differences in spacing within the range. In most cases the expression

$$\sigma(b_1) = 3{\cdot}47\sigma/\mathscr{R}(x)\sqrt{n} \tag{3}$$

will be an adequate approximation.

6.3.3 *Example*

In Table 6.3.3 are shown the mirror settings y (in units of 10^{-3} mm.) for successive positions of minimum visibility in a Michelson interferometer with a sodium light source. The calculation of the slope of the line fitting these observations is shown in the same table. The value obtained for a_1 is $(289{\cdot}77 \pm 0{\cdot}47) \times 10^{-3}$ mm.

TABLE 6.3.3

Positions of minimum visibility with sodium light (*Example* 6.3.3)

$\lvert 2\epsilon \rvert$	y_+	y_-
1	3648	3353
3	3965	3084
5	4230	2792
7	4546	2514
9	4817	2204
11	5123	1929
13	5421	1650
15	5694	1363
17	5993	1074
19	6278	763

$M_0 = \Sigma y = 70441$

$a_0 = M_0/n = 3522{\cdot}05$

$M_1 = \tfrac{1}{2}\Sigma 2\epsilon y = 192699{\cdot}5$ $\qquad \Sigma\epsilon_i^2 = 665$

$a_1 = M_1/\Sigma\epsilon_i^2 = 289{\cdot}7737$

Σy^2	303938669
$-M_0 a_0$	248096724
$-M_1^2/\Sigma\epsilon_i^2$	55839244
$= \Sigma v^2$	2701

$s = \{\Sigma v^2/(n-2)\}^{\frac{1}{2}} = 12{\cdot}2$

$s(a_1) = s/\sqrt{\Sigma\epsilon_i^2} = 0{\cdot}47$

The difference in wavelength of the two components of the sodium doublet is given by the formula

$$\lambda_1 - \lambda_2 = \lambda_1\lambda_2/2a_1.$$

Using the values 5896 A.U. and 5890 A.U. for λ_1 and λ_2, the value of $\lambda_1 - \lambda_2$ given by this experiment is $5{\cdot}992 \pm 0{\cdot}010$ A.U.

If n is odd the observations are still listed in two columns of equal length as in Table 6.3.3, starting at the bottom of the left column with the value y for the largest value of ϵ, but the central value corresponding to $\epsilon = 0$ is unpaired. To illustrate the calculating scheme when n is odd, in Table 6.3.3*a* a straight line is fitted to the 17 observations obtained by omitting the three largest observations from the set in Table 6.3.3.

TABLE 6.3.3*a*

The fitting of a straight line when n is odd

$\lvert\epsilon\rvert$	y_+	y_-
0	3084	
1	3353	2792
2	3648	2514
3	3965	2204
4	4230	1929
5	4546	1650
6	4817	1363
7	5123	1074
8	5421	763

$$M_0 = \Sigma y = 52476$$

$$a_0 = M_0/n = 3086{\cdot}82$$

$$M_1 = \Sigma \epsilon y = 118127 \qquad \Sigma\epsilon_i^2 = 408$$

$$a_1 = M_1/\Sigma\epsilon_i^2 = 289{\cdot}5270$$

Σy^2	196187700
$-M_0^2/n$	161984152
$-M_1^2/\Sigma\epsilon_i^2$	34200951
$= \Sigma v^2$	2597

$$s = \{\Sigma v^2/(n-2)\}^{\frac{1}{2}} = 13{\cdot}2$$

$$s(a_1) = s/\sqrt{\Sigma\epsilon_i^2} = 0{\cdot}65$$

6.3.4 *The estimation of slope from successive differences*

The expectation of the difference

$$\Delta y_i = y_{i+1} - y_i$$

is $B_1\Delta x = A_1$, where B_1 is the slope of the 'true' line. Hence it should be possible to obtain an estimate of A_1 from the difference values Δy_i.

Unfortunately, the mean difference $\Sigma\Delta y_i/(n-1)$ does not provide a satisfactory estimate, for the intermediate observations cancel out, and the value obtained is just $(y_n - y_1)/(n-1)$. This, being based on only two observations, is a very inefficient estimate. It will now be shown that, by ascribing suitable weights to the differences, the least-squares estimate a_1 can be obtained.

The estimate a_1 is

$$\sum_{-\frac{1}{2}(n-1)}^{\frac{1}{2}(n-1)} \epsilon y(\epsilon)/\Sigma\epsilon^2 = \frac{12}{n(n^2-1)} \sum_{m=1}^{n} \tfrac{1}{2}(2m-n-1)\,y(m),$$

where $$m = \epsilon + \tfrac{1}{2}(n+1).$$

Thus, as $$2m-n-1 = (m-1)\{n-(m-1)\} - m(n-m),$$

$$a_1 = \left[\sum_{m=2}^{n} (m-1)\{n-(m-1)\}\,y(m) - \sum_{m=1}^{n-1} m(n-m)\,y(m)\right] \Big/ [n(n^2-1)/6]$$

$$= \left[\sum_{m=1}^{n-1} m(n-m)\,\{y(m+1)-y(m)\}\right] \Big/ [n(n^2-1)/6],$$

or $$a_1 = \sum_{m=1}^{n-1} m(n-m)\,\Delta y(m)/\{n(n^2-1)/6\}. \qquad (1)$$

Thus a_1 is the weighted sum of the differences Δy. Also if

$$W(m) = m(n-m), \tag{2a}$$

$$\sum_1^{n-1} W(m) = n\Sigma m - \Sigma m^2 = n(n^2-1)/6. \tag{2b}$$

Hence (1) can be put in the form

$$a_1 = \Sigma W(m)\,\Delta y(m)/\Sigma W(m). \tag{3}$$

The weights $W(m)$ are $(n-1)\times 1, (n-2)\times 2, \ldots$. The weights $W(m)$ and their sums $\Sigma W(m)$ are listed in Table 6.7c for $n2(1)55$.

6.3.4.1 *Estimation of standard deviation.* The standard deviation σ can be estimated from the residuals

$$V_i = \Delta y_i - a_1. \tag{1}$$

For, as $E(V_i) = 0$,

$$\begin{aligned} E(V_i^2) &= \operatorname{var} V_i = \operatorname{var}(y_{i+1} - y_i - a_1) \\ &= \operatorname{var} y_i + \operatorname{var} y_{i+1} + \operatorname{var} a_1 + 2\operatorname{cov}(a_1, y_i) - 2\operatorname{cov}(a_1, y_{i+1}) \\ &= 2\sigma^2 - \sigma^2/\Sigma\epsilon^2. \end{aligned}$$

But $\Sigma\epsilon^2$ is $0(n^{-3})$, and so the second term can be neglected. Hence

$$E(V_i^2) = 2\sigma^2, \tag{2}$$

and

$$s_V^2 = \Sigma V_i^2/2(n-1) \tag{3}$$

will provide an unbiased estimate of σ^2.

The corresponding estimate of the standard deviation of the slope is, from (6.3.1,2a) and (6.3.4,2b),

$$s_V(a_1) = \{\Sigma V_i^2/(n-1)\,\Sigma W\}^{\frac{1}{2}} \doteqdot \sqrt{(6\Sigma V_i^2)}/n^2. \tag{4}$$

6.3.4.2 *Example*

The calculation of the spacing of the successive positions of minimum visibility for the example of § 6.3.3 is carried out using finite differences in Table 6.3.4. Since this method will often be employed when there is no calculating-machine available, the calculations in this example have all been done without the use of a machine. It is then advantageous to reduce the magnitudes of the differences Δy by subtracting a suitable constant—in this case 260. The weights W_1 and the value ΣW_1 are obtained from Table 6.7c.

The formation of the differences $\Delta y'$ can be checked by the equation

$$\Sigma\Delta y' + (n-1)\times \text{const.} = y(n-1) - y(0).$$

In this case $\quad 575 + 19\times 260 = 6278 - 763.$

Since $\Sigma W_1\,\Delta y'/\Sigma W_1 = 29{\cdot}77$, the residuals V_i may be taken as $\Delta y_i' - 30$. The final estimate for the slope is

$$a_1 = 289{\cdot}77 \pm 0{\cdot}43.$$

TABLE 6.3.4

Calculation of slope using differences (Example 6.3.4.2)

y	Δy	$\Delta y' = \Delta y -$ const. (260)	W_1	$W_1\,\Delta y'$	V	V^2
763	311	51	19	969	21	441
1074	289	29	36	1044	1	1
1363	287	27	51	1377	3	9
1650	279	19	64	1216	11	121
1929	275	15	75	1125	15	225
2204	310	50	84	4200	20	400
2514	278	18	91	1638	12	144
2792	292	32	96	3072	2	4
3084	269	9	99	891	21	441
3353	295	35	100	3500	5	25
3648	317	57	99	5643	27	729
3965	265	5	96	480	25	625
4230	316	56	91	5096	26	676
4546	271	11	84	924	19	361
4817	306	46	75	3450	16	256
5123	298	38	64	2432	8	64
5421	273	13	51	663	17	289
5694	299	39	36	1404	9	81
5993	285	25	19	475	5	25
6278	—					
		575	1330	39599		4917
		$\Sigma\Delta y'$	ΣW_1	$\Sigma W_1\,\Delta y'$		ΣV^2

$\log \Sigma W_1\,\Delta y'$	4·5977
$\log \Sigma W_1$	3·1239
diff.	1·4738
$\Sigma W_1\,\Delta y'/\Sigma W_1$	29·77

$$\begin{aligned} a_1 &= \text{const.} + \Sigma W_1\,\Delta y'/\Sigma W_1 \pm \sqrt{(6\Sigma V^2)}/n^2 \\ &= 260 \quad + 29{\cdot}77 \qquad\qquad \pm \sqrt{29502}/400 \\ &= 289{\cdot}77 \pm 0{\cdot}43 \end{aligned}$$

6.3.5 *Efficiency of* s_V

The variance of the estimate s_V^2 is

$$\operatorname{var} s_V^2 = E(s_V^2-\sigma^2)^2 = E(s_V^4)-\sigma^4 = \frac{1}{4(n-1)^2}E(\Sigma V_i^2)^2-\sigma^4. \quad (1)$$

If the deviations $y_i - Y_i$ are denoted by δy_i,

$$V_i = y_{i+1}-y_i-a_1 = \delta y_{i+1}-\delta y_i + A_1 - a_1.$$

Now the standard deviation of a_1 is of the order of $\sigma n^{-\frac{3}{2}}$, and so the deviations $A_1 - a_1$ can be neglected in comparison with the deviations δy_i. Hence the approximation

$$E(\Sigma V_i^2)^2 = E\sum_i\sum_j V_i^2 V_j^2 = E\sum_i\sum_j (\delta y_{i+1}-\delta y_i)^2(\delta y_{j+1}-\delta y_j)^2,$$

or
$$E(\Sigma V_i^2)^2 = E\sum_i\sum_j \{(\delta y_i)^2+(\delta y_{i+1})^2-2\delta y_i\,\delta y_{i+1}\}$$
$$\times\{(\delta y_j)^2+(\delta y_{j+1})^2-2\delta y_j\,\delta y_{j+1}\} \qquad (2)$$

is obtained.

This can only be evaluated if the form of the deviation law is known. If the deviations are assumed to follow a normal law,

$$E(\delta y_i)^4 = \mu_4 = 3\sigma^4, \quad E(\delta y_i)^2(\delta y_j)^2 = \sigma^4. \qquad (3)$$

Then

$$E(V_i^4) = 12\sigma^4, \quad E(V_i^2 V_{i+1}^2) = 6\sigma^4, \quad E(V_i^2 V_{i+1+q}^2) = 4\sigma^4. \qquad (4)$$

Of the $(n-1)^2$ terms on the left-hand side of (2), $n-1$ are of the first type, and $2(n-2)$ are of the second type, the remaining terms being of the third type. Hence

$$E(\Sigma V_i^2)^2 = 12(n-1)\sigma^4 + 12(n-2)\sigma^4$$
$$+4\{(n-1)^2-(n-1)-2(n-2)\}\sigma^4$$
$$= 4\sigma^4(n^2+n-3),$$

and so
$$\operatorname{var} s_V^2 = \frac{\sigma^4}{(n-1)^2}(3n-4) = \frac{3\sigma^4}{n-1}\left\{1-\frac{1}{3(n-1)}\right\}. \qquad (5)$$

The variance of the estimate s^2 is, from (2.5.3,2),

$$\operatorname{var} s^2 = 2\sigma^4/(n-2).$$

Hence the efficiency of s_V^2 is

$$\eta(s_V^2) = \frac{\operatorname{var} s^2}{\operatorname{var} s_V^2} = \tfrac{2}{3}\{1+\tfrac{4}{3}(n-1)^{-1}\}. \qquad (6)$$

The efficiency of the estimate s_V^2 of the variance of an observation is thus about 0·67. Since $\operatorname{var} s^2 = 4\sigma^2 \operatorname{var} s$, from (1.2,14*a*), the efficiency of the standard deviation estimate s_V will also be 0·67. These estimates are somewhat inefficient, and if a more accurate estimate is desired, it will be best to calculate a_0 and then calculate s from (6.3.1,5*a*–*d*).

6.3.6 *Calculation of slope by double summation*

In this method the observations are listed in decreasing order of x, and the column of y values summed from the top, recording the intermediate sums $\sum_{i=j}^{n} y_i$ and the final sum

$$M_0 = \sum_{i=1}^{n} y_i. \qquad (1)$$

These sums are now added to give the quantity

$$M_1 = \sum_{j=1}^{n} \sum_{i=j}^{n} y_i. \qquad (2a)$$

This can be rewritten

$$M_1 = \sum_{i=1}^{n} \sum_{j=1}^{i} y_i = \sum_{i=1}^{n} i y_i. \qquad (2b)$$

Now in terms of the equally-spaced variable $x \equiv i$ taking the values 1 to n at the points of observation, the slope is

$$b_1 = a_1 = \frac{\Sigma i y_i - \Sigma y_i \Sigma i/n}{\Sigma i^2 - (\Sigma i)^2/n}, \qquad (3)$$

and the denominator is

$$\Sigma(i - \bar{i})^2 = \Sigma \epsilon^2 = n(n^2 - 1)/12, \qquad (4)$$

which is tabulated in Table 6.7*b*. Hence

$$a_1 = \{M_1 - \tfrac{1}{2}(n+1)\, M_0\}/\Sigma\epsilon^2, \qquad (5a)$$

while

$$a_0 = M_0/n. \qquad (5b)$$

This method is useful when a printing-adding machine is employed, as the intermediate sums can then be printed automatically.

6.3.6.1 *Example*

The calculations by the double summation method for the observations given in Table 6.3.3 are carried out in Table 6.3.6. The observed values

TABLE 6.3.6

Double summation method (*Example* 6.3.6.1)

	(37872)		*(61458)*
6278	4230		2204
6278	*42102*		*63662*
5993	3965		1929
12271	*46067*		*65591*
5694	3648		1650
17965	*49715*		*67241*
5421	3353		1363
23386	*53068*		*68604*
5123	3084		1074
28509	*56152*		*69678*
4817	2792		763
33326	*58944*	M_0	*70441*
4546	2514		
37872	*61458*	M_1	*932330*

$n = 20$ $\qquad \Sigma\epsilon^2 = n(n^2 - 1)/12 = 665$

$a_1 = \{M_1 - \tfrac{1}{2}(n+1)\, M_0\}/\Sigma\epsilon^2 = 192699{\cdot}5/665 = 289{\cdot}77$

are listed in decreasing order of x, using double spacing, and the intermediate sums are entered in between the observations so that the quantities to be summed lie directly under one another. When the calculations are done mentally it is best to recheck the progressive total from time to time by direct summation of the last checked total and the intervening observations.

6.4 OTHER ESTIMATES OF THE SLOPE

It is clear that, if $W(x)$ is any function whatsoever for which the sum of the values $W(x_i) \equiv W_i$ at the points of observation is zero,

$$\Sigma W_i \equiv \Sigma W(x_i) = 0, \tag{1}$$

then

$$E[\Sigma W_i(y_i - B_0 - B_1 x_i)] = 0 = E[\Sigma W_i(y_i - B_1 x_i)],$$

where B_0 and B_1 are the coefficients of the regression line or functional relationship line. Hence

$$b_1 = \Sigma W_i y_i / \Sigma W_i x_i \tag{2}$$

will provide an unbiased estimate of B_1, the variance of which is given by

$$\sigma^2(b_1)/\sigma^2 = \Sigma W_i^2 / (\Sigma W_i x_i)^2. \tag{3}$$

When

$$W_i = w_i(x_i - \bar{x}) = w_i \xi_i,$$

the estimate is the least-squares estimate, which will be denoted by the symbol b_1^*. The efficiency of any other estimate b_1 will be given by

$$\eta[b_1] = \operatorname{var} b_1^* / \operatorname{var} b_1 = (\Sigma W_i x_i)^2 / \Sigma W_i^2 \, \Sigma w_i(x_i - \bar{x})^2. \tag{4}$$

The estimate b_1 will be less accurate than the least-squares estimate, but, if the number of observations is very large, or if the computing facilities are limited, it may be best to put up with a slight loss in accuracy in return for a considerable reduction in computing time.

6.4.1 *Step function methods for equally-spaced observations*

When the observations are equally spaced and of equal weight, it is convenient to replace the variable x by the variable ϵ defined in § 6.3. Then (6.4,2) and (6.4,4) become

$$b_1 = (\Sigma W_i y_i) / (\Sigma W_i \epsilon_i) \, \Delta x \tag{1a}$$

and

$$\eta(b_1) = 12(\Sigma W_i \epsilon_i)^2 / n(n^2 - 1) \, \Sigma W_i^2 \doteqdot 12 n^{-3} (\Sigma W_i \epsilon_i)^2 / \Sigma W_i^2. \tag{1b}$$

A step function is a function which is constant in magnitude over specified ranges of ϵ, the magnitude usually being different in different ranges. Now $W(\epsilon)$ must be an odd function of ϵ, from (6.4,1), and so, for any arbitrary function $f(\epsilon)$, if $W(\epsilon)$ is a step function $\sum_i W(\epsilon_i) f(\epsilon_i)$ will be of the form

$$\left[q_1\left(\sum_{0,\frac{1}{2}}^{\frac{1}{2}(n-1)} - \sum_{0,\frac{1}{2}}^{\frac{1}{2}(a_1-1)}\right) + q_2\left(\sum_{0,\frac{1}{2}}^{\frac{1}{2}(a_1-1)} - \sum_{0,\frac{1}{2}}^{\frac{1}{2}(a_2-1)}\right) + \dots \right.$$
$$\left. + q_m\left(\sum_{0,\frac{1}{2}}^{\frac{1}{2}(a_{m-1}-1)} - \sum_{0,\frac{1}{2}}^{\frac{1}{2}(a_m-1)}\right)\right][f(\epsilon_i) - f(-\epsilon_i)],$$

the numbers $\frac{1}{2}(a_j - 1)$ being the values of ϵ at the ends of the steps. The function will be referred to as a single-step function if $m = 1$, a double-step function if $m = 2$, etc. The total number of different steps, counting those corresponding to both positive and negative values of ϵ and the central step of zero weight, is $2m+1$. It is required to find, for any given m, the best values of the parameters a_j locating the steps and of the corresponding weights q_j.

The expression $\Sigma W_i \epsilon_i$ will consist of terms of the form

$$2\sum_{0}^{\frac{1}{2}(a_j-1)} \epsilon = (a_j^2 - 1)/4, \quad 2\sum_{\frac{1}{2}}^{\frac{1}{2}(a_j-1)} \epsilon = a_j^2/4. \tag{2}$$

If
$$\alpha_j = a_j/n, \quad \alpha_0 = 1, \tag{3}$$

then
$$\Sigma W_i \epsilon_i = \frac{n^2}{4}\sum_{k=1}^{m} q_k(\alpha_{k-1}^2 - \alpha_k^2). \tag{4a}$$

Similarly,
$$\Sigma W_i^2 = n\sum_{1}^{m} q_k^2(\alpha_{k-1} - \alpha_k). \tag{4b}$$

Thus, on substituting these expressions in (1b), the efficiency of the estimate b_1 is found to be

$$\eta(b_1) = \frac{3}{4}\left\{\sum_1^m q_k(\alpha_{k-1}^2 - \alpha_k^2)\right\}^2 \Big/ \sum_1^m q_k^2(\alpha_{k-1} - \alpha_k). \tag{5}$$

The expression on the right may be written as $\frac{3}{4}(F^2/D)$. To maximize $\eta(b_1)$, the fraction F^2/D is differentiated with respect to the q_k, α_k, and the resultant expressions equated to zero. On differentiating with respect to q_k,

$$\alpha_{k-1} + \alpha_k = (F/D)\, q_k.$$

It is clear that the weights may be multiplied by any arbitrary common factor without altering the efficiency. It is convenient to choose this factor so that (F/D) is unity. Then

$$\alpha_{k-1} + \alpha_k = q_k. \tag{6}$$

On differentiation with respect to α_k,

$$\left.\begin{aligned} q_k + q_{k+1} &= 4\alpha_k, \quad k = 1 \text{ to } m-1; \\ q_m &= 4\alpha_m. \end{aligned}\right\} \tag{7}$$

Combining (6) and (7),

$$\alpha_{k-1}+\alpha_{k+1} = 2\alpha_k,$$

and so $$\alpha_{k-1}-\alpha_k = \Delta\alpha = 1-\alpha_1, \text{ a constant.}$$

But $$\alpha_{m-1}+\alpha_m = q_m = 4\alpha_m,$$

and so $$\Delta\alpha = 2\alpha_m.$$

Thus $$\alpha_m = \alpha_1-(m-1)\,\Delta\alpha = 1-m\Delta\alpha,$$

and $$\alpha_m = 1/(2m+1). \tag{8a}$$

Hence $$\alpha_j = \frac{2(m-j)+1}{2m+1}, \tag{8b}$$

and, from (6), $$q_j = \frac{4(m-j+1)}{2m+1}. \tag{8c}$$

These give the optimum values of α_j and q_j. To find the corresponding efficiency, the numerator

$$F = \sum_{j=1}^{m} q_{m-j+1}(\alpha_{m-j}^2-\alpha_{m-j+1}^2)$$

must be evaluated. Using (8*b*) and (8*c*),

$$F = \sum_j 4j\{(2j+1)^2-(2j-1)^2\}/(2m+1)^3$$

$$= 32\sum_j j^2/(2m+1)^3 = \frac{32m(m+1)}{6(2m+1)^2}.$$

Thus, since $F = D$, the efficiency of the estimate b_1 is

$$\eta(b_1) = \tfrac{3}{4}F = 1-\frac{1}{(2m+1)^2}. \tag{9}$$

6.4.1.1 *Optimum weights and steps.* The calculations of the previous section can now be summarized. The number of observations in the jth step should be

$$\tfrac{1}{2}(a_{j-1}-1)-\tfrac{1}{2}(a_j-1) = \tfrac{1}{2}n(\alpha_{j-1}-\alpha_j) = n/(2m+1);$$

that is, the observations should be divided uniformly among the $2m+1$ steps. The weight q_j, from (6.4.1,8*c*), should be proportional to $m-j+1$; that is, the weights should be $m, m-1, \ldots, 1$. The efficiency is then given by (6.4.1,9); η is 0·89 for three steps, and 0·96 for five steps.

If the number of observations n is not an integral multiple of the number of steps, but

$$n = (2m+1)r+\nu \quad (|\nu| \leqslant m), \tag{1}$$

then the number of observations in some of the steps will be $r \pm 1$ instead of r. The methods of choosing the number n_j of observations in the step of weight j to give maximum efficiency when $m = 1$ and 2 are listed in Table 6.7*d*, together with the corresponding formulae for $\Sigma W_i \epsilon_i$.

For the cases $m = 1$ and $m = 2$, the values $\Sigma W_i \epsilon_i$ are listed in Table 6.7*e* for $n7(1)75$. It will be seen that, when $\nu = \pm 1$, the central group contains $r \pm 1$ observations, while when $\nu = \pm 2$ the numbers in the two groups on each side of the central one are $r \pm 1$. In forming $\Sigma W_i y_i$, the sum of the n_j observations in the step of weight j corresponding to negative values of ϵ is subtracted from the sum of the n_j observations in the step of weight j corresponding to positive values of ϵ, and the difference is multiplied by the weight j. Division of $\Sigma W_i y_i$ by $\Sigma W_i \epsilon_i$ gives b_1, and division by Δx gives the slope in terms of the variable x.

6.4.1.2 *Example*

Table 6.4.1 shows the calculations for the 20 observations of Example 6.3.3, using single-step and double-step functions. The values obtained for b_1 are 289·84 and 289·68, compared with the least-squares value of $289{\cdot}77 \pm 0{\cdot}47$.

TABLE 6.4.1

Solution of Example 6.3.3 by step function methods

Three-step Solution

$n = 20 \qquad r = 7 \qquad \nu = -1$

	Σy
(7)	37872
(6)	21072
(7)	11497

$\Sigma\epsilon_i^2 = 2r^2 - r = 91$

$b_1 = (37872 - 11497)/91 = 26375/91 = 289{\cdot}84$

$b_0 = 70441/20 = 3522{\cdot}05$

Five-step Solution

$n = 20 \qquad r = 4 \qquad \nu = 0$

	Σy
(4)	23386
(4)	18716
(4)	14050
(4)	9439
(4)	4850

$\Sigma\epsilon_i^2 = 10r^2 = 160$

$b_1 = (2 \times 23386 + 18716 - 9439 - 2 \times 4850) = 46349/160 = 289{\cdot}68$

$b_0 = 70441/20 = 3522{\cdot}05$

6.4.2 *Observations not equally-spaced*

When the values of the independent variable are not spaced at equal intervals, it is still true that any function W_i for which $\Sigma W_i x_i$ vanishes can be used to provide an estimate b_1 of the slope. However, the optimum function of a particular type will depend on the spacing of the observed values x_i, and will be different in different examples.

It is possible to describe the departure from uniform spacing in terms of two parameters κ_2 and κ_3, and to calculate the efficiencies in terms of these parameters. The procedure is complicated and will be left till § 8.5, but a brief summary of the results will be given here.

6.4.3 *Step function methods for unequally-spaced observations*

The only practicable procedure appears to be to use the same steps and weights as in the equally-spaced case (§ 6.4.1.1).

Table 6.7*f* gives the efficiencies, for selected values of the parameters κ_2 and κ_3 describing the departure from uniform spacing, when the number of steps $N = 2m+1$ is 3, 5, 7, and ∞. From this table it is seen that five steps ($m = 2$) are sufficient to give adequate values for the efficiency in all cases, while often three steps ($m = 1$) will suffice.

It will be realized that for the estimate

$$b_1 = \Sigma W_i y_i / \Sigma W_i x_i$$

both the values $\Sigma W_i y_i$ and $\Sigma W_i x_i$ must be calculated.

6.4.3.1 *Estimation of fitted values.* To obtain an estimate of the coefficient B_0, another equation is required. The obvious choice is the first least-squares equation

$$\Sigma(y_i - b_0 - b_1 x_i) = 0$$

and then
$$b_0 = (\Sigma y_i - b_1 \Sigma x_i)/n, \tag{1a}$$
or
$$b_0 = a_0 - b_1 \bar{x}, \tag{1b}$$
where
$$a_0 = \Sigma y_i / n. \tag{1c}$$

The fitted curve is then

$$u_1(x) = a_0 + b_1(x - \bar{x}). \tag{2}$$

Now
$$\begin{aligned} \operatorname{cov}(a_0, b_1) &= E\{\Sigma(y_i - Y_i)/n\}\{\Sigma W_i(y_i - Y_i)/\Sigma W_i x_i\} \\ &= \Sigma W_i\{E(y_i - Y_i)^2 / n\Sigma W_i x_i\} \end{aligned}$$

and so
$$\operatorname{cov}(a_0, b_1) = 0, \tag{3a}$$
and
$$\operatorname{var} u_1(x) = \operatorname{var} a_0 + (x - \bar{x})^2 \operatorname{var} b_1. \tag{3b}$$

The standard deviation can then be put in the form (cf. § 6.1.5)

$$\sigma[u_1(k')] = n^{-\frac{1}{2}}\sigma\rho_{10}(k'), \tag{4a}$$

where, from (6.4,3),

$$k' = (x-\bar{x})\{n\Sigma W_i^2/3(\Sigma W_i x_i)^2\}^{\frac{1}{2}}. \tag{4b}$$

The efficiency of the fitted value can be calculated from the ratio

$$\eta[u_1(k)] = \{\rho_{10}(k)/\rho_{10}(k')\}^2, \tag{5a}$$

where for the unequally-spaced case (6.1.5,2),

$$k = (x-\bar{x})\,n/\surd(3D), \tag{5b}$$

and for the equally-spaced case (6.3.1,4c)

$$k = 2\epsilon/n. \tag{5c}$$

Table 6.7g shows how the efficiencies vary with $|k|$ for selected values of κ_2 and κ_3, when the fitting is done by single-step and by double-step functions. The column $\kappa_2 = 0$, $\kappa_3 = 0$, corresponds to the equally-spaced case. The efficiency is close to unity when k is small, and approaches the value $\eta(b_1)$ when k is much greater than unity.

TABLE 6.4.3

Solutions of Example 6.1.4 *by step function methods*

Single-step Functions

$n/3 = 7$

x	Σy	Σx
3–14	6·478	62
16–33	12·099	133
35–49	26·068	295

$b_1 = (26{\cdot}068-6{\cdot}478)/(295-62) = 19{\cdot}590/233 = 0{\cdot}0840773$

$b_0 = (\Sigma y - b_1\,\Sigma x)/20 = 3{\cdot}447123/20 = 0{\cdot}172356$

Double-step Functions

$n/5 = 4$

x	Σy	Σx
3– 9	2·845	25
11–16	5·124	53
18–24	7·670	84
33–41	13·216	148
42–49	15·790	180

$b_1 = (2\times 15{\cdot}790+13{\cdot}216-5{\cdot}124-2\times 2{\cdot}845)/(2\times 180+148-53-2\times 25)$

$= 33{\cdot}982/405 = 0{\cdot}0839062$

$b_0 = (\Sigma y - b_1\,\Sigma x)/20 = 0{\cdot}176548$

6.4.3.2 *Example*

Table 6.4.3 shows the calculations for the observations listed in Table 6.1.4.

The values obtained using single-step, double-step, and least-squares methods are:

$$b_1: \quad 0\cdot0841, \quad 0\cdot0839, \quad 0\cdot0842 \pm 0\cdot0006;$$

$$b_0: \quad 0\cdot172, \quad 0\cdot177, \quad 0\cdot169 \ \pm 0\cdot018.$$

The variances of the estimates b_1 are $\{\Sigma W^2/(\Sigma Wx)^2\}\,\sigma^2$, while for the least-squares estimate the variance is $n\sigma^2/D$. Hence the efficiencies of the estimates b_1 are

(*a*) for the single-step, $233^2/14 \times 4319 = 0\cdot898$.

(*b*) for the double-step estimate, $405^2/40 \times 4319 = 0\cdot949$.

6.4.4 *Estimation of standard deviation*

The value Σv_i^2 can still be used to estimate the standard deviation of an observation when step function methods are used. For, if $\xi_i = x_i - \bar{x}$,

$$v_i = y_i - a_0 - b_1\,\xi_i. \tag{1}$$

Now

$$\operatorname{cov}(y_i, b_1) = E(y_i - Y_i)\,\Sigma W_i(y_i - Y_i)/\Sigma W_i\,\xi_i = (\sigma^2/\Sigma W_i\,\xi_i)\,W_i,$$

and, from (6.1.3,2*b*) and (6.4.3.1,3*a*),

$$\operatorname{cov}(y_i, a_0) = \operatorname{var} a_0, \quad \operatorname{cov}(a_0, b_1) = 0.$$

Hence

$$\operatorname{var} v_i = E(v_i^2) = \operatorname{var} y_i - \operatorname{var} a_0 + \xi_i^2 \operatorname{var} b_1 - 2\xi_i \operatorname{cov}(y_i, b_1)$$

$$= \sigma^2 - n^{-1}\sigma^2 + \eta_1^{-1}\,\xi_i^2 \operatorname{var} b_1^* - 2W_i\,\xi_i\,\sigma^2/\Sigma W_i\,\xi_i,$$

b_1^* being the least-squares estimate of slope. So

$$E(\Sigma v_i^2) = n\sigma^2 - 3\sigma^2 + \eta_1^{-1}\sigma^2 = (n-2)\,\sigma^2 + (\eta_1^{-1} - 1)\,\sigma^2. \tag{2}$$

It follows that, since η_1 is close to unity,

$$s = \{\Sigma v_i^2/(n-2)\}^{\frac{1}{2}} \tag{3}$$

will provide an estimate of σ, where the residuals v_i are calculated from (1). But there are no formulae comparable to those of (6.1.3,6*a*–*e*) for the least-squares case, since here $\Sigma y_i\,\xi_i \neq a_1\,\Sigma\xi_i^2$. The residuals v_i have to be calculated individually.

6.4.5 *Least-squares method with grouped observations*

The n observations y_i are supposed grouped into $N = n/r$ groups, the observations being grouped in order of x_i. Then, if y_{Nj} and x_{Nj} are the sums of the r values y_i and x_i in the jth group, estimates of the regression line may be obtained by fitting a least-squares line to these N grouped values. If

$$\xi_{Nj} = x_{Nj} - \bar{x}_N, \tag{1}$$

the estimated slope of the line is

$$a_{1N} = \Sigma y_{Nj}\,\xi_{Nj}/\Sigma\xi_{Nj}^2. \tag{2}$$

Now
$$E\Sigma y_{Nj}\,\xi_{Nj} = E\sum_j \xi_{Nj}\sum_r y_i$$

$$= \sum_j \xi_{Nj}\sum_r (A_0 + A_1\,\xi_i) = A_1\,\Sigma\xi_{Nj}^2. \tag{3}$$

Hence
$$Ea_{1N} = A_1, \tag{4}$$

and a_{1N} will provide an unbiased estimate of the slope of the line. The variance of this estimate will be

$$\operatorname{var} a_{1N} = \Sigma(\xi_{Nj}^2 \operatorname{var} y_{Nj})/(\Sigma\xi_{Nj}^2)^2$$

and as
$$\operatorname{var} y_{Nj} = r \operatorname{var} y_i = r\sigma^2, \tag{5a}$$

it follows that
$$\operatorname{var} a_{1N} = r\sigma^2/\Sigma\xi_{Nj}^2. \tag{5b}$$

The efficiency of this estimate is

$$\eta(a_{1N}) = \operatorname{var} a_1^*/\operatorname{var} a_{1N} = \Sigma\xi_{Nj}^2/r\Sigma\xi_i^2. \tag{5c}$$

The efficiency can be expressed in terms of κ_2 and κ_3, the values for $N = 3$, 5, and 7 being shown in Table 6.7*h*. It will be seen that in all cases (except $\kappa_2 = 0 = \kappa_3$) the efficiency is higher than for the step function method with the same value of N. The calculations, however, take a little longer. The value $N = 5$ will always give a satisfactory efficiency.

The case $\kappa_2 = 0 = \kappa_3$ corresponds to the equally-spaced case. In this case, if $N = 2m+1$, the values ξ_{Nj} are proportional to $m, m-1, \ldots 0, \ldots -m$, which are the same as the weights in the step function method. For the equally-spaced case the step function and least-squares grouped methods are identical.

In forming the groups, if n/N is not integral, but

$$n = rN + \nu, \quad |\nu| \leqslant (N-1)/2,$$

the ν observations should be included in (or omitted from, if ν is negative) the ν groups near the centre of the range of x. The corresponding sums Σx_i and Σy_i are multiplied by $r/(r+1)$ or $r/(r-1)$, according as ν is positive or negative.

6.4.5.1 *The fitted curve.* If the fitted curve is written in terms of grouped variables as

$$u_{1N}(x_N) = a_{0N} + a_{1N}(x_N - \bar{x}_N),$$

then
$$a_{0N} = \sum_j y_{jN}/N = \sum_i y_i/N = ra_0. \tag{1}$$

Hence the estimated curve in terms of the original variable x is

$$u_1(x) = a_0 + a_1(x-\bar{x}), \tag{2}$$

where

$$a_0 = r^{-1}a_{0N}, \quad a_1 = a_{1N}. \tag{3}$$

Since

$$\operatorname{var} a_0 = \sigma^2/n, \tag{4a}$$

it follows that

$$\sigma^2[u_1(x)] = n^{-1}\sigma^2[1+(x-\bar{x})^2 rn/\Sigma(x_N-\bar{x}_N)^2]. \tag{4b}$$

6.4.5.2 *Estimation of standard deviation.* The usual least-squares results hold for the residuals v_{jN}. That is,

$$E(\Sigma v_{jN}^2) = (N-2)\operatorname{var} y_N = r(N-2)\operatorname{var} y_n, \tag{1a}$$

and

$$\Sigma v_{jN}^2 = \Sigma y_N^2 - Na_{0N}^2 - \Sigma(x_N-\bar{x}_N)^2 a_{1N}^2. \tag{1b}$$

Hence the estimate

$$s_N = \{\Sigma v_{jN}^2/(n-2r)\}^{\frac{1}{2}} \tag{1c}$$

for σ can be rapidly calculated. But it is based only on $N-2$ d.f., and so will not be very accurate unless N is large. If a more accurate estimate of σ is required, it will be necessary to evaluate the individual residuals

$$v_i = y_i - a_0 - a_1(x_i - \bar{x}),$$

and use (6.4.4,3).

TABLE 6.4.5

Solution of Example 6.1.4 *by the least-squares grouped method*

Calculations for Five Groups

x	y_N	x_N
3– 9	2·845	25
11–16	5·124	53
18–24	7·670	84
33–41	13·216	148
42–49	15·790	180

Σy 44·645 $\quad$ Σxy 5785·145 $\quad$ Σy^2 517·165057
N 5 $\quad$ Σx 490 $\quad$ Σx^2 64794
$D = N\Sigma x^2 - (\Sigma x)^2 = 83870$
$b_1 = (N\Sigma xy - \Sigma x\Sigma y)/D = 7049{\cdot}675/D = 0{\cdot}0840548$
$b_0 = (-\Sigma x\Sigma xy + \Sigma x^2\,\Sigma y)/D = 0{\cdot}691631$
$b_0/4 = 0{\cdot}172908$
Σy^2 517·165057
$(\Sigma y)^2/N$ 398·635205
$(b_1 D)^2/ND$ 118·511786
$= \Sigma v^2$ 0·018066 $\quad$ $s_N^2 = \Sigma v^2/(N-2) = 0{\cdot}006022$ $\quad$ $s_N = 0{\cdot}0776$
$s[b_1] = s/\sqrt{(D/N)} = s/129{\cdot}5 = 0{\cdot}00060$

6.4.5.3 *Example*

Table 6.4.5 shows the calculations for the observations of Table 6.1.4, when the observations are formed into five groups. The estimate of the slope is $0{\cdot}08405 \pm 0{\cdot}00060$, compared with the full least-squares value $0{\cdot}08422 \pm 0{\cdot}00062$. The estimate of standard deviation in the grouped case is based on 3 d.f. only, and so will not be very accurate.

6.4.6 *Observations of different weight*

If the observations are of different weight, the observation of weight w_i is regarded as w_i observations all at the same value of x_i in forming the steps or groups. Thus Σw_i replaces n in the earlier discussions. In forming the steps the observations are divided into $N = 2m+1$ groups in such a way that the sums Σw_i are approximately equal.

6.5 THE INDEPENDENT VARIABLE SUBJECT TO ERROR

The cases discussed in this section are those in which the errors in the independent variable cause the experimental regression line and the functional relationship line to diverge to such an extent that the experimental regression line is not an adequate estimate of the functional relationship. Cases in which the experimental regression line is an adequate estimate of the error-free regression line or the functional relationship (cf. § 5.5) are of course treated as in earlier sections of this chapter.

6.5.1 *Estimation of regression lines*

The regression equations in terms of the error-free variable x' are

$$\Sigma w_i(y_i - b_0' - b_1' x_i')\, x_i'^k = 0, \quad k = 0 \text{ and } 1. \tag{1a}$$

These are unbiased estimating equations, in the sense that they are satisfied 'on the average' by the true values B_0' and B_1'; that is

$$E\Sigma w_i(y_i - B_0' - B_1' x_i')\, x_i'^k = 0. \tag{1b}$$

It is shown in § 5.3.2 that the solutions b_0' and b_1' are also unbiased estimates of B_0' and B_1', in the sense that

$$E(b_j') = B_j'. \tag{1c}$$

When the observed values $x_i = x_i' + \gamma_i$ are subject to error, unbiased estimating equations similar to (1) can be set up if the standard deviations of the error terms are known. These equations are

$$\Sigma w_i(y_i - b_0' - b_1' x_i) = 0, \tag{2a}$$

$$\Sigma w_i(y_i - b_0' - b_1' x_i)\, x_i = -b_1' \Sigma w_i \sigma_{\gamma i}^2. \tag{2b}$$

These are unbiased in the sense that if the true values B_0' and B_1' are substituted the expectations of the two sides are equal.

A more useful pair of equations is obtained by choosing the origins of both x and y at the weighted means. Now

$$E\Sigma w_i(x_i-\bar{x})^2 = E\{\Sigma w_i x_i^2 - (\Sigma w_i x_i)^2/\Sigma w_i\}$$

$$= E\Sigma w_i(x_i'-\overline{x_i'})^2 + E\{\Sigma w_i \gamma_i^2 - \Sigma w_i^2 \gamma_i^2/\Sigma w_i\},$$

and $$E\Sigma w_i(x_i-\bar{x})(y_i-\bar{y}) = E\Sigma w_i(x_i'-\overline{x'})(y_i'-\overline{y'}).$$

Thus the unbiased estimating equations for the regression of y on x when x and y denote the variables with the origins at the means are

$$b_{0y}' = 0 \tag{3a}$$

and $$b_{1y}'\{\Sigma w_i x_i^2 - \Sigma w_i \sigma_{\gamma i}^2(1-w_i/\Sigma w_i)\} = \Sigma w_i x_i y_i. \tag{3b}$$

The weights w_i should be inversely proportional to the variance, for fixed x_i', of $(y_i - B_0' - B_1' x_i)$. Thus

$$w_i \propto (\sigma_{yi'}^2 + \sigma_{\delta i}^2 + B_{1y}'^2 \sigma_{\gamma i}^2)^{-1}, \tag{4}$$

as in § 5.5. It will only be very occasionally that all these standard deviations will be known at all accurately. For the special case in which the standard deviations are the same at all points—or when they are assumed to be the same, in default of further information—w_i can be taken as unity and (3b) becomes

$$b_{1y}' = \Sigma x_i y_i/\{\Sigma x_i^2 - (n-1)\sigma_\gamma^2\}. \tag{5}$$

In a similar manner, the estimated slope of the error-free regression line of x' on y' is given by

$$b_{1x}' = \Sigma w_i x_i y_i/\{\Sigma w_i y_i^2 - \Sigma w_i \sigma_{\delta i}^2(1-w_i/\Sigma w_i)\}, \tag{6a}$$

with $$w_i \propto (\sigma_{xi'}^2 + \sigma_{\gamma i}^2 + B_{1x}'^2 \sigma_{\delta i}^2)^{-1}. \tag{6b}$$

If the standard deviations are the same at all the points of observation,

$$b_{1x}' = \Sigma x_i y_i/\{\Sigma y_i^2 - (n-1)\sigma_\delta^2\}. \tag{6c}$$

6.5.2 *The functional relationship*

When the error-free points x_i' and y_i' lie exactly on the straight line

$$y = B_0' + B_1' x, \tag{1}$$

for each value of x_i' there is a single value of y_i' given by (1), and for each value of y_i' there is a single value of x_i' given by the same equation. Hence the two error-free regression lines and the line of functional relationship all coincide in this case.

Since $\sigma_{xi'}$ and $\sigma_{yi'}$ in (6.5.1,4) and (6.5.1,6b) are now both zero, and

$$B'_{1y} = B'^{-1}_{1x} = B'_1,$$

the weights w_i in (6.5.1,3b) and (6.5.1,6a) can be made identical. Then

$$w_i = \sigma^2/(\sigma^2_{\delta i} + B'^2_1\,\sigma^2_{\gamma i}), \tag{2a}$$

and, if

$$k_i = \sigma^2_{\delta i}/\sigma^2_{\gamma i}, \tag{2b}$$

$$w_i = \sigma^2/\sigma^2_{\gamma i}(k_i + B'^2_1). \tag{2c}$$

A knowledge of the relative magnitudes of $\sigma_{\delta i}$ and $\sigma_{\gamma i}$, as well as an estimate of the slope B'_1, is required before the weights can be evaluated. Often accurate values for these quantities are not available. However, in the special case where $\sigma_{\delta i}$ and $\sigma_{\gamma i}$ are constant, the weights are constant and so may be made unity, and estimates of the standard deviations are not required for the calculation of the weights.

6.5.2.1 *Estimates of the slope.* The estimates of B'_1 obtained from the unbiased equations for the error-free regression lines are

$$b'_{1(y)} = \Sigma w_i x_i y_i/\{\Sigma w_i x^2_i - \Sigma w'_i\,\sigma^2_{\gamma i}\} \tag{1a}$$

and

$$b'_{1(x)} = \{\Sigma w_i y^2_i - \Sigma w'_i\,\sigma^2_{\delta i}\}/\Sigma w_i x_i y_i, \tag{1b}$$

where

$$w'_i = w_i(1 - w_i/\Sigma w_i). \tag{1c}$$

These equations may be combined to give a third form in terms of the ratio

$$k = \Sigma w'_i\,\sigma^2_{\delta i}/\Sigma w'_i\,\sigma^2_{\gamma i}. \tag{2}$$

Combining (1a) and (1b), the estimate b'_1 is given by

$$\Sigma w_i y^2_i - b'_1\,\Sigma w_i x_i y_i = k(\Sigma w_i x^2_i - b'^{-1}_1\,\Sigma w_i x_i y_i), \tag{3a}$$

or

$$b'^2_1\,\Sigma w_i x_i y_i - b'_1(\Sigma w_i y^2_i - k\Sigma w_i x^2_i) - k\Sigma w_i x_i y_i = 0. \tag{3b}$$

The solution of this quadratic equation is

$$b'_1 = m + \surd(m^2 + k), \tag{4a}$$

where

$$m = (\Sigma w_i y^2_i - k\Sigma w_i x^2_i)/2\Sigma w_i x_i y_i. \tag{4b}$$

This estimate can be expressed in terms of the estimated slopes b^{-1}_{1x} and b_{1y} of the experimental regression lines. For, on dividing (3a) by $\Sigma w_i x_i y_i$,

$$b^{-1}_{1x} - b'_1 = k(b^{-1}_{1y} - b'^{-1}_1), \tag{5a}$$

and so

$$b'_1 = \frac{kb_{1y} + (b_{1y}b'_1)\,b^{-1}_{1x}}{k + b_{1y}b'_1} \doteqdot \frac{kb_{1y} + b'^2_1\,b^{-1}_{1x}}{k + b'^2_1} \tag{5b}$$

An alternative form is

$$b'_1 \doteqdot b_{1y} + \frac{b^{-1}_{1x} - b_{1y}}{1 + k/b'^2_1} \doteqdot b_{1y} + \frac{b^{-1}_{1x} - b_{1y}}{1 + k/b^2_{1y}}. \tag{6}$$

6.5.2.2 *Ratio of standard deviations constant.* If the ratio

$$k_i = \sigma_{\delta i}^2/\sigma_{\gamma i}^2 \tag{1}$$

is a constant, then from (6.5.2,2*a*–*c*),

$$w_i = \sigma_\delta^2/\sigma_{\delta i}^2 = \sigma_\gamma^2/\sigma_{\gamma i}^2, \tag{2}$$

where σ_δ and σ_γ are constants corresponding to the standard deviation of an error term of unit weight. Hence

$$\Sigma w_i' \sigma_{\delta i}^2 = \Sigma w_i(1-w_i/\Sigma w_i)\,\sigma_{\delta i}^2 = \Sigma(1-w_i/\Sigma w_i)\,\sigma_\delta^2,$$

and so
$$\Sigma w_i' \sigma_{\delta i}^2 = (n-1)\,\sigma_\delta^2 \tag{3a}$$
and
$$\Sigma w_i' \sigma_{\gamma i}^2 = (n-1)\,\sigma_\gamma^2, \tag{3b}$$
while
$$k = \sigma_\delta^2/\sigma_\gamma^2. \tag{3c}$$

The estimating equations are then

$$b'_{1(y)} = \Sigma w_i x_i y_i/\{\Sigma w_i x_i^2 - (n-1)\,\sigma_\gamma^2\}, \tag{4a}$$

$$b'_{1(x)} = \{\Sigma w_i y_i^2 - (n-1)\,\sigma_\delta^2\}/\Sigma w_i x_i y_i, \tag{4b}$$

and
$$b'_{1(k)} = m + \sqrt{(m^2+k)} \tag{4c}$$
with
$$k = \sigma_\delta^2/\sigma_\gamma^2 = \sigma_{\delta i}^2/\sigma_{\gamma i}^2. \tag{4d}$$

It is shown in § 6.5.5 that the estimate (4*c*) can be derived from least-squares and maximum likelihood postulates.

6.5.3 *Choice of equation for estimating the slope*

The three estimating equations are

(*a*) (6.5.2.1,1*a*) and (6.5.2.2,4*a*), which require an accurate estimate of $\sigma_{\gamma i}$;

(*b*) (6.5.2.1,1*b*) and (6.5.2.2,4*b*), which require an accurate estimate of $\sigma_{\delta i}$;

(*c*) (6.5.2.1,4*a*) and (6.5.2.2,4*c*), which require an accurate estimate of $\sigma_{\delta i}/\sigma_{\gamma i}$.

Which equation will be used in any particular case depends on the information available. It is clear that some information about the errors is needed before an estimate of the slope of the error-free line can be found.

The estimate obtained from (*c*) will always lie between the estimates obtained from (*a*) and (*b*). For, combining (6.5.2.1,1*a*) and (6.5.2.1,1*b*),

$$(\Sigma w_i y_i^2 - b'_{1(x)}\,\Sigma w_i x_i y_i) = k(\Sigma w_i x_i^2 - b'^{-1}_{1(y)}\,\Sigma w_i x_i y_i), \tag{1}$$

and on comparing this with the equation (6.5.2.1,3*a*) giving b_1', it is clear that if $b_1' \lessgtr b'_{1(x)}$, then $b_1'^{-1} \lessgtr b'^{-1}_{1(y)}$, and so b_1' lies between $b'_{1(x)}$ and $b'_{1(y)}$.

6.5.3.1 *Example*

The values listed in Table 6.5.3 form a set artificially constructed to illustrate the calculation of the slope by the various methods. These calculations are carried out in the lower part of the table. The three values of b_1' obtained differ only slightly from one another. The standard deviation of the estimate b_{1y} of the regression line is 0·04, and this value will serve as an approximation to the standard deviations of the estimates b_1' (see § 6.5.4).

TABLE 6.5.3

Estimation of slope when the independent variable is subject to error

x	y	x	y
0·2	0·7	10·1	4·7
2·3	0·9	11·5	5·1
2·8	2·8	14·7	7·9
5·4	2·8	16·3	7·1
8·8	3·5	18·1	9·6

Σy 45·1	Σxy 567·71	Σy^2 282·31
n 10	Σx 90·2	Σx^2 1163·42

(*a*) Sums referred to the mean
$\Sigma(y-\bar{y})^2 = 78{\cdot}909 \quad \Sigma(y-\bar{y})(x-\bar{x}) = 160{\cdot}908 \quad \Sigma(x-\bar{x})^2 = 349{\cdot}816$

(*b*) Slopes of regression lines
$b_{1y} = \Sigma(y-\bar{y})(x-\bar{x})/\Sigma(x-\bar{x})^2 = 0{\cdot}4600$
$b_{1x}^{-1} = \Sigma(y-\bar{y})^2/\Sigma(y-\bar{y})(x-\bar{x}) = 0{\cdot}4904$

(*c*) Slope of line calculated by assuming $\sigma_\gamma = 1{\cdot}0$ (6.5.2.2,4*a*)
$b_{1(y)}' = 160{\cdot}908/(349{\cdot}816-9) = 0{\cdot}4721$

(*d*) Slope of line calculated by assuming $\sigma_\delta = 0{\cdot}5$ (6.5.2.2,4*b*)
$b_{1(x)}' = (78{\cdot}909-2{\cdot}25)/160{\cdot}908 = 0{\cdot}4764$

(*e*) Slope of line calculated by assuming $k = 0{\cdot}25$ (6.5.2.2,4*c*)
$b_1' = -0{\cdot}02655+\sqrt{0{\cdot}25070} = 0{\cdot}4742$

(*f*) Slope of line calculated from (6.5.2.1,6)
$b_1' = b_{1y}+(b_{1x}^{-1}-b_{1y})/(1+k/b_{1y}^2) = 0{\cdot}4600+0{\cdot}0304/2{\cdot}183 = 0{\cdot}4739$

(*g*) Variance of b_{1y}
$\Sigma v_y^2 = \Sigma(y-\bar{y})^2-b_{1y}^2\,\Sigma(x-\bar{x})^2 = 4{\cdot}894 \quad s_{y\,|\,x} = 0{\cdot}782$
$s(b_{1y}) = 0{\cdot}042$

6.5.4 *Estimation of standard deviations*

The standard deviation of an observation of unit weight can be estimated from the residuals for the *experimental* regression curve by the usual formula

$$s^2 = \Sigma w_i v_i^2/(n-2). \tag{1}$$

Also from (6.5.2,2*a*)

$$\sigma^2 = w_i(\sigma_{\delta i}^2+B'^2\sigma_{\gamma i}^2) = w_i\sigma_{\gamma i}^2(k_i+B_1'^2),$$

and so from the estimate s^2 of σ^2 one of the standard deviations $\sigma_{\gamma i}$ and $\sigma_{\delta i}$ can be estimated if the other is known, or both can be estimated if the ratio k_i is known.

The standard deviation of the slope of the experimental regression line is

$$s^2(b_{1y}) = \frac{\Sigma w_i(y_i-\bar{y})^2 - b_{1y}^2\,\Sigma w_i(x_i-\bar{x})^2}{(n-2)\,\Sigma w_i(x_i-\bar{x})^2} = \frac{b_{1y}}{n-2}(b_{1x}^{-1}-b_{1y}). \quad (2a)$$

Similarly,
$$s^2(b_{1x}) = \frac{b_{1x}}{n-2}(b_{1y}^{-1}-b_{1x}),$$

and so
$$s^2(b_{1x}^{-1}) = b_{1x}^{-4}\,s^2(b_{1x}) = (b_{1x}^{-1}/b_{1y})^2\,s^2(b_{1y}). \quad (2b)$$

Thus the standard deviations of b_{1y} and b_{1x}^{-1} are very nearly equal. Equation $(2a)$ can be written in the form

$$\frac{b_{1x}^{-1}-b_{1y}}{s(b_{1y})} = (n-2)\frac{s(b_{1y})}{b_{1y}}. \quad (3a)$$

The quantity on the left is the ratio of the difference of the slopes to the standard deviation. If this quantity is much less than unity the difference in slope of the two regression lines and the line of functional relationship can be neglected. Using the approximation (6.3.2.1,2) for $s(b_{1y})$, this becomes

$$\frac{b_{1x}^{-1}-b_{1y}}{s(b_{1y})} \doteqdot \sqrt{(12n)}\frac{\sigma}{\mathscr{R}(x)}\frac{1}{b_{1y}} \doteqdot \sqrt{(12n)}\frac{\sigma}{\mathscr{R}(y)}, \quad (3b)$$

showing that the ratio is proportional to $\sigma/\mathscr{R}(y)$.

The standard deviations of the estimates of the slopes of the error-free line are all very nearly equal to $\sigma(b_{1y})$. For the estimate $(6.5.2.1,1a)$,

$$b'_{1(y)} = b_{1y}\left(1-\frac{\Sigma w'_i\,\sigma_{yi}^2}{\Sigma w_i\,x_i^2}\right)^{-1},$$

and as σ_{yi} is supposed known,

$$\operatorname{var} b'_{1(y)} = (b'_{1(y)}/b_{1y})^2 \operatorname{var} b_{1y} \quad (4a)$$

to a very good approximation. Similarly, for the estimate $(6.5.2.1,1b)$,

$$\operatorname{var} b'_{1(x)} = (b'_{1(x)}/b_{1x}^{-1})^2 \operatorname{var} b_{1x}^{-1} = \operatorname{var} b'_{1(y)}. \quad (4b)$$

Since $(6.5.2.1,5b)$ expresses b'_1 as the weighted mean of b_{1y} and b_{1x}^{-1}, the variance of b'_1 would also be expected to be very nearly equal to the variance of these quantities. A formula for the variance of b'_1 in the case where the ratio k_i of the standard deviations is constant will now be derived. From $(6.5.2.1,3a)$,

$$b'_1 - k/b'_1 = \frac{\Sigma w_i\,y_i^2 - k\Sigma w_i\,x_i^2}{\Sigma w_i\,x_i\,y_i}. \quad (5)$$

Now

$$\operatorname{var}\frac{\Sigma w_i y_i^2 - k\Sigma w_i x_i^2}{\Sigma w_i x_i y_i}$$
$$= \sum_j \left\{\frac{2w_j y_j}{\Sigma w_i x_i y_i} - \frac{\Sigma w_i y_i^2 - k\Sigma w_i x_i^2}{(\Sigma w_i x_i y_i)^2} w_j x_j\right\}^2 \sigma_{\delta j}^2$$
$$+ \sum_j \left\{\frac{-2kw_j x_j}{\Sigma w_i x_i y_i} - \frac{\Sigma w_i y_i^2 - k\Sigma w_i x_i^2}{(\Sigma w_i x_i y_i)^2} w_j y_j\right\}^2 \sigma_{\gamma j}^2$$
$$= \frac{\sigma_\gamma^2}{\Sigma w_i x_i y_i}\{4kb_{1x}^{-1} - 4k(b_1' - k/b_1') + k(b_1' - k/b_1')^2 b_{1y}^{-1}$$
$$+ 4k^2 b_{1y}^{-1} + 4k(b_1' - k/b_1') + (b_1' - k/b_1')^2 b_{1x}^{-1}\}$$
$$= \frac{\sigma_\gamma^2}{\Sigma w_i x_i y_i}(b_1' + k/b_1')^2 (b_{1x}^{-1} + kb_{1y}^{-1}),$$

and
$$\operatorname{var}(b_1' - k/b_1') = (1 + k/b_1'^2)^2 \operatorname{var} b_1'.$$

Equating these two variances,

$$\operatorname{var} b_1' = \left(\frac{b_1'}{b_{1y}}\right)^2 \frac{\sigma_\gamma^2(k + b_{1y} b_{1x}^{-1})}{\Sigma w_i x_i^2},$$

or
$$\operatorname{var} b_1' = \left(\frac{b_1'}{b_{1y}}\right)^2 \frac{k + b_{1y} b_{1x}^{-1}}{k + b_1'^2} \operatorname{var} b_{1y}. \tag{6a}$$

Usually the approximation

$$\operatorname{var} b_1' \doteqdot \operatorname{var} b_{1y} \tag{6b}$$

will be sufficiently accurate.

6.5.5 *Least-squares postulates*

Estimates of the coefficients B_0' and B_1' may be obtained by minimizing the sum of squares of deviations. One treatment assumes that the coefficients b_0' and b_1' and the estimates $\hat{x}_i'$ of the error-free values x_i' corresponding to the observed values x_i are chosen to minimize

$$\Sigma(y_i - \hat{y}_i')^2/\sigma_{\delta i}^2 + \Sigma(x_i - \hat{x}_i')^2/\sigma_{\gamma i}^2, \tag{1a}$$

with
$$\hat{y}_i' = b_0' + b_1'\hat{x}_i'. \tag{1b}$$

Differentiation with respect to the parameters b_0', b_1', and $\hat{x}_i'$ leads to the equations

$$\Sigma\sigma_{\delta i}^{-2}(y_i - b_0' - b_1'\hat{x}_i') = 0, \tag{2a}$$
$$\Sigma\sigma_{\delta i}^{-2}(y_i - b_0' - b_1'\hat{x}_i')\hat{x}_i' = 0, \tag{2b}$$
$$\sigma_{\delta i}^{-2}(y_i - b_0' - b_1'\hat{x}_i')b_1' + \sigma_{\gamma i}^{-2}(x_i - \hat{x}_i') = 0. \tag{2c}$$

These equations can be solved for b_0', b_1', and $\hat{x}_i'$. The solutions are complicated unless it can be assumed that $k_i = \sigma_{\delta i}^2/\sigma_{\gamma i}^2$ is a constant. Then, as w_i is proportional to $\sigma_{\delta i}^{-2}$, summing (2c) over all values of i and using (2a) gives

$$\Sigma w_i x_i = \Sigma w_i \hat{x}_i'. \tag{3}$$

If the origins are chosen at the weighted means, it follows from (2a) and (3) that b_0' vanishes. Then (2c) gives

$$\hat{x}_i' = (k+b_1'^2)^{-1}(y_i b_1' + kx_i) \tag{4}$$

and substitution of (4) in (2b) leads to the quadratic equation (6.5.2.1,3a, b). Thus when the ratio k_i is constant this equation follows from the application of the least-squares principle to (1a).

Alternatively, it may be postulated that the coefficients should be chosen to minimize the expression

$$\Sigma w_i(y_i - b_0' - b_1' x_i)^2, \tag{5a}$$

where w_i also contains the parameter b_1'. Since

$$w_i = w_i''/(k_i + B_1'^2),$$

where w_i'' is proportional to $\sigma_{\delta i}^{-2}$, the expression to be minimized is

$$\Sigma w_i''(y_i - b_0' - b_1' x_i)^2/(k_i + b_1'^2). \tag{5b}$$

Differentiation with respect to b_0' gives the normal equation

$$\Sigma w_i(y_i - b_0' - b_1' x_i) = 0,$$

and so b_0' vanishes if the origins are at the weighted means. Differentiation with respect to b_1' leads to a complicated equation, but if k_i is assumed constant this equation also simplifies to the quadratic form (6.5.2.1,3a, b).

6.5.5.1 *Maximum likelihood estimates.* If it is assumed that the errors follow a normal law, then the probability of obtaining the observed values x_i and y_i is proportional to

$$\left\{(2\pi)^n \prod_i (\sigma_{\gamma i}\sigma_{\delta i})\right\}^{-1} \exp - \tfrac{1}{2}\left\{\sum_i (y_i - y_i')^2/\sigma_{\delta i}^2 + \sum_i (x_i - x_i')^2/\sigma_{\gamma i}^2\right\}, \tag{1a}$$

with

$$y_i' = B_0' + B_1' x_i'. \tag{1b}$$

The principle of maximum likelihood states that the estimates are to be chosen so that this probability is a maximum. If the errors are known, the estimates obtained will be identical with the least-squares estimates, as is obvious on comparing (1a) with (6.5.5,1a).

It should also be possible to estimate the standard deviations themselves, as well as the coefficients, by the method of maximum likelihood. Such estimates were first obtained by Dent (1935), who, for the case when the standard deviations are constant at all points, derived the equation

$$\hat{\sigma}_\delta^2/\hat{\sigma}_\gamma^2 = b_1'^2 \tag{2a}$$

for the estimates of σ_δ and σ_γ, and the equation

$$b_1' = \Sigma(y_i-\bar{y})^2/\Sigma(x_i-\bar{x})^2 \tag{2b}$$

for the estimate of the slope. But, as Lindley (1947) pointed out, (2*a*) is not 'consistent', for the slope B_1' of the line and the ratio $\sigma_\delta/\sigma_\gamma$ are not connected in any way. In other words, the maximum likelihood estimate of this ratio is completely unreliable, and the slope b_1' given by (2*b*) is no more accurate than that obtained by an arbitrary choice of this ratio. The method of maximum likelihood breaks down here.

6.5.6 *The method of grouping with both variables subject to error*

For an observed pair of values x_i and y_i,

$$y_i - B_0' - B_1'x_i = (y_i' - B_0' - B_1'x_i') + \delta_i - B_1'\gamma_i,$$

and so, if W_i is an arbitrary function for which

$$\Sigma W_i = 0, \tag{1}$$

$$E\Sigma W_i(y_i - B_1'x_i) = E\Sigma W_i(\delta_i - B_1'\gamma_i). \tag{2}$$

Thus, provided W_i is independent of γ_i and δ_i,

$$b_1'\Sigma W_i x_i = \Sigma W_i y_i \tag{3}$$

will be an unbiased estimating equation for B_1'. In this method, unlike the methods discussed in §§ 6.5.1 to 6.5.5, the standard deviations of the errors are not required.

The most convenient functions W_i are the step functions, and presumably the step functions used when the x_i are free from error will also be reasonably efficient in this case. When the x_i are free from error (i.e. when $x_i = x_i'$) the observations are listed in order of x_i' before grouping, and if reasonable efficiencies are required this should also be done when the x_i are subject to error. But the values x_i' are unknown, and the usual procedure is to list the values in order of x_i. This will be quite satisfactory provided the errors γ_i are not comparable with the spacing of the x_i near the boundaries of the groups. For then the group into which an observation goes—i.e. the value W_i in (2)—depends on the error γ_i, and so the right-hand side of (2) does not vanish and the estimate will be biased.

Of course, it is only in cases where the errors are comparable to the spacing that the difference between the regression curve and the curve of functional relationship is important. Often in the design of the experiment the spacing near the boundaries of the groups can be chosen so that there is practically no possibility of an observation going into the 'wrong' group. For example, all the observations could be taken near the extremes of the range of x, so that there would be no doubt about the group into which an observation should go. Sometimes it will be possible to order the observations in terms of a third variable. Thus if the independent variable is allowed to change with time, an unbiased estimate may be obtained by ordering the observations in time before grouping.

6.5.6.1 *Example*

For the example of § 6.5.3.1, using 3 groups,

$$b_1' = (24{\cdot}6-4{\cdot}4)/(49{\cdot}1-5{\cdot}3) = 20{\cdot}2/43{\cdot}8 = 0{\cdot}4612.$$

6.5.6.2 *Confidence limits in the method of grouping.* Bartlett (1949) has derived formulae giving confidence limits for B_1' in terms of the significance levels for the t distribution. If

$$z_i = y_i - B_0' - B_1' x_i,$$

then z_i is distributed normally about zero with variance

$$\sigma_z^2 = \sigma_\delta^2 + B'^2 \sigma_\gamma^2.$$

Hence
$$\frac{\Sigma W_i z_i}{\sigma_z \sqrt{(\Sigma W_i^2)}} = \frac{\Sigma W_i x_i (b_1' - B_1')}{\sigma_z \sqrt{(\Sigma W_i^2)}} \tag{1}$$

is a standardized normal variate, and if s_z is an estimate of σ_z whose distribution is independent of that of $\Sigma W_i z_i$,

$$t = \frac{\Sigma W_i x_i (b_1' - B_1')}{s_z \sqrt{(\Sigma W_i^2)}} \tag{2}$$

will follow a t distribution.

An estimate s_z can be found by considering the residuals from the means of the various steps. Thus for the jth step

$$\left\{\sum_i (y_{ji} - \bar{y}_j) - \sum_i B_1'(x_{ji} - \bar{x}_j)\right\}^2 \Big/ \sigma_z^2$$

is distributed as χ^2, independently of the mean $\bar{y}_j - B_1' \bar{x}_j$ and so independently of the sum $\sum_i W_{ji} z_i$. Hence

$$s_z^2 = \sum_j \left\{ \sum_i (y_{ji} - \bar{y}_j)^2 - 2B_1' \sum_i (y_{ji} - \bar{y}_j)(x_{ji} - \bar{x}_j) + B_1'^2 \sum_i (x_{ji} - \bar{x}_j)^2 \right\} \Big/ (n - N) \qquad (3)$$

will be an estimate of σ_z^2 based on $n - N$ d.f., where N is the total number of steps. If the three sums of the form

$$\sum_j \sum_i (y_{ji} - \bar{y}_j)^2/(n-N)$$

are denoted by s_{yy}, s_{yx}, and s_{xx}, then, from (2),

$$t^2 \Sigma W_i^2 (s_{yy} - 2B_1' s_{yx} + B_1'^2 s_{xx}) = (\Sigma W_i x_i)^2 (b_1' - B_1')^2, \qquad (4)$$

where the number of degrees of freedom for t is $n - N$.

For a given significance level—i.e. for a given value of t—this is a quadratic equation in B_1' whose roots give the confidence limits for B_1' at the assumed level of significance.

6.5.6.3 *Approximate estimate of standard deviation of the slope.* From (6.5.6.2,1),

$$\operatorname{var}\{(b_1' - B_1')(\Sigma W_i x_i)\} = \Sigma W_i^2 \sigma_z^2. \qquad (1)$$

If the experiment were repeated a number of times under the same conditions, the proportional variation in the term $b_1' - B_1'$ would be considerably greater than the proportional variation in $\Sigma W_i x_i$. Hence (1) can be written

$$\operatorname{var}(b_1' - B_1') \doteqdot \{\Sigma W_i^2/(\Sigma W_i x_i)^2\} \sigma_z^2,$$

and

$$s^2(b_1') = \{\Sigma W_i^2/(\Sigma W_i x_i)^2\} s_z^2 \qquad (2)$$

will provide an estimate of the standard deviation of b_1'.

6.6 NOTES AND REFERENCES

(6.2) For a test of significance for concurrent regression lines, see Tocher (1952) and Williams (1953).

(6.3) The method of fitting using successive differences has been extended by Birge (1947) to polynomials of higher degree.

(6.4) The fitting of a straight line by dividing the observations into three groups goes back to Eddington (cf. Jeffreys, 1948, p. 193).

(6.5) The estimate (6.5.2.1,4*a*, *b*) was apparently first discovered by Kummell in a neglected paper published in 1879 (*The Analyst*, Des Moines, **6**, 97). It was rediscovered by K. Pearson (1901) and Gini (1921); a discussion is given by Deming (1943). See also Lindley (1947).

Grouping methods when both variables are subject to error were considered by Wald (1940) and Bartlett (1949); see also Neyman and Scott (1951) and Smith (1956).

For the treatment of similar problems in economics, see Geary (1949), Reiersöl (1950), Koopmans (1950), and Hood (1953).

TABLE 6.7*a*

The function $\rho_{10}(k)$ *giving the standard deviation of the fitted value*

k	$\rho_{10}(k)$	k	$\rho_{10}(k)$	k	$\rho_{10}(k)$
0·00	1·00	1·0	2·00	2·0	3·61
0·05	1·00	1·1	2·15	2·1	3·77
0·10	1·01	1·2	2·31	2·2	3·94
0·15	1·03	1·3	2·46	2·3	4·11
0·20	1·06	1·4	2·62	2·4	4·28
0·25	1·09	1·5	2·78	2·5	4·44
0·3	1·13	1·6	2·95	2·6	4·61
0·4	1·22	1·7	3·11	2·7	4·78
0·5	1·32	1·8	3·27	2·8	4·95
0·6	1·44	1·9	3·44	2·9	5·12
0·7	1·57	2·0	3·61	3·0	5·29
0·8	1·71				
0·9	1·85			$k > 3$	$1·73k$
1·0	2·00				

TABLE 6.7*b*

The sums $S_{11} = \Sigma\epsilon_i^2$

n	S_{11}	n	S_{11}	n	S_{11}
6	17·5	26	1462·5	51	11050
7	28	27	1638	52	11713
8	42	28	1827	53	12402
9	60	29	2030	54	13117·5
10	82·5	30	2247·5	55	13860
11	110	31	2480	56	14630
12	143	32	2728	57	15428
13	182	33	2992	58	16254·5
14	227·5	34	3272·5	59	17110
15	280	35	3570	60	17995
16	340	36	3885	61	18910
17	408	37	4218	62	19855·5
18	484·5	38	4569·5	63	20832
19	570	39	4940	64	21840
20	665	40	5330	65	22880
21	770	41	5740	66	23952·5
22	885·5	42	6170·5	67	25058
23	1012	43	6622	68	26197
24	1150	44	7095	69	27370
25	1300	45	7590	70	28577·5
		46	8107·5	71	29820
		47	8648	72	31098
		48	9212	73	32412
		49	9800	74	33762·5
		50	10412·5	75	35150

TABLE 6.7*c*

W(m) and ΣW(m) for the fitting of a line using successive differences

$n =$	2	4	6	8	10	12	14	16	18	20	22	24	26	28
	1	3	5	7	9	11	13	15	17	19	21	23	25	27
	1	4	8	12	16	20	24	28	32	36	40	44	48	52
		10	9	15	21	27	33	39	45	51	57	63	69	75
			35	16	24	32	40	48	56	64	72	80	88	96
				84	25	35	45	55	65	75	85	95	105	115
					165	36	48	60	72	84	96	108	120	132
						286	49	63	77	91	105	119	133	147
							455	64	80	96	112	128	144	160
								680	81	99	117	135	153	171
									969	100	120	140	160	180
										1330	121	143	165	187
											1771	144	168	192
												2300	169	195
													2925	196
														3654

$n =$	30	32	34	36	38	40	42	44	46	48	50	52	54
	29	31	33	35	37	39	41	43	45	47	49	51	53
	56	60	64	68	72	76	80	84	88	92	96	100	104
	81	87	93	99	105	111	117	123	129	135	141	147	153
	104	112	120	128	136	144	152	160	168	176	184	192	200
	125	135	145	155	165	175	185	195	205	215	225	235	245
	144	156	168	180	192	204	216	228	240	252	264	276	288
	161	175	189	203	217	231	245	259	273	287	301	315	329
	176	192	208	224	240	256	272	288	304	320	336	352	368
	189	207	225	243	261	279	297	315	333	351	369	387	405
	200	220	240	260	280	300	320	340	360	380	400	420	440
	209	231	253	275	297	319	341	363	385	407	429	451	473
	216	240	264	288	312	336	360	384	408	432	456	480	504
	221	247	273	299	325	351	377	403	429	455	481	507	533
	224	252	280	308	336	364	392	420	448	476	504	532	560
	225	255	285	315	345	375	405	435	465	495	525	555	585
	4495	256	288	320	352	384	416	448	480	512	544	576	608
		5456	289	323	357	391	425	459	493	527	561	595	629
			6545	324	360	396	432	468	504	540	576	612	648
				7770	361	399	437	475	513	551	589	627	665
					9139	400	440	480	520	560	600	640	680
						10660	441	483	525	567	609	651	693
							12341	484	528	572	616	660	704
								14190	529	575	621	667	713
									16215	576	624	672	720
										18424	625	675	725
											20825	676	728
												23426	729
													26235

TABLE 6.7c (*continued*)

$n =$	3	5	7	9	11	13	15	17	19	21	23	25	27	29
	1	2	3	4	5	6	7	8	9	10	11	12	13	14
	$\overline{2}$	3	5	7	9	11	13	15	17	19	21	23	25	27
		$\overline{10}$	6	9	12	15	18	21	24	27	30	33	36	39
			$\overline{28}$	10	14	18	22	26	30	34	38	42	46	50
				$\overline{60}$	15	20	25	30	35	40	45	50	55	60
					$\overline{110}$	21	27	33	39	45	51	57	63	69
						$\overline{182}$	28	35	42	49	56	63	70	77
							$\overline{280}$	36	44	52	60	68	76	84
								$\overline{408}$	45	54	63	72	81	90
									$\overline{570}$	55	65	75	85	95
										$\overline{770}$	66	77	88	99
											$\overline{1012}$	78	90	102
												$\overline{1300}$	91	104
													$\overline{1638}$	105
														$\overline{2030}$

$n =$	31	33	35	37	39	41	43	45	47	49	51	53	55
	15	16	17	18	19	20	21	22	23	24	25	26	27
	29	31	33	35	37	39	41	43	45	47	49	51	53
	42	45	48	51	54	57	60	63	66	69	72	75	78
	54	58	62	66	70	74	78	82	86	90	94	98	102
	65	70	75	80	85	90	95	100	105	110	115	120	125
	75	81	87	93	99	105	111	117	123	129	135	141	147
	84	91	98	105	112	119	126	133	140	147	154	161	168
	92	100	108	116	124	132	140	148	156	164	172	180	188
	99	108	117	126	135	144	153	162	171	180	189	198	207
	105	115	125	135	145	155	165	175	185	195	205	215	225
	110	121	132	143	154	165	176	187	198	209	220	231	242
	114	126	138	150	162	174	186	198	210	222	234	246	258
	117	130	143	156	169	182	195	208	221	234	247	260	273
	119	133	147	161	175	189	203	217	231	245	259	273	287
	120	135	150	165	180	195	210	225	240	255	270	285	300
	$\overline{2480}$	136	152	168	184	200	216	232	248	264	280	296	312
		$\overline{2992}$	153	170	187	204	221	238	255	272	289	306	323
			$\overline{3570}$	171	189	207	225	243	261	279	297	315	333
				$\overline{4218}$	190	209	228	247	266	285	304	323	342
					$\overline{4940}$	210	230	250	270	290	310	330	350
						$\overline{5740}$	231	252	273	294	315	336	357
							$\overline{6622}$	253	275	297	319	341	363
								$\overline{7590}$	276	299	322	345	368
									$\overline{8648}$	300	324	348	372
										$\overline{9800}$	325	350	375
											$\overline{11050}$	351	377
												$\overline{12402}$	378
													$\overline{13860}$

TABLE 6.7*d*

Step functions for estimating the slope of a line;
n_j is the number of observations in step of weight j

Three steps: $n = 3r+\nu$				Five steps: $n = 5r+\nu$				
ν	n_1	n_0	$\Sigma W_i \epsilon_i$	ν	n_2	n_1	n_0	$\Sigma W_i \epsilon_i$
0	r	r	$2r^2$	0	r	r	r	$10r^2$
± 1	r	$r \pm 1$	$2r^2 \pm r$	± 1	r	r	$r \pm 1$	$10r^2 \pm 3r$
				± 2	r	$r \pm 1$	r	$10r^2 \pm 7r + 1$
$\eta(b_1) = 0{\cdot}889$				$\eta(b_1) = 0{\cdot}960$				

TABLE 6.7*e*

Values $\Sigma W_i \epsilon_i$ for single-step and double-step functions

n	$\Sigma W_i \epsilon_i$ Single	Double	n	$\Sigma W_i \epsilon_i$ Single	Double	n	$\Sigma W_i \epsilon_i$ Single	Double
7	10	18	26	153	265	51	578	1030
8	15	27	27	162	286	52	595	1071
9	18	34	28	171	319	53	630	1134
10	21	40	29	190	342	54	648	1177
			30	200	360	55	666	1210
11	28	46	31	210	378	56	703	1243
12	32	55	32	231	403	57	722	1288
13	36	70	33	242	442	58	741	1357
14	45	81	34	253	469	59	780	1404
15	50	90	35	276	490	60	800	1440
16	55	99	36	288	511	61	820	1476
17	66	112	37	300	540	62	861	1525
18	72	133	38	325	585	63	882	1600
19	78	148	39	338	616	64	903	1651
20	91	160	40	351	640	65	946	1690
21	98	172	41	378	664	66	968	1729
22	105	189	42	392	697	67	990	1782
23	120	216	43	406	748	68	1035	1863
24	128	235	44	435	783	69	1058	1918
25	136	250	45	450	810	70	1081	1960
			46	465	837	71	1128	2002
			47	496	874	72	1152	2059
			48	512	931	73	1176	2146
			49	528	970	74	1225	2205
			50	561	1000	75	1250	2250

TABLE 6.7*f*

Percentage efficiencies of b_1 using step functions

κ_2^2	0					0·5					1·0				
κ_3	−1·0	−0·5	0	0·5	1·0	−1·0	−0·5	0	0·5	1·0	−1·0	−0·5	0	0·5	1·0
$N = 3$	91·1	90·4	88·9	86·3	81·8	89·1	88·0	86·0	82·9	77·9	87·1	85·7	83·3	79·7	74·3
$N = 5$	96·1	96·4	96·0	94·6	91·6	94·0	93·8	92·9	90·9	87·2	91·9	91·3	90·0	87·5	83·2
$N = 7$	97·5	98·0	98·0	97·0	94·4	95·3	95·4	94·8	93·2	89·9	93·2	92·9	91·8	89·7	85·7
$N = \infty$	98·8	99·6	100	99·5	97·4	96·6	97·0	96·8	95·6	92·7	94·5	94·5	93·8	91·9	88·4

TABLE 6.7*g*

Efficiencies of fitted values using step functions

Single-step functions

$\lvert\kappa_2\rvert$	0					0·5					1·0				
$\lvert k\rvert$ \ κ_3	−1·0	−0·5	0	+0·5	+1·0	−1·0	−0·5	0	+0·5	+1·0	−1·0	−0·5	0	+0·5	+1·0
1·4	0·923	0·917	0·903	0·880	0·840	0·914	0·906	0·891	0·865	0·822	0·888	0·875	0·854	0·821	0·772
1·2	0·927	0·920	0·908	0·885	0·847	0·918	0·910	0·895	0·871	0·830	0·893	0·881	0·860	0·829	0·781
1·0	0·932	0·926	0·914	0·893	0·857	0·924	0·916	0·903	0·879	0·840	0·900	0·889	0·870	0·840	0·794
0·8	0·940	0·935	0·924	0·905	0·873	0·933	0·926	0·914	0·893	0·857	0·912	0·901	0·884	0·857	0·815
0·6	0·952	0·948	0·939	0·924	0·897	0·946	0·941	0·931	0·913	0·884	0·929	0·920	0·906	0·883	0·848
0·4	0·969	0·967	0·961	0·951	0·933	0·966	0·962	0·955	0·944	0·924	0·954	0·949	0·939	0·924	0·899
0·2	0·990	0·989	0·987	0·983	0·977	0·988	0·987	0·985	0·981	0·974	0·984	0·982	0·979	0·973	0·964
0	1	1	1	1	1	1	1	1	1	1	1	1	1	1	1

TABLE 6.7*g* (*cont.*)

Double-step functions

$\lvert\kappa_2\rvert$	0					0·5					1·0				
κ_3 / $\lvert k\rvert$	−1·0	−0·5	0	+0·5	+1·0	−1·0	−0·5	0	+0·5	+1·0	−1·0	−0·5	0	+0·5	+1·0
1·4	0·969	0·970	0·966	0·953	0·924	0·960	0·959	0·952	0·936	0·904	0·935	0·928	0·914	0·889	0·848
1·2	0·971	0·971	0·967	0·955	0·927	0·963	0·961	0·954	0·938	0·907	0·939	0·932	0·918	0·894	0·852
1·0	0·974	0·974	0·970	0·957	0·930	0·966	0·964	0·958	0·942	0·911	0·945	0·938	0·924	0·900	0·860
0·8	0·978	0·977	0·973	0·962	0·936	0·971	0·969	0·963	0·948	0·919	0·953	0·946	0·933	0·911	0·872
0·6	0·983	0·983	0·979	0·969	0·946	0·978	0·976	0·970	0·958	0·932	0·965	0·958	0·947	0·927	0·892
0·4	0·990	0·989	0·987	0·979	0·963	0·987	0·986	0·981	0·972	0·953	0·979	0·975	0·967	0·952	0·925
0·2	0·997	0·997	0·996	0·993	0·986	0·996	0·995	0·994	0·990	0·982	0·994	0·992	0·989	0·983	0·972
0	1	1	1	1	1	1	1	1	1	1	1	1	1	1	1

TABLE 6.7*h*

Efficiencies of the estimates b_1 obtained using least-squares grouped methods

κ_2^2	0					0·5					1·0				
κ_3	−1·0	−0·5	0	+0·5	+1·0	−1·0	−0·5	0	+0·5	+1·0	−1·0	−0·5	0	+0·5	+1·0
$N = 3$	0·911	0·904	0·889	0·863	0·818	0·902	0·893	0·876	0·848	0·803	0·893	0·882	0·864	0·835	0·788
$N = 5$	0·967	0·965	0·960	0·949	0·930	0·964	0·961	0·955	0·944	0·924	0·960	0·957	0·950	0·938	0·918
$N = 7$	0·983	0·982	0·980	0·974	0·963	0·981	0·980	0·977	0·971	0·960	0·980	0·978	0·975	0·968	0·958

PART III

POLYNOMIALS AND OTHER CURVES

CHAPTER 7

ESTIMATION OF THE POLYNOMIAL COEFFICIENTS

In this chapter and the next a full account will be given of the problems associated with the fitting of a polynomial curve to a series of observations. The fundamental ideas and postulates were discussed in § 5.3, and were shown to lead to the normal equations

$$\Sigma w_i(y_i - \Sigma b_{pj}\, x_i^j)\, x_i^k = 0.$$

The solution of these equations using the Gauss–Doolittle method will be developed in § 7.1. In § 7.2 the problem will be treated from the point of view of orthogonal polynomials, and the connection between these two approaches will be established. Then in § 7.3 the two treatments will be combined, using matrix notation. The matrix treatment summarizes in compact notation the results obtained. It would be possible to develop the theory directly in matrix notation, and in fact this approach might be preferred by the mathematical reader. However, it is felt that the approach given here, in which each step is set out in detail, is easier to follow, especially for readers not accustomed to matrix manipulation.

The remainder of the chapter is concerned with various special methods of procedure, particularly those applicable to equally-spaced observations. In § 12.1.2 there is a guide to enable the reader to select the scheme most suitable for any one of the various types of example commonly encountered in practice.

7.1 THE NORMAL EQUATIONS

The $p+1$ normal equations for the coefficients b_{pj} are obtained on the least-squares principle by differentiation of $\Sigma w_i(y_i - \Sigma b_{pj}\, x_i^j)^2$ with respect to b_{pk}. As in § 5.3, the resulting equations are

$$\sum_j b_{pj} \sum_i w_i\, x_i^j\, x_i^k = \sum_i w_i\, y_i\, x_i^k, \quad k = 0 \text{ to } p. \tag{1}$$

It is convenient to define the symbols

$$\phi_{jk} = \Sigma w_i\, x_i^j\, x_i^k = \phi_{kj} \tag{2a}$$

and

$$M_k = \Sigma w_i\, y_i\, x_i^k. \tag{2b}$$

In terms of these symbols the normal equations are

$$\sum_j \phi_{kj}\, b_{pj} = M_k, \quad k = 0 \text{ to } p. \tag{3}$$

The quantities M_k are called the moments. There are two distinct stages in the fitting of a least-squares curve: the calculation of the values ϕ_{kj} and M_k, and the solution of the normal equations.

7.1.1 *Moments and sums of powers*

The values ϕ_{kj} and M_k can be obtained by listing the quantities $w_i^{\frac{1}{2}} x_i^k, w_i^{\frac{1}{2}} y_i$ in columns, and forming the sums of the products of corresponding elements in two of the columns. This is done for all possible pairings of the columns. For example, the products for the columns $w_i^{\frac{1}{2}} x_i^k, w_i^{\frac{1}{2}} y_i$ give the moment M_k. When the observations are all of equal weight, the columns are just x_i^k, y_i. An alternative method which may be used when the weights are different is to form two sets of columns, one of values x_i^k, y_i, and the other of values $w_i x_i^k, w_i y_i$. The columns in one set are multiplied by the columns in the other. In each method the products of individual terms are not listed separately but are allowed to accumulate in the product register of the calculating machine. It is absolutely essential to check very carefully the formation of the columns and the final values for ϕ_{kj} and M_k.

7.1.1.1 *Example*

Table 7.1.1 shows a series of 67 observations of the mechanical equivalent of heat J made at different temperatures t by Jaeger and von Steinwehr (1921). The values x are the temperature readings, referred to an origin of 20° C. The values y are obtained from the observed values J by subtracting a constant of magnitude 4·17 and then multiplying by 10^4. That is,

$$x = t - 20, \quad y = (J - 4{\cdot}17) \times 10^4.$$

It is usually an advantage to choose the origin of x near the centre of the range, since the magnitudes of the powers and moments are then much smaller than when the origin is at one end of the range and greater accuracy is obtained for calculations carried out to a given number of significant figures.

The calculation of the moments and sums of powers is best done systematically, using the scheme of Table 7.1.1*a*. It will be assumed that the third degree polynomial is required. Powers of ten are removed from the values x_i and y_i to bring them to the order of unity. This enables the calculations to be checked by means of a 'check column'. When the number of observations is large it is best to subdivide them into groups of about twenty so that calculating mistakes can be detected more easily. Table 7.1.1*a* gives the calculations for the last twenty observations of Table 7.1.1. The steps are performed in the order described below.

(*a*) The values x, y are entered, and these columns summed to give the entries $[x^0 x]$ and $[x^0 y]$ at the bottom of the table (the symbol $[x^j x^k]$ will be used to denote $\Sigma x^j x^k$, the element in row x^j, column x^k in the lower section of the table). The entries are checked by summing the values in the original table of observations.

TABLE 7.1.1

Observations of Jaeger and von Steinwehr (Example 7.1.1.1)

x	y	x	y	x	y
−15·25	291	−2·59	133	+5·55	50
−14·35	315	−1·58	104	+6·24	71
−13·85	245	−1·58	97	+6·33	41
−13·62	259	−1·45	104	+7·76	66
−12·59	254	−1·18	117	+9·15	79
−11·49	230	−1·17	104	+9·19	58
−11·19	200	−0·25	106	+10·60	57
−11·05	207	−0·05	119	+11·49	88
−9·98	176	+0·01	99	+12·54	68
−8·67	202	+1·13	96	+13·98	45
−8·55	175	+1·15	89	+14·32	113
−8·39	153			+14·39	69
−7·94	187	+1·41	80	+15·75	83
−7·41	169	+1·80	83	+15·79	85
−7·40	165	+2·53	80	+16·64	173
−7·13	101	+3·24	71	+17·19	35
−6·97	137	+3·96	76	+19·41	66
−6·03	114	+4·11	66	+23·09	104
−6·00	201	+4·82	60	+24·34	103
−5·12	114	+5·36	70	+25·56	25
−5·01	118			+25·79	74
−3·69	130			+26·96	103
−3·61	137			+28·36	102
−2·67	90			+29·60	108

(*b*) Values x^2 and x^3 are entered from *Barlow's Tables*. The number of decimal places to be retained depends on the scatter of the observations from a smooth curve. Here the scatter is rather large, and six decimals are more than adequate, but if the curve fits the points closely eight decimals may be required. The columns of values x^2, x^3 are summed (entries $[x^0\,x^2]$, $[x^0\,x^3]$ in the x^0 row). The columns x, x^2 are intermultiplied (entry $[xx^2]$ in x row). If no copying or calculating mistakes have been made, this entry should be the same as $[x^0\,x^3]$, perhaps differing by 1 or 2 in the last figure.

(*c*) The quantities are now summed horizontally to give the check elements z_i:

$$z_i = \sum_{j=0}^{3} x_i^j + y_i.$$

(*d*) The column of values z_i is now summed. If all entries are correct,

$$\sum_j [x^0\,x^j] = [x^0\,z].$$

(*e*) The columns are now multiplied in turn by the x column, the sums being entered in the x row at the bottom of the table. The entry $[xx]$

equals $[x^0 x^2]$, and so need not be recalculated. The entry $[xx^0]$ equals $[x^0 x]$, and is usually omitted, as are the other elements below the diagonal in the lower table. If the calculations are free from mistakes,

$$\sum_j [xx^j] = [xz].$$

It will be noted that the element $[xx^0]$ omitted from the lower table by reason of symmetry must be included in this sum.

TABLE 7.1.1*a*

Calculation of moments and sums of powers

	Factors removed: x, 10^q, $q = 1$; y, 10^r, $r = 2$						
	x^0	x	x^2	x^3	y	z	
	1	+2·960	8·761600	+25·934336	1·08	39·735936	
	1	+2·836	8·042896	+22·809653	1·02	35·708549	
	1	+2·696	7·268416	+19·595650	1·03	31·590066	
	1	+2·579	6·651241	+17·153550	0·74	28·123791	
	1	+2·556	6·533136	+16·698696	0·25	27·037832	
	1	+2·434	5·924356	+14·419882	1·03	24·808238	
	1	+2·309	5·331481	+12·310390	1·04	21·990871	
	1	+1·941	3·767481	+7·312681	0·66	14·681162	
	1	+1·719	2·954961	+5·079578	0·35	11·103539	
	1	+1·664	2·768896	+4·607443	1·73	11·770339	
	1	+1·579	2·493241	+3·936828	0·85	9·859069	
	1	+1·575	2·480625	+3·906984	0·83	9·792609	
	1	+1·439	2·070721	+2·979768	0·69	8·179489	
	1	+1·432	2·050624	+2·936494	1·13	8·549118	
	1	+1·398	1·954404	+2·732257	0·45	7·534661	
	1	+1·254	1·572516	+1·971935	0·68	6·478451	
	1	+1·149	1·320201	+1·516911	0·88	5·866112	
	1	+1·060	1·123600	+1·191016	0·57	4·944616	
	1	+0·919	0·844561	+0·776152	0·58	4·119713	
	1	+0·915	0·837225	+0·766061	0·79	4·308286	Check
x^0	20	36·414	74·752182	168·636265	16·38	316·182447	447
x		74·752182	168·636263	406·894568	30·69715	717·394163	163
x^2			406·894564	1027·455221	64·492194	1742·230425	424
x^3				2674·462633	148·160059	4425·608746	746
y					15·4604	275·189803	803

(*f*) The columns are multiplied in turn by the x^2 column. The entry $[x^2 x^2]$ equals $[xx^3]$, and need not be recalculated. However, if it is recalculated the check column entry will agree more closely with the sum of the values in the x^2 row, since $\Sigma(x^2)^2$ will differ from Σxx^3 because of the rounding off to six decimals. In checking the entries against the value $[x^2 z]$, the values omitted by reason of symmetry must be included in the sum. Thus

$$[x^0 x^2]+[xx^2]+[x^2 x^2]+[x^2 x^3]+[x^2 y] = [x^2 z],$$

and $[x^2 z]$ is the sum of the entries in the x^2 column down to the diagonal and then along the x^2 row.

(*g*) The products of the x^3 column with the x^3, y and z columns are formed and checked.

(*h*) The products of the y column with the y and z columns are formed and checked.

The elements in the lower table are the quantities $\Sigma x_i^j x_i^k$, $\Sigma y_i x_i^k$, Σy_i^2, for the twenty observations. Similar tables are formed for the remainder of the observations, and these are summed to give the final table of values ϕ_{jk}, M_k, Σy^2, shown in Table 7.1.1*b*. This final table is checked by comparing the sums of the rows (including the elements omitted by reason of symmetry) with the check column values.

TABLE 7.1.1*b*

Moments and sums of powers for Example 7.1.1.1

ϕ_{jk}				M_j	C_j	Check
67	20·173	98·726409	146·564411	79·90	412·363820	820
	98·726409	146·564410	436·475630	−9·221	692·718449	449
		436·475625	991·695975	113·263288	1786·725708	707
			2722·811792	92·954958	4390·502768	766
			Σy^2	121·8016	398·698846	846

7.1.2 *The method of single division*

The most common methods of solving the normal equations are variants of what Dwyer (1951) has called the method of single division. The first step is the elimination of b_{p0} from the normal equations, leaving p equations in $b_{p1}, \ldots, b_{pp}$. Then b_{p1} is eliminated from this set of p equations. This is followed in turn by the elimination of $b_{p2}, b_{p3}, \ldots$, until finally a single equation in b_{pp} remains. The forward section of the method is then complete. The backward section begins with the substitution of b_{pp} in one of the second last pair of equations and the solution of this equation for $b_{p,p-1}$. The values $b_{pp}, b_{p,p-1}$ are then substituted in one of the third last trio of equations, and this equation is solved for $b_{p,p-2}$. The process is continued till all coefficients down to b_{p0} have been obtained.

Table 7.1.2 gives in symbols the detailed steps for the forward solution when p is 2. It is clear that some systematic notation is necessary for the discussion of the method. In one convenient notation, the unknowns which have been removed are indicated by writing the degree of the unknown last eliminated as a suffix separated from the other suffixes by a stop. Thus (2′) in Table 7.1.2 is written

$$\phi_{21\cdot0}\, b_{21} + \phi_{22\cdot0}\, b_{22} = M_{2\cdot0}.$$

In the general case, if the coefficients up to $b_{p,j-1}$ have been eliminated from the normal equations, the resulting equations are

$$\sum_{k=j}^{p} \phi_{rk.j-1} b_{pk} = M_{r.j-1}, \quad r = j \text{ to } p. \tag{1a}$$

TABLE 7.1.2

The method of single division

(0)	$\phi_{00} b_{20} + \phi_{01} b_{21} + \phi_{02} b_{22} = M_0$
(1)	$\phi_{10} b_{20} + \phi_{11} b_{21} + \phi_{12} b_{22} = M_1$
(2)	$\phi_{20} b_{20} + \phi_{21} b_{21} + \phi_{22} b_{22} = M_2$
$(0') = (0) \div \phi_{00}$	$b_{20} + \alpha_{01} b_{21} + \alpha_{02} b_{22} = a_0$
where	$\alpha_{01} = \phi_{01}/\phi_{00}, \quad \alpha_{02} = \phi_{02}/\phi_{00}, \quad a_0 = M_0/\phi_{00}$
$(1') = (1) - \phi_{10}(0')$	$(\phi_{11} - \phi_{10}\alpha_{01}) b_{21} + (\phi_{12} - \phi_{10}\alpha_{02}) b_{22} = M_1 - \phi_{10} a_0$
$(2') = (2) - \phi_{20}(0')$	$(\phi_{21} - \phi_{20}\alpha_{01}) b_{21} + (\phi_{22} - \phi_{20}\alpha_{02}) b_{22} = M_2 - \phi_{20} a_0$
$(1'') = (1') \div (\phi_{11} - \phi_{10}\alpha_{01})$	$b_{21} + \alpha_{12} b_{22} = a_1$
where	$\alpha_{12} = (\phi_{12} - \phi_{10}\alpha_{02})/(\phi_{11} - \phi_{10}\alpha_{01}), \quad a_1 = (M_1 - \phi_{10} a_0)/(\phi_{11} - \phi_{10}\alpha_{01})$
$(2'') = (2') - (\phi_{21} - \phi_{20}\alpha_{01})(1'')$	
$[\phi_{22} - \phi_{20}\alpha_{02} - (\phi_{21} - \phi_{20}\alpha_{01})\alpha_{12}] b_{22} = [M_2 - \phi_{20} a_0 - (\phi_{21} - \phi_{20}\alpha_{01}) a_1]$	

In eliminating b_{pj} from this set by the method of single division, the equation with $r = j$ is first divided by $\phi_{jj.j-1}$. The resulting equation is then multiplied by $\phi_{rj.j-1}$ and subtracted from (1*a*) for all values of r from $j+1$ to p. This gives the equations

$$\sum_{k=j+1}^{p} \phi_{rk.j} b_{pk} = M_{r.j}, \quad r = j+1 \text{ to } p. \tag{1b}$$

Hence
$$\phi_{rk.j} = \phi_{rk.j-1} - \phi_{rj.j-1}(\phi_{jk.j-1}/\phi_{jj.j-1}) \tag{2a}$$

and
$$M_{r.j} = M_{r.j-1} - \phi_{rj.j-1}(M_{j.j-1}/\phi_{jj.j-1}). \tag{2b}$$

If the symbols

$$S_{rj} = \phi_{rj.j-1} \quad (r \geqslant j), \tag{3a}$$

$$\alpha_{jk} S_{jj} = \phi_{jk.j-1} \quad (k \geqslant j), \tag{3b}$$

$$\mathscr{M}_j = a_j S_{jj} = M_{j.j-1}, \tag{3c}$$

are introduced for the elements in the first row and first column of the set (1*a*), then (2*a*, *b*) become

$$\phi_{rk.j} = \phi_{rk.j-1} - S_{rj}\alpha_{jk} \tag{4a}$$

and
$$M_{r.j} = M_{r.j-1} - S_{rj} a_j. \tag{4b}$$

The terms $\phi_{rk.j-1}$ and $M_{r.j-1}$ can be expanded in a similar way, and on continuing these expansions the following equations are obtained:

$$\phi_{rk.j} = \phi_{rk} - \sum_{q=0}^{j} S_{rq}\,\alpha_{qk}, \tag{5a}$$

$$M_{r.j} = M_r - \sum_{q=0}^{j} S_{rq}\,a_q. \tag{5b}$$

Hence any particular element can be calculated in terms of the quantities ϕ_{rk}, M_r, and the elements occurring in the first row and the first column of each set (1*a*).

The elements occurring in all the other rows and columns need not be recorded, for the whole solution may be carried out in terms of the quantities S_{rq}, α_{qk} and a_k. From (5*a*, *b*), these quantities satisfy the recurrence relations

$$S_{rk} = \phi_{rk} - \sum_{q=0}^{k-1} S_{rq}\,\alpha_{qk} \quad (r \geqslant k); \tag{6a}$$

$$S_{rr}\,\alpha_{rk} = \phi_{rk} - \sum_{q=0}^{r-1} S_{rq}\,\alpha_{qk} \quad (r \leqslant k); \tag{6b}$$

$$S_{rr}\,a_r = M_r - \sum_{q=0}^{r-1} S_{rq}\,a_q = \mathscr{M}_r. \tag{6c}$$

The final equation in the forward solution is

$$\phi_{pp.p-1}\,b_{pp} = M_{p.p-1}, \tag{7a}$$

i.e.

$$b_{pp} = \mathscr{M}_p/S_{pp} = a_p. \tag{7b}$$

The backward solution is, from (1*a*), with $r = j$,

$$\sum_{k=j}^{p} \alpha_{jk}\,S_{jj}\,b_{pk} = \mathscr{M}_j,$$

and as from (3) α_{jj} is unity,

$$b_{pj} = a_j - \sum_{k=j+1}^{p} \alpha_{jk}\,b_{pk}. \tag{8}$$

The values b_{pj} can then be built up in turn, using previously calculated values $b_{pk} (k > j)$.

7.1.3 *The Gauss–Doolittle method*

The theory of the method of single division, developed in the previous section, applies to a general set of equations. As shown there, the procedure can be simplified by recording only the leading row and column at each stage. For the normal equations a further simplification is possible, since in these equations the quantities ϕ_{jk} are symmetrical, $\phi_{jk} = \phi_{kj}$. The simplified method is due primarily to Gauss, but it was popularized by Doolittle and is often called the Doolittle method.

From (7.1.2, 2*a*), the quantities $\phi_{rk.j}$ will be symmetrical if the quantities $\phi_{rk.j-1}$ are symmetrical. Thus, by induction, if the original normal equations are symmetrical, so are all the other sets of equations obtained by the method of single division. In particular, from (7.1.2, 3*a*–*b*),

$$S_{rj} = \alpha_{jr} S_{jj} \quad (r \geqslant j). \tag{1}$$

The Gauss–Doolittle scheme uses (7.1.2, 6*a*), (7.1.2, 6*c*), and (7.1.2, 8)

$$S_{rj} = \phi_{rj} - \sum_{q=0}^{j-1} S_{rq} \alpha_{qj} = \alpha_{jr} S_{rr}, \tag{2a}$$

$$\mathscr{M}_r = M_r - \sum_{q=0}^{r-1} \alpha_{qr} \mathscr{M}_q = a_r S_{rr}, \tag{2b}$$

and

$$b_{pj} = a_j - \sum_{k=j+1}^{p} \alpha_{jk} b_{pk}. \tag{2c}$$

The scheme is illustrated in Table 7.1.3. Steps 1 to 11 constitute the forward solution, steps 12 to 14 the backward solution.

TABLE 7.1.3

The Gauss–Doolittle method

1. Enter	$\phi_{00} = S_{00}$	$\phi_{01} = S_{10}$	$\phi_{02} = S_{20}$	$M_0 = \mathscr{M}_0$
2. Divide by S_{00}	1	α_{01}	α_{02}	a_0
3. Enter		ϕ_{11}	ϕ_{12}	M_1
4. $\alpha_{01} \times (1)$		$S_{10}\alpha_{01}$	$S_{20}\alpha_{01}$	$\mathscr{M}_0\alpha_{01}$
5. Subtract		S_{11}	S_{21}	$\mathscr{M}_1$
6. Divide by S_{11}		1	α_{12}	a_1
7. Enter			ϕ_{22}	M_2
8. $\alpha_{02} \times (1)$			$S_{20}\alpha_{02}$	$\mathscr{M}_0\alpha_{02}$
9. $\alpha_{12} \times (5)$			$S_{21}\alpha_{12}$	$\mathscr{M}_1\alpha_{12}$
10. Subtract			S_{22}	$\mathscr{M}_2$
11. Divide by S_{22}			1	a_2
12. $b_{22} =$				a_2
13. $b_{21} =$			a_1	$-\alpha_{12} b_{22}$
14. $b_{20} =$		a_0	$-\alpha_{01} b_{21}$	$-\alpha_{02} b_{22}$

7.1.4 *The abbreviated Doolittle method*

Doolittle recognized that lines 3, 4, 7, 8, and 9 of Table 7.1.3 were subsidiary to the main calculations, and recommended that they be transferred to a separate working sheet. There seems to be little advantage in this, but when modern calculating machines are employed it is not necessary to record these lines explicitly. Their omission leads to the abbreviated Doolittle method shown in Table 7.1.4.

TABLE 7.1.4

Abbreviated Doolittle method

ϕ_{00}	ϕ_{01}	ϕ_{02}	M_0
	ϕ_{11}	ϕ_{12}	M_1
		ϕ_{22}	M_2
S_{00}	S_{10}	S_{20}	$\mathscr{M}_0$
1	α_{01}	α_{02}	a_0
	S_{11}	S_{21}	$\mathscr{M}_1$
	1	α_{12}	a_1
		S_{22}	$\mathscr{M}_2$
		1	a_2
b_{20}	b_{21}	b_{22}	

TABLE 7.1.4*a*

Selection of elements in the abbreviated Doolittle method

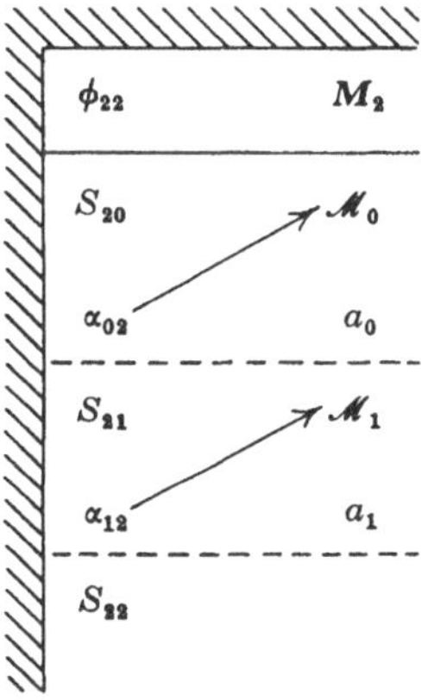

The selection of the correct elements in forming the products is simplified by covering unwanted elements with a pair of cards or a right-angled template. Thus to calculate the element S_{jk}, the rows of the ϕ_{rs} up to $k-1$ and the columns of the ϕ_{rs} and S_{rs} up to $k-1$ are covered. Then the element ϕ_{jk} is entered on the machine, and the products of the α_{rk} in the 'k' column and the S_{jr} (or $\mathscr{M}_r$) in the 'j' column are subtracted. The calculation of $\mathscr{M}_2$ is illustrated in Table 7.1.4*a*:

$$\mathscr{M}_2 = M_2 - \alpha_{02}\mathscr{M}_0 - \alpha_{12}\mathscr{M}_1.$$

The advantage of the method lies in the reduction of the number of entries. There is little, if any, saving in time. Because of its complexity, the method is not recommended for someone fitting curves only occasionally—the full Doolittle method, in which each step is set out separately, is much safer.

7.1.5 *The square root method*

In the square root method, instead of dividing S_{jk} by S_{kk} to give α_{kj}, S_{jk} is divided by $\sqrt{S_{kk}}$. The quantities

$$s_{jj} = \sqrt{S_{jj}}, \tag{1a}$$

$$s_{jk} = S_{jk}/\sqrt{S_{kk}} = \alpha_{kj}\sqrt{S_{kk}}, \tag{1b}$$

$$m_j = \mathscr{M}_j/\sqrt{S_{jj}} = a_j\sqrt{S_{jj}}, \tag{1c}$$

are recorded. From (7.1.3, 2*a*–*b*), the equations for the calculation of these quantities are

$$s_{rj} = \left(\phi_{rj} - \sum_{q=0}^{j-1} s_{rq}s_{jq}\right)\Big/ s_{jj}, \tag{2a}$$

$$s_{jj}^2 = \left(\phi_{jj} - \sum_{q=0}^{j-1} s_{jq}^2\right), \tag{2b}$$

$$m_r = \left(M_r - \sum_{q=0}^{r-1} s_{rq}m_q\right)\Big/ s_{rr}. \tag{2c}$$

The terms in (2*a*) are the products of the elements in the r and j rows of the quantities s. These are subtracted from ϕ_{rj} and divided by s_{jj}. s_{jj} itself is found from the square root of (2*b*). The calculating scheme is shown in Table 7.1.5.

TABLE 7.1.5

The square root method

ϕ_{00}	ϕ_{01}	ϕ_{02}	M_0
	ϕ_{11}	ϕ_{12}	M_1
		ϕ_{22}	M_2
s_{00}	s_{10}	s_{20}	m_0
	s_{11}	s_{21}	m_1
		s_{22}	m_2
b_{20}	b_{21}	b_{22}	

The extraction of the square root is simply accomplished by dividing the value in the product register [i.e. the right-hand side of (2*b*)] by an approximate 4 or 5 figure value obtained from *Barlow's Tables*, and taking the mean of the divisor and the quotient (cf. the introduction to *Barlow's Tables*, p. xi).

The backward solution (7.1.3, 2*c*) becomes

$$b_{pj} = \left(m_j - \sum_{j+1}^{p} s_{kj}b_{pk}\right)\Big/ s_{jj}, \tag{2d}$$

and it is necessary to divide by s_{jj} in the backward solution as well as in the forward solution.

Though the number of quantities recorded is reduced to the absolute minimum, the square root method is even more involved than the abbreviated Doolittle method, and is perhaps best left to the professional computor.

For 3 or 4 unknowns there is no significant difference between the times required for the calculations in the three methods, but if the number of unknowns is large the square root method does enable the calculations to be performed more rapidly. Laderman (1948) reports that a system of ten equations was solved in under three hours by the square root method at the Computation Laboratory of the National Bureau of Standards.

7.1.6 *The check column*

It is extremely difficult to solve the normal equations without making an arithmetical mistake. Hence the provision of a check column becomes almost essential if the correct values are to be obtained in a reasonable time. The check column is made up of quantities

$$C_j = \sum_{k=0}^{p} \phi_{jk} + M_j \tag{1}$$

which are the sum of all the elements in the jth normal equation, including those elements omitted from the Doolittle scheme by reason of symmetry. The C_j are the quantities used in checking Table 7.1.1*b*. They are operated on in just the same way as the quantities ϕ_{jk} and M_j, giving rise to values

$$c_r = \left(C_r - \sum_{q=0}^{r-1} S_{rq} c_q\right) \bigg/ S_{rr} = C_{r.r-1}/S_{rr}. \tag{2}$$

The individual components of C_r will change under the operations according to the rules (7.1.2, 5*a*, *b*), and so

$$C_{r.r-1} = \sum_{q=0}^{p} \phi_{rq.r-1} + M_{r.r-1}.$$

Now $\phi_{rq.r-1}$ vanishes for q less than r, and so, from (7.1.2,3*a*–*c*),

$$c_r = \sum_{q=r}^{p} \alpha_{rq} + a_r. \tag{3}$$

Hence the values c_r will provide a check on the accuracy of the calculations up to that stage.

The quantities c_q can also be used to check the arithmetic in the backward solution. They are operated on in the same way as the a_q to produce quantities

$$d_{pq} = c_q - \sum_{q+1}^{p} \alpha_{qr} d_{pr}. \tag{4}$$

It can be shown that

$$d_{pq} = 1 + b_{pq}. \tag{5}$$

Perhaps the simplest proof of (5) is by induction. If this equation holds for values of the suffix from $q+1$ to p, then

$$d_{pq} = c_q - \sum_{q+1}^{p} \alpha_{qr} d_{pr} = a_q + \sum_{q}^{p} \alpha_{qr} - \sum_{q+1}^{p} \alpha_{qr} - \sum_{q+1}^{p} \alpha_{qr} b_{pr} = 1 + b_{pq},$$

from (7.1.3,2c), the value α_{qq} being unity. Hence (5) holds for all q.

When the equations are non-symmetrical, the columns of values ϕ_{qr} can be summed to produce a row of values

$$C'_r = \sum_{q=0}^{p} \phi_{qr}. \tag{6a}$$

These are operated on in the same way as the other rows, and the operation gives elements

$$C'_{r.r-1} = \sum_{q=0}^{p} \phi_{qr.r-1} = \sum_{q=r}^{p} S_{qr}, \tag{6b}$$

and these sums provide a check on the values S_{qr}. This check is not of interest in the present chapter, since the normal equations are symmetrical.

7.1.7 *Changes of scale*

If the check columns are to provide an adequate check on the calculations, and if troubles due to misplacing the decimal point are to be avoided, it is almost essential that the quantities ϕ_{jk} and M_k be all of the same order of magnitude. When the calculations of the moments and sums of powers are done by the scheme suggested in § 7.1.1.1, the quantities obtained will be all of the same order. Often in the solution of the normal equations it is desired to invert the matrix of quantities ϕ_{jk}, M_k, and then it is best to take out a further power of ten to bring the quantities to the order of unity. If the factors removed from x and y are 10^q and 10^r, and a further factor 10^s is removed from all the elements to bring them to the order of unity, Table 7.1.7 lists the factors by which the quantities given by the Doolittle scheme must be multiplied to return to the original variables.

TABLE 7.1.7

Removal of powers of ten in the Doolittle scheme

x is divided by 10^q.
y is divided by 10^r.
Elements ϕ_{jk}, M_k are divided by a further factor 10^s.
To return to the original variables:

b_{pj}, $s(b_{pj})$, a_j	are multiplied by	$10^r\,10^{-qj}$;
Σv_p^2	is multiplied by	$10^{2r}\,10^s$;
s_p	is multiplied by	$10^r\sqrt{10^s}$;
α_{kj}, β_{kj}	are multiplied by	$10^{q(j-k)}$;
χ_{jk}	is multiplied by	$10^{-q(j+k)}\,10^{-s}$.

7.1.8 *Example using the Doolittle scheme*

Table 7.1.8 shows the calculations using the full Doolittle scheme for the example of Table 7.1.1. To bring the quantities in Table 7.1.1*b* to the order of unity, a factor 10^2 is removed. The quantities $\phi_{jk}(=\Sigma x^{j+k})$ and $M_k(=\Sigma yx^k)$ are divided by this factor and rounded to six decimal places, and then entered in lines 1, 3, 7, and 12 of the scheme. The sums C_j of the elements in these lines are then formed, the elements missing by reason of symmetry being included. Thus the elements are summed from the top of the column to the diagonal and then along the row—

$$C_2 = \phi_{02}+\phi_{12}+\phi_{22}+\phi_{23}+M_2.$$

The quantities C_j are then checked against the original values in Table 7.1.1*b*.

The calculations in each of the four subsections of the scheme are performed in order. The value c_j at the end of each subsection is checked ($c_j = a_j+\Sigma\alpha_{jk}$) before proceeding with the next section. The column at the far right is used in standard deviation calculations and for checking the fitted curve, and will not be considered for the present.

After a_3 has been calculated, the coefficients b_{3j} are evaluated in the lower section of the scheme. Finally, the correctness of the solution is checked by comparing $\Sigma b_{3j}\phi_{j3}$ with M_3.

To bring the coefficients back to the original scale, they have to be multiplied by $10^r\,10^{-qj} = 10^2\,10^{-j}$ (Table 7.1.7). Thus

$$b_{30} = 92{\cdot}3894, \qquad b_{31} = -5{\cdot}56669,$$

$$b_{32} = 0{\cdot}406555, \quad b_{33} = -0{\cdot}0074431.$$

It will be observed that S_{33} and $\mathcal{M}_3$ are small fractions of ϕ_{33} and M_3, so that it is necessary to retain a fairly large number of significant figures throughout the calculations.

TABLE 7.1.8

The Doolittle scheme

Factors removed: x, 10^q, $q=1$; y, 10^r, $r=2$; elements divided by 10^s, $s=2$							Σy^2 1·218016
1. Enter	ϕ_{00} 0·67	ϕ_{01} 0·20173	ϕ_{02} 0·987264	ϕ_{03} 1·465644	M_0 0·7990	C_0 4·123638	$M_0 a_0$ 0·952837
2. $\div\phi_{00}$	α_{00} 1	α_{01} 0·301090	α_{02} 1·473528	α_{03} 2·187528	a_0 1·192537	c_0 6·154684	Σv_0^2 0·265179
Check						683	
	3. Enter	ϕ_{11} 0·987264	ϕ_{12} 1·465644	ϕ_{13} 4·364756	M_1 −0·09221	C_1 6·927184	$\mathscr{M}_1 a_1$ 0·119525
	4. $1\times\alpha_{01}$	0·060739	0·297255	0·441291	0·240571	1·241586	Σv_1^2 0·145654
	5. Subtract	S_{11} 0·926525	S_{21} 1·168389	S_{31} 3·923465	$\mathscr{M}_1$ −0·332781	$\mathscr{C}_1$ 5·685598	(s_1)
	6. $\div S_{11}$	α_{11} 1	α_{12} 1·261044	α_{13} 4·234602	a_1 −0·359171	c_1 6·136476	
	Check					475	
		7. Enter	ϕ_{22} 4·364756	ϕ_{23} 9·915960	M_2 1·132633	C_2 17·867257	$\mathscr{M}_2 a_2$ 0·097853
		8. $1\times\alpha_{02}$	1·454761	2·159667	1·177349	6·076296	Σv_2^2 0·047801
		9. $5\times\alpha_{12}$	1·473390	4·947662	−0·419651	7·169789	(s_2) 0·02733
		10. Subtract	S_{22} 1·436605	S_{32} 2·809631	$\mathscr{M}_2$ 0·374935	$\mathscr{C}_2$ 4·621172	
		11. $\div S_{22}$	α_{22} 1	α_{23} 1·955744	a_2 0·260987	c_2 3·216731	
		Check				731	
			12. Enter	ϕ_{33} 27·228118	M_3 0·929550	C_3 43·905028	$\mathscr{M}_3 a_3$ 0·010597
			13. $1\times\alpha_{03}$	3·206137	1·747835	9·020574	Σv_3^2 0·037204
			14. $5\times\alpha_{13}$	16·614313	−1·409195	24·076245	(s_3) 0·02430
			15. $10\times\alpha_{23}$	5·494919	0·733277	9·037829	
			16. Subtract	S_{33} 1·912749	$\mathscr{M}_3$ −0·142367	$\mathscr{C}_3$ 1·770380	
			17. $\div S_{33}$	α_{33} 1	a_3 −0·074431	c_3 0·925568	
			Check			569	
	$-\alpha_{j3} b_{33}$	+0·162820	+0·315186	+0·145568	$a_3 = b_{33}$ −0·074431		
				$+a_2$ 0·260987			
	$-\alpha_{j2} b_{32}$	−0·599070	−0·512684	$=b_{32}$ 0·406555			
			$+a_1$ −0·359171				
	$-\alpha_{j1} b_{31}$	+0·167607	$=b_{31}$ −0·556669				
		$+a_0$ +1·192537					
		$=b_{30}$ +0·923894					
	Check	$\Sigma b_{3j}\phi_{j3}$	0·929549				

TABLE 7.1.8.1

The abbreviated Doolittle scheme

Factors removed: x, 10^q, $q = 1$; y, 10^r, $r = 2$; elements divided by 10^s, $s = 2$

ϕ_{00}	0·670000	ϕ_{01}	0·201730	ϕ_{02}	0·987264	ϕ_{03}	1·465644	M_0	0·799000	C_0	4·123638		
		ϕ_{11}	0·987264	ϕ_{12}	1·465644	ϕ_{13}	4·364756	M_1	−0·092210	C_1	6·927184		
				ϕ_{22}	4·364756	ϕ_{23}	9·916960	M_2	1·132633	C_2	17·867257		
						ϕ_{33}	27·228118	M_3	0·929550	C_3	43·905028		
												Σy^2	1·218016
S_{00}	0·670000	S_{10}	0·201730	S_{20}	0·987264	S_{30}	1·465644	$\mathscr{M}_0$	0·799000	$\mathscr{C}_0$	4·123638	$a_0 \mathscr{M}_0$	0·952837
α_{00}	1	α_{01}	0·301090	α_{02}	1·473528	α_{03}	2·187528	a_0	1·192537	c_0	6·154684	Σv_0^2	0·265179
											Check 683		
		S_{11}	0·926525	S_{21}	1·168389	S_{31}	3·923465	$\mathscr{M}_1$	−0·332781	$\mathscr{C}_1$	5·685598	$a_1 \mathscr{M}_1$	0·119525
		α_{11}	1	α_{12}	1·261044	α_{13}	4·234603	a_1	−0·359171	c_1	6·136475	Σv_1^2	0·145654
											Check 476		
				S_{22}	1·436605	S_{32}	2·809631	$\mathscr{M}_2$	0·374936	$\mathscr{C}_2$	4·621172	$a_2 \mathscr{M}_2$	0·097853
				α_{22}	1	α_{23}	1·955743	a_2	0·260987	c_2	3·216731	Σv_2^2	0·047801
											Check 730		
						S_{33}	1·912748	$\mathscr{M}_3$	−0·142370	$\mathscr{C}_3$	1·770379	$a_3 \mathscr{M}_3$	0·010597
						α_{33}	1	a_3	−0·074432	c_3	0·925568	Σv_3^2	0·037204
											Check 568		
b_{30}	0·923893	b_{31}	−0·556667	b_{32}	0·406557	b_{33}	−0·074432	Check $\Sigma b_{3j} \phi_{j3}$	0·929549			s_2	0·02733
d_{30}	1·923893	d_{31}	0·443330	d_{32}	1·406558	d_{33}	0·925568					s_3	0·02430

TABLE 7.1.8.2

The square root scheme

Factors removed: x, 10^q, $q = 1$; y, 10^r, $r = 2$; elements divided by 10^s, $s = 2$

0·670000	0·201730	0·987264	1·465644	0·799000	4·123638	
	0·987264	1·465644	4·364756	−0·092210	6·927184	
		4·364756	9·916960	1·132633	17·867257	
			27·228118	0·929550	43·905028	
						Σy^2 1·218016
0·818535	0·246453	1·206135	1·790570	0·976134	5·037827	Σv_0^2 0·265178
	0·962562	1·213832	4·076064	−0·345724	5·906733	Σv_1^2 0·145653
		1·198585	2·344126	0·312814	3·855526	Σv_2^2 0·047800
			1·383023	−0·102940	1·280084	Σv_3^2 0·037203
0·923896	−0·556667	0·406554	−0·074431	Check	$\Sigma b_{3j}\phi_{j3}$ 0·929551	
1·923896	0·443329	1·406553	0·925570			

7.1.8.1 *The abbreviated Doolittle scheme.* Table 7.1.8.1 shows the evaluation of the coefficients using the abbreviated Doolittle scheme. The values d_{3j} have been evaluated in the back solution as a check on the formation of the coefficients b_{3j}. Comparison with Table 7.1.8 shows that the quantities calculated in the two forms of the Doolittle scheme are the same.

7.1.8.2 *The square root scheme.* Table 7.1.8.2 shows the calculation of the coefficients using the square root scheme. The check column is used in the same way as in the Doolittle scheme.

7.2 ORTHOGONAL POLYNOMIALS

The orthogonal polynomials $T_j(x)$, of degree j in x, for a set of points x_i with associated weights w_i, are defined by the equations

$$\sum_i w_i T_j(x_i) T_k(x_i) = 0, \quad j \neq k. \tag{1}$$

These equations determine the polynomials completely, except for an arbitrary constant factor. The factor will be chosen so that the coefficient of x^j in the expansion of $T_j(x)$ in powers of x is unity. This expansion will be written

$$T_j(x) = \sum_{k=0}^{j} \beta_{kj} x^k, \tag{2a}$$

and the corresponding expansion of x^j as a series in $T_k(x)$ will be written

$$x^j = \sum_{k=0}^{j} \alpha_{kj} T_k(x), \tag{2b}$$

with

$$\beta_{jj} = \alpha_{jj} = 1. \tag{2c}$$

Equation (2*b*) may be regrouped in the form

$$T_j(x) = x^j - \sum_{k=0}^{j-1} \alpha_{kj} T_k(x). \tag{2d}$$

To find the relation between α_{kj} and β_{kj}, the expression (2*a*) for $T_j(x)$ is substituted in (2*b*) to give

$$x^j = \sum_{k=0}^{j} \alpha_{kj} \sum_{m=0}^{k} \beta_{mk} x^m = \sum_{m=0}^{j} \sum_{k=m}^{j} \beta_{mk} \alpha_{kj} x^m,$$

or

$$\sum_{k=m}^{j} \beta_{mk} \alpha_{kj} = \delta_{mj} \begin{Bmatrix} = 1, & m = j \\ = 0, & m \neq j \end{Bmatrix}, \tag{3a}$$

where δ_{mj} is the Kronecker delta. Thus

$$\beta_{mj} = -\sum_{k=m}^{j-1} \beta_{mk} \alpha_{kj}. \tag{3b}$$

Alternatively, (2*b*) can be substituted in (2*a*), giving

$$T_j(x) = \sum_{m=0}^{j} \sum_{k=m}^{j} \alpha_{mk}\beta_{kj} T_m(x),$$

and so
$$\sum_{k=m}^{j} \alpha_{mk}\beta_{kj} = \delta_{mj} \tag{3c}$$

and
$$\beta_{mj} = -\sum_{m+1}^{j} \alpha_{mk}\beta_{kj}. \tag{3d}$$

Hence the quantities β_{mj} can be calculated if the quantities α_{kj} are known.

From (2*d*), by reason of the orthogonal property,

$$\Sigma w_i T_j(x_i) T_k(x_i) = \Sigma w_i x_i^j T_k(x_i) - \alpha_{kj} \Sigma w_i T_k^2(x_i) = 0,$$

or
$$\alpha_{kj} = S_{jk}/S_{kk}, \tag{4a}$$

where
$$S_{jk} = \Sigma w_i x_i^j T_k(x_i), \tag{4b}$$

$$S_{kk} = \Sigma w_i x_i^k T_k(x_i) = \Sigma w_i T_k^2(x_i). \tag{4c}$$

Now the S_{jk} can be calculated by the recurrence relation

$$S_{jk} = \Sigma w_i x_i^j \left\{ x_i^k - \sum_{m=0}^{k-1} \alpha_{mk} T_m(x_i) \right\},$$

i.e.
$$S_{jk} = \phi_{jk} - \Sigma \alpha_{mk} S_{jm} = \alpha_{kj} S_{kk} \tag{4d}$$

with
$$\phi_{jk} = \Sigma w_i x_i^j x_i^k. \tag{4e}$$

Hence the quantities S_{jk}, α_{kj}, are identical with the corresponding quantities defined by (7.1.3,2*a*) which occur in the Doolittle scheme. Thus the α_{kj} can be calculated by this scheme, and the coefficients β_{mj} by (3*b*).

7.2.1 *Independence of origin*

It can be shown that the value of $T_j(x)$ is independent of the origin of the variable x. If the variable is changed to

$$x' = x + \xi,$$

then for the new variable the orthogonal polynomial will be

$$\begin{aligned} T_j'(x') &= x'^j - \sum_k \alpha_{kj}' T_k'(x') \\ &= x'^j - \sum_k \left\{ \sum_i w_i x_i'^j T_k'(x_i') / \Sigma w_i T_k'^2(x_i') \right\} T_k'(x'). \end{aligned}$$

If it is assumed that $T_k'(x') = T_k(x)$ for all $k < j$, this becomes

$$T_j'(x') = (x+\xi)^j - \sum_k \left[\sum_i w_i (x_i + \xi)^j T_k(x_i) / \Sigma w_i T_k^2(x_i) \right] T_k(x). \tag{1}$$

But $(x+\xi)^j - x^j$ can be expanded in powers of $T_k(x)$, in the form

$$(x+\xi)^j - x^j = \sum_k \alpha''_{kj} T_k(x),$$

where $$\sum_i w_i T_k(x_i)\{(x_i+\xi)^j - x_i^j\} = \alpha''_{kj} \Sigma w_i T_k^2(x_i).$$

So on substituting $$x^j + \sum_k \alpha''_{kj} T_k(x)$$

for $(x+\xi)^j$ in (1),

$$T'_j(x') = x^j - \sum_k \left[\sum_i w_i x_i^j T_k(x_i)/\Sigma w_i T_k^2(x_i)\right] T_k(x) = T_j(x),$$

and the equality of the values of the polynomial in the two different variables follows by induction.

7.2.2 *The fitted curve in terms of orthogonal polynomials*

The fitted curve will be written as

$$u_p(x) = \sum_{j=0}^{p} a_j T_j(x) = \sum_{j=0}^{p} b_{pj} x^j. \tag{1}$$

Substituting for x^j the expression (7.2,2*b*),

$$u_p(x) = \sum_{k=0}^{p} b_{pk} \sum_{j=0}^{k} \alpha_{jk} T_j(x) = \sum_{j=0}^{p} \sum_{k=j}^{p} \alpha_{jk} b_{pk} T_j(x),$$

and so $$a_j = \sum_{k=j}^{p} \alpha_{jk} b_{pk}. \tag{2}$$

Comparing this with (7.1.3,2*c*), the a_j are identical with the corresponding quantities in the Doolittle scheme.

If (7.2,2*a*) is substituted in (1),

$$u_p(x) = \sum_{k=0}^{p} a_k \sum_{j=0}^{k} \beta_{jk} x^j = \sum_{j=0}^{p} \sum_{k=j}^{p} \beta_{jk} a_k x^j,$$

and so $$b_{pj} = \sum_{k=j}^{p} \beta_{jk} a_k. \tag{3}$$

One advantage of the orthogonal form is that the coefficients a_j are independent of the degree of the polynomial, so that, if the degree is changed from p to $p+1$, the only effect is to add an additional term $a_{p+1} T_{p+1}(x)$ to $u_p(x)$. With the power-series representation all the b_{pj} are altered to new values $b_{p+1,j}$. A second advantage arises from the independence of the coefficients a_j, which considerably simplifies the discussion of standard deviations.

The quantities $\mathscr{M}_j$ occurring in the Doolittle scheme are, from (7.1.3,2*b*), related to the moments M_j by the equation

$$M_j = \sum_{k=0}^{j} \alpha_{kj} \mathscr{M}_k. \tag{4a}$$

TABLE

The Doolittle scheme

Factors removed: x, 10^q, $q = 1$; y, 10^r, $r = 2$; elements divided by 10^s, $s = 2$

β_{00} 1	1. Enter	ϕ_{00} 0·67	ϕ_{01} 0·20173
R_{00} 1·492537	2. $\div\phi_{00}$	α_{00} 1	α_{01} 0·301090
$R_{00}\,\phi_{00}$	Check		
		3. Enter	ϕ_{11} 0·987264
α_{01} 0·301090		4. $1 \times \alpha_{01}$	0·060739
β_{01} −0·301090	β_{11} 1	5. Subtract	S_{11} 0·926525
R_{10} −0·324967	R_{11} 1·079302	6. $\div S_{11}$	α_{11} 1
$\Sigma R_{1j}\,\phi_{0j}$		Check	
			7. Enter
α_{02} 1·473528			8. $1 \times \alpha_{02}$
−0·379688	α_{12} 1·261044		9. $5 \times \alpha_{12}$
β_{02} −1·093840	β_{12} −1·261044	β_{22} 1	10. Subtract
R_{20} −0·761406	R_{21} −0·877795	R_{22} 0·696086	11. $\div S_{22}$
$\Sigma R_{2j}\,\phi_{0j}$			Check
α_{03} 2·187528			
−1·274996	α_{13} 4·234602		
−2·139271	−2·466279	α_{23} 1·955744	
β_{03} +1·226739	β_{13} −1·768323	β_{23} −1·955744	β_{33} 1
R_{30} +0·641349	R_{31} −0·924493	R_{32} −1·022478	R_{33} 0·522808
$\Sigma R_{3j}\,\phi_{0j}$			

$a_0 =$	b_{00} 1·192537			
$+\beta_{j1}\,a_1$	0·108143	a_1 −0·359171		
	b_{10} 1·300680	b_{11} −0·359171		
Check	$\Sigma b_{1j}\,\phi_{j1}$			
$+\beta_{j2}\,a_2$	−0·285478	−0·329116	a_2 0·260987	
	b_{20} 1·015202	b_{21} −0·688287	b_{22} 0·260987	
Check	$\Sigma b_{2j}\,\phi_{j2}$			
$+\beta_{j3}\,a_3$	−0·091307	+0·131618	+0·145568	a_3 −0·074431
	b_{30} 0·923895	b_{31} −0·556669	b_{32} +0·406555	b_{33} −0·074431
Check	$\Sigma b_{3j}\,\phi_{j3}$			

		$j = 0$			
	p	χ_{00}	χ_{01}	χ_{02}	χ_{03}
R_{00}	0	1·492537			
$+\beta_{j1}\,R_{1k}$		+0·097844	−0·324967		
	1	1·590381	−0·324967		
$+\beta_{j2}\,R_{2k}$		+0·832856	+0·960167	−0·761406	
	2	2·423237	+0·635200	−0·761406	
$+\beta_{j3}\,R_{3k}$		+0·786768	−1·134112	−1·254314	+0·641349
	3	3·210005	−0·498912	−2·015720	+0·641349
		Check	$\sum_k \phi_{jk}\,\chi_{jk}$		

7.2.3

using the coefficients β_{kj}

								Σy^2	1·218016
ϕ_{02}	0·987264	ϕ_{03}	1·465644	M_0	0·7990	C_0	4·123638	$M_0\,a_0$	0·952837
α_{02}	1·473528	α_{03}	2·187528	a_0	1·192537	c_0	6·154684	Σv_0^2	0·265179
							683		
ϕ_{12}	1·465644	ϕ_{13}	4·364756	M_1	−0·09221	C_1	6·927184	$\mathscr{M}_1\,a_1$	0·119525
	0·297255		0·441291		0·240571		1·241586	Σv_1^2	0·145654
S_{21}	1·168389	S_{31}	3·923465	$\mathscr{M}_1$	−0·332781	$\mathscr{C}_1$	5·685598	(s_1)	
α_{12}	1·261044	α_{13}	4·234602	a_1	−0·359171	c_1	6·136476		
							475		
ϕ_{22}	4·364756	ϕ_{23}	9·916960	M_2	1·132633	C_2	17·867257	$\mathscr{M}_2\,a_2$	0·097853
	1·454761		2·159667		1·177349		6·076296	Σv_2^2	0·047801
	1·473390		4·947662		−0·419651		7·169789	(s_2)	0·02733
S_{22}	1·436605	S_{32}	2·809631	$\mathscr{M}_2$	0·374935	$\mathscr{C}_2$	4·621172		
α_{22}	1	α_{23}	1·955744	a_2	0·260987	c_2	3·216731		
							731		
12. Enter		ϕ_{33}	27·228118	M_3	0·929550	C_3	43·905028	$\mathscr{M}_3\,a_3$	0·010597
13. $1\times\alpha_{03}$			3·206137		1·747835		9·020574	Σv_3^2	0·037204
14. $5\times\alpha_{13}$			16·614313		−1·409195		24·076245	(s_3)	0·02430
15. $10\times\alpha_{23}$			5·494919		0·733277		9·037829		
16. Subtract		S_{33}	1·912749	$\mathscr{M}_3$	−0·142367	$\mathscr{C}_3$	1·770380		
17. $\div S_{33}$		α_{33}	1	a_3	−0·074431	c_3	0·925568		
Check							569		

$j = 1$			$j = 2$		$j = 3$
χ_{11}	χ_{12}	χ_{13}	χ_{22}	χ_{23}	χ_{33}
1·079302					
1·079302					
+1·106938	−0·877795		0·696086		
2·186240	−0·877795		0·696086		
+1·634802	+1·808071	−0·924493	+1·999705	−1·022478	0·522808
3·821042	+0·930276	−0·924493	2·695791	−1·022478	0·522808

TABLE

The abbreviated Doolittle scheme

Factors removed: x, 10^q, $q = 1$; y, 10^r, $r = 2$; elements divided by 10^s, $s = 2$

		ϕ_{00}	0·670000	ϕ_{01}	0·201730	ϕ_{02}	0·987264	ϕ_{03}	1·465644
				ϕ_{11}	0·987264	ϕ_{12}	1·465644	ϕ_{13}	4·364756
						ϕ_{22}	4·364756	ϕ_{23}	9·916960
								ϕ_{33}	27·228118
β_{00}	1	S_{00}	0·670000	S_{10}	0·201730	S_{20}	0·987264	S_{30}	1·465644
R_{00}	1·492537	α_{00}	1	α_{01}	0·301090	α_{02}	1·473528	α_{03}	2·187528
	Check								
β_{01}	−0·301090	β_{11}	1	S_{11}	0·926525	S_{21}	1·168389	S_{31}	3·923465
R_{10}	−0·324967	R_{11}	1·079302	α_{11}	1	α_{12}	1·261044	α_{13}	4·234603
	Check								
β_{02}	−1·093840	β_{12}	−1·261044	β_{22}	1	S_{22}	1·436605	S_{32}	2·809631
R_{20}	−0·761406	R_{21}	−0·877795	R_{22}	0·696086	α_{22}	1	α_{23}	1·955743
	Check								
β_{03}	1·226739	β_{13}	−1·768325	β_{23}	−1·955743	β_{33}	1	S_{33}	1·912748
R_{30}	0·641349	R_{31}	−0·924494	R_{32}	−1·022478	R_{33}	0·522808	α_{33}	1
	Check								
b_{00}	1·192537	Check							
b_{10}	1·300680	b_{11}	−0·359171	Check					
b_{20}	1·015202	b_{21}	−0·688287	b_{22}	0·260987	Check			
b_{30}	0·923893	b_{31}	−0·556667	b_{32}	0·406557	b_{33}	−0·074432		Check

Degree p	χ_{00}	χ_{01}	χ_{02}	χ_{03}	χ_{11}
0	1·492537				
1	1·590381	−0·324967			1·079302
2	2·423238	+0·635199	−0·761406		2·186240
3	3·210005	−0·498913	−2·015719	+0·641349	3·821046
	Check	$\Sigma\phi_{0k}\chi_{0k}$			Check

Hence

$$\sum_{j=0}^{q} \beta_{jq} M_j = \sum_{j=0}^{q} \beta_{jq} \sum_{k=0}^{j} \alpha_{kj} \mathscr{M}_k = \sum_{k=0}^{q} \sum_{j=k}^{q} \alpha_{kj} \beta_{jq} \mathscr{M}_k,$$

and so, from (7.2,3*c*),

$$\sum_{j=0}^{q} \beta_{jq} M_j = \mathscr{M}_q = a_q S_{qq}. \tag{4b}$$

If the values $\Sigma w_i y_i x_i^j$ are substituted for M_j in (4*b*),

$$\mathscr{M}_q = \sum_i w_i y_i T_q(x_i). \tag{5}$$

Hence the quantities $\mathscr{M}_q$ occurring in the Doolittle scheme are the orthogonal moments.

7.2.3*a*

using the coefficients β_{kj}

M_0	0·799000	C_0	4·123638		
M_1	−0·092210	C_1	6·927184		
M_2	1·132633	C_2	17·867257		
M_3	0·929550	C_3	43·905028		
				Σy^2	1·218016
$\mathscr{M}_0$	0·799000	$\mathscr{C}_0$	4·123638	$a_0 \mathscr{M}_0$	0·952837
a_0	1·192537	c_0	6·154684	Σv_0^2	0·265179
			683		
$\mathscr{M}_1$	−0·332781	$\mathscr{C}_1$	5·685598	$a_1 \mathscr{M}_1$	0·119525
a_1	−0·359171	c_1	6·136475	Σv_1^2	0·145654
			476		
$\mathscr{M}_2$	0·374936	$\mathscr{C}_2$	4·621172	$a_2 \mathscr{M}_2$	0·097853
a_2	0·260987	c_2	3·216731	Σv_2^2	0·047801
			730		
$\mathscr{M}_3$	−0·142370	$\mathscr{C}_3$	1·770379	$a_3 \mathscr{M}_3$	0·010597
a_3	−0·074432	c_3	0·925568	Σv_3^2	0·037204
			568		

χ_{12}	χ_{13}	χ_{22}	χ_{23}	χ_{33}
−0·877795		0·696086		
0·930278	−0·924494	2·695790	−1·022478	0·522808
$\Sigma\phi_{1k}\chi_{1k}$		Check	$\Sigma\phi_{2k}\chi_{2k}$	$\Sigma\phi_{3k}\chi_{3k}$

7.2.3 *Example*

The various forms of the Doolittle scheme can be extended to include the calculations of the quantities β_{kj}. Table 7.2.3 shows the calculations for the full Doolittle scheme. The β_{kj} are evaluated from the relation (7.2,3*b*) in the triangular region at the left of the scheme. The instructions are the same as for the forward scheme, and the calculations may be done at the same time. It should be noted, however, that the two sections are completely separate; thus in line 14, products $\alpha_{13} S_{11}$ do not occur. The checking of the β_{kj} is also done separately. If the quantities $R_{kj} = \beta_{jk}/S_{jj}$ are formed for use in later standard deviation calculations, the values may be checked by evaluating

$$\sum_j R_{kj}\phi_{0j} = 0, \quad k > 0. \tag{1a}$$

Otherwise the equation

$$\sum_j \beta_{jk}\phi_{0j} = 0 \tag{1b}$$

may be used. The proof of these two equations is given in § 7.3.1.

The back section of the Doolittle scheme is replaced by a scheme using the values β_{jk} in which the coefficients b_{pj} for all values of the degree p from 0 to 3 are obtained.

The lowest section of Table 7.2.3 contains the calculations for the elements χ_{jk} of the inverse matrices. These calculations are discussed in § 7.3.4.

The abbreviated Doolittle scheme can be modified in exactly the same way. The scheme is shown in Table 7.2.3a. There is no advantage in calculating the quantities β_{jk} at the same time as the forward section of the Doolittle scheme, and the calculations of the β_{jk} may be left till the forward section is completed. The lowest section of the table gives the calculations for the inverse matrices.

7.2.4 *The square root method*

In the square root method a quantity

$$s_{jk} = \alpha_{kj}\, s_{kk} = S_{jk}/s_{kk} \tag{1}$$

replaces α_{kj} and S_{jk}. Similarly, a quantity

$$r_{jk} = \beta_{kj}/s_{jj} = s_{jj}\, R_{jk} \tag{2}$$

will replace the quantities β_{kj} and R_{jk}.

The equation (7.2,3a) relating α_{kj} and β_{kj} becomes

$$\sum_{k=m}^{j} \beta_{mk}\, \alpha_{kj} = \sum_{k=m}^{j} (r_{km}\, s_{kk})\,(s_{jk}/s_{kk}) = \sum_{k=m}^{j} s_{jk}\, r_{km} = \delta_{jm}. \tag{3}$$

The equation for r_{jm} is then

$$r_{jm} = -\sum_{k=m}^{j-1} s_{jk}\, r_{km}/s_{jj}. \tag{4}$$

The equation (7.2.2,3),

$$b_{pj} = \sum_{k=j}^{p} \beta_{jk}\, a_k,$$

for the power-series coefficients becomes

$$b_{pj} = \sum_{k=j}^{p} r_{kj}\, m_k. \tag{5}$$

The orthogonal polynomial $T_j(x)$ is

$$T_j(x) = \Sigma\beta_{kj}\, x^k = s_{jj}\, \Sigma r_{jk}\, x^k,$$

and so $$\sum_i w_i\left(\sum_k r_{jk}\, x_i^k\right)^2 = s_{jj}^{-2} \sum_i w_i\, T_j^2(x_i) = 1.$$

The polynomials $\Sigma r_{jk}\, x^k$ are then the orthogonal polynomials for which the arbitrary constants are chosen so that the weighted sums of the squares of the values at the points of observation are unity.

7.2.4.1 *Example*

The calculations of r_{jk} and b_{pk} for the previous example are shown in Table 7.2.4. The r_{jk} are formed in a triangular array at the left of the scheme.

The formation of the r_{jk} may be checked by the equation

$$\Sigma r_{kj}\,\phi_{0j} = 0, \tag{1}$$

which is obtained by substituting r_{kj} for β_{jk} in (7.2.3,1*b*).

The lowest section gives the elements of the inverse matrices for different values of p (§ 7.3.4.1).

7.2.4.2 *Orthogonal polynomials with high-speed computers.* When an automatic computer is available, it becomes feasible to calculate the orthogonal polynomials

$$t_j(x) = \sum_{k=0}^{j} r_{jk}\,x^k \tag{1}$$

at each point of observation, and thence to obtain the fitted curve in the form

$$u_p(x) = \sum_{j=0}^{p} a_j\,t_j(x), \tag{2a}$$

where

$$a_j = \sum_i w_i\,y_i\,t_j(x_i), \tag{2b}$$

the sum $\Sigma w_i\,t_j^2(x_i)$ being unity.

In the scheme developed by Davis and Rabinowitz (1954), $t_j(x)$ is calculated from the inverse of (1),

$$x^j = \sum_{k=0}^{j} s_{jk}\,t_k(x),$$

that is,

$$t_j(x) = \left\{x^j - \sum_{k=0}^{j-1} s_{jk}\,t_k(x)\right\} \Big/ s_{jj}, \tag{3a}$$

where

$$s_{jj}^2 = \sum_i w_i\,x_i^{2j} - \sum_{k=0}^{j-1} s_{jk}^2 \tag{3b}$$

and

$$s_{jk} = \sum_i w_i\,x_i^j\,t_k(x_i). \tag{3c}$$

The calculation of the orthogonal polynomials by means of the equations (3*a*–*c*) is often referred to as the Gram–Schmidt orthonormalization process. Some detailed applications of this process to various problems, using the SEAC computer at the National Bureau of Standards, are given in the reference cited.

An alternative method of calculating the orthogonal polynomials with a high-speed computer is described in § 7.2.6.

TABLE

The square root scheme

Factors removed: x, $10^q, q = 1$; y, $10^r, r = 2$; elements divided by $10^s, s = 2$				
		0·670000	0·201730 0·987264	0·987264 1·465644 4·364756
	1·221695	0·818535	0·246453	1·206135
	−0·312801	1·038894	0·962562	1·213832
	−0·912610	−1·052110	0·834317	1·198585
	0·886998	−1·278590	−1·414108	0·723054
Degree p	b_{p0}	b_{p1}	b_{p2}	b_{p3}
0	1·192538			
1	1·300681	−0·359171		
2	1·015204	−0·688285	0·260986	
3	0·923896	−0·556667	0·406554	−0·074431
Degree p	χ_{00}	χ_{01}	χ_{02}	χ_{03}
0	1·492539			
1	1·590383	−0·324967		
2	2·423240	+0·635199	−0·761406	
3	3·210006	−0·498908	−2·015717	+0·641347

7.2.5 *The use of residuals as a check on the arithmetical calculations*

The residuals v_i are the quantities

$$v_i = y_i - u_p(x_i) = y_i - \sum_{j=0}^{p} a_j T_j(x_i). \tag{1}$$

Now, from (7.2.2,4*b*) and (7.2.2,5),

$$\mathscr{M}_j = \Sigma w_i y_i T_j(x_i) = a_j S_{jj} = a_j \Sigma w_i T_j^2(x_i), \tag{2}$$

and so, using this equation and the orthogonal property, $\Sigma w_i v_i^2$ can be reduced to the forms

$$\Sigma w_i v_i^2 = \Sigma w_i y_i^2 - \sum_{j=0}^{p} a_j \mathscr{M}_j; \tag{3a}$$

$$\Sigma w_i v_i^2 = \Sigma w_i y_i^2 - \sum_{j=0}^{p} a_j^2 S_{jj}; \tag{3b}$$

$$\Sigma w_i v_i^2 = \Sigma w_i y_i^2 - \sum_{j=0}^{p} \mathscr{M}_j^2 / S_{jj}. \tag{3c}$$

7.2.4

using the coefficients r_{jk}

1·465644	0·799000	4·123638	
4·364756	−0·092210	6·927184	
9·916960	1·132633	17·867257	
27·228118	0·929550	43·905028	
			Σy^2 1·218016
1·790570	0·976134	5·037827	Σv_0^2 0·265178
4·076064	−0·345724	5·906733	Σv_1^2 0·145653
2·344126	0·312814	3·855526	Σv_2^2 0·047800
1·383023	−0·102940	1·280084	Σv_3^2 0·037203

χ_{11}	χ_{12}	χ_{13}	χ_{22}	χ_{23}	χ_{33}
1·079301					
2·186236	−0·877793		0·696085		
3·821029	+0·930271	−0·924490	2·695786	−1·022476	0·522807

The form (3*a*) is that used in the Doolittle calculating schemes. A fourth form

$$\Sigma w_i v_i^2 = \Sigma w_i y_i^2 - \sum_{j=0}^{p} b_{pj} M_j \tag{3d}$$

follows from the identity

$$\sum_{j=0}^{p} b_{pj} M_j = \sum_{j=0}^{p} a_j \mathcal{M}_j. \tag{4}$$

This identity is verified by noting that each side is equal to $\Sigma w_i y_i u_p(x_i)$; for

$$\sum_i w_i y_i u_p(x_i) = \sum_i w_i y_i \sum_j b_{pj} x_i^j = \sum_i w_i y_i \sum_j a_j T_j(x_i).$$

If the residuals are also calculated by evaluating $y_i - u_p(x_i)$ for all points of observation x_i, the agreement of the values for $\Sigma w_i v_i^2$ obtained by direct calculation and by use of (3*a*) will provide an excellent check on the arithmetical calculations.

7.2.5.1 *Example*

A suggested method of calculating the fitted values and residuals is shown in Table 7.2.5. The calculations are for the 20 points of Table 7.1.1*a*. The values b_{pj} are entered at the top of the columns from Table 7.1.8 or Table 7.2.3. The entries in the columns are the products of b_{pj} and the values x^j given in Table 7.1.1*a*. As a check, each column is summed and the sum compared with the product of b_{pj} and the value Σx^j; these should agree except for rounding-off errors.

TABLE 7.2.5

Calculation of residuals

b_{3j}	0·924	− 0·5567	+ 0·4066	− 0·07443		
	$b_{30}\, x_i^0$	$b_{31}\, x_i$	$b_{32}\, x_i^2$	$b_{33}\, x_i^3$	$u_3\, (x_i)$	v_{3i}
	0·924	− 1·648	+ 3·562	− 1·930	0·908	+0·172
		− 1·579	+ 3·270	− 1·698	0·917	+0·103
		− 1·501	+ 2·955	− 1·459	0·919	+0·111
		− 1·436	+ 2·704	− 1·277	0·915	−0·175
		− 1·423	+ 2·656	− 1·243	0·914	−0·664
		− 1·355	+ 2·409	− 1·073	0·905	+0·125
		− 1·285	+ 2·168	− 0·916	0·891	+0·149
		− 1·081	+ 1·532	− 0·544	0·831	−0·171
		− 0·957	+ 1·202	− 0·378	0·791	−0·441
		− 0·926	+ 1·126	− 0·343	0·781	+0·949
		− 0·879	+ 1·014	− 0·293	0·766	+0·084
		− 0·877	+ 1·009	− 0·291	0·765	+0·065
		− 0·801	+ 0·842	− 0·222	0·743	−0·053
		− 0·797	+ 0·834	− 0·219	0·742	+0·388
		− 0·778	+ 0·795	− 0·203	0·738	−0·288
		− 0·698	+ 0·639	− 0·147	0·718	−0·038
		− 0·640	+ 0·537	− 0·113	0·708	+0·172
		− 0·590	+ 0·457	− 0·089	0·702	−0·132
		− 0·512	+ 0·343	− 0·058	0·697	−0·117
		− 0·509	+ 0·340	− 0·057	0·698	+0·092
Sum Check	18·480	−20·272	30·394	−12·553	16·049	0·331

The rows of values $b_{pj}\, x_i^j$ are summed to give the fitted values $u_p(x_i)$. If there are no mistakes, $\sum_i u_p(x_i)$ should be identical with

$$\sum_j \left\{ b_{pj} \sum_i x_i^j \right\}.$$

The residuals v_i are next calculated by subtracting the fitted values $u_p(x_i)$ from the original observations y_i. These are checked by summing the v_i column and using the identity

$$\Sigma v_i = \Sigma y_i - \Sigma u_p(x_i),$$

Σy_i being obtained from Table 7.1.1*a*.

The residuals are calculated in a similar way for the two other sub-groups into which the observations were divided. The final sum Σv_i for the full set of observations should be zero, except for rounding-off errors. In the present example a value $-0{\cdot}012$ is obtained. The sum of the squares of the residuals is then formed. The value $3{\cdot}720544$ agrees very well with the value $0{\cdot}037204 \times 10^2$ given in Table 7.1.8 by use of (7.2.5,3*a*). Hence it seems reasonably certain that no arithmetical mistakes have been made.

7.2.6 *Recurrence relations*

Consider the sums

$$\sum_i w_i T_k(x_i)\, x_i T_j(x_i) \tag{1}$$

for a fixed value of j. The product $xT_k(x)$ can be expressed as the sum of polynomials $T_q(x)$ of degrees up to $k+1$. Hence when $k<j-1$, the sum (1) vanishes because of the orthogonal property. Also $xT_j(x)$ can be expressed in terms of the polynomials $T_q(x)$ of degrees up to $j+1$, and so (1) vanishes when $k>j+1$. Thus

$$\Sigma w_i T_k(x_i)\,[x_i T_j(x_i)] = 0, \quad k>j+1,\ k<j-1,$$

and so the expansion of $xT_j(x)$ in terms of orthogonal polynomials must be of the form

$$xT_j(x) = T_{j+1}(x) + \pi_j T_j(x) + \rho_j T_{j-1}(x). \tag{2a}$$

This gives a recurrence equation for the orthogonal polynomials of the form

$$T_{j+1}(x) = xT_j(x) - \pi_j T_j(x) - \rho_j T_{j-1}(x). \tag{2b}$$

To find π_j, (2*a*) is multiplied by $w_i T_j(x_i)$ for each value $x = x_i$ and the products summed, giving

$$\pi_j = \Sigma w_i x_i T_j^2(x_i)/\Sigma w_i T_j^2(x_i). \tag{3}$$

Similarly, multiplication by $w_i T_{j-1}(x_i)$ leads to the equation

$$\Sigma w_i T_j(x_i)\, x_i T_{j-1}(x_i) = \rho_j \Sigma w_i T_{j-1}^2(x_i).$$

But the left-hand side of this equation is

$$\Sigma w_i T_j(x_i)\,[x_i T_{j-1}(x_i)] = \Sigma w_i T_j^2(x_i),$$

and so

$$\rho_j = \Sigma w_i T_j^2(x_i)/\Sigma w_i T_{j-1}^2(x_i). \tag{4}$$

A recurrence relation for the coefficients β_{kj} can be established by substituting $T_q(x) = \Sigma \beta_{kq} x^k$ in (2*b*) and equating corresponding powers of x on each side. It is found that

$$\beta_{k+1,\,j+1} = \beta_{kj} - \pi_j \beta_{k+1,\,j} - \rho_j \beta_{k+1,\,j-1}. \tag{5}$$

When a high-speed computer is used, the recurrence relation (2*b*) enables the orthogonal polynomials to be calculated at each point of observation. Forsythe (1957) has given a discussion of this method. He also recommends in this case that the variable x should be re-scaled so that the values x_i cover the range $+2$ to -2.

It is shown in § 7.7.3 that for equally-spaced observations of equal weight ρ_j is approximately $n^2/16$ when x_i covers the range $+\frac{1}{2}(n-1)$ to $-\frac{1}{2}(n-1)$. Since $T_j(x)$ varies as x^j when the scale of x is changed, ρ_j will, from (4), vary as x^2 and so ρ_j will be close to unity when x_i covers the range $+2$ to -2. The re-scaling will then ensure that $T_j(x)$ and $T_{j-1}(x)$ will be of the same order of magnitude.

7.3 MATRIX NOTATION

The solution of the normal equations by the Doolittle method can be considered in terms of the factorization into two triangular matrices of the square matrix whose elements are ϕ_{jk}. If $\boldsymbol{\phi}$ is the $(p+1)\times(p+1)$ matrix whose elements are ϕ_{jk}, $\mathbf{b}$ and $\mathbf{M}$ the column vectors whose elements are b_{pj} and M_j, then the quantities

$$\sum_j \phi_{kj} b_{pj}$$

are just the terms of the product $\boldsymbol{\phi}\mathbf{b}$, and so the normal equations (7.1,3) are

$$\boldsymbol{\phi}\mathbf{b} = \mathbf{M}. \tag{1}$$

The symbol $\mathbf{S}$ denotes the lower triangular matrix with elements $S_{rq}, r \geqslant q$, and zeros above the principal diagonal, $S_{rq} = 0$, $r < q$; and the symbol $\boldsymbol{\alpha}$ denotes the upper triangular matrix with elements $\alpha_{qk}, q \leqslant k$, and zeros below the principal diagonal, $\alpha_{qk} = 0, q > k$. That is

$$\mathbf{S} = \begin{bmatrix} S_{00} & 0 & 0 & \cdots \\ S_{10} & S_{11} & 0 & \cdots \\ S_{20} & S_{21} & S_{22} & \cdots \\ \cdots & \cdots & \cdots & \cdots \end{bmatrix}, \quad \boldsymbol{\alpha} = \begin{bmatrix} \alpha_{00} & \alpha_{01} & \alpha_{02} & \cdots \\ 0 & \alpha_{11} & \alpha_{12} & \cdots \\ 0 & 0 & \alpha_{22} & \cdots \\ \cdots & \cdots & \cdots & \cdots \end{bmatrix}.$$

Then (7.1.2,6*a*) and (7.1.2,6*b*) give

$$\phi_{rk} = \sum_{q=0}^{k,\,r} S_{rq}\alpha_{qk} = \sum_{q=0}^{p} S_{rq}\alpha_{qk},$$

i.e.

$$\boldsymbol{\phi} = \mathbf{S}\boldsymbol{\alpha}. \tag{2}$$

Thus the Doolittle process effectively factorizes $\boldsymbol{\phi}$ into two triangular matrices $\mathbf{S}$ and $\boldsymbol{\alpha}$.

Similarly, (7.1.2,6*c*) is

$$M_r = \sum_{q=0}^{r} S_{rq} a_q,$$

i.e.

$$\mathbf{M} = \mathbf{S}\mathbf{a}. \tag{3}$$

Then, from (1),

$$\mathbf{S\alpha b} = \mathbf{Sa}$$

or

$$\mathbf{\alpha b} = \mathbf{a}, \tag{4}$$

which is just (7.1.2,8). If the solution of these equations is written

$$\mathbf{b} = \mathbf{\beta a}, \tag{5}$$

then

$$\mathbf{\alpha\beta} = \mathbf{I}, \tag{6}$$

where $\mathbf{I}$ is the unit matrix. These are (7.2.2,3) and (7.2,3*a*–*d*) respectively. Since, when $j \geqslant m$,

$$\sum_{k=j}^{p} \alpha_{jk}\beta_{km} = \delta_{jm},$$

it follows by taking in turn $j = p, p-1, \ldots, m+1, m$, that

$$\beta_{pm}, \beta_{p-1,m}, \ldots, \beta_{m+1,m}$$

all vanish and $\beta_{mm} = 1$. Hence β_{km} is an upper triangular matrix.

7.3.1 *The check column*

If $\mathbf{i}$ is a unit column vector, the check elements $\mathbf{C}$ constitute a column vector

$$\mathbf{C} = \mathbf{\phi i} + \mathbf{M}. \tag{1}$$

Hence, as $\mathbf{C}$ is operated on to give $\mathbf{c}$ in the same way as $\mathbf{M}$ is operated on to give $\mathbf{a}$, from (7.3,3),

$$\mathbf{Sc} = \mathbf{C},$$

or

$$\mathbf{c} = \mathbf{S}^{-1}\mathbf{\phi i} + \mathbf{S}^{-1}\mathbf{M} = \mathbf{\alpha i} + \mathbf{a}. \tag{2}$$

This is just (7.1.6,3). Similarly, the elements d_{pj} form a vector

$$\mathbf{d} = \mathbf{\alpha}^{-1}\mathbf{c} = \mathbf{i} + \mathbf{\alpha}^{-1}\mathbf{a} = \mathbf{i} + \mathbf{b}, \tag{3}$$

which is (7.1.6,5).

The formulae (7.2.3,1) used for checking the calculation of β_{jk} and R_{kj} are established from the equation

$$\mathbf{S\alpha} = \mathbf{\phi},$$

or

$$\mathbf{S} = \mathbf{\phi\beta}.$$

Since $\mathbf{S}$ is a lower triangular matrix,

$$\sum_j \beta_{jk}\phi_{0j} = S_{0k} = 0, \quad k > 0,$$

and, on dividing by S_{jj},

$$\sum_j R_{kj}\phi_{0j} = 0.$$

These are (7.2.3,1*b*) and (7.2.3,1*a*).

7.3.2 *The square root method*

The quantities s_{jk} form a lower triangular matrix. From (7.1.5,2*a*),

$$\boldsymbol{\phi} = \mathbf{s}\mathbf{s}^T, \tag{1}$$

where $\mathbf{s}^T$ is the transpose of $\mathbf{s}$, an upper triangular matrix obtained by interchanging the rows and columns of s, so that $s_{jk} = s^T_{kj}$.

Equation (7.1.5,2*c*) gives

$$\mathbf{M} = \mathbf{sm}, \tag{2}$$

and so the normal equations become

$$\mathbf{s}^T\mathbf{b} = \mathbf{m}, \tag{3}$$

which is (7.1.5,2*d*).

7.3.3 *The inverse matrix*

The matrix which is the inverse of $\boldsymbol{\phi}$ will be denoted by $\boldsymbol{\chi}$,

$$\boldsymbol{\chi} = \boldsymbol{\phi}^{-1}. \tag{1}$$

If the inverse matrix is known, the coefficients b_{pj} are given directly in terms of the moments by the equations

$$b_{pj} = \sum_{k=0}^{p} \chi_{jk} M_k. \tag{2}$$

Hence if a number of curves with the same values of x_i are to be fitted, it may be an advantage to calculate the inverse matrix. The inverse matrix may also be useful in standard error calculations.

7.3.4 *Calculation of the inverse matrix from the values β_{kj}*

The inverse of the matrix $\boldsymbol{\alpha}$ is $\boldsymbol{\beta}$. If the inverse of $\mathbf{S}$ is denoted by $\mathbf{R}$, so that

$$\mathbf{RS} = \mathbf{I}, \tag{1}$$

then

$$\Sigma R_{jk} S_{km} = \delta_{jm}.$$

Now for the symmetrical normal equations,

$$S_{km} = \alpha_{mk} S_{mm},$$

and so

$$\Sigma \alpha_{mk} R_{jk} S_{mm} = \delta_{jm}.$$

But

$$\Sigma \alpha_{mk} \beta_{kj} = \delta_{jm},$$

and so

$$R_{jk} = \beta_{kj}/S_{jj}. \tag{2}$$

Hence the elements of $\mathbf{R}$ are readily calculated (see Tables 7.2.3 and 7.2.3*a*).

The inverse matrix is

$$\boldsymbol{\chi} = \boldsymbol{\phi}^{-1} = \boldsymbol{\alpha}^{-1}\mathbf{S}^{-1} = \boldsymbol{\beta}\mathbf{R}, \tag{3a}$$

or

$$\chi_{jk} = \sum_{q=0}^{p} \beta_{jq} R_{qk} = \sum_{q=j,k}^{p} \beta_{jq}\beta_{kq}/S_{qq}, \tag{3b}$$

the lower limit being the larger of the two values j and k.

7.3.4.1 *Example*

The values R_{jk} were calculated in Tables 7.2.3 and 7.2.3*a*. The lowest section of Table 7.2.3 shows the calculations of the inverse matrices for various values of p, and Table 7.2.3*a* shows the abbreviated form of this calculation. The matrices are symmetrical, and only the elements in the upper half are shown. The elements are formed by adding $\beta_{jp} R_{pk}$ to the element in the matrix of lower degree.

The arithmetical calculations should be checked by forming the sums $\sum_k \phi_{jk}\chi_{jk}$ for $j = 0, 1, 2, 3$. The values should be unity in each case.

When the calculations are done by the square root method,

$$\chi_{jk} = \Sigma r_{qj} r_{qk}. \tag{1}$$

The calculation of the inverse matrices for the square root scheme is shown at the bottom of Table 7.2.4.

7.3.5 *Calculation of the inverse matrix from the values* α_{kj}

The elements of the inverse matrix can also be calculated directly from the equation

$$\boldsymbol{\alpha}\boldsymbol{\chi} = \mathbf{S}^{-1}, \tag{1}$$

although the formulae and the arithmetical calculations are much more complicated than in § 7.3.4. From (7.3.4,2), $\mathbf{R} = \mathbf{S}^{-1}$ is a lower triangular matrix and so, when $k \leqslant j$,

$$S^{-1}_{kj} = \delta_{kj} S^{-1}_{jj}.$$

Hence when $k \leqslant j$,

$$\sum_{m=k}^{p} \alpha_{km}\chi_{mj} = \delta_{kj} S^{-1}_{jj},$$

or

$$\chi_{kj} = \delta_{kj} S^{-1}_{jj} - \sum_{k+1}^{p} \alpha_{km}\chi_{mj}. \tag{2}$$

Thus the values χ_{kj} can be built up in turn from the previously calculated values χ_{mj}, $m > k$.

7.3.5.1 *Example*

Table 7.3.5 shows the calculation of the inverse matrix using the values of α from the Doolittle scheme of Table 7.1.8. In each section the calculations start from right of the page. For example, χ_{13} and χ_{12} are entered using the previously calculated values χ_{31} and χ_{21}; χ_{11} and then χ_{10} are worked out as shown. The calculations are checked by the equation

$$\sum_k \chi_{jk}\phi_{jk} = 1.$$

This scheme is more complicated than that of § 7.3.4, but does not use the quantities β_{kj}. However, since the calculation of the quantities only requires a few extra minutes, it is considered preferable to use the scheme of § 7.3.4 because the possibility of error is much smaller.

TABLE 7.3.5

The inverse matrix using the coefficients α_{kj}

				S_{33}^{-1} =
$\times -\alpha_{j3}$	−1·143657	−2·213884	−1·022479	χ_{33} 0·522808
$\times -\alpha_{j2}$	+1·506651	+1·289391	χ_{32} −1·022479	
$\times -\alpha_{j1}$	+0·278356	χ_{31} −0·924493		
	χ_{30} +0·641350			
Check	$\Sigma\chi_{3j}\phi_{j3}$			
			S_{22}^{-1} 0·696086	
$\times -\alpha_{j3}$	+2·236701	+4·329792	+1·999707	χ_{23} −1·022479
$\times -\alpha_{j2}$	−3·972326	−3·399514	χ_{22} +2·695793	
$\times -\alpha_{j1}$	−0·280097	χ_{21} +0·930278		
	χ_{20} −2·015722			
Check	$\Sigma\chi_{2j}\phi_{j2}$			
		S_{11}^{-1} 1·079302		
$\times -\alpha_{j3}$	+2·022354	+3·914860		χ_{13} −0·924493
$\times -\alpha_{j2}$	−1·370791	−1·173121	χ_{12} +0·930278	
$\times -\alpha_{j1}$	−1·150477	χ_{11} +3·821041		
	χ_{10} −0·498914			
Check	$\Sigma\chi_{1j}\phi_{j1}$			
	S_{00}^{-1} 1·492537			
$\times -\alpha_{03}$	−1·402971			χ_{03} +0·641350
$\times -\alpha_{02}$	+2·970223		χ_{02} −2·015722	
$\times -\alpha_{01}$	+0·150218	χ_{01} −0·498914		
	χ_{00} +3·210007			
Check	$\Sigma\chi_{0j}\phi_{j0}$			

7.3.6 *The omission of observations*

It is occasionally of interest to determine what changes would be brought about by the omission of one or more of the series of observations. Plackett (1950) has developed formulae giving the coefficients and the inverse matrix in terms of the corresponding quantities for the full series of observations. These formulae require the inverse of a $r \times r$ matrix, where r is the number of observations dropped, and so are only useful when r is small. If r is greater than 2 it is probably simplest to recompute ϕ_{jk} and M_j and carry out the calculations afresh using the Doolittle scheme.

A double prime superscript will be used for quantities relating to the r observations which are to be dropped, and a single prime superscript will be used for quantities relating to the new curve fitted to the remaining $n-r$ observations. For example, the new matrix $\boldsymbol{\phi}'$ is

$$\boldsymbol{\phi}' = \boldsymbol{\phi} - \mathbf{X}''\mathbf{W}''\mathbf{X}''^T,$$

where $\mathbf{X}''$ is the $p+1 \times r$ matrix whose elements are $x_i''^j$ and $\mathbf{W}''$ is the $r \times r$ matrix whose elements are w_i''. The elements of the product $\mathbf{X}''\mathbf{W}''\mathbf{X}''^T$ are $\Sigma w_i'' x_i''^j x_i''^k$. The formulae for the new polynomial coefficients, inverse matrix, and sum of the squares of the residuals are

$$\mathbf{b}' = \mathbf{b} - \boldsymbol{\chi}\mathbf{X}''\mathbf{G}\mathbf{v}'', \tag{1a}$$

$$\boldsymbol{\chi}' = \boldsymbol{\chi} + \boldsymbol{\chi}\mathbf{X}''\mathbf{G}\mathbf{X}''^T\boldsymbol{\chi}, \tag{1b}$$

and

$$\mathbf{v}'^T\mathbf{W}'\mathbf{v}' = \mathbf{v}^T\mathbf{W}\mathbf{v} - \mathbf{v}''^T\mathbf{G}\mathbf{v}'', \tag{1c}$$

where $\mathbf{G}$ is the $r \times r$ matrix defined by

$$\mathbf{G} = \mathbf{W}''(\mathbf{I} - \mathbf{X}''^T\boldsymbol{\chi}\mathbf{X}''\mathbf{W}'')^{-1}. \tag{1d}$$

The proof of these results will be deferred to § 7.3.6.2.

If only one observation is to be omitted, $\mathbf{G}$ is the single quantity

$$G = w''(1 - w''\Sigma\Sigma\chi_{jk}x''^j x''^k)^{-1}. \tag{2a}$$

If two observations are to be omitted, the elements of $\mathbf{G}$ are

$$G_{11} = w_1''(1 - w_2''\Sigma\Sigma\chi_{jk}x_2''^j x_2''^k)/D, \quad G_{12} = w_1''w_2''\Sigma\Sigma\chi_{jk}x_1''^j x_2''^k/D, \tag{2b}$$

$$G_{21} = w_1''w_2''\Sigma\Sigma\chi_{jk}x_1''^j x_2''^k/D, \quad G_{22} = w_2''(1 - w_1''\Sigma\Sigma\chi_{jk}x_1''^j x_1''^k)/D, \tag{2c}$$

where

$$D = (1 - w_1''\Sigma\Sigma\chi_{jk}x_1''^j x_1''^k)(1 - w_2''\Sigma\Sigma\chi_{jk}x_2''^j x_2''^k) - w_1''w_2''(\Sigma\Sigma\chi_{jk}x_1''^j x_2''^k)^2. \tag{2d}$$

The following forms of (1a) to (1c) are suitable for purposes of calculation:

$$b_j' = b_j - \sum_i (\Sigma\chi_{jk}x_i''^k)(\Sigma G_{im}v_m''); \tag{3a}$$

$$\chi_{jk}' = \chi_{jk} + \sum_i (\Sigma\chi_{jr}x_i''^r)(\Sigma G_{im}\Sigma\chi_{kr}x_m''^r); \tag{3b}$$

$$\Sigma w_i' v_i'^2 = \Sigma w_i v_i^2 - \sum_i v_i''(\Sigma G_{im}v_m''). \tag{3c}$$

7.3.6.1 *Example*

Table 7.3.6 shows the calculation of the inverse matrix and polynomial coefficients when the two observations at $x = +16{\cdot}64$ and $x = +25{\cdot}56$ are omitted from Example 7.1.1.1. It is necessary to restore the factor 10^{-8} to the elements χ_{jk} in Table 7.2.3*a*.

As a check on the calculations, the values of the moments

$$M'_j = M_j - \Sigma y''_i x''^j_i \tag{1}$$

are calculated. The values

$$b'_j = \Sigma \chi'_{jk} M'_k \tag{2}$$

should agree with those given by (7.3.6,3*a*).

7.3.6.2 *Derivation of equations.* The new matrix is

$$\boldsymbol{\phi}' = \boldsymbol{\phi} - \mathbf{X}''\mathbf{W}''\mathbf{X}''^T. \tag{1}$$

Now $(\mathbf{I} - \mathbf{X}''^T\boldsymbol{\chi}\mathbf{X}''\mathbf{W}'')\mathbf{X}''^T\boldsymbol{\chi}' = \mathbf{X}''^T\boldsymbol{\chi}' - \mathbf{X}''^T\boldsymbol{\chi}(\boldsymbol{\phi} - \boldsymbol{\phi}')\boldsymbol{\chi}',$

or

$$(\mathbf{I} - \mathbf{X}''^T\boldsymbol{\chi}\mathbf{X}''\mathbf{W}'')\mathbf{X}''^T\boldsymbol{\chi}' = \mathbf{X}''^T\boldsymbol{\chi}, \tag{2}$$

and so, if

$$\mathbf{G} = \mathbf{W}''(\mathbf{I} - \mathbf{X}''^T\boldsymbol{\chi}\mathbf{X}''\mathbf{W}'')^{-1}, \tag{3}$$

then

$$\boldsymbol{\chi}\mathbf{X}''\mathbf{G}\mathbf{X}''^T\boldsymbol{\chi} = \boldsymbol{\chi}\mathbf{X}''\mathbf{W}''\mathbf{X}''^T\boldsymbol{\chi}' = \boldsymbol{\chi}(\boldsymbol{\phi} - \boldsymbol{\phi}')\boldsymbol{\chi}' = \boldsymbol{\chi}' - \boldsymbol{\chi}, \tag{4}$$

which is (7.3.6,1*b*).

The moments are connected by the equation

$$\boldsymbol{\phi}\mathbf{b} = \boldsymbol{\phi}'\mathbf{b}' + \mathbf{X}''\mathbf{W}''\mathbf{y}''. \tag{5}$$

By use of (1) and (5),

$$(\mathbf{I} - \mathbf{X}''^T\boldsymbol{\chi}\mathbf{X}''\mathbf{W}'')(\mathbf{y}'' - \mathbf{X}''^T\mathbf{b}') = \mathbf{y}'' - \mathbf{X}''^T\boldsymbol{\chi}(\boldsymbol{\phi}\mathbf{b} - \boldsymbol{\phi}'\mathbf{b}') - \mathbf{X}''^T\mathbf{b}' + \mathbf{X}''^T\boldsymbol{\chi}(\boldsymbol{\phi} - \boldsymbol{\phi}')\mathbf{b}',$$

or

$$(\mathbf{I} - \mathbf{X}''^T\boldsymbol{\chi}\mathbf{X}''\mathbf{W}'')(\mathbf{y}'' - \mathbf{X}''^T\mathbf{b}') = \mathbf{y}'' - \mathbf{X}''^T\mathbf{b} = \mathbf{v}'', \tag{6}$$

and so

$$\boldsymbol{\chi}\mathbf{X}''\mathbf{G}\mathbf{v}'' = \boldsymbol{\chi}\mathbf{X}''\mathbf{W}''(\mathbf{y}'' - \mathbf{X}''^T\mathbf{b}') = \boldsymbol{\chi}(\boldsymbol{\phi}\mathbf{b} - \boldsymbol{\phi}'\mathbf{b}') - \boldsymbol{\chi}(\boldsymbol{\phi} - \boldsymbol{\phi}')\mathbf{b}',$$

or

$$\boldsymbol{\chi}\mathbf{X}''\mathbf{G}\mathbf{v}'' = \mathbf{b} - \mathbf{b}', \tag{7}$$

which is (7.3.6,1*a*).

The equation (7.2.5,3*d*) for the sum of the squares of the residuals is

$$\mathbf{v}^T\mathbf{W}\mathbf{v} = \mathbf{y}^T\mathbf{W}\mathbf{y} - \mathbf{b}^T\boldsymbol{\phi}\mathbf{b}. \tag{8a}$$

The new sum will be given by the similar form

$$\mathbf{v}'^T\mathbf{W}'\mathbf{v}' = \mathbf{y}'^T\mathbf{W}'\mathbf{y}' - \mathbf{b}'^T\boldsymbol{\phi}'\mathbf{b}', \tag{8b}$$

and so

$$\mathbf{v}'^T\mathbf{W}'\mathbf{v}' = \mathbf{v}^T\mathbf{W}\mathbf{v} - \mathbf{y}''^T\mathbf{W}''\mathbf{y}'' - \mathbf{b}'^T\boldsymbol{\phi}'\mathbf{b}' + \mathbf{b}^T\boldsymbol{\phi}\mathbf{b}. \tag{8c}$$

Now, using (6),

$$\begin{aligned}\mathbf{v}''^T\mathbf{G}\mathbf{v}'' &= (\mathbf{y}'' - \mathbf{X}''^T\mathbf{b})^T\mathbf{W}''(\mathbf{y}'' - \mathbf{X}''^T\mathbf{b}') \\ &= \mathbf{y}''^T\mathbf{W}''\mathbf{y}'' - \mathbf{b}^T(\boldsymbol{\phi}\mathbf{b} - \boldsymbol{\phi}'\mathbf{b}') - (\boldsymbol{\phi}\mathbf{b} - \boldsymbol{\phi}'\mathbf{b}')^T\mathbf{b}' \\ &\qquad + \mathbf{b}^T(\boldsymbol{\phi} - \boldsymbol{\phi}')\mathbf{b}' \\ &= \mathbf{y}''^T\mathbf{W}''\mathbf{y}'' - \mathbf{b}^T\boldsymbol{\phi}\mathbf{b} + \mathbf{b}'^T\boldsymbol{\phi}'\mathbf{b}',\end{aligned}$$

and so $$\mathbf{v}'^T\mathbf{W}'\mathbf{v}' = \mathbf{v}^T\mathbf{W}\mathbf{v} - \mathbf{v}''^T\mathbf{G}\mathbf{v}'', \tag{9}$$

which is (7.3.6,1*c*).

7.3.6.3 *The fitted values at the omitted points.* If $\mathbf{v}''$ denotes the residuals at the points which are to be omitted and $\mathbf{u}'$ denotes the new fitted values at these points, then, from (7.3.6.2,6),

$$\mathbf{u}' = \mathbf{y}'' - \mathbf{W}''^{-1}\mathbf{G}\mathbf{v}''. \tag{1}$$

The most important case is that in which only one observation y_i is omitted. The new value u_i' then gives the fitted value at x_i as determined from all the observations except the one at x_i. Dropping the double prime superscript, (1) becomes in this case

$$u_i' = y_i - v_i(1 - w_i\Sigma\Sigma\chi_{jk}x_i^jx_i^k)^{-1}. \tag{2a}$$

TABLE 7.3.6

The omission of two observations from Example 7.1.1.1

b_j	+0·923893	−0·556667	+0·406557	−0·074432
χ_{jk}	+0·03210005			
	−0·00498913	+0·03821046		
	−0·02015719	+0·00930278	+0·02695790	
	+0·00641349	−0·00924494	−0·01022478	+0·00522808
$x_1''^j$	1	+1·664	+2·768896	+4·607443
$x_2''^j$	1	+2·556	+6·533136	+16·698696
$\Sigma\chi_{jk}x_1''^k$	−0·00246524	+0·04175597	+0·02285617	−0·01319336
$\Sigma\chi_{jk}x_2''^k$	−0·00524491	−0·00092531	+0·00899985	+0·00328566
$\Sigma\Sigma\chi_{jk}x_1''^jx_i''^k$	+0·06951540	+0·03327358		
$\Sigma\Sigma\chi_{jk}x_2''^jx_i''^k$	+0·03327358	+0·10605348		
(7.3.6,2*d*) D	0·930485 × 0·893947 − 0·001107 = 0·830697			
(7.3.6,2*b*) G_{1m}	+1·076141	+0·040055		
(7.3.6,2*c*) G_{2m}	+0·040055	+1·120126		
v_m''	+0·949628	−0·664227		
$\Sigma G_{im}v_m''$	+0·995328	−0·705981		
(7.3.6,3*a*) b_j'	+0·922644	−0·598881	+0·390161	−0·058981
$\Sigma G_{1m}(\Sigma\chi_{jk}x_m''^k)$	−0·00286303	+0·04489825	+0·02495695	−0·01406631
$\Sigma G_{2m}(\Sigma\chi_{jk}x_m''^k)$	−0·00597371	+0·00063607	+0·01099647	+0·00315189
(7.3.6,3*b*) χ_{jk}'	+0·03213844			
	−0·00510315	+0·04008464		
	−0·02027639	+0·01033471	+0·02762729	
	+0·00643163	−0·00983521	−0·01051792	+0·00542402
(7.3.6,3*c*) $\Sigma v_i'^2$	3·720544 − 1·414123 = 2·306421			
Check				
M_j	+79·9000	−9·2210	+113·2633	+92·9550
y_i''	+1·73	+0·25		
M_j'	+77·9200	−12·7387	+106·8398	+80·8094
b_j'	+0·922645	−0·598883	+0·390160	−0·058980

Cohen (1956) has used this result in discussing the estimation of the best values for the atomic constants. The form he uses is

$$u_i' = \frac{u_i \operatorname{var} y_i - y_i \operatorname{var} u_i}{\operatorname{var} y_i - \operatorname{var} u_i}. \tag{2b}$$

It is shown in § 8.1 that

$$\operatorname{var} u_i = \sigma^2 \Sigma\Sigma \chi_{jk} x_i^j x_i^k,$$

and so (2a) and (2b) are equivalent.

7.4 CHANGES OF ORIGIN

It is occasionally necessary to change from the variable x to a new variable with a different origin. Suppose that the new variable has its origin at $x = g$,

$$z = x - g. \tag{1}$$

The fitted curve will be written

$$u_p(z) = \sum_{j=0}^{p} c_{pj} z^j. \tag{2}$$

The relation between the coefficients b_{pj} and c_{pj} is found from the equation

$$u_p(z) = \sum_{j=0}^{p} b_{pj}(z+g)^j.$$

Thus

$$c_{pj} = \sum_{q=j}^{p} \binom{q}{j} g^{q-j} b_{pq}. \tag{3}$$

Alternatively, the coefficients c_{pj} may be expressed in terms of the orthogonal coefficients a_j in the form

$$c_{pj} = \sum_{k=j}^{p} \gamma_{jk} a_k. \tag{4}$$

Then, using (3),

$$c_{pj} = \sum_{q=j}^{p} \binom{q}{j} g^{q-j} \sum_{k=q}^{p} \beta_{qk} a_k = \sum_{k=j}^{p} \sum_{q=j}^{k} \binom{q}{j} g^{q-j} \beta_{qk} a_k,$$

and so

$$\gamma_{jk} = \sum_{q=j}^{k} \binom{q}{j} g^{q-j} \beta_{qk}. \tag{5}$$

The γ_{kj} are useful in the evaluation of the standard deviations of the c_{pj}. If these standard deviations are not required, the coefficients c_{pj} can be obtained from the b_{pj} by (3). Table 7.4 gives the explicit formulae for polynomials up to the fifth degree. It will be observed from (5) that the formulae for γ_{jk} in terms of β_{jk} are also of the form shown in Table 7.4.

TABLE 7.4

Relation between the power-series coefficients b and c

$$
\begin{aligned}
c_0 &= b_0 + gb_1 + g^2b_2 + g^3b_3 + g^4b_4 + g^5b_5 \\
c_1 &= b_1 + 2gb_2 + 3g^2b_3 + 4g^3b_4 + 5g^4b_5 \\
c_2 &= b_2 + 3gb_3 + 6g^2b_4 + 10g^3b_5 \\
c_3 &= b_3 + 4gb_4 + 10g^2b_5 \\
c_4 &= b_4 + 5gb_5 \\
c_5 &= b_5
\end{aligned}
$$

7.4.1 *Example*

Table 7.4.1 shows the detailed calculations when the origin in the scheme of Table 7.2.3 is transferred to the point $x = -2$. This corresponds to a change of origin to a temperature of 0° C. in the original data.

Section (*a*) shows the evaluation of the c_{3j} directly from the b_{pj}. This is the most rapid method when these coefficients only are required. If coefficients of polynomials of lower degree, or elements of the inverse matrix, are to be obtained, the values γ_{jk} and $Q_{kj} = \gamma_{jk}/S_{kk}$ may be evaluated first, as in section (*b*) of the table. In the scheme shown all the intermediate products have been written down, but the scheme may be abbreviated by accumulating these intermediate products in the register of the calculating machine and not recording them separately. The coefficients are then calculated as in section (*c*) of the table.

The scales of the variables used in the illustrative schemes differ from the original variables x and y by the factors listed in Table 7.1.8. Hence, from § 7.1.7, if t is the temperature ($= x + 20$),

$$u_t = \Sigma 10^r\, 10^{-qj} c_{pj}\, t^j,$$

where $q = 1, r = 2$. Thus

$$
\begin{aligned}
u_t &= \Sigma 10^2 (10^{-j} c_{pj})\, t^j \\
&= 425{\cdot}89 - 30{\cdot}761t + 0{\cdot}85314t^2 - 0{\cdot}007443t^3,
\end{aligned}
$$

and $$J_t = 4{\cdot}17 + 10^{-4} u_t.$$

7.5 ITERATIVE METHODS FOR THE SOLUTION OF THE NORMAL EQUATIONS

When the number of unknowns is large, it may be preferable to solve the normal equations by iterative procedures. This is specially true if high-speed automatic calculating machinery is being used, since repetitive processes are easily programmed on such machines. When desk calculators are being used, the two methods given below may be of advantage in certain cases; for example, when each unknown occurs in only a few of the normal equations.

TABLE 7.4.1

Change of origin

(*a*) Evaluation of c_{pj}

$g = -2$

b_{30} +0·923895 b_{31} −0·556669 b_{32} +0·406555 b_{33} −0·074431

×	1 +0·923895	g +1·113338	g^2 +1·626220	g^3 +0·595448	c_{30} +4·258901
×		1 −0·556669	$2g$ −1·626220	$3g^2$ −0·893172	c_{31} −3·076061
×			1 +0·406555	$3g$ +0·446586	c_{32} +0·853141
×				1 −0·074431	c_{33} −0·074431
				Check $\Sigma c_{3j}(-g)^j = b_{30}$	

(*b*) Evaluation of γ_{jk}

$g = -2$

					γ_{00} 1	Q_{00} +1·492537
	β_{01} −0·301090	β_{11} 1				$\div S_{11}$ 0·926525
×	1 −0·301090	g −2			γ_{01} −2·301090	Q_{10} −2·483570
×		1 1			γ_{11} 1	Q_{11} +1·079302
	β_{02} −1·093840	β_{12} −1·261044	β_{22} 1			$\div S_{22}$ 1·436605
×	1 −1·093840	g +2·522088	g^2 +4		γ_{02} +5·428248	Q_{20} +3·778525
×		1 −1·261044	$2g$ −4		γ_{12} −5·261044	Q_{21} −3·662137
×			1 1		γ_{22} 1	Q_{22} +0·696086
				Check	$\Sigma Q_{2j}(-g)^j = \beta_{02}/S_{22} = R_{20}$	
	β_{03} +1·226739	β_{13} −1·768323	β_{23} −1·955744	β_{33} 1		$\div S_{33}$ 1·912749
×	1 +1·226739	g +3·536646	g^2 −7·822976	g^3 −8	γ_{03} −11·059591	Q_{30} −5·782040
×		1 −1·768323	$2g$ +7·822976	$3g^2$ 12	γ_{13} +18·054653	Q_{31} +9·439113
×			1 −1·955744	$3g$ −6	γ_{23} −7·955744	Q_{32} −4·159325
×				1 1	γ_{33} 1	Q_{33} +0·522808
				Check	$\Sigma Q_{3j}(-g)^j = \beta_{03}/S_{33} = R_{30}$	

(*c*) Evaluation of c_{pj} using the values γ_{jp}

a_0 1·192537			
$\gamma_{01}a_1$ +0·826485	a_1 −0·359171		
c_{10} +2·019022	c_{11} −0·359171		

$\gamma_{02}a_2$ +1·416702	$\gamma_{12}a_2$ −1·373064	a_2 +0·260987	
c_{20} +3·435724	c_{21} −1·732235	c_{22} +0·260987	

Check $\Sigma c_{2j}(-g)^j = \sum_0^2 \beta_{0j}a_j = b_{20}$

$\gamma_{03}a_3$ +0·823176	$\gamma_{13}a_3$ −1·343826	$\gamma_{23}a_3$ +0·592154	a_3 −0·074431
c_{30} +4·258900	c_{31} −3·076061	c_{32} +0·853141	c_{33} −0·074431

Check $\Sigma c_{3j}(-g)^j = b_{20} + \beta_{03}a_3 = b_{30}$

7.5.1 *Von Seidel's method*

In von Seidel's method, the value b_{pj} is determined from the jth normal equation,

$$b_{pj} = \left\{M_j - \sum_{k \neq j} \phi_{jk} b_{pk}\right\} \Big/ \phi_{jj}. \tag{1a}$$

The earlier approximations b_{pk} are used on the right-hand side. The initial approximations by which the process is started are usually taken as

$$b_{pk} = M_k/\phi_{kk}. \tag{1b}$$

The values $b_{p0}, b_{p1}, \ldots$ are corrected in turn, using (1*a*). For example, in calculating the second approximation to b_{pj}, the second approximation to the values b_{pk} $(k<j)$ and the first approximation to the values b_{pq} $(q>j)$ are used in (1*a*). The process is continued till b_{pp} is reached, and then started all over again by determining the third approximation to b_{p0}. The iterative procedure is continued until the desired accuracy is obtained.

7.5.2 *Relaxation method*

The relaxation method (the terminology comes from structural engineering) is similar to the von Seidel method, but the coefficients are not adjusted in any fixed order, the order being selected by the computer. Usually the equation which has the largest residual

$$R_j = M_j - \sum_{k=0}^{p} \phi_{jk} b_{pk} \tag{1}$$

will be selected. Hence the residuals are reduced more rapidly than in the von Seidel method.

It is probably best to draw up a systematic scheme, consisting of an 'operations table' (Table 7.5.2) and a 'relaxation table' (Table 7.5.2*a*). Line j in the operations table gives the decreases in the residuals for a change in Δb_j of one unit. Hence when a change Δb_j is made in the relaxation table, the residuals are reduced by Δb_j times the factors in line j of the operations table.

TABLE 7.5.2

Operations table for three variables

	Δb_0	Δb_1	Δb_2	ΔR_0	ΔR_1	ΔR_2
0.	1	0	0	ϕ_{00}	ϕ_{10}	ϕ_{20}
1.	0	1	0	ϕ_{01}	ϕ_{11}	ϕ_{21}
2.	0	0	1	ϕ_{02}	ϕ_{12}	ϕ_{22}

TABLE 7.5.2*a*

Relaxation table for three variables

	Coefficients			Residuals		
A.	b_0^0	b_1^0	b_2^0	R_0^0	R_1^0	R_2^0
B.		Δb_1		(A)	$-\Delta b_1 \times (1)$	
C.	Δb_0			(B)	$-\Delta b_0 \times (0)$	
		etc.			etc.	
Add	b_0^1	b_1^1	b_2^1		Check	

The zero-order approximations b_j^0 obtained from (7.5.1,1*b*) and the corresponding residuals R_j^0 obtained from (1) are written down in line A of the relaxation table. Suppose that R_1^0 is the largest residual. Then a relaxation Δb_1 is made in b_1 so that the residual R_1 is very small. Clearly

$$\Delta b_1 \sim R_1^0/\phi_{11}.$$

The new residuals are obtained by subtracting from those in line A the product of Δb_1 and line (1) of the operations table. If the largest residual is now R_0, a relaxation

$$\Delta b_0 \sim R_0/\phi_{00}$$

is made, and the new residuals calculated. The relaxations are continued, the residuals becoming smaller and smaller. A halt is called at a suitable point, and the corrected values b_j^1 obtained by summing the columns of the relaxation table. These values are then substituted in (1). If the arithmetic is free from mistakes, the residuals obtained should agree with those in the last line of the relaxation table. Then if necessary the process can be carried further until the required accuracy is obtained.

7.5.2.1 *Example*

In Table 7.5.2*b* are shown the calculations of the second-degree coefficients for the example previously used in this chapter.

The first section of the table shows the operations table for this case. The first line A in the second section shows the initial approximations M_j/ϕ_{jj}, and the residuals obtained using these approximations. At the end of this section the approximations 1·02, −0·69, +0·26 are obtained. The residuals are calculated afresh using these approximations, and the procedure carried a stage further in the third section of the table.

It should be emphasized that such a procedure takes a much longer time than the solution by the Doolittle scheme, and it would only be employed in exceptional cases. However, iterative schemes are very useful with modern automatic computers, where a large number of similar operations can be rapidly performed.

TABLE 7.5.2*b*

Relaxation method for the solution of normal equations

0.	1	0	0	0·670000	0·201730	0·987264
1.	0	1	0	0·201730	0·987264	1·465644
2.	0	0	1	0·987264	1·465644	4·364756
A.	1·19	−0·09	+0·26	−0·2368	−0·6245	−1·0451
		−0·63		−0·1097	−0·0025	−0·1217
	−0·16			−0·0025	+0·0298	+0·0363
		+0·03		−0·0086	+0·0002	−0·0077
	−0·01			−0·0019	+0·0022	+0·0022
					$\times 10^3$	
	1·02	−0·69	+0·26	−1·895	+2·170	+2·082
		$\times 10^3$				
		+2·2		−2·339	−0·002	−1·142
	−3·0			−0·329	+0·603	+1·820
			+0·5	−0·823	−0·130	−0·362
	−1·4			+0·115	+0·152	+1·020
			+0·3	−0·181	−0·288	−0·289
		−0·4		−0·100	+0·107	+0·297
	−0·2			+0·034	+0·147	+0·494
			+0·1	−0·065	0	+0·058
	−0·1			+0·002	+0·020	+0·157
					$\times 10^3$	
	1·0153	−0·6882	+0·2609	+0·0024	+0·0221	+0·1552

7.5.3 *Group relaxations*

The simple operations of the operations table may be combined to give more complicated group relaxations, where several of the coefficients b_{pj} are varied simultaneously. The 'groups' are selected so that the relaxation causes a large change in one residual, and very small changes in the other residuals. An operations table entry is of the form

$$\alpha_0 \quad \alpha_1 \quad \alpha_2 \qquad \psi_0 \quad \psi_1 \quad \psi_2,$$

where (say) ψ_0 and ψ_2 are very small. The ψ_k are given by

$$\psi_k = \Sigma \alpha_j \phi_{kj}.$$

If the residual to be reduced is R_1, the relaxations are $\alpha_j r_1$, where

$$r_1 \sim R_1/\psi_1.$$

When single relaxations are used, a decrease in R_1 may lead to an increase in R_0 and R_2. With group relaxations, R_0 and R_2 are practically unaffected.

7.5.3.1 *Example*

The top portion of Table 7.5.3 shows the method of obtaining group relaxations for the example of § 7.5.2.1. The calculation of the approximate solutions is shown in the lower section.

TABLE 7.5.3

Solution of normal equations by group relaxations

0′.	1	0	0	0·670000	0·201730	0·987264
1′.	0	1	0	0·201730	0·987264	1·465644
2′.	0	0	1	0·987264	1·465644	4·364756
1″ : 5 × 0′ − 1′	5	−1	0	3·148270	0·021386	3·470676
2″ : 7 × 0′ − 2′	7	0	−1	3·702736	−0·053534	2·546092
0 : 7 × 2″ − 5 × 1″	24	+5	−7	10·177802	−0·481668	0·469264
1 : 3 × 1′ − 2′	0	+3	−1	−0·382074	1·496148	0·032176
2 : 6 × 1″ − 5 × 2″	−5	−6	+5	0·375940	0·395986	8·093596
	1·19	−0·09	0·26	−0·2368	−0·6245	−1·0451
1 × −0·42	0	−1·26	+0·42	−0·3973	+0·0039	−1·0316
0 × −0·040	−0·96	−0·20	+0·28	+0·0098	−0·0154	−1·0128
2 × −0·12	+0·60	+0·72	−0·60	+0·0549	+0·0321	−0·0416
1 × 0·021	0	+0·063	−0·021	+0·0629	+0·0007	−0·0422
0 × 0·006	+0·144	+0·030	−0·042	+0·0018	+0·0036	−0·0450
2 × −0·006	+0·030	+0·036	−0·030	+0·0040	+0·0060	+0·0036
1 × 0·004	0	+0·012	−0·004	+0·0055	0	+0·0035
0 × 0·00055	+0·013	+0·003	−0·004	−0·0001	+0·0003	+0·0033
2 × 0·0004	−0·002	−0·002	+0·002	−0·0002	+0·0001	+0·0001
	1·015	−0·688	+0·261	+0·000064	−0·000261	−0·000278

7.5.4 *The method of steepest descent*

Relaxation methods require the exercise of judgment, and so they are not very suitable for automatic computers, where a definite programme is desirable. Various iterative methods suitable for such machines have been described by Householder (1953) and Booth (1955). One of these methods, called the method of steepest descent, will now be briefly outlined.

The normal equations are in matrix notation

$$\mathbf{M} - \boldsymbol{\phi}\mathbf{b} = \mathbf{0}, \tag{1}$$

and if $\boldsymbol{\beta}$ is an approximate solution of these equations

$$\mathbf{M} - \boldsymbol{\phi}\boldsymbol{\beta} = \mathbf{r}, \tag{2}$$

where $\mathbf{r}$ represents the residuals. The quantity

$$S = \tfrac{1}{2}\mathbf{r}^T \boldsymbol{\phi}^{-1}\mathbf{r} \tag{3}$$

is a function of the β_j, and it is always positive except for $\boldsymbol{\beta} = \mathbf{b}$, when it vanishes. Hence an iterative process can be based on the minimizing of this function. The variations of the coefficients β_j to give new approximations β'_j which are considered are those of the form

$$\beta'_j = \beta_j + \lambda u_j,$$

or

$$\boldsymbol{\beta}' = \boldsymbol{\beta} + \lambda \mathbf{u}. \tag{4}$$

For a given set u_j, the value of λ which minimizes S is found from the equation

$$\partial S'/\partial\lambda = \partial(\tfrac{1}{2}\mathbf{r}'^T \boldsymbol{\phi}^{-1} \mathbf{r}')/\partial\lambda = 0. \tag{5}$$

Now

$$\mathbf{r}' = \mathbf{M} - \boldsymbol{\phi}\boldsymbol{\beta}' = \mathbf{r} - \lambda\boldsymbol{\phi}\mathbf{u},$$

and so (5) gives

$$-\mathbf{r}^T \mathbf{u} + \lambda \mathbf{u}^T \boldsymbol{\phi}\mathbf{u} = 0,$$

or

$$\lambda = \mathbf{r}^T \mathbf{u}/\mathbf{u}^T \boldsymbol{\phi}\mathbf{u}. \tag{6}$$

To find the best set u_j, the steepest descent criterion states that the values are to be chosen so that the decrease of S is most rapid. Now the vector $\mathbf{u}$ which gives the most rapid decrease is the set u_j whose elements are proportional to $-\partial S/\partial\beta_j$. This vector corresponds to $-\operatorname{grad} S$ in a space of $p+1$ dimensions. Then

$$u_j = -\frac{\partial S}{\partial \beta_j} = -\frac{\partial}{\partial \beta_j}\{\tfrac{1}{2}(\mathbf{M} - \boldsymbol{\phi}\boldsymbol{\beta})^T \boldsymbol{\phi}^{-1}(\mathbf{M} - \boldsymbol{\phi}\boldsymbol{\beta})\},$$

and when this is differentiated it is found that

$$\mathbf{u} = \mathbf{M} - \boldsymbol{\phi}\boldsymbol{\beta} = \mathbf{r}.$$

Hence the method of steepest descent gives for the next approximation

$$\boldsymbol{\beta}' = \boldsymbol{\beta} + \lambda\mathbf{r}, \tag{7a}$$

where

$$\lambda = \mathbf{r}^T \mathbf{r}/\mathbf{r}^T \boldsymbol{\phi}\mathbf{r} = \Sigma r_j^2 \Big/ \sum_{j,k} r_j r_k \phi_{jk}. \tag{7b}$$

The method of steepest descent can also be based on the function

$$R = \tfrac{1}{2}\mathbf{r}^T \mathbf{r}$$

instead of on the function S, but the formulae are more complicated.

7.5.4.1 *Example*

The steps in the application of the method of steepest descent based on S to the example used in this section are shown in Table 7.5.4. Of course this example is only intended to illustrate the method, as iterative procedures would never be applied when the number of unknowns is so small.

Starting with the initial approximation

$$\beta_j = M_j/\phi_{jj},$$

the residuals

$$r_j = M_j - \Sigma\beta_j \phi_{jk}$$

TABLE 7.5.4

The method of steepest descent

				ϕ_{jk}		M_k	
			0·670000	0·201730	0·987264	0·799000	
			0·201730	0·987264	1·465644	−0·092210	
			0·987264	1·465644	4·364756	1·132633	
	β_j and $\Delta\beta_j$			r_j and $\Sigma r_k \phi_{jk}$		Σr_j^2 and $\Sigma r_j \Sigma r_k \phi_{jk}$	λ
1·19	−0·09	0·26	−0·236833	−0·624482	−1·045140	1·538385	
−0·0476	−0·1255	−0·2101	−1·316484	−2·196108	−5·710866	7·651871	0·201046
1·1424	−0·2155	0·0499	+0·027800	−0·183047	+0·102828	0·044853	
+0·0443	−0·2915	+0·1638	+0·083218	−0·024398	+0·207983	0·028166	1·5925
1·1867	−0·5070	+0·2137	−0·104790	−0·144268	−0·228620	0·084061	
−0·0217	−0·0299	−0·0473	−0·325021	−0·498645	−1·312771	0·406123	0·206984
1·1650	−0·5369	+0·1664	−0·037522	−0·041047	+0·043079	0·0049486	
−0·0629	−0·0688	+0·0722	+0·009110	+0·015045	+0·090825	0·0029533	1·6756
1·1021	−0·6057	+0·2386	−0·052780	−0·066253	−0·109121	0·019083	
−0·0109	−0·0137	−0·0226	−0·156459	−0·235989	−0·625498	0·092148	0·207086
1·0912	−0·6194	+0·2160	−0·020401	−0·017406	+0·020363	0·0011338	
−0·0344	−0·0294	+0·0344	+0·002924	+0·008545	+0·043227	0·0006718	1·6877
1·0568	−0·6488	+0·2504	−0·025384	−0·031859	−0·052733	0·0044401	
−0·0052	−0·0066	−0·0109	−0·075496	−0·113862	−0·301921	0·0214651	0·206853
1·0516	−0·6554	+0·2395	−0·009808	−0·008318	+0·009650	0·00025851	
−0·0169	−0·0143	+0·0166	+0·001278	+0·003953	+0·020246	0·00014996	1·7238
1·0347	−0·6697	0·2561					
	b_j						
1·0152	−0·6883	0·2610					

are calculated, and from these in turn

$$\Sigma r_k \phi_{jk}, \quad \Sigma r_j(\Sigma r_k \phi_{jk}), \quad \Sigma r_j^2.$$

Then

$$\lambda = \Sigma r_j^2 / \Sigma r_j(\Sigma r_k \phi_{jk}),$$

$$\Delta\beta_j = \lambda r_j,$$

and the next approximation is

$$\beta_j' = \beta_j + \Delta\beta_j.$$

It is seen that the values β_j tend towards the solution b_j. It is interesting to observe that the quantities in alternate approximations are similar in form, as if the descent is following a zigzag path.

7.6 EQUALLY-SPACED OBSERVATIONS OF EQUAL WEIGHT

When the interval Δx between successive observations is constant, the calculations are best performed in terms of the variable

$$\epsilon = (x - \bar{x})/\Delta x. \tag{1}$$

The values ϵ_i at the points of observation are the integers or half-integers from $-\frac{1}{2}(n-1)$ to $+\frac{1}{2}(n-1)$.

7.6.1 *The orthogonal polynomials* $T_j(\epsilon)$

The orthogonal polynomial $T_j(\epsilon)$ is related to the powers ϵ^k by the equations

$$T_j(\epsilon) = \sum_{k=0}^{j} \beta_{kj}\, \epsilon^k, \tag{1a}$$

$$\epsilon^j = \sum_{k=0}^{j} \alpha_{kj}\, T_k(\epsilon). \tag{1b}$$

It will now be shown that

$$T_j(\epsilon) = (-)^j\, T_j(-\epsilon), \tag{2a}$$

and

$$\alpha_{kj} = 0 = \beta_{kj}, \quad k+j \text{ odd}. \tag{2b}$$

For if (2a) holds for all values up to $j-1$, then from (7.2,4a–e),

$$\alpha_{kj} = \sum_i \epsilon_i^j\, T_k(\epsilon_i) / \sum_i T_k^2(\epsilon_i) = (-)^{j+k} \sum_i (-\epsilon_i)^j\, T_k(-\epsilon_i) / \sum_i T_k^2(\epsilon_i).$$

Now the values ϵ_i are symmetrical about zero, and so

$$\sum_i \epsilon_i^j\, T_k(\epsilon_i) = \sum_i (-\epsilon_i)^j\, T_k(-\epsilon_i),$$

and α_{kj} vanishes when $j+k$ is odd. Hence

$$T_j(\epsilon) = \epsilon^j - \sum_{\substack{j+k \\ \text{even}}} \alpha_{kj}\, T_k(\epsilon),$$

and

$$T_j(-\epsilon) = (-\epsilon)^j - \sum_{\substack{j+k \\ \text{even}}} \alpha_{kj}(-)^k\, T_k(\epsilon) = (-)^j\, [\epsilon^j - \Sigma \alpha_{kj}\, T_k(\epsilon)],$$

and so (2*a*) follows by induction. Finally, since

$$T_j(-\epsilon) = \Sigma\beta_{kj}(-\epsilon)^k = (-)^j\,\Sigma\beta_{kj}\,\epsilon^k,$$

β_{kj} also vanishes when $k+j$ is odd.

It is possible to derive formulae for α_{kj}, β_{kj}, and $S_{jj} = \Sigma T_j^2(\epsilon_i)$ from the values $\sum_i \epsilon_i^{j+k}$, but these formulae are obtained more simply when the orthogonal polynomials are expressed in terms of the factorial powers $\binom{x}{j}$. The formulae are listed in Table 7.10*a* for polynomials up to the 9th degree. Proof of these formulae will be postponed until § 7.7.

7.6.2 *Fitting by power moments*

The power moments are simply calculated and, since the values ϵ_i do not depend on n, the only quantities to be tabulated are β_{kj} and S_{jj}. Then, from (7.2.2,4*b*) and (7.2.2,3),

$$M_k = \Sigma\epsilon_i^k\,y_i, \tag{1a}$$

$$a_j = \Sigma\beta_{kj}\,M_k/S_{jj}, \tag{1b}$$

and

$$b_{pj} = \Sigma\beta_{jk}\,a_k. \tag{1c}$$

Values of β_{jk} and S_{jj} for polynomials up to the fifth degree are listed in Table 7.10*b* for $n6(1)75$.

When n is even, the values ϵ_i are half-integers. It is then simplest to calculate M_k by means of the equation

$$2^k\,M_k = \sum_i (2\epsilon_i)^k\,y_i.$$

Hence different calculating schemes are used for the cases when n is even (§ 7.6.2.1) and when n is odd (§ 7.6.2.2).

7.6.2.1 *Example—n even*

The calculation of the moments when n is even is done by entering the observations in the scheme shown in Table 7.6.2.1. In the present example the values y represent changes in sugar prices over a period of 62 years. This example was used by Anderson and Houseman (1942), who smoothed the values by fitting a cubic curve.

The observations are entered in the y columns, starting from the lower portion of the y_+ column and entering the values in decreasing order of ϵ. The moments M_j may then be calculated by summing the products of the corresponding values in the columns y and $(2\epsilon)^j$,

$$M_j = \Sigma(2\epsilon)^j\,\{y(+\epsilon)+(-)^j\,y(-\epsilon)\}/2^j.$$

The products for the observations of y corresponding to negative values of ϵ are subtracted if j is odd and added if j is even. If the number of observations is large it may be convenient to form first the sums and

TABLE 7.6.2.1

Calculation of moments—n even

$$M_j = \Sigma \epsilon^j (y_+ + (-)^j y_-)$$

$32\epsilon^5$	$8\epsilon^3$	2ϵ	Diff.	y_+	y_-	Sum	$4\epsilon^2$	$16\epsilon^4$
1	1	1	−8	5	13	18	1	1
243	27	3	−2	6	8	14	9	81
3125	125	5	+4	10	6	16	25	625
16807	343	7	+3	8	5	13	49	2401
59049	729	9	0	10	10	20	81	6561
161051	1331	11	0	13	13	26	121	14641
371293	2197	13	+1	10	9	19	169	28561
759375	3375	15	−7	3	10	13	225	50625
1,419857	4913	17	+2	7	5	12	289	83521
2,476099	6859	19	+11	16	5	21	361	130321
4,084101	9261	21	+27	29	2	31	441	194481
6,436343	12167	23	+36	37	1	38	529	279841
9,765625	15625	25	+30	38	8	46	625	390625
14,348907	19683	27	+47	50	3	53	729	531441
20,511149	24389	29	+68	74	6	80	841	707281
28,629151	29791	31	0	22	22	44	961	923521
39,135393	35937	33	−17	19	36	55	1089	1,185921
52,521875	42875	35	+14	44	30	74	1225	1,500625
69,343957	50653	37	+15	35	20	55	1369	1,874161
90,224199	59319	39	−6	15	21	36	1521	2,313441
115,856201	68921	41	−9	15	24	39	1681	2,825761
147,008443	79507	43	−10	18	28	46	1849	3,418801
184,528125	91125	45	−30	15	45	60	2025	4,100625
229,345007	103823	47	−42	10	52	62	2209	4,879681
282,475249	117649	49	−51	6	57	63	2401	5,764801
345,025251	132651	51	−52	4	56	60	2601	6,765201
418,195493	148877	53	−48	0	48	48	2809	7,890481
503,284375	166375	55	−52	3	55	58	3025	9,150625
601,692057	185193	57	−72	1	73	74	3249	10,556001
714,924299	205379	59	−62	3	65	68	3481	12,117361
844,596301	226981	61	−60	7	67	74	3721	13,845841
992,436543	250047	63					3969	15,752961
1160,290625	274625	65					4225	17,850625
1350,125107	300763	67					4489	20,151121
1564,031349	328509	69					4761	22,667121
1804,229351	357911	71					5041	25,411681
2073,071593	389017	73					5329	28,398241
	Sum		−270	533	803	1336	Check	

NOTE that the values found by multiplying the columns must be divided by 2^j to give M_j.

differences of the observations of equal $|\epsilon|$; this halves the number of multiplications. The sums and differences are checked by adding the columns, the total of the sums being $\Sigma y_+ + \Sigma y_-$ and the total of the differences $\Sigma y_+ - \Sigma y_-$. It is advisable to repeat the calculation of the moments before proceeding further.

The coefficients for the third-degree curve are evaluated in Table 7.6.2.1*a*. The values Σv^2 are for use in calculations of standard deviations (§ 8.4) and in checking the calculations (§ 7.6.2.3). The fitted curve is

$$u_3(\epsilon) = 12{\cdot}378 + 0{\cdot}9798\epsilon + 0{\cdot}028635\epsilon^2 - 0{\cdot}0025868\epsilon^3.$$

TABLE 7.6.2.1*a*

Evaluation of coefficients—third-degree curve

	S_{00} 62 S_{11} 19855·5 S_{22} 5,083008 S_{33} 1253,143008 β_{02} −320·25 β_{13} −576·25	
$a_j = \mathscr{M}_j/S_{jj}$	Σy^2 54038	$a_3 = b_{33}$ −0·002586832
$M_0 = \mathscr{M}_0$ +1336 a_0 +21·54839	$a_0\,\mathscr{M}_0$ 28789 Σv_0^2 25249	$\beta_{13}\,a_3$ +1·490662 $+a_1$ −0·5108408 $= b_{p1}$ +0·979821
$M_1 = \mathscr{M}_1$ −10143 a_1 −0·5108408	$a_1\,\mathscr{M}_1$ 5181 Σv_1^2 20068	$a_2 = b_{p2}$ +0·02863462
M_2 +573404 $+\beta_{02}\,M_0$ −427854 $= \mathscr{M}_2$ +145550 a_2 +0·02863462	$a_2\,\mathscr{M}_2$ 4168 Σv_2^2 15900 (s_2)	$\beta_{02}\,a_2$ −9·17024 $+a_0$ +21·54839 $= b_{p0}$ +12·37815
M_3 −9086574·75 $+\beta_{13}\,M_1$ +5844903·75 $= \mathscr{M}_3$ −3241671 a_3 −0·002586832	$a_3\,\mathscr{M}_3$ 8386 Σv_3^2 7514 (s_3) 11·38	

7.6.2.2 *Example—n odd*

The scheme for calculating the moments when n is odd is shown in Table 7.6.2.2. The observations are the measurements of the frequencies of the first 25 lines in one of the bands of the CuH spectrum (Birge and Shea, 1927). The frequencies vary from 22330·52 cm^{-1} to 23295·47 cm^{-1}. The constant 22300 is subtracted from each figure before entering it in the calculating scheme. The observations, in contrast to those of Example 7.6.2.1, are very accurate, and it is necessary to retain a large number of figures throughout the calculations.

The scheme for the evaluation of the coefficients for a curve of the fourth or fifth degree is given in Table 7.6.2.2*a*. In the present example a fourth-degree curve is fitted, the curve obtained being

$$u_4(\epsilon) = 647{\cdot}254 + 40{\cdot}8346\epsilon - 0{\cdot}93435\epsilon^2 - 0{\cdot}004376\epsilon^3 + 0{\cdot}00001383\epsilon^4.$$

TABLE 7.6.2.2

Calculation of moments—n odd

$$M_j = \Sigma \epsilon^j (y_+ + (-)^j y_-)$$

ϵ^5	ϵ^3	ϵ	Diff.	y_+	y_-	Sum	ϵ^2	ϵ^4
0	0	0		647·29			0	0
1	1	1	81·67	687·15	605·48	1292·63	1	1
32	8	2	163·32	725·15	561·83	1286·98	4	16
243	27	3	244·85	761·27	516·42	1277·69	9	81
1024	64	4	326·17	795·39	469·22	1264·61	16	256
3125	125	5	407·25	827·54	420·29	1247·83	25	625
7776	216	6	488·11	857·71	369·60	1227·31	36	1296
16807	343	7	568·68	885·85	317·17	1203·02	49	2401
32768	512	8	648·88	911·94	263·06	1175·00	64	4096
59049	729	9	728·55	935·95	207·40	1143·35	81	6561
100000	1000	10	807·88	957·86	149·98	1107·84	100	10000
161051	1331	11	886·74	977·79	91·05	1068·84	121	14641
248832	1728	12	964·95	995·47	30·52	1025·99	144	20736
371293	2197	13					169	28561
537824	2744	14					196	38416
759375	3375	15					225	50625
1,048576	4096	16					256	65536
1,419857	4913	17					289	83521
1,889568	5832	18					324	104976
2,476099	6859	19					361	130321
3,200000	8000	20					400	160000
4,084101	9261	21					441	194481
5,153632	10648	22					484	234256
6,436343	12167	23					529	279841
7,962624	13824	24					576	331776
9,765625	15625	25					625	390625
11,881376	17576	26					676	456976
14,348907	19683	27					729	531441
17,210368	21952	28					784	614656
20,511149	24389	29					841	707281
24,300000	27000	30					900	810000
28,629151	29791	31					961	923521
33,554432	32768	32					1024	1,048576
39,135393	35937	33					1089	1,185921
45,435424	39304	34					1156	1,336336
52,521875	42875	35					1225	1,500625
60,466176	46656	36					1296	1,679616
69,343957	50653	37					1369	1,874161
	Sum		6317·05	10319·07	4002·02	14321·09	Check	
				y_0 647·29				

TABLE 7.6.2.2a

Evaluation of coefficients—fourth-degree curve

S_{00}	25	S_{44}	82,409184
S_{11}	1300	β_{24}	−133
S_{22}	53820	β_{04}	+2059·2
S_{33}	2,131272	S_{55}	
β_{02}	−52	β_{35}	
β_{13}	−93·4	β_{15}	

$a_j = \mathcal{M}_j/S_{jj}$		Σy^2	11,133456·4770		
$M_0 = \mathcal{M}_0$	14968·38	$a_0 \mathcal{M}_0$	8,962095·9930		
a_0	+598·7352	Σv_0^2	2,171360·4840		
$M_1 = \mathcal{M}_1$	52553·54	$a_1 \mathcal{M}_1$	2,124518·8973		
a_1	+40·4258	Σv_1^2	46841·5867	$a_4 = b_{p4}$	+0·0000138302061
M_2	728167·96	$a_2 \mathcal{M}_2$	46800·7296	$\beta_{24} a_4$	−0·001839417
$+\beta_{02} M_0$	−778355·76	Σv_2^2	40·8571	$+a_2$	−0·932512077
$=\mathcal{M}_2$	−50187·80	(s_2)		$=b_{p2}$	−0·934351494
a_2	−0·932512077				
M_3	4899173·36	$a_3 \mathcal{M}_3$	40·8198	$\beta_{04} a_4$	+0·028479
$+\beta_{13} M_1$	−4908500·636	Σv_3^2	0·0373	$\beta_{02} a_2$	+48·490628
$=\mathcal{M}_3$	−9327·276	(s_3)		$+a_0$	+598·7352
a_3	−0·00437638931			$=b_{p0}$	+647·254307
M_4	66024590·32	$a_4 \mathcal{M}_4$	0·0158	$a_5 = b_{55}$	
$+\beta_{24} M_2$	−96846338·68	Σv_4^2	0·0215	$\beta_{35} a_5$	
$+\beta_{04} M_0$	+30822888·096	(s_4)	0·0328	$+a_3$	−0·00437638931
$=\mathcal{M}_4$	+1139·736			$=b_{p3}$	
a_4	+0·0000138302061				
M_5		$a_5 \mathcal{M}_5$		$\beta_{15} a_5$	
$+\beta_{35} M_3$		Σv_5^2		$+\beta_{13} a_3$	+0·4087548
$+\beta_{15} M_1$		(s_5)		$+a_1$	+40·4258
$=\mathcal{M}_5$				$=b_{p1}$	+40·8345548
a_5					

7.6.2.3 *Evaluation of residuals.* It is very desirable to check the calculations by forming the residuals v_i and comparing the value Σv_i^2 with that obtained using formula (7.2.5,3a).

When n is even (Example 7.6.2.1), the coefficients b_{pj} are divided by 2^j, and the products $(b_{pj}/2^j)(2\epsilon_i)^j$ are evaluated using the columns of values $(2\epsilon_i)^j$ in Table 7.6.2.1. The sum u' of the terms containing even powers and the sum u'' of the terms containing odd powers are evaluated separately in the two extreme columns of Table 7.6.2.3a. Two results which are useful

TABLE 7.6.2.3*a*

Evaluation of residuals—n even

Odd powers					Even powers	
$b_{31}/2$	$+0{\cdot}4899$			b_{30}	$12{\cdot}378$	
$b_{33}/8$	$-0{\cdot}0003234$			$b_{32}/4$	$0{\cdot}007159$	

2ϵ	u''	u_+	v_+	v_-	u_-	u'
1	0·5	12·9	−7·9	+1·1	11·9	12·4
3	1·5	13·9	−7·9	−2·9	10·9	12·4
5	2·4	15·0	−5·0	−4·2	10·2	12·6
7	3·3	16·0	−8·0	−4·4	9·4	12·7
9	4·2	17·2	−7·2	+1·2	8·8	13·0
11	5·0	18·2	−5·2	+4·8	8·2	13·2
13	5·7	19·3	−9·3	+1·1	7·9	13·6
15	6·3	20·3	−17·3	+2·3	7·7	14·0
17	6·7	21·1	−14·1	−2·7	7·7	14·4
19	7·1	22·1	−6·1	−2·9	7·9	15·0
21	7·3	22·8	+6·2	−6·2	8·2	15·5
23	7·3	23·5	+13·5	−7·9	8·9	16·2
25	7·2	24·1	+13·9	−1·7	9·7	16·9
27	6·9	24·5	+25·5	−7·7	10·7	17·6
29	6·3	24·7	+49·3	−6·1	12·1	18·4
31	5·6	24·9	−2·9	+8·3	13·7	19·3
33	4·5	24·7	−5·7	+20·3	15·7	20·2
35	3·3	24·4	+19·6	+12·2	17·8	21·1
37	1·7	23·9	+11·1	−0·5	20·5	22·2
39	−0·1	23·2	−8·2	−2·4	23·4	23·3
41	−2·2	22·2	−7·2	−2·6	26·6	24·4
43	−4·6	21·0	−3·0	−2·2	30·2	25·6
45	−7·4	19·5	−4·5	+10·7	34·3	26·9
47	−10·6	17·6	−7·6	+13·2	38·8	28·2
49	−14·0	15·6	−9·6	+13·4	43·6	29·6
51	−17·9	13·1	−9·1	+7·1	48·9	31·0
53	−22·2	10·3	−10·3	−6·7	54·7	32·5
55	−26·9	7·1	−4·1	−5·9	60·9	34·0
57	−32·0	3·6	−2·6	+5·4	67·6	35·6
59	−37·5	−0·2	+3·2	−9·8	74·8	37·3
61	−43·5	−4·5	+11·5	−15·5	82·5	39·0
Sums	−126·1	542·0	−9·0	+8·8	794·2	668·1

TABLE 7.6.2.3*b*

Evaluation of residuals—n odd

Odd powers				Even powers	
b_{41}	40·834555			b_{40}	647·2543
b_{43}	−0·00437639			b_{42}	−0·934351
				b_{44}	+0·00001383

ϵ	u''	u_+	v_+	v_-	u_-	u'
0		647·2543	+0·0357			
1	40·8302	687·1502	−0·0002	−0·0098	605·4898	646·3200
2	81·6341	725·1512	−0·0012	−0·0530	561·8830	643·5171
3	122·3855	761·2318	+0·0382	−0·0408	516·4608	638·8463
4	163·0581	795·3663	+0·0237	−0·0301	469·2501	632·3082
5	203·6257	827·5299	+0·0101	+0·0115	420·2785	623·9042
6	244·0620	857·6976	+0·0124	+0·0264	369·5736	613·6356
7	284·3408	885·8451	+0·0049	+0·0065	317·1635	601·5043
8	324·4357	911·9482	−0·0082	−0·0168	263·0768	587·5125
9	364·3206	935·9831	−0·0331	+0·0581	207·3419	571·6625
10	403·9692	957·9267	−0·0667	−0·0083	149·9883	553·9575
11	443·3551	977·7554	+0·0346	+0·0048	91·0452	534·4003
12	482·4523	995·4468	+0·0232	−0·0222	30·5422	512·9945
Sums	3158·4693	10319·0323	+0·0377	−0·0737	4002·0937	7160·5630

in checking the formation of these columns are

$$2\Sigma u'(+u_0, \text{ if } n \text{ odd}) = M_0,$$

$$2\Sigma \epsilon u'' = M_1.$$

Here the first sum is 1336·2 and the second −10147·1.

The fitted values u_+ for ϵ positive are the sums $u'+u''$, the fitted values u_- for ϵ negative are the differences $u'-u''$. As a check on the formation of these sums and differences, the columns themselves are summed. If no mistakes have been made, Σu_+ is the sum of $\Sigma u'$ and $\Sigma u''$, and Σu_- the difference of these totals.

The residuals are now formed by subtracting the fitted values from the observed values y_i. Again the sums of the columns are formed as a check, using the identity

$$\Sigma v = \Sigma y - \Sigma u,$$

the value Σy being obtained from Table 7.6.2.1.

If there have been no mistakes, Σv_i will be zero, except for rounding-off errors. In the present example Σv_i is −0·2. The squares of the values v_i are then summed, giving a value 7515·44.

This compares very well with the value 7514 obtained in Table 7.6.2.1*a*, and so the calculations are free from mistakes.

When n is odd (Example 7.6.2.2), the procedure is similar. The calculations are given in Table 7.6.2.3*b*. The u and v values for $\epsilon = 0$ are not included in either the + or − columns, but are left separate. The checks on the formation of the values u' and u'' are

$$2\Sigma u' + u_0 = 14968{\cdot}3803,$$

$$2\Sigma \epsilon u'' = 52553{\cdot}5402.$$

The sum of the residuals is $-0{\cdot}0003$, and the sum of their squares $0{\cdot}02151359$. The value given in Table 7.6.2.2*a* is $0{\cdot}0215$.

7.6.3 *Fitting by orthogonal moments*

If the values $T_j(\epsilon_i)$ at the points of observation are known, it is possible to calculate the orthogonal moments

$$\Sigma T_j(\epsilon_i)\, y_i, \tag{1a}$$

and hence to find the coefficients a_j from the equation

$$a_j = \mathscr{M}_j / S_{jj}. \tag{1b}$$

Although this method is no more rapid than that using power moments, at least for polynomials of degree less than 6, cases arise in which the use of orthogonal moments may be preferred.

In general the values $T_j(\epsilon_i)$ are not integers, while it is clearly an advantage both in tabulating and in computing to work with integral values. As a consequence, the polynomial usually tabulated is that multiple of $T_j(\epsilon)$ for which the values at the points ϵ_i are the smallest possible set of integers. These polynomials have been denoted by the symbols $\xi'_j(\epsilon)$ (Fisher), $V_j(\epsilon)$ (Birge), $\phi_j(\epsilon)$ (van der Reyden), and $P_{j,n}(\epsilon)$ (Milne). Here the symbol $T'_j(\epsilon)$ will be used,

$$T'_j(\epsilon) = \beta'_{jj}\, T_j(\epsilon). \tag{2}$$

If the fitted polynomial is written

$$u_p(\epsilon) = \Sigma a'_j\, T'_j(\epsilon), \tag{3}$$

then
$$a_j = \beta'_{jj}\, a'_j. \tag{4}$$

Hence
$$a'_j = \Sigma \beta'_{jj}\, y_i\, T_j(\epsilon_i) / \Sigma \beta'^2_{jj}\, T^2_j(\epsilon_i),$$

or
$$a'_j = \Sigma y_i\, T'_j(\epsilon_i) / \Sigma T'^2_j(\epsilon_i) = \mathscr{M}'_j / S'_{jj}, \tag{5}$$

where $\mathscr{M}'_j$ is the orthogonal moment, which may be calculated from tables of $T'_j(\epsilon_i)$.

If the fitted values are required only at the points of observation, then only the coefficients a'_j are needed. If the fitted values are required at other points it is simplest to expand the curve in power-series form. The coefficient b_{pj} is given by

$$b_{pj} = \sum_{k=j}^{p} \beta'_{jk} a'_k, \tag{6}$$

where, from (7.6.2,1c) and (4),

$$\beta'_{jk} = \beta'_{jj} \beta_{jk}. \tag{7}$$

7.6.3.1 *Tables of orthogonal polynomials.* Tables of $T'_j(\epsilon_i)$ and $\Sigma T'^2_j(\epsilon_i)$ have been prepared by various authors. A summary of the tables most readily available is given below.

Author	j to	n to
1. van der Reyden (1943)	9	52
2. Fisher and Yates (1948)	5	75
3. *Biometrika Tables* (1954)	6	52
4. Anderson and Houseman (1942)	5	104
5. Birge (1947)	5	30
6. Milne (1949)	5	21

As far as is known, the only errors in these tables are those for the Anderson and Houseman table listed by Sherman in *Mathematical Tables and Other Aids to Computation*, **5**, 81 (1951). Sherman also prepared I.B.M. punched cards for the table. A table of values β'_{kj} has been prepared by Guest (1952) for $n6(1)104$ and j to 5. The values in the range 6(1)75 are reproduced in Table 7.10*c*. A table for the range 3(1)30 has been given by Birge (1947).

7.6.3.2 *Example*

In order to calculate the orthogonal moments the observations should be entered on a slip of paper in two columns, with observations corresponding to positive and negative values of ϵ side by side. The spacing should be the same as in the Table of $T'_j(\epsilon)$ to facilitate multiplication of corresponding entries. The formation of sum and difference columns, as in Tables 7.6.2.1 and 7.6.2.2, may be an advantage in preventing arithmetical mistakes. The orthogonal polynomial tables only list values $T'_j(\epsilon)$ for ϵ positive. The even moments are obtained by multiplying the sum column by the values $T'_j(\epsilon)$, the odd moments by multiplying the difference column by $T'_j(\epsilon)$. The values $T'_1(\epsilon)$ are often omitted from the tables; they are simply the quantities 0, 1, 2, ... for n odd, and 1, 3, 5, ... for n even. When the number of observations is large it is best to record the progressive total in the register of the calculating machine at a few points throughout the range, so that, if on checking the calculations a different value is obtained, the mistake can be readily located.

Table 7.6.3 shows the calculation of the coefficients from the moments $\mathcal{M}'_j$. Values S'_{jj} are given in the orthogonal polynomial tables. The power-series coefficients are calculated, using the values β'_{jk} from Table 7.10*c*, in the lower section of the scheme.

If the fitted values are required only at the points of observation, it is not essential to calculate b_{pj}. The fitted values may be calculated from the equation

$$u_p(\epsilon_i) = \sum_{j=0}^{p} a'_j T'_j(\epsilon_i),$$

using the tables of orthogonal polynomial values. Thus

$$u_3(-\tfrac{7}{2}) = a'_0 - 7a'_1 - 154a'_2 + 658a'_3 = 9{\cdot}4105.$$

TABLE 7.6.3

Fitting a third-degree curve using tables of orthogonal polynomials

$S'_{jj}(=\Sigma T'^2_j)$ and β'_{kj}

S'_{00} 62	S'_{11} 79422	S'_{22} 1270752	S'_{33} 139238112
β'_{11} 2	β'_{33} 0·3r	β'_{13} −192·083r	
	β'_{22} 0·5	β'_{02} −160·125	

		Σy^2 54038
$\mathscr{M}'_0$	1336	28789
a'_0	21·548387	Σv_0^2 25249
$\mathscr{M}'_1$	−20286	5181
a'_1	−0·25542041	Σv_1^2 20068
$\mathscr{M}'_2$	72775	4168
a'_2	0·057269239	Σv_2^2 15900
$\mathscr{M}'_3$	−1080557	8386
a'_3	−0·007760497	Σv_3^2 7514

$b_{33} = \beta'_{33} a'_3$	−0·00258683
$b_{31} = \beta'_{13} a'_3 + \beta'_{11} a'_1$	+0·979822
$b_{32} = \beta'_{22} a'_2$	0·0286346
$b_{30} = \beta'_{02} a'_2 + a'_0$	12·37815

7.6.3.3 *Calculation of fitted values for polynomials of different degrees.* The tables of orthogonal polynomials are specially useful when it is desired to calculate the fitted values at the points of observation for two or more values of the degree p. This may happen when the polynomial is merely being used to smooth the observations and the degree p can be chosen to give the most suitable curve. The fitted values are calculated from the formula

$$u(\epsilon_i) = a'_0 + a'_1 T'_1(\epsilon_i) + a'_2 T'_2(\epsilon_i) + \dots.$$

Table 7.6.3.1 shows the calculations of the fitted values for the third- and fourth-degree polynomials in the example of Table 7.6.2.1. The value a'_4 is first calculated as in Table 7.6.3:

$\mathscr{M}'_4$	−7599201	$a'_4 \mathscr{M}'_4$	557
a'_4	−0·00007332336	Σv_4^2	6957.

TABLE

Fitted values using

j	3	1	0	2	4
a'_j	$-7{\cdot}760 \times 10^{-3}$	$-0{\cdot}2554$	$21{\cdot}5$	$+0{\cdot}05727$	$-7{\cdot}332 \times 10^{-5}$
2ϵ					
1	+0·7	−0·3	21·5	−9·2	−3·4
3	+2·2	−0·8		−9·1	−3·3
5	+3·7	−1·3		−9·0	−3·2
7	+5·1	−1·8		−8·8	−3·0
9	+6·5	−2·3		−8·6	−2·7
11	+7·8	−2·8		−8·3	−2·4
13	+9·0	−3·3		−8·0	−2·0
15	+10·1	−3·8		−7·6	−1·5
17	+11·1	−4·3		−7·1	−1·1
19	+11·9	−4·9		−6·6	−0·5
21	+12·7	−5·4		−6·0	0
23	+13·2	−5·9		−5·4	+0·5
25	+13·6	−6·4		−4·7	+1·1
27	+13·8	−6·9		−4·0	+1·6
29	+13·7	−7·4		−3·1	+2·1
31	+13·5	−7·9		−2·3	+2·6
33	+13·0	−8·4		−1·4	+3·0
35	+12·2	−8·9		−0·4	+3·4
37	+11·2	−9·4		+0·6	+3·7
39	+9·9	−10·0		+1·7	+3·8
41	+8·3	−10·5		+2·9	+3·9
43	+6·3	−11·0		+4·1	+3·8
45	+4·1	−11·5		+5·3	+3·5
47	+1·5	−12·0		+6·6	+3·0
49	−1·5	−12·5		+8·0	+2·3
51	−4·9	−13·0		+9·4	+1·4
53	−8·6	−13·5		+10·9	+0·2
55	−12·8	−14·0		+12·5	−1·2
57	−17·4	−14·6		+14·1	−3·0
59	−22·4	−15·1		+15·7	−5·1
61	−27·9	−15·6		+17·5	−7·7
	+119·6	−245·5	+666·5	−0·3	−0·2

7.6.3.1

orthogonal polynomials

ε positive		ε negative			
$p = 3$	$p = 4$	$p = 3$	$p = 4$		
u_{3+}	u_{4+}	u_{3-}	u_{4-}	v_{4+}	v_{4-}
12·7	9·3	11·9	8·5	−4·3	+4·5
13·8	10·5	11·0	7·7	−4·5	+0·3
14·9	11·7	10·1	6·9	−1·7	−0·9
16·0	13·0	9·4	6·4	−5·0	−1·4
17·1	14·4	8·7	6·0	−4·4	+4·0
18·2	15·8	8·2	5·8	−2·8	+7·2
19·2	17·2	7·8	5·8	−7·2	+3·2
20·2	18·7	7·6	6·1	−15·7	+3·9
21·2	20·1	7·6	6·5	−13·1	−1·5
21·9	21·4	7·9	7·4	−5·4	−2·4
22·8	22·8	8·2	8·2	+6·2	−6·2
23·4	23·9	8·8	9·3	+13·1	−8·3
24·0	25·1	9·6	10·7	+12·9	−2·7
24·4	26·0	10·6	12·2	+24·0	−9·2
24·7	26·8	12·1	14·2	+47·2	−8·2
24·8	27·4	13·6	16·2	−5·4	+5·8
24·7	27·7	15·5	18·5	−8·7	+17·5
24·4	27·8	17·8	21·2	+16·2	+8·8
23·9	27·6	20·3	24·0	+7·4	−4·0
23·1	26·9	23·3	27·1	−11·9	−6·1
22·2	26·1	26·6	30·5	−11·1	−6·5
20·9	24·7	30·3	34·1	−6·7	−6·1
19·4	22·9	34·2	37·7	−7·9	+7·3
17·6	20·6	38·6	41·6	−10·6	+10·4
15·5	17·8	43·5	45·8	−11·8	+11·2
13·0	14·4	48·8	50·2	−10·4	+5·8
10·3	10·5	54·5	54·7	−10·5	−6·7
7·2	6·0	60·8	59·6	−3·0	−4·6
3·6	0·6	67·6	64·6	+0·4	+8·4
−0·3	−5·4	74·7	69·6	+8·4	−4·6
−4·5	−12·2	82·5	74·8	+19·2	−7·8
540·3	540·1	792·1	791·9	−7·1	+11·1

Then columns of values $a_j' T_j'(\epsilon_i)$ are formed by multiplying the values in the orthogonal polynomial table by a_j'. These may be checked by summing the column and comparing with the product of a_j' and the value $\sum_{+} T_j'(\epsilon_i)$. This latter value may be obtained by summing the values $T_j'(\epsilon_i)$ in the table of polynomials. It will be observed that

$$\sum_{+} T_j'(\epsilon_i) = 0, \qquad j \text{ even};$$

$$\sum_{+} T_1'(\epsilon_i) = \begin{cases} n^2/4, & n \text{ even}, \\ (n^2-1)/8, & n \text{ odd}, \end{cases}$$

the + under the summation sign indicating that the sum is for the positive value of ϵ for which the polynomial is tabulated.

The rows are then summed to give the fitted values for ϵ positive. The differences of the terms for even and odd values of j give the fitted values for ϵ negative. The calculation of the fitted values is checked using the sums of the columns.

As a final check, the residuals for the fourth-degree curve are formed. Since only one decimal has been retained, the rounding-off errors are rather large, and this accounts for the value $+4{\cdot}0$ for Σv_4. The value obtained for Σv^2 is 6965·82, which compares well with the value 6957 obtained using the products $a_j' \mathscr{M}_j'$.

7.6.4 *Other tables*

In special cases use may be found for certain other tables, of which those due to Kerawala and to Davis are the most important. Kerawala (1941) effectively combines the two equations

$$a_j = \Sigma y_i T_j(\epsilon_i)/\Sigma T_j^2(\epsilon_i), \quad b_{pj} = \Sigma \beta_{jk} a_k,$$

into a single equation

$$b_{pj} = \Sigma z_{pj}(\epsilon_i)\, y_i / Z_{pj}, \tag{1}$$

the quantities $z_{pj}(\epsilon_i)$ being the smallest possible set of integers. Values of $z_{pj}(\epsilon_i)$ and Z_{pj} are listed for polynomials of degrees up to the 5th and for n up to 30. This provides the most rapid method of fitting a polynomial if the degree is known, but the coefficients a_j are not determined and hence the polynomials of different degrees cannot be obtained simply.

Davis (1935) writes his equations in the form

$$b_{pj} = \Sigma \chi_{jk,\,p} M_k, \tag{2}$$

and tabulates the elements $\chi_{jk,\,p}$ of the inverse matrix. The values are given to ten significant figures, but only for odd values of n. A similar table is given by Cox and Matuschak (1941) for both odd and even values of n. Birge has suggested that there may be a considerable loss of accuracy due to the fact that the quantities $\chi_{jk,p}$ are non-terminating decimals, and it is certainly desirable to retain all ten significant figures in the individual terms of (2). Calculation of the polynomials of different degrees requires the recalculation of the sums (2) for each degree—strictly, only half the sums, as $b_{pj} = b_{p+1,j}$ for $p+j$ even.

7.6.5 *The fitted curve in terms of factorials*

The factorials are the quantities

$$x^{(j)} = x(x-1)\dots(x-j+1) = x!/(x-j)!, \tag{1a}$$

and the reduced factorials are the binomial coefficients

$$\binom{x}{j} = x^{(j)}/j! = x!/(x-j)!\,j!. \tag{1b}$$

The fitted curve may be expanded in terms of the factorials instead of the powers, in the form

$$u_p(x) = \sum_{j=0}^{p} b_{p(j)}\, x^{(j)} = \sum_{j=0}^{p} b_{p[j]}\binom{x}{j}. \tag{2}$$

The orthogonal polynomials $T_j(x)$ can also be expanded in terms of factorials,

$$T_j(x) = \Sigma\beta_{(kj)}\, x^{(k)} = \Sigma\beta_{[kj]}\binom{x}{k}. \tag{3}$$

Then the orthogonal coefficients a_j are given by

$$a_j = \mathscr{M}_j/S_{jj} = \Sigma y_i\, T_j(x_i)/S_{jj},$$

or

$$a_j = \Sigma\beta_{(kj)}\, M_{(k)}/S_{jj} = \Sigma\beta_{[kj]}\, M_{[k]}/S_{jj}, \tag{4a}$$

where $M_{(k)}$ and $M_{[k]}$ are the factorial moments,

$$M_{(k)} = \sum_i y_i\, x_i^{(k)}, \quad M_{[k]} = \Sigma y_i\binom{x_i}{k} = M_{(k)}/k!. \tag{4b}$$

Also the fitted curve is

$$\Sigma a_j\, T_j(x) = \Sigma a_j\, \Sigma\beta_{(kj)}\, x^{(k)} = \Sigma a_j\, \Sigma\beta_{[kj]}\binom{x}{k},$$

and on comparing this with (2),

$$b_{p(j)} = \Sigma\beta_{(jk)}\, a_k, \quad b_{p[j]} = \Sigma\beta_{[jk]}\, a_k. \tag{5}$$

Thus if the factorial moments are evaluated, the fitted curve can be found by means of (4a) and (5), provided the coefficients are known. Table 7.10d gives the values of $\beta_{[kj]}$ for n 2(1)75 and for polynomials up to the fifth degree. The explicit formulae for

$\beta_{[kj]}$ are given in Table 7.10e, and the proofs are given in § 7.7. The quantities $S_{jj} = \Sigma T_j^2(\epsilon_i)$ are given in Table 7.10b.

When factorial moments are employed the range of the independent variable x will be taken to be from 0 to $n-1$. Factorial methods are only of use when the observations are equally-spaced.

7.6.5.1 *Calculation of factorial moments.* The factorial moments can be obtained by repeated summation, without multiplication. The values y_i are entered in a column in decreasing order of x, the progressive sums being recorded in an adjoining column. If an adding machine is being used, the progressive sums are obtained by printing the sub-totals after each addition. These progressive sums are then treated in the same way, as illustrated in Table 7.6.5.

TABLE 7.6.5

Calculation of factorial moments

Column / Row	0	1	2
$n-1$	y_{n-1}	y_{n-1}	y_{n-1}
$n-2$	$y_{n-2}+y_{n-1}$	$y_{n-2}+2y_{n-1}$	$y_{n-2}+3y_{n-1}$
$n-3$	$y_{n-3}+y_{n-2}+y_{n-1}$	$y_{n-3}+2y_{n-2}+3y_{n-1}$	$y_{n-3}+3y_{n-2}+6y_{n-1}$
...	...	...	...
2	$y_2+y_3+\ldots+y_{n-1}$	$y_2+2y_3+\ldots$	$^*y_2+3y_3+\ldots+\frac{1}{2}(n-1)(n-2)\,y_{n-1}$
1	$y_1+y_2+\ldots+y_{n-1}$	$^*y_1+2y_2+\ldots+(n-1)\,y_{n-1}$	omitted
0	$^*y_0+y_1+\ldots+y_{n-1}$	omitted	omitted

The entry in row k, column k, will be proved to be $M_{[k]}$. More generally, the entry in row j, column k is

$$y_{jk} = \sum_{s=j}^{n-1} \binom{s+k-j}{k} y_s. \qquad (1)$$

For the entry y_{jk} is

$$y_{jk} = y_{j,\,k-1} + y_{j+1,\,k},$$

and if (1) holds for the earlier entries $y_{j,\,k-1}$ and $y_{j+1,\,k}$,

$$y_{jk} = \sum_{s=j}^{n-1} \binom{s+k-1-j}{k-1} y_s + \sum_{s=j+1}^{n-1} \binom{s+k-j-1}{k} y_s$$

$$= y_j + \sum_{s=j+1}^{n-1} \left\{ \binom{s+k-j-1}{k-1} + \binom{s+k-j-1}{k} \right\} y_s,$$

and so
$$y_{jk} = \sum_{s=j}^{n-1} \binom{s+k-j}{k} y_s$$

and (1) follows by induction. Hence
$$y_{kk} = \sum_{s=k}^{n-1} \binom{s}{k} y_s = M_{[k]} = M_{(k)}/k!, \tag{2}$$

since $\binom{s}{k}$ vanishes for values of s from 0 to $k-1$.

7.6.5.2 *Calculation of fitted values from the differences of zero.* The finite differences of $\binom{x}{k}$ are

$$\Delta\binom{x}{k} = \binom{x+1}{k} - \binom{x}{k}$$
$$= \frac{x!}{(k-1)!(x-k+1)!}\left[\frac{x+1}{k} - \frac{x-k+1}{k}\right] = \binom{x}{k-1},$$

and so
$$\Delta^q\binom{x}{k} = \binom{x}{k-q}. \tag{1}$$

Hence the finite differences of the fitted polynomial are
$$\Delta^q[u_p(x)] = \sum_{k=0}^{p} b_{p[k]}\,\Delta^q\binom{x}{k},$$

and at $x = 0$,
$$\Delta^q[u_p(0)] = \sum_{k=0}^{p} b_{p[k]}\binom{0}{k-q} = b_{p[q]} = q!\,b_{p(q)}, \tag{2}$$

as $\binom{0}{r}$ is zero except for $r = 0$.

The fitted values for integral values of x can be built up from the differences at $x = 0$ by summation. This is illustrated in Table 7.6.5.2.

7.6.5.3 *Example*

Table 7.6.5*a* shows (in condensed form) the method of calculating the moments $M_{[j]}$ for the example of § 7.6.2.1. It is not really necessary to record the progressive sums in the last column unless there is a possibility that a polynomial of higher degree may be required.

Table 7.6.5.1 gives the calculation of the orthogonal coefficients a_j, and the reduced factorial coefficients $b_{p[j]}$.

The building up of the fitted values by repeated summation is shown in Table 7.6.5.2. The differences of zero $\Delta^j[u(0)]$ are simply the coefficients $b_{p[j]}$ obtained in Table 7.6.5.1. The values $\Delta^j[u(x)]$ are obtained in turn as $\Delta^j[u(x-1)] + \Delta^{j+1}[u(x-1)]$, commencing with high values of j and finally finishing with the fitted values corresponding to $j = 0$.

TABLE 7.6.5*a*

Calculation of factorial moments (*Example* 7.6.5.3)

y	0	1	2	3
7	7	7	7	7
3	10	17	24	31
1	11	28	52	83
...	...	...	...	...
56	1028	25925	498976	7161225
48	1076	27001	525977	7687202
55	1131	28132	554109	8241311
73	1204	29336	583445	
65	1269	30605		
67	1336			

TABLE 7.6.5.1

Fitted curve in terms of factorials

n 62

(*a*) $\beta_{[kj]}$ (from Table 7.10*d*)

$\beta_{[00]}$ 1

$\beta_{[01]}$ $-30\cdot5$ $\beta_{[11]}$ 1

$\beta_{[02]}$ 610 $\beta_{[12]}$ -60 $\beta_{[22]}$ 2

$\beta_{[03]}$ -10797 $\beta_{[13]}$ 2124 $\beta_{[23]}$ -177 $\beta_{[33]}$ 6

(*b*) S_{jj} (from Table 7.10*b*)

S_{00} 62 S_{11} 19855·5 S_{22} 5,083008 S_{33} 1253,143008

(*c*) $M_{[j]}$ (from Table 7.6.5*a*)

$M_{[0]}$ 1336 $M_{[1]}$ 30605 $M_{[2]}$ 583445 $M_{[3]}$ 8,241311

$$a_j = \mathscr{M}_j/S_{jj} = \sum_{k=0}^{j} \beta_{[kj]} M_{[k]}/S_{jj}; \qquad \Sigma v^2 = \Sigma y^2 - \Sigma a_j \mathscr{M}_j$$

Σy^2 54038

$a_0 = 1336/62 = 21\cdot548387$ $\qquad \Sigma v_0^2$ 25249

$a_1 = -10143/19855\cdot5 = -0\cdot5108408$ $\qquad \Sigma v_1^2$ 20068

$a_2 = +145550/5{,}083008 = +0\cdot02863462$ $\qquad \Sigma v_2^2$ 15900

$a_3 = -3241671/1253{,}143008 = -0\cdot002586832$ $\qquad \Sigma v_3^2$ 7514

$$b_{[3j]} = \sum_{k=j}^{3} \beta_{[jk]} a_k$$

$b_{[30]}$ 82·526175 $b_{[31]}$ $-7\cdot723349$ $b_{[32]}$ $+0\cdot5151385$ $b_{[33]}$ $-0\cdot01552099$

TABLE 7.6.5.2

Fitted values obtained from the differences of zero

x	u	Δu	$\Delta^2 u$	$\Delta^3 u$
0	82·5			
		−7·723		
1	74·8		+0·51514	
		−7·208		−0·0155210
2	67·6		+0·49962	
		−6·708		−0·0155210
3	60·9		+0·48410	
		−6·224		−0·0155210
4	54·6		+0·46858	
		−5·756		−0·0155210
5	48·9		+0·45306	
		−5·302		−0·0155210
6	43·6		+0·43754	
		−4·865		−0·0155210
7	38·7		+0·42201	
		−4·443		−0·0155210
8	34·3		+0·40649	
		−4·036		−0·0155210
...	...		...	
		...		...

7.7 PROPERTIES OF THE ORTHOGONAL POLYNOMIALS FOR THE EQUALLY-SPACED CASE

The orthogonal polynomials are, in terms of the variable ϵ,

$$T_j(\epsilon) = \sum_{k=0}^{j} \beta_{kj}\, \epsilon^k, \tag{1}$$

while in terms of the variable x which takes the value 0 to $n-1$ at the points of observation

$$T_j(x) = \sum_{k=0}^{j} \beta_{(kj)}\, x^{(k)}. \tag{2}$$

It was shown in § 7.2.1 that the values of the orthogonal polynomials are independent of the origin, so that $T_j(x)$ and $T_j(\epsilon)$ are identical. It is convenient to introduce another orthogonal polynomial $P_j(x)$ for which the constant term is unity,

$$P_j(x) = T_j(x)/\beta_{(0j)} = \Sigma \pi_{kj}\, x^{(k)}. \tag{3a}$$

Then $$\pi_{kj} = \beta_{(kj)}/\beta_{(0j)}, \quad \pi_{0j} = 1, \quad \beta_{(kj)} = \pi_{kj}/\pi_{jj}. \tag{3b}$$

7.7.1 *The factorial coefficients* $\beta_{(kj)}$

A general formula for $\beta_{(kj)}$ will now be developed. The two formulae

$$(x+q)^{(q)}\, x^{(k)} = (x+q)^{(q+k)} \tag{1a}$$

and $$\sum_{x=0}^{n-1} (x+q)^{(r)} = (n+q)^{(r+1)}/(r+1) - q^{(r+1)}/(r+1) \qquad (1b)$$

will be needed in the discussion. The formula (1*a*) is obvious when both sides are written out as products of factors $(x+j)$. The second formula can be derived from the easily verified relation

$$kq^{(k-1)} + q^{(k)} = (q+1)^{(k)}.$$

On replacing k by $r+1$ and q by $n+q-1$,

$$(n+q)^{(r+1)}/(r+1) = (n-1+q)^{(r)} + (n-1+q)^{(r+1)}/(r+1).$$

Continued expansion of the last term in this equation leads to (1*b*). $q^{(r+1)}/(r+1)$ vanishes when $r \geqslant q$.

From the orthogonal property,

$$\sum_{x=0}^{n-1} (x+q)^{(q)} P_j(x) = 0, \quad q<j, \qquad (2a)$$

and so, using (1*a*) and (1*b*),

$$\sum_{k=0}^{j} \pi_{kj}(n+q)^{(q+k+1)}/(q+k+1) = 0, \quad q<j. \qquad (2b)$$

Division by the common factor $(n+q)^{(q+1)}$ gives

$$\sum_{k=0}^{j} \pi_{kj}(n-1)^{(k)}/(q+k+1) = 0, \quad q<j. \qquad (2c)$$

The solutions π_{kj} of these $j-1$ equations can be found in the following way. (2*c*) is written

$$\sum_{k=0}^{j} \pi_{kj}(n-1)^{(k)}/(q+k+1) = \sum_{k=0}^{j} z_k/(q+k+1)$$

$$= \phi(q)/(q+j+1)^{(j+1)}, \qquad (3)$$

where $$z_k = \pi_{kj}(n-1)^{(k)}$$

and $\phi(q)$ is a polynomial of degree j which is to be determined. Now, from (2*c*), $\phi(q)$ vanishes at $q = 0, 1, \ldots, j-1$, and so is of the form $Cq^{(j)}$. To determine the constant C, (3) is multiplied by $q+1$, giving

$$z_0 + \sum_{k=1}^{j} z_k(q+1)/(q+k+1) = \phi(q)/(q+j+1)^{(j)}.$$

When q is set equal to -1,

$$\phi(-1) = C(-1)^{(j)} = j^{(j)} z_0 = j^{(j)},$$

since z_0 is unity. So $$C = (-)^j$$

and $$\sum_{k=0}^{j} z_k/(q+k+1) = (-)^j q^{(j)}/(q+j+1)^{(j+1)}. \qquad (4a)$$

The value z_k is found by multiplying (4a) by $(q+k+1)$, and setting $q = -(k+1)$. Thus

$$z_k = (-)^j(-k-1)^{(j)}/(j-k)!\,(-)^k k!$$
$$= (-)^k(j+k)!/(j-k)!\,k!\,k!, \tag{4b}$$

and
$$\pi_{kj} = (-)^k\binom{j+k}{k}\binom{j}{k}\bigg/(n-1)^{(k)}. \tag{4c}$$

On dividing this by π_{jj}, the formula

$$\beta_{(kj)} = (-)^{j-k}\frac{\binom{j+k}{k}\binom{j}{k}}{\binom{2j}{j}}\frac{(n-1)^{(j)}}{(n-1)^{(k)}} \tag{5}$$

is obtained for the coefficient $\beta_{(kj)}$.

7.7.2 *The sum of the squares of the orthogonal polynomial values*

An expression for $\Sigma P_j^2(x_i)$ will first be derived. Since

$$(x+j)^{(j)} = x^{(j)} + \text{factorials of lower degree},$$
$$\sum_i P_j^2(x_i) = \pi_{jj}\sum_i x_i^{(j)}P_j(x_i) = \pi_{jj}\sum_i (x_i+j)^{(j)}P_j(x_i),$$

or
$$\Sigma P_j^2(x_i) = \pi_{jj}\sum_{k=0}^{j}\pi_{kj}\sum_i x_i^{(k)}(x_i+j)^{(j)}$$
$$= \pi_{jj}\sum_{k=0}^{j}\pi_{kj}(n+j)^{(j+k+1)}/(j+k+1).$$

Using (7.7.1,3),

$$\Sigma P_j^2(x_i) = \pi_{jj}(n+j)^{(j+1)}\sum_{k=0}^{j} z_k/(j+k+1),$$

and, substituting for the sum from (7.7.1,4a),

$$\Sigma P_j^2(x_i) = \pi_{jj}(n+j)^{(j+1)}(-)^j j!/(2j+1)^{(j+1)}. \tag{1}$$

The sum of the squares for the polynomial $T_j(x)$ will be given by

$$\Sigma T_j^2(x_i) = \Sigma P_j^2(x_i)/\pi_{jj}^2,$$

and so, from (7.7.1,4c) and (1),

$$\Sigma T_j^2(x_i) = (n+j)^{(j+1)}(n-1)^{(j)}j!/(2j+1)^{(j+1)}\binom{2j}{j}.$$

This is equivalent to the forms

$$\Sigma T_j^2(x_i) = j!^2\binom{n+j}{2j+1}\bigg/\binom{2j}{j} \tag{2a}$$

and
$$\Sigma T_j^2(x_i) = \frac{j!^4}{(2j)!\,(2j+1)!}n(n^2-1)\ldots(n^2-j^2). \tag{2b}$$

Since the orthogonal polynomials are independent of the choice of origin, these equations also give the sums $\Sigma T_j^2(\epsilon_i)$. The values

$$R_j = (2j)!\,(2j+1)!/j!^4 \tag{2c}$$

of the inverse of the numerical coefficient are given in Table 7.10*a* for the first ten polynomials.

7.7.3 *Recurrence relations*

For the equally-spaced case the quantity π_j in (7.2.6,2*a*) vanishes and

$$\epsilon T_j(\epsilon) = T_{j+1}(\epsilon) + \rho_j T_{j-1}(\epsilon), \tag{1a}$$

or

$$T_{j+1}(\epsilon) = \epsilon T_j(\epsilon) - \rho_j T_{j-1}(\epsilon), \tag{1b}$$

with

$$\rho_j = \Sigma T_j^2(\epsilon_i)/\Sigma T_{j-1}^2(\epsilon_i). \tag{2}$$

On substituting for $\Sigma T_j^2(\epsilon_i)$ from (7.7.2,2*b*),

$$\rho_j = j^2(n^2-j^2)/4(4j^2-1). \tag{3}$$

The expressions ρ_j are listed in Table 7.10*a* for polynomials up to the ninth degree.

The recurrence relation for the coefficients β_{kj} is

$$\beta_{k+1,\,j+1} = \beta_{kj} - \rho_j \beta_{k+1,\,j-1}. \tag{4}$$

The values β_{kj} can then be built up in turn. The expressions for polynomials up to the ninth degree are given in Table 7.10*a*.

7.8 EQUALLY-SPACED OBSERVATIONS WITH DIFFERENT WEIGHTS

The usual method of procedure here is to calculate the sums $\Sigma w_i y_i x_i^k$ and $\Sigma w_i x_i^{j+k}$, and to solve the resulting normal equations by the Doolittle method. The scale and origin of the independent variable may be altered so that the values x_i are successive integers and the origin is near the centre of the range. Since the quantities β_{kj} and S_{jj} will depend on the weights, they cannot be tabulated but have to be obtained from the Doolittle scheme. The procedure is identical with that given in earlier sections, and need not be discussed further.

The method of curve fitting using factorials can also be modified for use in the present case. This modification will now be considered.

7.8.1 *Factorial form of solution*

If the fitted curve is required in the factorial form

$$u_p(x) = \sum_{j=0}^{p} b_{p[j]}\binom{x}{j}, \tag{1}$$

the least-squares condition for the coefficients $b_{p[j]}$ leads to the normal equations

$$\Sigma b_{p[j]} \Sigma w_i \binom{x_i}{j}\binom{x_i}{k} = \Sigma w_i y_i \binom{x_i}{k}, \tag{2a}$$

or

$$\Sigma b_{p[j]} \phi_{[jk]} = M_{[k]}. \tag{2b}$$

The moments $M_{[k]}$ can be calculated by repeated summation of the values $w_i y_i$, as in § 7.6.5.1, but the quantities $\phi_{[jk]}$ cannot be directly calculated in this way. However, it will be shown in § 7.8.1.2 below that

$$\binom{x}{j}\binom{x}{k} = \sum_{q=0}^{k} \binom{k}{q}\binom{j+q}{k}\binom{x}{j+q}, \tag{3a}$$

and so it follows that the quantities $\phi_{[jk]}$ can be expressed as linear sums of the values

$$W_{[r]} = \Sigma w_i \binom{x_i}{r}.$$

These latter values can be obtained by repeated summation of the quantities w_i. The linear sums are

$$\phi_{[jk]} = \sum_{q=0}^{k} \binom{k}{q}\binom{j+q}{k} W_{[j+q]}. \tag{3b}$$

These formulae are written out in detail in Table 7.10*f* for polynomials up to the fifth degree.

Once the $\phi_{[jk]}$ have been calculated, the equations (2*a*, *b*) are solved by the standard Doolittle method.

7.8.1.1 *Example*

Table 7.8.1 shows a series of observations y which are spaced at equal intervals of the variable x. Observations at some of the values of x are missing. This example can be treated by the method of § 7.8.1 by setting w_i equal to 1 for the observations which are present and equal to 0 for the observations which are missing. Table 7.8.1.1 shows a portion of the scheme for the calculation of $M_{[k]}$ and $W_{[k]}$ by repeated summation.

The values $\phi_{[jk]}$ are then evaluated in Table 7.8.1.2, using the formulae of Table 7.10*f*. Finally the normal equations are solved by the Doolittle technique. The fitted curve is obtained by multiplying the coefficients in the Doolittle scheme by the factors $10^r\, 10^{-qj} = 10 \times 10^{-j}$. The curve is

$$u(x) = 0{\cdot}4875 + 0{\cdot}4253x - 0{\cdot}02950\tbinom{x}{2} + 0{\cdot}003516\tbinom{x}{3}.$$

It will be observed that the number of significant figures is rather small, because of the small value of S_{33}.

TABLE 7.8.1

Equally-spaced observations with some of the set missing

x	y	x	y	x	y
0	0·4				
1	0·9	11	—	21	7·7
2	1·4	12	4·1	22	8·3
3	—	13	4·6	23	—
4	1·9	14	—	24	9·6
5	2·5	15	5·2	25	10·6
6	—	16	—	26	11·3
7	2·9	17	5·9	27	—
8	3·4	18	6·6	28	12·5
9	—	19	7·0	29	13·7
10	4·2	20	7·8	30	14·7

TABLE 7.8.1.1

Calculation of factorials (*Example* 7.8.1.1)

x	wy	0	1	2	3
...	...	...	...	...	...
7	2·9	140·1	2265·8	22299·3	162700·1
6	—	140·1	2405·9	24705·2	187405·3
5	2·5	142·6	2548·5	27253·7	214659·0
4	1·9	144·5	2693·0	29946·7	244605·7
3	—	144·5	2837·5	32784·2	277389·9
2	1·4	145·9	2983·4	35767·6	
1	0·9	146·8	3130·2		
0	0·4	147·2			

x	w	0	1	2	3	4	5	6
...	...	...	...	...	...	...	...	...
7	1	18	236	2104	14425	81679	398600	1723998
6	0	18	254	2358	16783	98462	497062	2221060
5	1	19	273	2631	19414	117876	614938	
4	1	20	293	2924	22338	140214		
3	0	20	313	3237	25575			
2	1	21	334	3571				
1	1	22	356					
0	1	23						

TABLE 7.8.1.2

Calculation of fitted curve in terms of factorials

(*a*) Quantities $W_{[j]}$ (Table 7.8.1.1)

$W_{[0]}$	23	$W_{[1]}$	356	$W_{[2]}$	3571	$W_{[3]}$	25575
		$W_{[4]}$	140214	$W_{[5]}$	614938	$W_{[6]}$	2221060

(*b*) Quantities $M_{[j]}$ (Table 7.8.1.1)

$M_{[0]}$	147·2	$M_{[1]}$	3130·2	$M_{[2]}$	35767·6	$M_{[3]}$	277389·9

(*c*) Elements of matrix (Table 7.10*f*)

$\phi_{[00]}$	23	$\phi_{[01]}$	356	$\phi_{[02]}$	3571	$\phi_{[03]}$	25575
		$\phi_{[11]}$	7498	$\phi_{[12]}$	83867	$\phi_{[13]}$	637581
				$\phi_{[22]}$	998305	$\phi_{[23]}$	7908673
						$\phi_{[33]}$	64577483

(*d*) Abbreviated Doolittle scheme

Factors removed: $\phi_{[jk]}$, $10^s\, 10^{q(j+k)}$; $M_{[j]}$, $10^s\, 10^r\, 10^{qj}$; $q = 1$, $r = 1$, $s = 1$

2·300000	3·560000	3·571000	2·557500	1·472000	13·460500
	7·498000	8·386700	6·375810	3·130200	28·950710
		9·983050	7·908673	3·576760	33·426183
			6·457748	2·773899	26·073630
2·300000	3·560000	3·571000	2·557500	1·472000	13·460500
1	1·547826	1·552609	1·111957	0·640000	5·852391
					392
	1·987739	2·859412	2·417243	0·851800	8·116198
	1	1·438525	1·216077	0·428527	4·083131
					129
		0·325348	0·460609	0·065985	0·851941
		1	1·415744	0·202813	2·618553
					557
			0·022260	0·007828	0·030091
			1	0·351645	1·351802
					645
0·048752	0·425302	−0·295026	0·351645		
1·048716	1·425440	0·704747	1·351802		

7.8.1.2 *Calculation of factorial products by summation.* In this section the relation (7.8.1,3*a*) will be established. It will first be shown that

$$x^{(j)}\, x^{(k)} = \left[\sum_{q=0}^{r} \binom{r}{q} j^{(r-q)}\, x^{(j+q)}\right] (x-r)^{(k-r)}. \tag{1}$$

If (1) is true for a particular value of r, then

$$x^{(j)}x^{(k)} = \sum_{q=0}^{r}\binom{r}{q} j^{(r-q)} x^{(j+q)} (x-r)(x-r-1)^{(k-r-1)}$$

$$= \left[\sum_{q=0}^{r}\binom{r}{q} j^{(r-q)} \{x^{(j+q+1)} + (j+q-r)\,x^{(j+q)}\}\right] \times (x-r-1)^{(k-r-1)}$$

$$= \left[\sum_{q=1}^{r}\left(x^{(j+q)} j^{(r+1-q)}\left\{\binom{r}{q}+\binom{r}{q-1}\right\}\right) + x^{(j+r+1)} + j^{(r+1)}x^{(j)}\right] \times (x-r-1)^{(k-r-1)}.$$

Since $$\binom{r}{q}+\binom{r}{q-1} = \binom{r+1}{q},$$

it follows that

$$x^{(j)}x^{(k)} = \left[\sum_{q=0}^{r+1}\binom{r+1}{q} j^{(r+1-q)} x^{(j+q)}\right](x-r-1)^{(k-r-1)},$$

and so (1) follows by induction.

For the value $r = k$, (1) takes the form

$$x^{(j)}x^{(k)} = \sum_{q=0}^{k}\binom{k}{q} j^{(k-q)} x^{(j+q)}. \tag{2}$$

Then

$$\binom{x}{j}\binom{x}{k} = \sum_{q=0}^{k}\binom{k}{q}\frac{j!}{(j-k+q)!}\frac{x!}{(x-j-q)!}\frac{1}{j!\,k!}\frac{(j+q)!}{(j+q)!},$$

and the terms on the right can be regrouped to give

$$\binom{x}{j}\binom{x}{k} = \sum_{q=0}^{k}\binom{k}{q}\binom{j+q}{k}\binom{x}{j+q},$$

which is (7.8.1,3a).

7.8.1.3 *Factorial sums with the origin near the centre of the range.* If the number of observations is large, the sums may be reduced in magnitude by choosing the origin of x near the centre of the range of x. The observations are divided into two groups, each group being summed from the extreme values of x towards the origin, as illustrated in Table 7.8.1.3. It may be shown that

$$M_{[k]} = \sum_{i=-r}^{s-1} y_i \binom{i}{k} = y_{kk} + (-)^k y_{-1,\,k}, \tag{1a}$$

where y_{jk} is the element in row j and column k. The sums in the two halves are added for even values of k and subtracted for odd values.

The proof of (1a) follows on the same lines as that in § 7.6.5.1. It was shown there that

$$\sum_{i=0}^{s-1} y_i \binom{i}{k} = y_{kk}. \tag{1b}$$

Similarly, from (7.6.5.1,1),

$$y_{-1,k} = \sum_{q=1}^{r} \binom{q+k-1}{k} y_{-q}.$$

When q is replaced by $-i$,

$$\binom{q+k-1}{k} = (q+k-1)(q+k-2)\ldots q/k! = (-)^k \binom{i}{k},$$

and so

$$y_{-1,k} = (-)^k \sum_{i=-1}^{-r} \binom{i}{k} y_i. \tag{1c}$$

On combining (1b) and (1c), (1a) is established,

The choice of origin near the centre of the range is only worthwhile when the observations are weighted and the sums $\Sigma w_i \binom{x_i}{j}$ have to be calculated. If the observations are all of equal weight, the method of § 7.6.5.1 should be used.

TABLE 7.8.1.3

Factorial sums with the origin near the centre of the range

	(0)	(1)	(2)
↓ y_{-r}	y_{-r}	y_{-r}	y_{-r}
y_{-r+1}	$y_{-r+1}+y_{-r}$	$y_{-r+1}+2y_{-r}$	$y_{-r+1}+3y_{-r}$
$\cdots$	$\cdots$	$\cdots$	$\cdots$
y_{-1}	$*y_{-1}+y_{-2}+\ldots+y_{-r}$	$*y_{-1}+2y_{-2}+\ldots+ry_{-r}$	$*y_{-1}+3y_{-2}+\ldots$
↑ y_0	$*y_0+y_1+\ \ldots+y_{s-1}$	omitted	omitted
y_1	$y_1+y_2+\ \ldots+y_{s-1}$	$*y_1+2y_2+\ \ldots$	omitted
y_2	$y_2+y_3+\ \ldots+y_{s-1}$	$y_2+2y_3+\ \ldots$	$*y_2+3y_3+\ \ldots$
$\cdots$	$\cdots$	$\cdots$	$\cdots$
y_{s-1}	y_{s-1}	y_{s-1}	y_{s-1}

7.8.2 *Observations of equal weight, but some of the series missing*

It often happens that, although the interval between successive observations is constant and they are of equal weight, several of the series have not been recorded. For example, the observations may have been made at equal intervals of time, but poor conditions may have prevented the taking of some of the set.

TABLE 7.8.2

Equally-spaced case, observations missing: range of x, $-r'$ to $+r$

M_3	M_1	M_0	M_2
$\Sigma x^3(L-R)$	$\Sigma x(L-R)$	$\Sigma(L+R)+y(0)$	$\Sigma x^2(L+R)$
138217·2	922·2	147·2	13879·4

	Σx^5	Σx^3	Σx		n		Σx^2	Σx^4	Σx^6
$\sum_0^r$				$r+1$	16	$\sum_0^r$	1240	178312	30,482920
$-\sum_0^{r'}$				$+r'$	15	$+\sum_0^{r'}$	1240	178312	30,482920
	+ terms marked			− no. missing	− 8		+ terms marked		
	35081	497	11		23		1993	302941	54,149533

		y			
		L	R		
		$x+$ missing terms marked	$x-$ missing terms marked		
−					−
+					−

x^5	x^3	x	L	R	x^2	x^4	x^6
		0	5·2				
± 1	± 1	± 1	—	—	= 1	= 1	= 1
32	8	2	5·9	4·6	4	16	64
243	27	3	6·6	4·1	9	81	729
+ 1024	+ 64	+ 4	7·0	—	− 16	− 256	− 4096
3125	125	5	7·8	4·2	25	625	15625
+ 7776	+ 216	+ 6	7·7	—	− 36	− 1296	− 46656
16807	343	7	8·3	3·4	49	2401	117649
− 32768	− 512	− 8	—	2·9	− 64	− 4096	− 262144
+ 59049	+ 729	+ 9	9·6	—	− 81	− 6561	− 531441
100000	1000	10	10·6	2·5	100	10000	1,000000
161051	1331	11	11·3	1·9	121	14641	1,771561
± 248832	± 1728	± 12	—	—	= 144	= 20736	= 2,985984
371293	2197	13	12·5	1·4	169	28561	4,826809
537824	2744	14	13·7	0·9	196	38416	7,529536
759375	3375	15	14·7	0·4	225	50625	11,390625
1,048576	4096	16			256	65536	16,777216
1,419857	4913	17			289	83521	24,137569
1,889568	5832	18			324	104976	34,012224
2,476099	6859	19			361	130321	47,045881
3,200000	8000	20			400	160000	64,000000
4,084101	9261	21			441	194481	85,766121
5,153632	10648	22			484	234256	113,379904
6,436343	12167	23			529	279841	148,035889
7,962624	13824	24			576	331776	191,102976
9,765625	15625	25			625	390625	244,140625
11,881376	17576	26			676	456976	308,915776
14,348907	19683	27			729	531441	387,420489
17,210368	21952	28			784	614656	481,890304
20,511149	24389	29			841	707281	594,823321
24,300000	27000	30			900	810000	729,000000

If more than a third of the observations are missing, then no special technique can produce any great saving in time. But if less than a third of the observations are missing, the scheme of Table 7.8.2 is worth using. Here the elements $\phi_{jk} = \Sigma x^{j+k}$ are calculated by subtracting the contributions of the missing elements from a tabulated value $\sum_0^r x^{j+k}$. The procedure is illustrated in Example 7.8.2.1.

The method of repeated summation (§ 7.8.1) can also be used, with the weights of the missing observations being taken as zero, as in Example 7.8.1.1.

7.8.2.1 *Example*

Table 7.8.2 shows the calculation of ϕ_{jk} and M_j for the observations listed in Table 7.8.1. The origin is chosen near the centre of the range, so that the range of x is from $-r'$ to $+r$. Here r and r' are both 15. The moments are calculated in the usual way. The quantities Σx^j are calculated by subtracting from $\sum_0^r x^j + (-)^j \sum_0^{r'} x^j$ the contributions due to the missing terms. These missing terms are obtained by marking the values of x^j in the calculating scheme, the sums Σx^j from Table 7.10g.

The solution of the normal equations is then carried out by the standard Doolittle procedure in Table 7.8.2.1.

TABLE 7.8.2.1

Solution of normal equations (*Example* 7.8.2.1)

Factors removed: x, $10^q, q = 1$; y, $10^r, r = 1$; elements divided by $10^s, s = 1$

2·300000	0·110000	1·993000	0·049700	1·472000	5·924700
	1·993000	0·049700	3·029410	0·922200	6·104310
		3·029410	0·035081	1·387940	6·495131
			5·414953	1·382172	9·911316
2·300000	0·110000	1·993000	0·049700	1·472000	5·924700
1	0·047826	0·866522	0·021609	0·640000	2·575956
					957
	1·987739	−0·045617	3·027033	0·851800	5·820955
	1	−0·022949	1·522852	0·428527	2·928430
					430
		1·301385	0·061482	0·131968	1·494837
		1	0·047244	0·101406	1·148651
					650
			0·801251	0·046964	0·848215
			1	0·058613	1·058614
					613
0·536928	0·341532	0·098637	0·058613		
1·536926	1·341530	1·098638	1·058614		

7.8.2.2 *Hartley's method.* The subscript i will be used to denote one of the set of n possible points of observation. Points for which an observation was obtained will be denoted by the subscript o, and points for which an observation was missed by the subscript m. The fitted curve will be written in the form

$$u_p(\epsilon) = \sum_j a_j T_j(\epsilon),$$

where the polynomials $T_j(\epsilon)$ are orthogonal over the n points ϵ_i. The least-squares principle states that the coefficients a_j are to be chosen so that

$$\sum_o \left\{y_o - \sum_j a_j T_j(\epsilon_o)\right\}^2$$

is minimized, and so the normal equations are

$$\sum_o \left\{y_o - \sum_j a_j T_j(\epsilon_o)\right\} T_k(\epsilon_o) = 0. \qquad (1a)$$

If these equations were solved for the a_j, and the fitted values

$$y_m \equiv u_p(\epsilon_m) = \sum_j a_j T_j(\epsilon_m) \qquad (1b)$$

at the missing points calculated, then the normal equations (1a) can be written as

$$\sum_i \left\{y_i - \sum_j a_j T_j(\epsilon_i)\right\} T_k(\epsilon_i) = 0. \qquad (1c)$$

The value y_i is the observed value y_o or, if the particular observation is missing, the fitted value y_m. Since the polynomials are orthogonal over the ϵ_i,

$$a_j = \sum_i y_i T_j(\epsilon_i) / \sum_i T_j^2(\epsilon_i) = \left\{\sum_o y_o T_j(\epsilon_o) + \sum_m y_m T_j(\epsilon_m)\right\} \Big/ \Sigma T_j^2(\epsilon_i). \qquad (2)$$

The value $\Sigma T_j^2(\epsilon_i)$ can be obtained from the tables.

In practice, the polynomials $T'_j(\epsilon)$ introduced in § 7.6.3 are used, and

$$a'_j = \left\{\sum_o y_o T'_j(\epsilon_o) + \sum_m y_m T'_j(\epsilon_m)\right\} \Big/ \sum_i T'^2_j(\epsilon_i), \qquad (3a)$$

with

$$y_m = \sum_j a'_j T'_j(\epsilon_m). \qquad (3b)$$

Equations (3a) and (3b) are solved by an iterative process. Reasonable values $y_m^{\{1\}}$ are first assumed, and the values $a_j'^{\{1\}}$ calculated from (3a):

$$a_j'^{\{1\}} = \{\Sigma y_o T'_j(\epsilon_o) + \Sigma y_m^{\{1\}} T'_j(\epsilon_m)\} / \Sigma T'^2_j(\epsilon_i).$$

Second approximations $y_m^{\{2\}}$ are obtained by substituting $a_j'^{\{1\}}$ in (3*b*). These are substituted back in the second term of (3*a*) to give a second approximation $a_j'^{\{2\}}$. The iterative process can be continued, better and better approximations to the values a_j' and y_m being obtained.

If only the fitted curve and not the standard deviations are required, it will usually be satisfactory to stop at the second approximation $a_j'^{\{2\}}$. If the sum of the squares of the residuals is to be calculated by the formulae of § 7.2.5, it is necessary to proceed at least to the third approximation. The approximations approach a_j' from one side only, and it is often possible to get a better approximation by extrapolating the trend of the values $a_j'^{\{1\}}, a_j'^{\{2\}}, a_j'^{\{3\}}$.

It is possible to work with the corrections $\Delta a_j'$ and Δy_m rather than with the full values a_j' and y_m. This is the procedure suggested by Hartley. However, if only the corrections are used, a mistake in one of the earlier approximations may be undetected, while if the full values are used all earlier mistakes are automatically corrected by the iterative process.

7.8.2.3 *Example*

Table 7.8.2.2 shows the application of the Hartley method to the observations of Table 7.8.1. There are 23 observations, and 8 of the set are missing. The moments $\Sigma y_0 T_j'$ for the observations actually made are calculated, using the tabulated orthogonal polynomials for $n = 31$. The values T_j' for the missing observations are entered in the scheme, and values $y_m^{\{1\}}$ are assumed. The products $y_m^{\{1\}} T_j'$ are added to $\Sigma y_o T_j'$, and the sum is divided by $\Sigma T_j'^2$ to give the first approximations $a_j'^{\{1\}}$. The second approximations $y_m^{\{2\}}$ are then calculated from these values $a_j'^{\{1\}}$, and the second approximations $a_j'^{\{2\}}$ from the values $y_m^{\{2\}}$.

When the third approximations have been calculated, new values $a_j'^{\{4\}}$ are assumed from the trend of the previous values. For example,

$$a_0'^{\{2\}} - a_0'^{\{1\}} = 0{\cdot}0082, \quad a_0'^{\{3\}} - a_0'^{\{2\}} = 0{\cdot}0015,$$

and so the next correction might be of the order of $\frac{15}{82} \times 0{\cdot}0015$ or $0{\cdot}0003$, the following correction $\frac{15}{82} \times 0{\cdot}0003$ or $0{\cdot}0001$. Thus a reasonable value for $a_0'^{\{4\}}$ is $6{\cdot}1580 + 0{\cdot}0003 + 0{\cdot}0001$, or $6{\cdot}1584$. The values $a_j'^{\{4\}}$ may then be used to give the next two approximations. The process is continued until the required number of significant figures has been obtained.

It will be noted that the coefficients a_j' are for the polynomials with $n = 31$. If the curve is required in power-series form, the values β_{kj}' for $n = 31$ must be used.

7.8.2.4 *Direct calculation of values at missed points.* If the missing values are ignored in (7.8.2.2,3*a*), the coefficients

$$a_{j0}' = \Sigma y_0 T_j'(\epsilon_0)/\Sigma T_j'^2(\epsilon_i) = \Sigma y_0 T_j'(\epsilon_0)/S_{jj}'$$

TABLE 7.8.2.2

Hartley's iterative method

	T'_0	T'_1	T'_2	T'_3
$\Sigma T'^2_j$	31	2480	158224	6724520
$\Sigma y_0 T'_j$	147·2	922·2	2103·4	4670·7

Missing values							
$y_m^{\{1\}}$	$y_m^{\{2\}}$	$y_m^{\{3\}}$	$y_m^{\{4\}}$	$y_m^{\{5\}}$	T'_1	T'_2	T'_3
11·9	11·8880	11·8968	11·9015	11·90096	+12	+64	+2
8·9	8·9975	9·0262	9·0332	9·03301	+8	−16	−532
5·5	5·6987	5·7173	5·7213	5·72127	+1	−79	−119
4·9	5·0238	5·0347	5·0370	5·03700	−1	−79	+119
4·1	4·1242	4·1236	4·1234	4·12339	−4	−64	+426
3·8	3·5566	3·5501	3·5484	3·54848	−6	−44	+539
2·7	2·6798	2·6699	2·6668	2·66705	−9	+1	+471
1·6	1·6841	1·6809	1·6778	1·67833	−12	+64	−2

$a_j'^{\{1\}}$ (from $y_m^{\{1\}}$)	6·148387	0·425040	0·00996372	0·00073635
$a_j'^{\{2\}}$ (from $y_m^{\{2\}}$)	6·156539	0·425543	0·00987977	0·00070789
$a_j'^{\{3\}}$ (from $y_m^{\{3\}}$)	6·158048	0·425750	0·00986640	0·00070424
$a_j'^{\{4\}}$ (from trend)	6·1584	0·42587	0·009864	0·0007036
$a_j'^{\{5\}}$ (from $y_m^{\{4\}}$)	6·158368	0·425826	0·0098637	0·00070329
$a_j'^{\{6\}}$ (from $y_m^{\{5\}}$)	6·158371	0·425819	0·00986374	0·00070333

Power-series coefficients

β'_{11} 1	β'_{33} 0·83r	β'_{13} −119·83r	β'_{22} 1	β'_{02} −80
	b_{33} 0·00058611		b_{32} 0·00986374	
	b_{31} 0·341537		b_{30} 5·369272	

and the fitted values

$$u_0(\epsilon_m) = \Sigma a'_{j0}\, T'_j(\epsilon_m)$$

are obtained. Then (7.8.2.2,3*b*) becomes

$$y_r = u_0(\epsilon_r) + \Sigma\Sigma y_m\, T'_j(\epsilon_m)\, T'_j(\epsilon_r)/S'_{jj},$$

and so

$$\sum_m \left(\delta_{rm} - \sum_j T'_j(\epsilon_m)\, T'_j(\epsilon_r)/S'_{jj}\right) y_m = u_0(\epsilon_r), \tag{1}$$

δ_{rm} being the Kronecker delta. To solve these equations for y_m it is necessary to invert the matrix whose elements are the quantities in brackets on the left-hand side. Thus a direct calculation of y_m is really only practicable when one or two observations are missing. If one observation is missing,

$$y_m = u_0(\epsilon_m) \Big/ \left\{1 - \sum_j T'^2_j(\epsilon_m)/S'_{jj}\right\}. \tag{2}$$

If two observations are missing,

$$y_{m1} = G_{11}\,u_0(\epsilon_{m1}) + G_{12}\,u_0(\epsilon_{m2}), \quad y_{m2} = G_{21}\,u_0(\epsilon_{m1}) + G_{22}\,u_0(\epsilon_{m2}), \tag{3}$$

where

$$G_{11} = \{1 - \Sigma T_j'^2(\epsilon_{m2})/S_{jj}'\}/D, \quad G_{22} = \{1 - \Sigma T_j'^2(\epsilon_{m1})/S_{jj}'\}/D, \tag{4a}$$

$$G_{12} = G_{21} = \{\Sigma T_j'(\epsilon_{m1})\,T_j'(\epsilon_{m2})/S_{jj}'\}/D, \tag{4b}$$

and
$$D = \{1 - \Sigma T_j'^2(\epsilon_{m1})/S_{jj}'\}\{1 - \Sigma T_j'^2(\epsilon_{m2})/S_{jj}'\} - \{\Sigma T_j'(\epsilon_{m1})\,T_j'(\epsilon_{m2})/S_{jj}'\}^2. \tag{4c}$$

When the y_m have been calculated, the coefficients a_j' are given by

$$a_j' = a_{j0}' + \sum_m y_m\,T_j'(\epsilon_m)/S_{jj}'. \tag{5}$$

As a check on the calculations, the values $\Sigma a_j'\,T_j'(\epsilon_m)$ may be calculated. These values should equal y_m.

7.8.2.5 *Example*

In Table 7.8.2.5 a cubic curve is fitted to the 23 observations obtained by omitting the values at $\epsilon = +7$ and $\epsilon = -3$ from the set listed in Table 7.6.2.2. The moments $\mathscr{M}_{j0}'$ are calculated using a table of orthogonal polynomials, omitting the contributions at $\epsilon = +7$ and $\epsilon = -3$.

TABLE 7.8.2.5

Direct calculation of values at missed points

	$\epsilon_{m1}+7,\quad \epsilon_{m2}-3$			
$\mathscr{M}_{j0}'$	13566·11	47901·85	−25324·19	112697·80
S_{jj}'	25	1300	53820	1,480050
a_{j0}'	542·6444	36·847577	−0·4705349	+0·07614459
$T_j'(\epsilon_{m1})$	1	+7	−3	−259
$T_j'(\epsilon_{m2})$	1	−3	−43	+211
$1-\Sigma T_j'^2(\epsilon_{m2})/S_{jj}'$	0·88864092			
$1-\Sigma T_j'^2(\epsilon_{m1})/S_{jj}'$	0·87681700			
$\Sigma T_j'(\epsilon_{m1})\,T_j'(\epsilon_{m2})/S_{jj}'$	−0·01068072			
(7.8.2.4,4a–c) D	0·77906139			
G_{11}	1·14065583			
G_{12}	−0·01370973			
G_{22}	1·12547870			
$u_0(\epsilon_{mi})$	782·26759	468·40118		
(7.8.2.4,3) y_{mi}	885·87643	516·45087		
$\Sigma y_{mi}\,T_j'(\epsilon_{mi})/S_{jj}'$	56·09309	3·578294	−0·4620033	−0·08139648
a_j'	598·73749	40·425871	−0·9325382	−0·00525189
Check $\Sigma a_j'\,T_j'(\epsilon_{mi})$	885·87644	516·45087		

7.9 NOTES AND REFERENCES

(7.1) The Gauss–Doolittle method and its variants have been rediscovered many times. An account of the various forms of solution of linear equations is given by Dwyer (1951).

(7.2) The relations between the orthogonal polynomials and the quantities occurring in the Doolittle scheme were discussed by Guest (1950*a*). A calculating scheme similar to Table 7.2.3 is given by Wishart and Metakides (1953).

(7.3) The use of matrix theory in the discussion of least-squares fitting is well presented by Hayes and Vickers (1951).

(7.4) The discussion of changes of origin is based on that of Birge (1947).

(7.5) A very large number of papers on the solution of linear equations have appeared in recent years. Full bibliographies are given by Bodewig (1947), Paige and Taussky (1953), and Taussky (1954); see also Forsythe (1953*a*). A general bibliography of papers on numerical analysis has been prepared by Householder (1953, 1956).

The normal equations are often ill-conditioned (Booth, 1955; Riley, 1955). A simple account of the troubles encountered with ill-conditioned equations is given by Deming (1937).

(7.6) A useful treatment of the fitting of polynomials to equally-spaced observations, with a historical account of earlier work, is given by Birge (1947).

The advantage of fitting by power moments rather than by orthogonal moments is that less quantities require tabulation. It is convenient to prepare mimeographed sheets of the powers ϵ^j in Tables 7.6.2.1 and 7.6.2.2, so that the observations can be entered directly beside the powers. The difference in the times required for the two methods is so slight that the choice of method is largely a matter of personal preference.

(7.7) The treatment of the properties of the orthogonal polynomials for the equally-spaced case is based on Milne (1949), Birge (1947), and Allan (1930).

(7.8) The factorial method is discussed by Fisher (1948), Aitken (1933*b*), and Guest (1953*b*). The 'missing plot' method of § 7.8.2.2 is due to Hartley (1951), and is the quickest method if only a small number of observations are missing. It should also be very suitable for use with high-speed automatic computers. However, the usual formulae given in Chapter 8 for the standard deviations of the coefficients and fitted values are not valid when this method is used.

7.10 TABLES

TABLE 7.10*a*

Orthogonal polynomials for equally-spaced observations

(*a*) $S_{jj} = \sum_i T_j^2(\epsilon_i) = n(n^2-1)(n^2-4)\dots(n^2-j^2)/R_j$

Values of $R_j = (2j)!\,(2j+1)!/j!^4$

j	R_j	Factors	j	R_j	Factors
0	1		6	11,099088	$2^4\,3^2\,7^2\,11^2\,13$
1	12	$2^2\,3$	7	176,679360	$2^6\,3^3\,5\;\;11^2\,13^2$
2	180	$2^2\,3^2\,5$	8	2815,827300	$2^2\,3^4\,5^2\,11^2\,13^2\,17$
3	2800	$2^4\,5^2\,7$	9	44914,183600	$2^4\,5^2\,11^2\,13^2\,17^2\,19$
4	44100	$2^2\,3^2\,5^2\,7^2$			
5	698544	$2^4\,3^4\,7^2\,11$			

TABLE 7.10a (cont.)

(b) $\rho_j = S_{jj}/S_{j,\,j-1} = j^2(n^2-j^2)/4(4j^2-1)$

j	ρ_j	j	ρ_j
1	$(n^2-1)/12$	6	$9(n^2-36)/143$
2	$(n^2-4)/15$	7	$49(n^2-49)/780$
3	$9(n^2-9)/140$	8	$16(n^2-64)/255$
4	$4(n^2-16)/63$	9	$81(n^2-81)/1292$
5	$25(n^2-25)/396$		

(c) Values β_{kj} $[T_j(\epsilon) = \Sigma\beta_{kj}\,\epsilon^k]$

(i) $\beta_{02} = -(n^2-1)/12$

(ii) $\beta_{13} = -(3n^2-7)/20$

(iii) $\beta_{24} = -(3n^2-13)/14$ $\qquad \beta_{04} = +3(n^2-1)(n^2-9)/560$

(iv) $\beta_{35} = -5(n^2-7)/18$ $\qquad \beta_{15} = +(15n^4-230n^2+407)/1008$

(v) $\beta_{46} = -5(3n^2-31)/44$ $\qquad \beta_{26} = +(5n^4-110n^2+329)/176$

$\beta_{06} = -5(n^2-1)(n^2-9)(n^2-25)/14784$

(vi) $\beta_{57} = -7(3n^2-43)/52$ $\qquad \beta_{37} = +7(15n^4-450n^2+2051)/2288$

$\beta_{17} = -(35n^6-1645n^4+17297n^2-27207)/27456$

(vii) $\beta_{68} = -7(n^2-19)/15$ $\qquad \beta_{48} = +7(3n^4-118n^2+763)/312$

$\beta_{28} = -(105n^6-6405n^4+91679n^2-231491)/34320$

$\beta_{08} = +7(n^2-1)(n^2-9)(n^2-25)(n^2-49)/329472$

(viii) $\beta_{79} = -3(3n^2-73)/17$ $\qquad \beta_{59} = +21(3n^4-150n^2+1307)/680$

$\beta_{39} = -(21n^6-1617n^4+30387n^2-112951)/3536$

$\beta_{19} = +3(105n^8-11060n^6+334054n^4-2973140n^2$
$+4370361)/3111680$

(d) Values α_{kj} $[\epsilon^j = \Sigma\alpha_{kj}\,T_k(\epsilon)]$

(i) $\alpha_{02} = (n^2-1)/12$

(ii) $\alpha_{13} = (3n^2-7)/20$

(iii) $\alpha_{24} = (3n^2-13)/14$ $\qquad \alpha_{04} = (n^2-1)(3n^2-7)/240$

(iv) $\alpha_{35} = 5(n^2-7)/18$ $\qquad \alpha_{15} = (3n^4-18n^2+31)/112$

(v) $\alpha_{46} = 5(3n^2-31)/44$ $\qquad \alpha_{26} = (5n^4-50n^2+157)/112$

$\alpha_{06} = (3n^4-18n^2+31)(n^2-1)/1344$

(vi) $\alpha_{57} = 7(3n^2-43)/52$ $\qquad \alpha_{37} = 7(15n^4-230n^2+1127)/1584$

$\alpha_{17} = (5n^6-55n^4+239n^2-381)/960$

(vii) $\alpha_{68} = 7(n^2-19)/15$ $\qquad \alpha_{48} = 21(5n^4-110n^2+769)/1144$

$\alpha_{28} = (5n^6-85n^4+611n^2-1731)/528$

$\alpha_{08} = (5n^6-55n^4+239n^2-381)(n^2-1)/11520$

(viii) $\alpha_{79} = 3(3n^2-73)/17$ $\qquad \alpha_{59} = 21(3n^4-90n^2+847)/520$

$\alpha_{39} = 5(21n^6-525n^4+5635n^2-24187)/6864$

$\alpha_{19} = (3n^8-52n^6+410n^4-1636n^2+2555)/2816$

TABLE 7.10*b*

Numerical values of S_{jj} and β_{kj}

For non-terminating decimals r indicates that the last figure is repeated indefinitely; s indicates that the figures 142857 are repeated, beginning with the digit immediately before the s

$n = S_{00}$	S_{11}	S_{22}	S_{33}	β_{02}	β_{13}
6	17·5	37·3r	64·8	−2·916r	−5·05
7	28	84	216	−4	−7
8	42	168	594	−5·25	−9·25
9	60	308	1425·6	−6·6r	−11·8
10	82·5	528	3088·8	−8·25	−14·65
11	110	858	6177·6	−10	−17·8
12	143	1334·6r	11583	−11·916r	−21·25
13	182	2002	20592	−14	−25
14	227·5	2912	35006·4	−16·25	−29·05
15	280	4125·3r	57283·2	−18·6r	−33·4
16	340	5712	90698·4	−21·25	−38·05
17	408	7752	139536	−24	−43
18	484·5	10336	209304	−26·916r	−48·25
19	570	13566	306979·2	−30	−53·8
20	665	17556	441282·6	−33·25	−59·65
21	770	22432·6r	622987·2	−36·6r	−65·8
22	885·5	28336	865260	−40·25	−72·25
23	1012	35420	1,184040	−44	−79
24	1150	43853·3r	1,598454	−47·916r	−86·05
25	1300	53820	2,131272	−52	−93·4
26	1462·5	65520	2,809404	−56·25	−101·05
27	1638	79170	3,664440	−60·6r	−109
28	1827	95004	4,733235	−65·25	−117·25
29	2030	113274	6,058540·8	−70	−125·8
30	2247·5	134250·6r	7,689686·4	−74·916r	−134·65
31	2480	158224	9,683308·8	−80	−143·8
32	2728	185504	12,104136	−85·25	−153·25
33	2992	216421·3r	15,025824	−90·6r	−163
34	3272·5	251328	18,531849·6	−96·25	−173·05
35	3570	290598	22,716460·8	−102	−183·4
36	3885	334628	27,685686·6	−107·916r	−194·05
37	4218	383838	33,558408	−114	−205
38	4569·5	438672	40,467492	−120·25	−216·25
39	4940	499598·6r	48,560990·4	−126·6r	−227·8
40	5330	567112	58,003405·2	−133·25	−239·65

TABLE 7.10*b* (cont.)

$n = S_{00}$	S_{11}	S_{22}	S_{33}	β_{02}	β_{13}
41	5740	641732	68,977022·4	−140	−251·8
42	6170·5	724005·3r	81,683316	−146·916r	−264·25
43	6622	814506	96,344424	−154	−277
44	7095	913836	113,204698·2	−161·25	−290·05
45	7590	1,022626	132,532329·6	−168·6r	−303·4
46	8107·5	1,141536	154,621051·2	−176·25	−317·05
47	8648	1,271256	179,791920	−184	−331
48	9212	1,412506·6r	208,395180	−191·916r	−345·25
49	9800	1,566040	240,812208	−200	−359·8
50	10412·5	1,732640	277,457544	−208·25	−374·65
51	11050	1,913123·3r	318,781008	−216·6r	−389·8
52	11713	2,108340	365,269905	−225·25	−405·25
53	12402	2,319174	417,451320	−234	−421
54	13117·5	2,546544	475,894504·8	−242·916r	−437·05
55	13860	2,791404	541,213358·4	−252	−453·4
56	14630	3,054744	614,069002·8	−261·25	−470·05
57	15428	3,337590·6r	695,172456	−270·6r	−487
58	16254·5	3,641008	785,287404	−280·25	−504·25
59	17110	3,966098	885,233073·6	−290	−521·8
60	17995	4,314001·3r	995,887207·8	−299·916r	−539·65
61	18910	4,685898	1118,189145·6	−310	−557·8
62	19855·5	5,083008	1253,143008	−320·25	−576·25
63	20832	5,506592	1401,820992	−330·6r	−595
64	21840	5,957952	1565,366774·4	−341·25	−614·05
65	22880	6,438432	1744,999027·2	−352	−633·4
66	23952·5	6,949418·6r	1942,015046·4	−362·916r	−653·05
67	25058	7,492342	2157,794496	−374	−673
68	26197	8,068676	2393,803269	−385·25	−693·25
69	27370	8,679939·3r	2651,597467·2	−396·6r	−713·8
70	28577·5	9,327696	2932,827501·6	−408·25	−734·65
71	29820	10,013556	3239,242315·2	−420	−755·8
72	31098	10,739176	3572,693730	−431·916r	−777·25
73	32412	11,506260	3935,140920	−444	−799
74	33762·5	12,316560	4328,655012	−456·25	−821·05
75	35150	13,171876·6r	4755,423816	−468·6r	−843·4

TABLE 7.10*b* (cont.)

n	S_{44}	β_{24}	β_{04}
6	82·2s	−6·78s	+5·0625
7	452·5s	−9·5s	+10·2s
8	1810·2s	−12·78s	+18·5625
9	5883·4s	−16·4s	+30·8s
10	16473·6	−20·5	+48·2625
11	41184	−25	+72
12	94134·8s	−29·92s	+103·41964s
13	200036·5s	−35·2s	+144
14	400073·1s	−41·07s	+195·34821s
15	760138·97s	−47·2s	+259·2
16	1,382070·8s	−53·92s	+337·41964s
17	2,418624	−61	+432
18	4,093056	−68·5	+545·0625
19	6,724306·2s	−76·4s	+678·8s
20	10,758890·05s	−84·78s	+835·7625
21	16,810765·7s	−93·5s	+1018·2s
22	25,710582·8s	−102·78s	+1229·0625
23	38,565874·2s	−112·4s	+1470·8s
24	56,833920	−122·5	+1746·5625
25	82,409184	−133	+2059·2
26	117,727405·7s	−143·92s	+2411·91964s
27	165,888617·1s	−155·2s	+2808
28	230,801554·2s	−167·07s	+3250·84821s
29	317,352137·1s	−179·2s	+3744
30	431,598906·51s	−191·92s	+4291·11964s
31	580,998528	−205	+4896
32	774,664704	−218·5	+5562·5625
33	1023,664073·1s	−232·4s	+6294·8s
34	1341,352923·4s	−246·78s	+7097·0625
35	1743,758800·45s	−261·5s	+7973·48s
36	2250,011355·4s	−276·78s	+8928·5625
37	2882,827049·1s	−292·4s	+9966·8s
38	3669,052608	−308·5	+11093·0625
39	4640,272416	−325	+12312
40	5833,485322·97s	−341·92s	+13628·61964s

TABLE 7.10*b* (cont.)

n	S_{44}	β_{24}	β_{04}
41	7291,856653·7s	−359·2s	+15048
42	9065,551515·4s	−377·07s	+16575·34821s
43	11212,655821·7s	−395·2s	+18216
44	13800,191780·5s	−413·92s	+19975·41964s
45	16905,234931·2	−433	+21859·2
46	20616,140160	−452·5	+23873·0625
47	25033,884480	−472·4s	+26022·8s
48	30273,534720	−492·78s	+28314·5625
49	36465,848640	−513·5s	+30754·2s
50	43759,018368	−534·78s	+33348·2625
51	52320,565440	−556·4s	+36102·8s
52	62339,397120	−578·5	+39024·5625
53	74028,034080	−601	+42120
54	87625,019931·4s	−623·92s	+45395·91964s
55	103397,523519·08s	−647·2s	+48859·2
56	121644,145316·5s	−671·07s	+52516·84821s
57	142697,939698·2s	−695·2s	+56376
58	166929,665307·4s	−719·92s	+60443·91964s
59	194751,276192	−745	+64728
60	226619,666841·6	−770·5	+69235·7625
61	263040,684726·8s	−796·4s	+73974·8s
62	304573,424420·5s	−822·78s	+78953·0625
63	351834,817865·1s	−849·5s	+84178·2s
64	405504,535844·5s	−876·78s	+89658·5625
65	466330,216221·25s	−904·4s	+95402·05s
66	535133,035008	−932·5	+101417·0625
67	612813,636864	−961	+107712
68	700358,442130·2s	−989·92s	+114295·41964s
69	798846,348054·8s	−1019·2s	+121176
70	909455,842400·91s	−1049·07s	+128362·54821s
71	1,033472,548182·8s	−1079·2s	+135864
72	1,172297,218834·2s	−1109·92s	+143689·41964s
73	1,327454,203680	−1141	+151848
74	1,500600,404160	−1172·5	+160349·0625
75	1,693534,741837·7s	−1204·4s	+169202·05s

TABLE 7.10*b* (cont.)

n	S_{55}	β_{35}	β_{15}	
6	57·1s	−8·05r	+11·47519841	(479/1008)
7	685·7s	−11·6r	+24·95238095	(20/21)
8	4457·1s	−15·83r	+46·75297619	(253/336)
9	20800	−20·5r	+79·5r	(5/9)
10	78000	−25·83r	+126·39583r	(19/48)
11	249600	−31·6r	+190·6r	(2/3)
12	707200	−38·05r	+276·11805r	(17/144)
13	1,818514·2s	−45	+386·8s	(6/7)
14	4,318971·4s	−52·5	+527·34821s	(117/336)
15	9,597714·2s	−60·5r	+702·4126984	(26/63)
16	20,155200	−69·16r	+917·22916r	(11/48)
17	40,310400	−78·3r	+1177·3r	(1/3)
18	77,261600	−88·05r	+1488·61805r	(89/144)
19	142,636800	−98·3r	+1857·3r	(1/3)
20	254,708571·4s	−109·16r	+2290·086310	(29/336)
21	441,494857·1s	−120·5r	+2793·841270	(53/63)
22	745,022571·4s	−132·5	+3375·91964s	(103/112)
23	1227,096000	−145	+4044	
24	1976,988000	−158·05r	+4806·11805r	(17/144)
25	3121,560000	−171·6r	+5670·6r	(2/3)
26	4838,418000	−185·83r	+6646·39583r	(19/48)
27	7372,827428·5s	−200·5r	+7742·412698	(26/63)
28	11059,241142·8s	−215·83r	+8968·181548	(61/336)
29	16348,443428·5s	−231·6r	+10333·52381	(11/21)
30	23841,480000	−248·05r	+11848·61805r	(89/144)
31	34331,731200	−265	+13524	
32	48856,694400	−282·5	+15370·5625	(9/16)
33	68761,273600	−300·5r	+17399·5r	(5/9)
34	95774,631085·7s	−319·16r	+19622·58631	(197/336)
35	132102,939428·5s	−338·3r	+22051·61905	(13/21)
36	180540,683885·7s	−358·05r	+24698·97520	(983/1008)
37	244603,507200	−378·3r	+27577·3r	(1/3)
38	328685,962800	−399·16r	+30699·72916r	(35/48)
39	438247,950400	−420·5r	+34079·5r	(5/9)
40	580034,052000	−442·5	+37730·5625	(9/16)

TABLE 7.10*b* (cont.)

n	S_{55}	β_{35}	β_{15}	
41	762330,468342·8s	−465	+41666·8s	(6/7)
42	995264,778114·2s	−488·05r	+45902·90377	(911/1008)
43	1,291154,306742·8s	−511·6r	+50453·52381	(11/21)
44	1,664909,500800	−535·83r	+55333·89583r	(43/48)
45	2,134499,360000	−560·5r	+60559·5r	(5/9)
46	2,721486,684000	−585·83r	+66146·39583r	(19/48)
47	3,451641,648000	−611·6r	+72110·6r	(2/3)
48	4,355643,032000	−638·05r	+78468·97520	(983/1008)
49	5,469877,296000	−665	+85238·2s	(2/7)
50	6,837346,620000	−692·5	+92435·91964s	(103/112)
51	8,508698,016000	−720·5r	+100079·5r	(5/9)
52	10,543386,672000	−749·16r	+108187·22916r	(11/48)
53	13,010987,808000	−778·3r	+116777·3r	(1/3)
54	15,992672,514000	−808·05r	+125868·61805r	(89/144)
55	19,582864,302857·1s	−838·3r	+135480·1905	(4/21)
56	23,891094,449485·7s	−869·16r	+145631·5149	(173/336)
57	29,044075,605257·1s	−900·5r	+156342·4127	(26/63)
58	35,188014,675600	−932·5	+167633·0625	(1/16)
59	42,491187,532800	−965	+179524	
60	51,146799,808000	−998·05r	+192036·11805r	(17/144)
61	61,376159,769600	−1031·6r	+205190·6r	(2/3)
62	73,432191,152914·2s	−1065·83r	+219009·2530	(85/336)
63	87,603315,761371·4s	−1100·5r	+233513·8413	(53/63)
64	104,217737,716114·2s	−1135·83r	+248726·7530	(253/336)
65	123,648163,392000	−1171·6r	+264670·6r	(2/3)
66	146,316993,347200	−1208·05r	+281368·61805r	(89/144)
67	172,702024,934400	−1245	+298844	
68	203,342706,777600	−1282·5	+317120·5625	(9/16)
69	238,846988,913371·4s	−1320·5r	+336222·4127	(26/63)
70	279,898815,132857·1s	−1359·16r	+356174·0149	(5/336)
71	327,266306,924571·4s	−1398·3r	+377000·1905	(4/21)
72	381,810691,412000	−1438·05r	+398726·11805r	(17/144)
73	444,496028,808000	−1478·3r	+421377·3r	(1/3)
74	516,399798,174000	−1519·16r	+444979·72916r	(35/48)
75	598,724403,680000	−1560·5r	+469559·5r	(5/9)

TABLE 7.10c

The coefficients β'_{kj}

n	β'_{11}	β'_{22}	β'_{02}	β'_{33}	β'_{13}
6	2	1·5	−4·375	1·6r	−8·416r
7	1	1	−4	0·16r	−1·16r
8	2	1	−5·25	0·6r	−6·16r
9	1	3	−20	0·83r	−9·83r
10	2	0·5	−4·125	1·6r	−24·416r
11	1	1	−10	0·83r	−14·83r
12	2	3	−35·75	0·6r	−14·16r
13	1	1	−14	0·16r	−4·16r
14	2	0·5	−8·125	1·6r	−48·416r
15	1	3	−56	0·83r	−27·83r
16	2	1	−21·25	3·3r	−126·83r
17	1	1	−24	0·16r	−7·16r
18	2	1·5	−40·375	0·3r	−16·083r
19	1	1	−30	0·83r	−44·83r
20	2	1	−33·25	3·3r	−198·83r
21	1	3	−110	0·83r	−54·83r
22	2	0·5	−20·125	0·3r	−24·083r
23	1	1	−44	0·16r	−13·16r
24	2	3	−143·75	3·3r	−286·83r
25	1	1	−52	0·83r	−77·83r
26	2	0·5	−28·125	1·6r	−168·416r
27	1	3	−182	0·16r	−18·16r
28	2	1	−65·25	0·6r	−78·16r
29	1	1	−70	0·83r	−104·83r
30	2	1·5	−112·375	1·6r	−224·416r
31	1	1	−80	0·83r	−119·83r
32	2	1	−85·25	0·6r	−102·16r
33	1	3	−272	0·16r	−27·16r
34	2	0·5	−48·125	1·6r	−288·416r
35	1	1	−102	0·83r	−152·83r
36	2	3	−323·75	3·3r	−646·83r
37	1	1	−114	0·16r	−34·16r
38	2	0·5	−60·125	0·3r	−72·083r
39	1	3	−380	0·83r	−189·83r
40	2	1	−133·25	3·3r	−798·83r

TABLE 7.10c (cont.)

n	β'_{11}	β'_{22}	β'_{02}	β'_{33}	β'_{13}
41	1	1	−140	0·83r	−209·83r
42	2	1·5	−220·375	0·3r	−88·083r
43	1	1	−154	0·16r	−46·16r
44	2	1	−161·25	3·3r	−966·83r
45	1	3	−506	0·83r	−252·83r
46	2	0·5	−88·125	1·6r	−528·416r
47	1	1	−184	0·16r	−55·16r
48	2	3	−575·75	0·6r	−230·16r
49	1	1	−200	0·83r	−299·83r
50	2	0·5	−104·125	1·6r	−624·416r
51	1	3	−650	0·83r	−324·83r
52	2	1	−225·25	0·6r	−270·16r
53	1	1	−234	0·16r	−70·16r
54	2	1·5	−364·375	1·6r	−728·416r
55	1	1	−252	0·83r	−377·83r
56	2	1	−261·25	3·3r	−1566·83r
57	1	3	−812	0·16r	−81·16r
58	2	0·5	−140·125	0·3r	−168·083r
59	1	1	−290	0·83r	−434·83r
60	2	3	−899·75	3·3r	−1798·83r
61	1	1	−310	0·83r	−464·83r
62	2	0·5	−160·125	0·3r	−192·083r
63	1	3	−992	0·16r	−99·16r
64	2	1	−341·25	3·3r	−2046·83r
65	1	1	−352	0·83r	−527·83r
66	2	1·5	−544·375	1·6r	−1088·416r
67	1	1	−374	0·16r	−112·16r
68	2	1	−385·25	0·6r	−462·16r
69	1	3	−1190	0·83r	−594·83r
70	2	0·5	−204·125	1·6r	−1224·416r
71	1	1	−420	0·83r	−629·83r
72	2	3	−1295·75	0·6r	−518·16r
73	1	1	−444	0·16r	−133·16r
74	2	0·5	−228·125	1·6r	−1368·416r
75	1	3	−1406	0·83r	−702·83r

TABLE 7.10c (cont.)

n	β'_{44}	β'_{24}	β'_{04}
6	0·583r	−3·9583r	+2·953125
7	0·583r	−5·583r	+6
8	0·583r	−7·4583r	+10·828125
9	0·583r	−9·583r	+18
10	0·416r	−8·5416r	+20·109375
11	0·083r	−2·083r	+6
12	0·2916r	−8·72916r	+30·1640625
13	0·583r	−20·583r	+84
14	0·583r	−23·9583r	+113·953125
15	2·916r	−137·916r	+756
16	0·583r	−31·4583r	+196·828125
17	0·083r	−5·083r	+36
18	0·083r	−5·7083r	+45·421875
19	0·583r	−44·583r	+396
20	1·4583r	−123·64583r	+1218·8203125
21	0·583r	−54·583r	+594
22	0·583r	−59·9583r	+716·953125
23	0·583r	−65·583r	+858
24	0·083r	−10·2083r	+145·546875
25	0·416r	−55·416r	+858
26	0·583r	−83·9583r	+1406·953125
27	0·583r	−90·583r	+1638
28	0·2916r	−48·72916r	+948·1640625
29	0·583r	−104·583r	+2184
30	2·916r	−559·7916r	+12515·765625
31	0·083r	−17·083r	+408
32	0·083r	−18·2083r	+463·546875
33	0·583r	−135·583r	+3672
34	0·583r	−143·9583r	+4139·953125
35	2·916r	−762·916r	+23256
36	0·2916r	−80·72916r	+2604·1640625
37	0·583r	−170·583r	+5814
38	0·083r	−25·7083r	+924·421875
39	0.083r	−27·083r	+1026
40	2·916r	−997·2916r	+39750·140625

TABLE 7.10*c* (cont.)

n	β'_{44}	β'_{24}	β'_{04}
41	0·583r	−209·583r	+8778
42	0·583r	−219·9583r	+9668·953125
43	0·583r	−230·583r	+10626
44	0·2916r	−120·72916r	+5826·1640625
45	0·416r	−180·416r	+9108
46	0·083r	−37·7083r	+1989·421875
47	0·583r	−275·583r	+15180
48	0·583r	−287·4583r	+16516·828125
49	0·583r	−299·583r	+17940
50	2·916r	−1559·7916r	+97265·765625
51	0·583r	−324·583r	+21060
52	0·0416r	−24·10416r	+1626·0234375
53	0·083r	−50·083r	+3510
54	0·583r	−363·9583r	+26480·953125
55	2·916r	−1887·916r	+142506
56	0·583r	−391·4583r	+30634·828125
57	0·583r	−405·583r	+32886
58	0·583r	−419·9583r	+35258·953125
59	0·083r	−62·083r	+5394
60	0·2083r	−160·52083r	+14424·1171875
61	0·583r	−464·583r	+43152
62	0·583r	−479·9583r	+46055·953125
63	0·583r	−495·583r	+49104
64	0·583r	−511·4583r	+52300·828125
65	2·916r	−2637·916r	+278256
66	0·083r	−77·7083r	+8451·421875
67	0·083r	−80·083r	+8976
68	0·2916r	−288·72916r	+33336·1640625
69	0·583r	−594·583r	+70686
70	2·916r	−3059·7916r	+374390·765625
71	0·583r	−629·583r	+79254
72	0·583r	−647·4583r	+83818·828125
73	0·083r	−95·083r	+12654
74	0·083r	−97·7083r	+13362·421875
75	2·916r	−3512·916r	+493506

TABLE 7.10*c* (cont.)

n	β'_{55}	β'_{35}	β'_{15}
6	2·1	−16·916r	+24·097916r
7	0·35	−4·083r	+8·73r
8	0·7	−11·083r	+32·727083r
9	0·15	−3·083r	+11·93r
10	0·1	−2·583r	+12·639583r
11	0·025	−0·7916r	+4·76r
12	0·15	−5·7083r	+41·4177083r
13	0·0583r	−2·625	+22·56r
14	0·23r	−12·25	+123·047916r
15	1·05	−63·583r	+737·53r
16	0·1	−6·916r	+91·722916r
17	0·05	−3·916r	+58·86r
18	0·3	−26·416r	+446·585416r
19	0·025	−2·4583r	+46·43r
20	0·35	−38·2083r	+801·5302083r
21	0·525	−63·2916r	+1466·76r
22	0·23r	−30·916r	+787·714583r
23	0·016r	−2·416r	+67·4
24	0·3	−47·416r	+1441·835416r
25	0·05	−8·583r	+283·53r
26	0·1	−18·583r	+664·639583r
27	0·525	−105·2916r	+4064·76r
28	0·35	−75·5416r	+3138·8635416r
29	0·175	−40·5416r	+1808·36r
30	0·3	−74·416r	+3554·585416r
31	0·016r	−4·416r	+225·4
32	0·03r	−9·416r	+512·352083r
33	0·15	−45·083r	+2609·93r
34	0·7	−223·416r	+13735·810416r
35	0·175	−59·2083r	+3859·03r
36	1·05	−375·9583r	+25933·9239583r
37	0·025	−9·4583r	+689·43r
38	0·1	−39·916r	+3069·972916r
39	0·15	−63·083r	+5111·93r
40	0·03r	−14·75	+1257·685416r

TABLE 7.10*c* (cont.)

n	β'_{55}	β'_{35}	β'_{15}
41	0·116r	−54·25	+4861·13r
42	2·1	−1024·916r	+96396·097916r
43	0·175	−89·5416r	+8829·36r
44	0·05	−26·7916r	+2766·6947916r
45	0·075	−42·0416r	+4541·96r
46	0·1	−58·583r	+6614·639583r
47	0·05	−30·583r	+3605·53r
48	2·1	−1339·916r	+164784·847916r
49	0·116r	−77·583r	+9944·46r
50	0·23r	−161·583r	+21568·38125
51	0·075	−54·0416r	+7505·96r
52	0·05	−37·4583r	+5409·3614583r
53	0·025	−19·4583r	+2919·43r
54	0·3	−242·416r	+37760·585416r
55	0·35	−293·416r	+47418·06r
56	0·7	−608·416r	+101942·060416r
57	1·05	−945·583r	+164159·53r
58	0·03r	−31·083r	+5587·76875
59	0·0083r	−8·0416r	+1496·03r
60	0·15	−149·7083r	+28805·4177083r
61	0·025	−25·7916r	+5129·76r
62	0·7	−746·083r	+153306·477083r
63	1·05	−1155·583r	+245189·53r
64	0·7	−795·083r	+174108·727083r
65	0·05	−58·583r	+13233·53r
66	0·3	−362·416r	+84410·585416r
67	0·0083r	−10·375	+2490·36r
68	0·016r	−21·375	+5285·3427083r
69	0·525	−693·2916r	+176516·76r
70	0·7	−951·416r	+249321·810416r
71	0·35	−489·416r	+131950·06r
72	0·3	−431·416r	+119617·835416r
73	0·05	−73·916r	+21068·86r
74	0·1	−151·916r	+44497·972916r
75	0·075	−117·0416r	+35216·96r

TABLE 7.10*d*

The coefficients $\beta_{[kj]}$

r indicates that the last figure is repeated; s that the set 142857 is repeated; t that the decimal is a fraction of 21

n	$\beta_{[01]}$	$\beta_{[12]}$	$\beta_{[02]}$	$\beta_{[23]}$	$\beta_{[13]}$	$\beta_{[03]}$
1						
2	−0·5					
3	−1	−1	0·3r			
4	−1·5	−2	1	−3	1·2	−0·3
5	−2	−3	2	−6	3·6	−1·2
6	−2·5	−4	3·3r	−9	7·2	−3
7	−3	−5	5	−12	12	−6
8	−3·5	−6	7	−15	18	−10·5
9	−4	−7	9·3r	−18	25·2	−16·8
10	−4·5	−8	12	−21	33·6	−25·2
11	−5	−9	15	−24	43·2	−36
12	−5·5	−10	18·3r	−27	54	−49·5
13	−6	−11	22	−30	66	−66
14	−6·5	−12	26	−33	79·2	−85·8
15	−7	−13	30·3r	−36	93·6	−109·2
16	−7·5	−14	35	−39	109·2	−136·5
17	−8	−15	40	−42	126	−168
18	−8·5	−16	45·3r	−45	144	−204
19	−9	−17	51	−48	163·2	−244·8
20	−9·5	−18	57	−51	183·6	−290·7
21	−10	−19	63·3r	−54	205·2	−342
22	−10·5	−20	70	−57	228	−399
23	−11	−21	77	−60	252	−462
24	−11·5	−22	84·3r	−63	277·2	−531·3
25	−12	−23	92	−66	303·6	−607·2
26	−12·5	−24	100	−69	331·2	−690
27	−13	−25	108·3r	−72	360	−780
28	−13·5	−26	117	−75	390	−877·5
29	−14	−27	126	−78	421·2	−982·8
30	−14·5	−28	135·3r	−81	453·6	−1096·2
31	−15	−29	145	−84	487·2	−1218
32	−15·5	−30	155	−87	522	−1348·5
33	−16	−31	165·3r	−90	558	−1488
34	−16·5	−32	176	−93	595·2	−1636·8
35	−17	−33	187	−96	633·6	−1795·2

TABLE 7.10*d* (cont.)

n	$\beta_{[01]}$	$\beta_{[12]}$	$\beta_{[02]}$	$\beta_{[23]}$	$\beta_{[13]}$	$\beta_{[03]}$
36	−17·5	−34	198·3r	−99	673·2	−1963·5
37	−18	−35	210	−102	714	−2142
38	−18·5	−36	222	−105	756	−2331
39	−19	−37	234·3r	−108	799·2	−2530·8
40	−19·5	−38	247	−111	843·6	−2741·7
41	−20	−39	260	−114	889·2	−2964
42	−20·5	−40	273·3r	−117	936	−3198
43	−21	−41	287	−120	984	−3444
44	−21·5	−42	301	−123	1033·2	−3702·3
45	−22	−43	315·3r	−126	1083·6	−3973·2
46	−22·5	−44	330	−129	1135·2	−4257
47	−23	−45	345	−132	1188	−4554
48	−23·5	−46	360·3r	−135	1242	−4864·5
49	−24	−47	376	−138	1297·2	−5188·8
50	−24·5	−48	392	−141	1353·6	−5527·2
51	−25	−49	408·3r	−144	1411·2	−5880
52	−25·5	−50	425	−147	1470	−6247·5
53	−26	−51	442	−150	1530	−6630
54	−26·5	−52	459·3r	−153	1591·2	−7027·8
55	−27	−53	477	−156	1653·6	−7441·2
56	−27·5	−54	495	−159	1717·2	−7870·5
57	−28	−55	513·3r	−162	1782	−8316
58	−28·5	−56	532	−165	1848	−8778
59	−29	−57	551	−168	1915·2	−9256·8
60	−29·5	−58	570·3r	−171	1983·6	−9752·7
61	−30	−59	590	−174	2053·2	−10266
62	−30·5	−60	610	−177	2124	−10797
63	−31	−61	630·3r	−180	2196	−11346
64	−31·5	−62	651	−183	2269·2	−11913·3
65	−32	−63	672	−186	2343·6	−12499·2
66	−32·5	−64	693·3r	−189	2419·2	−13104
67	−33	−65	715	−192	2496	−13728
68	−33·5	−66	737	−195	2574	−14371·5
69	−34	−67	759·3r	−198	2653·2	−15034·8
70	−34·5	−68	782	−201	2733·6	−15718·2
71	−35	−69	805	−204	2815·2	−16422
72	−35·5	−70	828·3r	−207	2898	−17146·5
73	−36	−71	852	−210	2982	−17892
74	−36·5	−72	876	−213	3067·2	−18658·8
75	−37	−73	900·3r	−216	3153·6	−19447·2

TABLE 7.10*d* (cont.)

n	$\beta_{[34]}$	$\beta_{[24]}$	$\beta_{[14]}$	$\beta_{[04]}$
5	−12	+5·1s	−1·7s	+0·34s
6	−24	+15·4s	−6·8s	+1·7s
7	−36	+30·8s	−17·1s	+5·1s
8	−48	+51·4s	−34·2s	+12
9	−60	+77·1s	−60	+24
10	−72	+108	−96	+43·2
11	−84	+144	−144	+72
12	−96	+185·1s	−205·7s	+113·1s
13	−108	+231·4s	−282·8s	+169·7s
14	−120	+282·8s	−377·1s	+245·1s
15	−132	+339·4s	−490·2s	+343·2
16	−144	+401·1s	−624	+468
17	−156	+468	−780	+624
18	−168	+540	−960	+816
19	−180	+617·1s	−1165·7s	+1049·1s
20	−192	+699·4s	−1398·8s	+1328·91s
21	−204	+786·8s	−1661·1s	+1661·1s
22	−216	+879·4s	−1954·2s	+2052
23	−228	+977·1s	−2280	+2508
24	−240	+1080	−2640	+3036
25	−252	+1188	−3036	+3643·2
26	−264	+1301·1s	−3469·7s	+4337·1s
27	−276	+1419·4s	−3942·8s	+5125·7s
28	−288	+1542·8s	−4457·1s	+6017·1s
29	−300	+1671·4s	−5014·2s	+7020
30	−312	+1805·1s	−5616	+8143·2
31	−324	+1944	−6264	+9396
32	−336	+2088	−6960	+10788
33	−348	+2237·1s	−7705·7s	+12329·1s
34	−360	+2391·4s	−8502·8s	+14029·7s
35	−372	+2550·8s	−9353·1s	+15900·34s
36	−384	+2715·4s	−10258·2s	+17952
37	−396	+2885·1s	−11220	+20196
38	−408	+3060	−12240	+22644
39	−420	+3240	−13320	+25308
40	−432	+3425·1s	−14461·7s	+28200·34s

TABLE 7.10*d* (cont.)

n	$\beta_{[34]}$	$\beta_{[24]}$	$\beta_{[14]}$	$\beta_{[04]}$
41	−444	+3615·4s	−15666·8s	+31333·7s
42	−456	+3810·8s	−16937·1s	+34721·1s
43	−468	+4011·4s	−18274·2s	+38376
44	−480	+4217·1s	−19680	+42312
45	−492	+4428	−21156	+46543·2
46	−504	+4644	−22704	+51084
47	−516	+4865·1s	−24325·7s	+55949·1s
48	−528	+5091·4s	−26022·8s	+61153·7s
49	−540	+5322·8s	−27797·1s	+66713·1s
50	−552	+5559·4s	−29650·2s	+72643·2
51	−564	+5801·1s	−31584	+78960
52	−576	+6048	−33600	+85680
53	−588	+6300	−35700	+92820
54	−600	+6557·1s	−37885·7s	+100397·1s
55	−612	+6819·4s	−40158·8s	+108428·91s
56	−624	+7086·8s	−42521·1s	+116933·1s
57	−636	+7359·4s	−44974·2s	+125928
58	−648	+7637·1s	−47520	+135432
59	−660	+7920	−50160	+145464
60	−672	+8208	−52896	+156043·2
61	−684	+8501·1s	−55729·7s	+167189·1s
62	−696	+8799·4s	−58662·8s	+178921·7s
63	−708	+9102·8s	−61697·1s	+191261·1s
64	−720	+9411·4s	−64834·2s	+204228
65	−732	+9725·1s	−68076	+217843·2
66	−744	+10044	−71424	+232128
67	−756	+10368	−74880	+247104
68	−768	+10697·1s	−78445·7s	+262793·1s
69	−780	+11031·4s	−82122·8s	+279217·7s
70	−792	+11370·8s	−85913·1s	+296400·34s
71	−804	+11715·4s	−89818·2s	+314364
72	−816	+12065·1s	−93840	+333132
73	−828	+12420	−97980	+352728
74	−840	+12780	−102240	+373176
75	−852	+13145·1s	−106621·7s	+394500·34s

TABLE 7.10*d* (cont.)

n	$\beta_{[45]}$	$\beta_{[35]}$	$\beta_{[25]}$	$\beta_{[15]}$	$\beta_{[05]}$
6	−60	+26·6r	−10	+2·8s	−0·476190t (10/21)
7	−120	+80	−40	+14·2s	−2·8s
8	−180	+160	−100	+42·8s	−10
9	−240	+266·6r	−200	+100	−26·6r
10	−300	+400	−350	+200	−60
11	−360	+560	−560	+360	−120
12	−420	+746·6r	−840	+600	−220
13	−480	+960	−1200	+942·8s	−377·1s
14	−540	+1200	−1650	+1414·2s	−612·8s
15	−600	+1466·6r	−2200	+2042·8s	−953·3r
16	−660	+1760	−2860	+2860	−1430
17	−720	+2080	−3640	+3900	−2080
18	−780	+2426·6r	−4550	+5200	−2946·6r
19	−840	+2800	−5600	+6800	−4080
20	−900	+3200	−6800	+8742·8s	−5537·1s
21	−960	+3626·6r	−8160	+11074·2s	−7382·8s
22	−1020	+4080	−9690	+13842·8s	−9690
23	−1080	+4560	−11400	+17100	−12540
24	−1140	+5066·6r	−13300	+20900	−16023·3r
25	−1200	+5600	−15400	+25300	−20240
26	−1260	+6160	−17710	+30360	−25300
27	−1320	+6746·6r	−20240	+36142·8s	−31323·809523t (17/21)
28	−1380	+7360	−23000	+42714·2s	−38442·8s
29	−1440	+8000	−26000	+50142·8s	−46800
30	−1500	+8666·6r	−29250	+58500	−56550
31	−1560	+9360	−32760	+67860	−67860
32	−1620	+10080	−36540	+78300	−80910
33	−1680	+10826·6r	−40600	+89900	−95893·3r
34	−1740	+11600	−44950	+102742·8s	−113017·1s
35	−1800	+12400	−49600	+116914·2s	−132502·8s
36	−1860	+13226·6r	−54560	+132502·8s	−154586·6r
37	−1920	+14080	−59840	+149600	−179520
38	−1980	+14960	−65450	+168300	−207570
39	−2040	+15866·6r	−71400	+188700	−239020
40	−2100	+16800	−77700	+210900	−274170

TABLE 7.10*d* (cont.)

n	$\beta_{[45]}$	$\beta_{[35]}$	$\beta_{[25]}$	$\beta_{[15]}$	$\beta_{[05]}$
41	−2160	+17760	−84360	+235002·8s	−313337·1s
42	−2220	+18746·6r	−91390	+261114·2s	−356856·190476t (4/21)
43	−2280	+19760	−98800	+289342·8s	−405080
44	−2340	+20800	−106600	+319800	−458380
45	−2400	+21866·6r	−114800	+352600	−517146·6r
46	−2460	+22960	−123410	+387860	−581790
47	−2520	+24080	−132440	+425700	−652740
48	−2580	+25226·6r	−141900	+466242·8s	−730447·1s
49	−2640	+26400	−151800	+509614·2s	−815382·8s
50	−2700	+27600	−162150	+555942·8s	−908040
51	−2760	+28826·6r	−172960	+605360	−1,008933·3r
52	−2820	+30080	−184240	+658000	−1,118600
53	−2880	+31360	−196000	+714000	−1,237600
54	−2940	+32666·6r	−208250	+773500	−1,366516·6r
55	−3000	+34000	−221000	+836642·8s	−1,505957·1s
56	−3060	+35360	−234260	+903574·2s	−1,656552·8s
57	−3120	+36746·6r	−248040	+974442·8s	−1,818960
58	−3180	+38160	−262350	+1,049400	−1,993860
59	−3240	+39600	−277200	+1,128600	−2,181960
60	−3300	+41066·6r	−292600	+1,212200	−2,383993·3r
61	−3360	+42560	−308560	+1,300360	−2,600720
62	−3420	+44080	−325090	+1,393242·8s	−2,832927·1s
63	−3480	+45626·6r	−342200	+1,491014·2s	−3,081429·523809t (11/21)
64	−3540	+47200	−359900	+1,593842·8s	−3,347070
65	−3600	+48800	−378200	+1,701900	−3,630720
66	−3660	+50426·6r	−397110	+1,815360	−3,933280
67	−3720	+52080	−416640	+1,934400	−4,255680
68	−3780	+53760	−436800	+2,059200	−4,598880
69	−3840	+55466·6r	−457600	+2,189942·8s	−4,963870·476190t (10/21)
70	−3900	+57200	−479050	+2,326814·2s	−5,351672·8s
71	−3960	+58960	−501160	+2,470002·8s	−5,763340
72	−4020	+60746·6r	−523940	+2,619700	−6,199956·6r
73	−4080	+62560	−547400	+2,776100	−6,662640
74	−4140	+64400	−571550	+2,939400	−7,152540
75	−4200	+66266·6r	−596400	+3,109800	−7,670840

TABLE 7.10*e*

Formulae for the factorial coefficients $\beta_{[kj]}$

Zero degree

$\beta_{[00]} = 1$

1*st degree*

$\beta_{[11]} = 1$
$\beta_{[01]} = -(n-1)/2$

2*nd degree*

$\beta_{[22]} = 2$
$\beta_{[12]} = -(n-2)$
$\beta_{[02]} = +(n-1)(n-2)/6$

3*rd degree*

$\beta_{[33]} = 6$
$\beta_{[23]} = -3(n-3)$
$\beta_{[13]} = +3(n-2)(n-3)/5$
$\beta_{[03]} = -(n-1)(n-2)(n-3)/20$

4*th degree*

$\beta_{[44]} = 24$
$\beta_{[34]} = -12(n-4)$
$\beta_{[24]} = +18(n-3)(n-4)/7$
$\beta_{[14]} = -2(n-2)(n-3)(n-4)/7$
$\beta_{[04]} = +(n-1)(n-2)(n-3)(n-4)/70$

5*th degree*

$\beta_{[55]} = 120$
$\beta_{[45]} = -60(n-5)$
$\beta_{[35]} = +40(n-4)(n-5)/3$
$\beta_{[25]} = -5(n-3)(n-4)(n-5)/3$
$\beta_{[15]} = +5(n-2)(n-3)(n-4)(n-5)/42$
$\beta_{[05]} = -(n-1)(n-2)(n-3)(n-4)(n-5)/252$

TABLE 7.10*f*

Formulae for $\phi_{[jk]}$

Zero degree

$\phi_{[00]} = W_{[0]}$

1*st degree*

$\phi_{[10]} = W_{[1]}$
$\phi_{[11]} = 2W_{[2]} + W_{[1]}$

2*nd degree*

$\phi_{[20]} = W_{[2]}$
$\phi_{[21]} = 3W_{[3]} + 2W_{[2]}$
$\phi_{[22]} = 6W_{[4]} + 6W_{[3]} + W_{[2]}$

3*rd degree*

$\phi_{[30]} = W_{[3]}$
$\phi_{[31]} = 4W_{[4]} + 3W_{[3]}$
$\phi_{[32]} = 10W_{[5]} + 12W_{[4]} + 3W_{[3]}$
$\phi_{[33]} = 20W_{[6]} + 30W_{[5]} + 12W_{[4]} + W_{[3]}$

4*th degree*

$\phi_{[40]} = W_{[4]}$
$\phi_{[41]} = 5W_{[5]} + 4W_{[4]}$
$\phi_{[42]} = 15W_{[6]} + 20W_{[5]} + 6W_{[4]}$
$\phi_{[43]} = 35W_{[7]} + 60W_{[6]} + 30W_{[5]} + 4W_{[4]}$
$\phi_{[44]} = 70W_{[8]} + 140W_{[7]} + 90W_{[6]} + 20W_{[5]} + W_{[4]}$

5*th degree*

$\phi_{[50]} = W_{[5]}$
$\phi_{[51]} = 6W_{[6]} + 5W_{[5]}$
$\phi_{[52]} = 21W_{[7]} + 30W_{[6]} + 10W_{[5]}$
$\phi_{[53]} = 56W_{[8]} + 105W_{[7]} + 60W_{[6]} + 10W_{[5]}$
$\phi_{[54]} = 126W_{[9]} + 280W_{[8]} + 210W_{[7]} + 60W_{[6]} + 5W_{[5]}$
$\phi_{[55]} = 252W_{[10]} + 630W_{[9]} + 560W_{[8]} + 210W_{[7]} + 30W_{[6]} + W_{[5]}$

TABLE 7.10*g*

Sums $\sum_{x=0}^{r} x^j$ *of the powers* x^j

j / r	1	2	3	4	5	6
1	1	1	1	1	1	1
2	3	5	9	17	33	65
3	6	14	36	98	276	794
4	10	30	100	354	1300	4890
5	15	55	225	979	4425	20515
6	21	91	441	2275	12201	67171
7	28	140	784	4676	29008	184820
8	36	204	1296	8772	61776	446964
9	45	285	2025	15333	120825	978405
10	55	385	3025	25333	220825	1,978405
11	66	506	4356	39974	381876	3,749966
12	78	650	6084	60710	630708	6,735950
13	91	819	8281	89271	1,002001	11,562759
14	105	1015	11025	127687	1,539825	19,092295
15	120	1240	14400	178312	2,299200	30,482920
16	136	1496	18496	243848	3,347776	47,260136
17	153	1785	23409	327369	4,767633	71,397705
18	171	2109	29241	432345	6,657201	105,409929
19	190	2470	36100	562666	9,133300	152,455810
20	210	2870	44100	722666	12,333300	216,455810
21	231	3311	53361	917147	16,417401	302,221931
22	253	3795	64009	1,151403	21,571033	415,601835
23	276	4324	76176	1,431244	28,007376	563,637724
24	300	4900	90000	1,763020	35,970000	754,740700
25	325	5525	105625	2,153645	45,735625	998,881325
26	351	6201	123201	2,610621	57,617001	1307,797101
27	378	6930	142884	3,142062	71,965908	1695,217590
28	406	7714	164836	3,756718	89,176276	2177,107894
29	435	8555	189225	4,463999	109,687425	2771,931215
30	465	9455	216225	5,273999	133,987425	3500,931215

CHAPTER 8

STANDARD DEVIATIONS OF THE ESTIMATES

8.1 FORMULAE FOR VARIANCES

Since the estimate

$$a_j = \sum_i w_i T_j(x_i) y_i / \sum_i w_i T_j^2(x_i) = \sum_i w_i T_j(x_i) y_i / S_{jj}$$

is a linear function of the observations y_i,

$$\operatorname{var} a_j = \sum_i \left[w_i T_j(x_i) / \sum_i w_i T_j^2(x_i) \right]^2 \operatorname{var} y_i,$$

and as $$\operatorname{var} y_i = \sigma^2 / w_i,$$

$$\operatorname{var} a_j = \sigma^2 / \sum_i w_i T_j^2(x_i) = \sigma^2 / S_{jj}. \tag{1}$$

If the true values (or population means) of y_i are denoted by Y_i, then the true values of a_j are

$$A_j = \Sigma w_i T_j(x_i) Y_i / \Sigma w_i T_j^2(x_i),$$

and so

$$\begin{aligned} \operatorname{cov}(a_j, a_k) &= E(a_j - A_j)(a_k - A_k) \\ &= E\left\{ \sum_i w_i (y_i - Y_i) T_j(x_i) \right\} \left\{ \sum_i w_i (y_i - Y_i) T_k(x_i) \right\} \Big/ S_{jj} S_{kk} \\ &= \sum_i w_i^2 T_j(x_i) T_k(x_i) \operatorname{var} y_i / S_{jj} S_{kk}. \end{aligned}$$

Because of the orthogonal property the numerator vanishes and

$$\operatorname{cov}(a_j, a_k) = 0. \tag{2a}$$

The covariance of y_h, a_j is given by

$$\operatorname{cov}(y_h, a_j) = E(y_h - Y_h) \sum_i w_i (y_i - Y_i) T_j(x_i) / S_{jj},$$

or $$\operatorname{cov}(y_h, a_j) = T_j(x_h) \operatorname{var} a_j = T_j(x_h) \sigma^2 / S_{jj}. \tag{2b}$$

If z is any linear sum of the coefficients a_j,

$$z = \Sigma \lambda_j a_j, \tag{3a}$$

it follows from (2a) that

$$\operatorname{var} z = \Sigma \lambda_j^2 \operatorname{var} a_j = \Sigma (\lambda_j^2 / S_{jj}) \sigma^2. \tag{3b}$$

For the power-series coefficients b_{pj}, from (7.2.2,3),

$$\operatorname{var} b_{pj} = \sum_{k=j}^{p} \beta_{jk}^2 \operatorname{var} a_k = \sum_{k=j}^{p} (\beta_{jk}^2 / S_{kk}) \sigma^2 = \sum_{k=}^{p} \beta_{jk} R_{kj} \sigma^2. \tag{4}$$

For the power-series coefficients c_{pj} when the variable is changed from x to z, (7.4,4) gives

$$\operatorname{var} c_{pj} = \sum_{k=j}^{p} \gamma_{jk}^2 \operatorname{var} a_k = \sum_{k=j}^{p} (\gamma_{jk}^2/S_{kk})\,\sigma^2 = \sum_{k=j}^{p} \gamma_{jk} Q_{kj}\,\sigma^2. \tag{5}$$

For the fitted value $u_p(x)$, from (7.2.2,1),

$$\operatorname{var} u_p(x) = \sum_{j=0}^{p} T_j^2(x) \operatorname{var} a_j = \sum_{j=0}^{p} [T_j^2(x)/S_{jj}]\,\sigma^2. \tag{6}$$

The coefficients b_{pj}, unlike the a_j, are not statistically independent. If B_{pj}, B_{pk} are the true values, the covariance of b_{pj}, b_{pk} is given by

$$\begin{aligned}\operatorname{cov}[b_{pj}, b_{pk}] &= E[b_{pj}-B_{pj}]\,[b_{pk}-B_{pk}] \\ &= E[\Sigma\beta_{jq}(a_q-A_q)]\,[\Sigma\beta_{kr}(a_r-A_r)],\end{aligned}$$

and so

$$\operatorname{cov}[b_{pj}, b_{pk}] = \sum_{j,\,k}^{p} \beta_{jq}\beta_{kq} \operatorname{var} a_q = \sum_{j,\,k}^{p} (\beta_{jq}\beta_{kq}/S_{qq})\,\sigma^2. \tag{7a}$$

The term in brackets is, from (7.3.4,3*b*), the element χ_{jk} of the inverse matrix, and so

$$\operatorname{cov}[b_{pj}, b_{pk}] = \chi_{jk}\,\sigma^2, \tag{7b}$$

while, from (4),

$$\operatorname{var} b_{pj} = \chi_{jj}\,\sigma^2. \tag{7c}$$

The variance of the fitted value $u_p(x)$ can also be expressed in terms of the elements of the inverse matrix. For

$$\begin{aligned}\operatorname{var}\left[\sum_j b_{pj}x^j\right] &= E\left[\sum_j (b_{pj}-B_{pj})\,x^j\right]\left[\sum_k (b_{pk}-B_{pk})\,x^k\right] \\ &= \sum_{q=0}^{2p}\left\{\sum_{j+k=q} E(b_{pj}-B_{pj})\,(b_{pk}-B_{pk})\right\} x^q,\end{aligned}$$

and, on using (7*b*) and (7*c*),

$$\operatorname{var}[u_p(x)] = \sigma^2 \sum_{q=0}^{2p}\left\{\sum_{j+k=q} \chi_{jk}\right\} x^q. \tag{8}$$

The elements of the inverse matrix in the sum lie along a line parallel to the backward diagonal.

8.1.1 *Estimation of σ from the residuals*

The residual v_i is given by

$$v_i = y_i - \sum_j a_j T_j(x_i),$$

and, as $E(v_i) = 0$,

$$E(v_i^2) = \operatorname{var} v_i = \operatorname{var} y_i - 2\sum \operatorname{cov}(y_i, a_j)\,T_j(x_i) + \sum_j T_j^2(x_i) \operatorname{var} a_j.$$

On using (8.1,2b),

$$E(v_i^2) = \operatorname{var} y_i - \sum_j T_j^2(x_i) \operatorname{var} a_j$$

$$= (\sigma^2/w_i)\left(1 - \sum_{j=0}^{p} w_i T_j^2(x_i)/S_{jj}\right). \quad (1)$$

It follows that $$E \sum_i w_i v_i^2 = (n-p-1)\sigma^2,$$

and so $$s^2 = (\Sigma w_i v_i^2)/(n-p-1) \quad (2)$$

will provide an unbiased estimate of σ^2.

The value $\Sigma w_i v_i^2$ can be obtained without calculating the individual residuals, as was shown in § 7.2.5.

8.1.2 *Example*

In the column at the extreme right of the Doolittle scheme given in Table 7.2.3, the calculations of Σv^2 are performed using equation (7.2.5,3a). For the third-degree polynomial, Σv_3^2 is 0·037204, and so from (8.1.1,2)

$$s_3 = \sqrt{(0{\cdot}037204/63)} = 0{\cdot}02430$$

is an estimate of σ. The standard deviation of a_3 is then estimated as

$$s[a_3] = s_3/\sqrt{S_{33}} = 0{\cdot}0176.$$

The inverse matrix for this example is calculated at the bottom of Table 7.2.3. The diagonal elements give the standard deviations of the power-series coefficients. Thus

$$s[b_{30}] = s_3\sqrt{\chi_{00}} = 0{\cdot}0435$$
$$s[b_{31}] = s_3\sqrt{\chi_{11}} = 0{\cdot}0475$$
$$s[b_{32}] = s_3\sqrt{\chi_{22}} = 0{\cdot}0399$$
$$s[b_{33}] = s_3\sqrt{\chi_{33}} = 0{\cdot}0176$$

The standard deviations of the fitted values $u_p(x)$ are calculated for a number of selected values of x in Table 8.1.2, using (8.1,8).

When the origin of x is changed to the value -2 (Table 7.4.1), the standard deviations of the new coefficients are found from the sums of the products $\gamma_{jk} Q_{kj}$. From Table 7.4.1,

$$s[c_{30}] = s\sqrt{\Sigma\gamma_{0k} Q_{k0}} = s\sqrt{91{\cdot}665} = 0{\cdot}233,$$
$$s[c_{31}] = s\sqrt{\Sigma\gamma_{1k} Q_{k1}} = s\sqrt{190{\cdot}77} = 0{\cdot}336,$$
$$s[c_{32}] = s\sqrt{\Sigma\gamma_{2k} Q_{k2}} = s\sqrt{33{\cdot}787} = 0{\cdot}141,$$
$$s[c_{33}] = s\sqrt{\Sigma\gamma_{3k} Q_{k3}} = s\sqrt{0{\cdot}5228} = 0{\cdot}0176.$$

When the scale of the variable is changed, the standard deviations are multiplied by the same factors as the coefficients. The values derived in this section for $s[a_j]$, $s[b_{pj}]$, $s[c_{pj}]$, have to be multiplied by the factors $10^r\,10^{-qj} = 10^2\,10^{-j}$ (§ 7.1.7) to give the standard deviations of the coefficients when the curve is expressed in terms of the original variable x. For the fitted values, $j = 0$ and the factor is $10^r = 10^2$. For the standard deviation s_p the factor is $10^r\sqrt{10^s} = 10^3$.

TABLE 8.1.2

Calculation of standard deviations of fitted values

Factors removed: x, 10^q, $q = 1$; y, 10^r, $r = 2$; elements of Doolittle scheme divided by 10^s, $s = 2$

x	χ_{00} 3·2100 $\times 1$	$2\chi_{01}$ −0·9978 $\times x$	$2\chi_{02}+\chi_{11}$ −0·2104 $\times x^2$	$2\chi_{03}+2\chi_{12}$ 3·1432 $\times x^3$	$2\chi_{13}+\chi_{22}$ +0·8468 $\times x^4$	$2\chi_{23}$ −2·0450 $\times x^5$	χ_{33} +0·5228 $\times x^6$	sum = L	s_3 0·02430 $s[u_3(x)] = s\sqrt{L}$
−2	3·2100	1·9956	−0·8416	−25·1456	13·5488	+65·4400	33·4592	91·666	0·233
−1·5		1·4967	−0·4734	−10·6083	4·2869	+15·5292	5·9550	19·396	0·107
−1		0·9978	−0·2104	−3·1432	0·8468	+2·0450	0·5228	4·2688	0·0502
−0·5		0·4989	−0·0526	−0·3929	0·0529	+0·0639	0·0082	3·3884	0·0447
0		0	0	0	0	0	0	3·2100	0·0435
0·5		−0·4989	−0·0526	0·3929	0·0529	−0·0639	0·0082	3·0486	0·0424
1·0		−0·9978	−0·2104	3·1432	0·8468	−2·0450	0·5228	4·4696	0·0514
1·5		−1·4967	−0·4734	10·6083	4·2869	−15·5292	5·9550	6·5609	0·0622
2·0		−1·9956	−0·8416	25·1456	13·5488	−65·4400	33·4592	7·0864	0·0647
2·5		−2·4945	−1·3150	49·1125	33·0781	−199·7070	127·6367	9·5208	0·0750
3		−2·9934	−1·8936	84·8664	68·5908	−496·9350	381·1212	35·966	0·146

8.1.3 *Least-squares theory in matrix notation*

The symbols listed in Table 8.1.3 will be used for the various matrices and vectors. As is customary, a row vector will be represented as the transpose of a column vector. The transpose of a vector or a matrix will be indicated by the superscript T.

TABLE 8.1.3
Matrix symbols

Symbol	*Order*	*Element*
$\mathbf{x}$	$(p+1)\times 1$	x^j
$\mathbf{X}$	$(p+1)\times n$	x_i^j
$\mathbf{b}$	$(p+1)\times 1$	b_{pj}
$\mathbf{B}$	$(p+1)\times 1$	B_{pj} (the true value)
$\mathbf{y}$	$n\times 1$	y_i
$\mathbf{v}$	$n\times 1$	v_i
$\boldsymbol{\delta}$	$n\times 1$	$y_i - Y_i$
$\mathbf{W}$	$n\times n$	$W_{ii} = w_i, W_{ij} = 0$

8.1.3.1 *Normal equations.* In matrix notation, the residuals are given by the equation

$$\mathbf{v} = \mathbf{y} - \mathbf{X}^T\mathbf{b}. \tag{1}$$

The expression $\mathbf{v}^T\mathbf{W}\mathbf{v}$ is a 1×1 matrix—that is, a scalar—and its value is $\Sigma w_i v_i^2$. The least-squares principle calls for the minimization of

$$\Sigma w_i v_i^2 = \mathbf{v}^T\mathbf{W}\mathbf{v} = (\mathbf{y}^T - \mathbf{b}^T\mathbf{X})\mathbf{W}(\mathbf{y} - \mathbf{X}^T\mathbf{b}). \tag{2}$$

If any scalar ψ is of the form

$$\psi = \mathbf{b}^T\mathbf{z} = \mathbf{z}^T\mathbf{b} = \sum_j z_j b_{pj},$$

then

$$\frac{\partial\psi}{\partial b_{pj}} = z_j,$$

and the differentials can be represented either as a column vector $\mathbf{z}$ or a row vector $\mathbf{z}^T$. If (2) is multiplied out and differentiated, and the differentials written as column vectors,

$$\partial(\mathbf{v}^T\mathbf{W}\mathbf{v})/\partial\mathbf{b} = -2\mathbf{X}\mathbf{W}\mathbf{y} + 2\mathbf{X}\mathbf{W}\mathbf{X}^T\mathbf{b},$$

and hence the normal equations are

$$\mathbf{X}\mathbf{W}\mathbf{X}^T\mathbf{b} = \mathbf{X}\mathbf{W}\mathbf{y}. \tag{3a}$$

These are identical with the equations (7.1,3), since it is easily verified that

$$\boldsymbol{\phi} = \mathbf{X}\mathbf{W}\mathbf{X}^T, \quad \mathbf{M} = \mathbf{X}\mathbf{W}\mathbf{y}. \tag{3b}$$

$\mathbf{X}$ is not a square matrix, and, as division (i.e. multiplication by $\mathbf{X}^{-1}$) is only possible when the matrix is square, $\mathbf{X}$ cannot be divided out of (3*a*).

8.1.3.2 *Standard deviation formulae.* The coefficients b_{pj} form the vector

$$\mathbf{b} = \boldsymbol{\phi}^{-1}\mathbf{XWy},$$

and the deviations of the b_{pj} from the true values B_{pj} form the vector

$$\mathbf{b}-\mathbf{B} = \boldsymbol{\phi}^{-1}\mathbf{XW}\boldsymbol{\delta}.$$

Hence the square matrix $(\mathbf{b}-\mathbf{B})(\mathbf{b}^T-\mathbf{B}^T)$ is equal to

$$\boldsymbol{\phi}^{-1}\mathbf{XW}\boldsymbol{\delta}\boldsymbol{\delta}^T\mathbf{W}^T\mathbf{X}^T\boldsymbol{\phi}^{-1T}$$

The expectation of an element of this matrix is

$$E(b_{pj}-B_{pj})(b_{pk}-B_{pk}) = \operatorname{cov}(b_{pj}, b_{pk}),$$

and so the covariance matrix is

$$E(\mathbf{b}-\mathbf{B})(\mathbf{b}^T-\mathbf{B}^T) = \boldsymbol{\phi}^{-1}\mathbf{XW}(E\boldsymbol{\delta}\boldsymbol{\delta}^T)\mathbf{W}^T\mathbf{X}^T\boldsymbol{\phi}^{-1T} \tag{1}$$

Now the (h,j) element of the product $E(\boldsymbol{\delta}\boldsymbol{\delta}^T\mathbf{W}^T)$ is

$$\sum_i E(y_h-Y_h)(y_i-Y_i)W_{ji}.$$

If the deviations $\boldsymbol{\delta}$ are uncorrelated $E(y_h-Y_h)(y_i-Y_i)$ vanishes unless $h=i$, when it has the value σ^2/w_i. Similarly, $\mathbf{W}$ is a diagonal matrix, and W_{ij} vanishes unless $i=j$. Thus the product is just the diagonal matrix whose elements are σ^2. Hence the covariance matrix is

$$\boldsymbol{\phi}^{-1}\mathbf{XW}(\sigma^2\mathbf{I})\mathbf{X}^T\boldsymbol{\phi}^{-1T} = \sigma^2\boldsymbol{\phi}^{-1}[\mathbf{XWX}^T\boldsymbol{\phi}^{-1}],$$

or, on using (8.1.3.1,3*b*),

$$E(\mathbf{b}-\mathbf{B})(\mathbf{b}^T-\mathbf{B}^T) = \sigma^2\boldsymbol{\phi}^{-1}. \tag{2a}$$

Thus
$$\operatorname{cov}(b_{pj}, b_{pk}) = \sigma^2(\phi^{-1})_{jk}, \tag{2b}$$

which is identical with *(8.1,7b)* and *(8.1,7c)*.

The fitted value is

$$u_p(x) = \mathbf{b}^T\mathbf{x},$$

and its variance is given by

$$\operatorname{var} u_p(x) = \mathbf{x}^T E(\mathbf{b}-\mathbf{B})(\mathbf{b}^T-\mathbf{B}^T)\mathbf{x},$$

or
$$\operatorname{var} u_p(x) = \mathbf{x}^T\boldsymbol{\phi}^{-1}\mathbf{x}\sigma^2. \tag{3}$$

This is the matrix form of (8.1,8).

8.2 RESULTS BASED ON THE NORMAL LAW

8.2.1 *The distribution of s^2*

When the deviations of the observations y from the values Y on the true curve follow a normal law, the probability of obtaining a set in the ranges dy_i about the values y_i is

$$dP(y_i) = C\left\{\exp - \sum_i (y_i - Y_i)^2/2\sigma_i^2\right\} \Pi\, dy_i. \tag{1}$$

If the equation of the true curve on which the Y_i lie is

$$Y = \sum_{j=0}^{p} A_j\, T_j(x), \tag{2}$$

then

$$\Sigma(y_i - Y_i)^2/\sigma_i^2 = \sum_i w_i \left\{y_i - \sum_j A_j\, T_j(x_i)\right\}^2 \Big/ \sigma^2$$

$$= \sum_i w_i \left\{y_i - \sum_j a_j\, T_j(x_i) + \sum_j (a_j - A_j)\, T_j(x_i)\right\}^2 \Big/ \sigma^2.$$

On using the orthogonal properties of the polynomials and the equations

$$\Sigma w_i \{y_i - \Sigma a_j\, T_j(x_i)\}\, T_k(x_i) = 0,$$

it follows that

$$\Sigma(y_i - Y_i)^2/\sigma_i^2 = \Sigma w_i\, v_i^2/\sigma^2 + \sum_j (a_j - A_j)^2\, \Sigma w_i\, T_j^2(x_i)/\sigma^2. \tag{3}$$

A change of scale

$$y_i^0 = y_i/\sigma_i = w_i^{\frac{1}{2}}\, y_i/\sigma \tag{4a}$$

transforms $\Sigma(y_i - Y_i)^2/\sigma_i^2$ according to the equation

$$\Sigma(y_i - Y_i)^2/\sigma_i^2 = \Sigma(y_i^0 - Y_i^0)^2. \tag{4b}$$

The expression on the right corresponds to the separation in an n-dimensional space of the point whose coordinates are y_i^0 from the point whose coordinates are Y_i^0. If these coordinates are now changed by the transformation

$$\left.\begin{array}{l} z_j = \sum_i \{w_i^{\frac{1}{2}}\, T_j(x_i)/[\Sigma w_i\, T_j^2(x_i)]^{\frac{1}{2}}\}\, y_i^0 = [\Sigma w_i\, T_j^2(x_i)]^{\frac{1}{2}}\, a_j/\sigma, \\ j = 0 \text{ to } p,\ z_{p+1}, \ldots, z_{n-1} \text{ orthogonal to each other and} \\ \quad \text{to the } z_j, \end{array}\right\} \tag{5a}$$

this separation will be preserved provided the transformation is an orthogonal one, corresponding to a simple rotation of the axes. But, from the properties of the polynomials $T_j(x)$,

$$\Sigma\, w_i^{\frac{1}{2}}\, T_j(x_i)\, w_i^{\frac{1}{2}}\, T_k(x_i)/\{\Sigma w_i\, T_j^2(x_i)\}^{\frac{1}{2}}\, \{\Sigma w_i\, T_k^2(x_i)\}^{\frac{1}{2}} = \delta_{jk},$$

and so from § 2.5.1 the new coordinate axes are at right angles. Hence

$$\Sigma(y_i - Y_i)^2/\sigma_i^2 = \Sigma(z_i - Z_i)^2. \tag{5b}$$

The distance of the point Y_i^0 from the origin is

$$\Sigma Z_i^2 = \Sigma\, Y_i^{02} = \Sigma w_i\, Y_i^2/\sigma^2,$$

or, on substituting for Y_i from (2),

$$\sum_{i=0}^{n-1} Z_i^2 = \sum_{j=0}^{p} A_j^2 \sum_i w_i T_j^2(x_i)/\sigma^2.$$

But, from (5*a*),

$$\sum_{j=0}^{p} A_j^2 \Sigma w_i T_j^2(x_i)/\sigma^2 = \sum_{i=0}^{p} Z_i^2, \tag{5c}$$

and so $Z_{p+1}, \ldots, Z_{n-1}$ vanish and

$$\Sigma(y_i - Y_i)^2/\sigma_i^2 = \sum_{j=0}^{p} (z_j - Z_j)^2 + \sum_{p+1}^{n-1} z_i^2,$$

or, on substituting for z_j from (5*a*),

$$\Sigma(y_i - Y_i)^2/\sigma_i^2 = \sum_{j=0}^{p} (a_j - A_j)^2 \Sigma w_i T_j^2(x_i)/\sigma^2 + \sum_{p+1}^{n-1} z_i^2. \tag{6}$$

Hence the probability distribution (1) transforms into

$$dP(a_j, z_i) = C \exp\left\{-\tfrac{1}{2}\sum_{j=0}^{p}(a_j - A_j)^2 \Sigma w_i T_j^2(x_i)/\sigma^2\right\} \Pi\, da_j$$

$$\times \exp\left\{-\tfrac{1}{2}\sum_{p+1}^{n-1} z_i^2\right\} \Pi\, dz_i. \tag{7a}$$

Comparison of (3) and (6) shows that

$$\sum_{p+1}^{n-1} z_i^2 = \sum_{i=0}^{n-1} w_i v_i^2/\sigma^2. \tag{7b}$$

Therefore the a_j are normally distributed about the A_j with variance $\sigma^2/\Sigma w_i T_j^2(x_i)$, $\Sigma w_i v_i^2/\sigma^2$ is distributed as χ^2 with $\nu = n-p-1$ d.f., and the distributions of the a_j and of $\Sigma w_i v_i^2$ are all independent of one another.

The quantity

$$s^2 = \Sigma w_i v_i^2/(n-p-1) \tag{8a}$$

will then provide an unbiased estimate of σ^2, and $\nu s^2/\sigma^2$ is distributed as χ^2 with $\nu = n-p-1$ d.f. Hence, as in § 2.5.3,

$$\operatorname{var} s^2 = \sigma^4/\tfrac{1}{2}(n-p-1), \quad \operatorname{var} s = \sigma^2/2(n-p-1). \tag{8b}$$

The χ^2 tables can be used to test the significance of the difference between an observed value s and an expected value σ.

8.2.2 *Tests of significance*

The theorem proved in § 3.1.4 shows that the ratio

$$t = \frac{a_j - A_j}{s(a_j)} = \frac{(a_j - A_j)\sqrt{S_{jj}}}{s} \tag{1a}$$

is distributed as t with $\nu = n-p-1$ d.f. This ratio can be used to test the significance of the departure from an assumed true value. In particular, the ratio

$$t = a_{p+1}\sqrt{S_{p+1,\,p+1}}/s \tag{1b}$$

can be used to test whether the coefficient a_{p+1} is significantly different from zero, and hence to provide a guide in the choice of the degree of the polynomial if this is not already known.

Similarly, the significance of the difference between two separate determinations a'_j and a''_j can be tested by the ratio

$$\frac{a'_j - a''_j - (A'_j - A''_j)}{s(a'_j - a''_j)}, \tag{2a}$$

where the standard deviation of the difference is given by

$$s^2(a'_j - a''_j) = s^2(a'_j) + s^2(a''_j) = (S'^{-1}_{jj} + S''^{-1}_{jj})\, s_p^2, \tag{2b}$$

with $$s_p^2 = (\Sigma w'_i v'^2_i + \Sigma w''_i v''^2_i)/(n' + n'' - 2p - 2). \tag{2c}$$

The power-series coefficients b_{pj} and the fitted values $u_p(x)$ are linear functions of the a_j, and so t-tests can be used for these quantities also. Thus for two different estimates of the same curve

$$t = \{u'_p(x) - u''_p(x)\}/s_p\{\sigma^2[u'_p(x)]/\sigma^2 + \sigma^2[u''_p(x)]/\sigma^2\}^{\frac{1}{2}} \tag{3}$$

will test the significance of the difference between the two fitted values.

If the two curves are obtained by different experimental methods, so that the estimated quantities may have different standard deviations the ratio of which is unknown, the quantities

$$\frac{a'_j - a''_j - (A'_j - A''_j)}{\{s^2(a'_j) + s^2(a''_j)\}^{\frac{1}{2}}}, \quad \frac{u'_p(x) - u''_p(x)}{\{s^2[u'_p(x)] + s^2[u''_p(x)]\}^{\frac{1}{2}}} \tag{4}$$

should be used to test the significance of the differences, either by Behrens' test (§ 3.5.3) or Welch's test (§ 3.5).

8.2.2.1 *F-test for the degree of the polynomial.* If the degree of the polynomial is not known, it is customary to examine the sums of the squares of the residuals calculated from

$$\sum_i w_i v_{ki}^2 = \sum_i w_i y_i^2 - \sum_{j=0}^{k} a_j \mathcal{M}_j. \tag{1}$$

If the curve is really of degree p, the sum should decrease rapidly as k increases until the value p is reached, after which the decrease should be very slow.

If the degree of the curve is in fact p, A_{p+q} is zero and a_{p+q} is distributed normally about zero with variance $\sigma^2/S_{p+q,\,p+q}$. Hence

$$\sum_{q=1}^{j} a_{p+q}^2 S_{p+q,\,p+q}/\sigma^2 = \sum_{q=1}^{j} a_{p+q}\mathscr{M}_{p+q}/\sigma^2 \tag{2a}$$

is distributed as χ^2 with j d.f., while

$$\Sigma w_i v_{p+j,\,i}^2/\sigma^2 = (n-p-j-1)\,s_{p+j}^2/\sigma^2 \tag{2b}$$

is distributed as χ^2 with $n-p-j-1$ d.f. Hence, on the hypothesis that all the A_{p+q} are zero, the ratio

$$F = \frac{\sum_{q=1}^{j} a_{p+q}\mathscr{M}_{p+q}}{j} \Bigg/ \frac{\sum_i w_i v_{p+j,\,i}^2}{n-p-j-1}$$

$$= \left\{\frac{\Sigma w_i v_{p,\,i}^2}{\Sigma w_i v_{p+j,\,i}^2} - 1\right\}\left\{\frac{n-p-j-1}{j}\right\} \tag{3}$$

is distributed as F with $(j, n-p-j-1)$ d.f. The significance of a whole series of coefficients can thus be tested simultaneously.

More commonly, the significance of a_{p+1} alone would be tested. Since $F = t^2$ when $\nu_1 = 1$ (§ 3.1.3), the test reduces to the t-test discussed in § 8.2.2.

8.2.2.2 *Example*

For the example of Table 7.2.3,

$$a_3 = -0{\cdot}0744, \quad s_3 = 0{\cdot}0243, \quad \sqrt{S_{33}} = 1{\cdot}383.$$

Hence the value t given by (8.2.2,1b) is

$$t = -0{\cdot}0744 \times 1{\cdot}383/0{\cdot}0243 = -4{\cdot}23 \text{ (63 d.f.)},$$

which is well below the 0·1% level. It would then be expected that the third-degree coefficient would be significant.

It would therefore seem desirable to extend the calculations to obtain the fourth-degree coefficient a_4. When this is done, it is found that

$$S_{44} = 2{\cdot}2941, \quad a_4 = 0{\cdot}002619, \quad s_4 = 0{\cdot}0244.$$

Then $$t = 0{\cdot}002619 \times 1{\cdot}515/0{\cdot}0244 = 0{\cdot}16 \text{ (62 d.f.)},$$

and this coefficient is certainly negligible.

If the Doolittle scheme has been carried through to the fourth degree, it is possible to test whether a second-degree polynomial would suffice by using (8.2.2.1,3). Thus

$$\Sigma w_i v_{2i}^2 = 0{\cdot}04780, \quad \Sigma w_i v_{4i}^2 = 0{\cdot}03719,$$

and $$F = \left\{\frac{0{\cdot}04780}{0{\cdot}03719} - 1\right\}\frac{62}{2} = 8{\cdot}84,$$

while the 0·1% level is 7·8.

8.2.2.3 *Analysis of variance table.* Statisticians often rewrite the last column of Table 7.2.3 in the form of an analysis of variance table. A typical scheme is shown in Table 8.2.2 (Goulden, 1952).

The entry in the variance column is the sum of squares divided by the degrees of freedom, and the F value is the ratio of the two variances. The F-test here is identical with the t-test of § 8.2.2.2. As F is by definition always greater than unity, the ratio for $p = 4$ is error variance divided by regression variance.

TABLE 8.2.2

Analysis of variance table for Example 7.2.3

Degree of fitting		Sums of squares	Degrees of freedom	Variance	F	5% Point
	Total (Σy^2)	1·218016	67			
0	Regression ($M_0 a_0$)	0·952837	1	0·952837		
	Error (Σv_0^2)	0·265179	66	0·004018	237	4
1	Regression ($\mathscr{M}_1 a_1$)	0·119525	1	0·119525		
	Error (Σv_1^2)	0·145654	65	0·002241	53	4
2	Regression ($\mathscr{M}_2 a_2$)	0·097853	1	0·097853		
	Error (Σv_2^2)	0·047801	64	0·000747	131	4
3	Regression ($\mathscr{M}_3 a_3$)	0·010597	1	0·010597		
	Error(Σv_3^2)	0·037204	63	0·000591	18	4
4	Regression ($\mathscr{M}_4 a_4$)	0·000016	1	0·000016		
	Error (Σv_4^2)	0·037188	62	0·000600	38	252

8.2.3 *Test for homogeneity*

Suppose that r separate sets of observations have been taken, giving rise to r sets of coefficients $b_{pj,q}$. If the true values $B_{pj,q}$ are all the same, and σ is the same for each set, the $b_{pj,q}$ will be distributed normally with variance

$$\sigma^2(b_{pj,q}) = \sigma^2/W_{j,q}, \tag{1}$$

where $W_{j,q}$ is a known function of $x_{i,q}$ and $w_{i,q}$. The weighted mean of the estimates b_{pj} is

$$\bar{b}_{pj} = \sum_q W_{j,q} b_{pj,q} / \sum_q W_{j,q}. \tag{2}$$

As regards the residuals from the mean $\bar{b}_{pj}$, the sum

$$\sum_q W_{j,q}(b_{pj,q}-\bar{b}_{pj})^2/\sigma^2 \tag{3}$$

will be distributed as χ^2 with $r-1$ d.f. Also

$$\sum_q \sum_i w_{i,q}\, v_{i,q}^2/\sigma^2 \tag{4}$$

will be distributed as χ^2 with $n-r(p+1)$ d.f., where n is the total number of observations. Hence the ratio

$$F = \frac{\Sigma W_{j,q}(b_{pj,q}-\bar{b}_{pj})^2}{r-1} \Bigg/ \frac{\Sigma\Sigma w_{i,q}\, v_{i,q}^2}{n-r(p+1)} \tag{5}$$

will be distributed as F with $(r-1, n-rp-r)$ d.f., if the values $b_{pj,q}$ are homogeneous and σ is the same for each set.

It will be clear that, since the forms of the orthogonal polynomials depend on the values $x_{i,q}$ and $w_{i,q}$, the coefficients $a_{j,q}$ would not be expected to be homogeneous unless the values x_i and w_i were the same for all sets.

An example on the testing for homogeneity of the slopes of straight lines is given in § 6.2.4.1.

8.3 MINIMUM VARIANCE ESTIMATES

In this section it will be shown that the unbiased estimate b_{pj} whose variance has the smallest possible value is identical with the least-squares estimate. This result is often referred to as the Markoff theorem, although it was originally proved by Gauss.

If the quantity

$$b_{pr} = \sum_i z_{ri}\, y_i \tag{1}$$

is to be an unbiased estimate of B_{pr}, then

$$E(b_{pr}) = \Sigma z_{ri}\, E(y_i) = \sum_i z_{ri} \sum_k B_{pk}\, x_i^k$$

must equal B_{pr}, and so

$$\sum_i z_{ri}\, x_i^k = \delta_{rk}. \tag{2}$$

The variance of the estimate (1) is

$$\operatorname{var} b_{pr} = \Sigma z_{ri}^2 \operatorname{var} y_i,$$

and so, for small variations Δz_{ri},

$$\Delta(\operatorname{var} b_{pr}/2\sigma^2) = \Sigma(z_{ri}/w_i)\, \Delta z_{ri}. \tag{3}$$

The variance will be a minimum when this expression vanishes. However, the variations Δz_{ri} cannot be independent, since, from (2),

$$\Sigma x_i^j\, \Delta z_{ri} = 0 \quad (p+1 \text{ equations}). \tag{4}$$

It would be possible to eliminate $p+1$ of the variations Δz_{ri} from (3) by solving the set (4) for these variations, and then to obtain the minimum variance conditions by equating the coefficients of the remaining $n-p-1$ variations in (3) to zero. However, this would produce an awkward unsymmetrical set of equations, and it is better to introduce $p+1$ independent ('Lagrangian') multipliers λ_{rj}. Thus if each of the equations (4) is multiplied by λ_{rj} and subtracted from (3), the minimum variance conditions are

$$\sum_i \left\{(z_{ri}/w_i) - \sum_{j=0}^{p} \lambda_{rj} x_i^j\right\} \Delta z_{ri} = 0. \tag{5}$$

Now $n-p-1$ of the Δz_{ri} can be chosen arbitrarily, so that their coefficients must vanish. Further, the remaining $p+1$ coefficients can be made to vanish by suitable choice of the $p+1$ multipliers λ_{rj}. Hence the symmetrical set of conditions

$$z_{ri} = \sum_{j=0}^{p} \lambda_{rj} w_i x_i^j \tag{6}$$

is obtained.

The multipliers λ_{rj} are determined by the substitution of (6) in the conditions (2) for the estimates to be unbiased. This gives

$$\sum_{j=0}^{p} \lambda_{rj} \sum_i w_i x_i^{j+k} = \delta_{rk}$$

or
$$\sum_{j=0}^{p} \lambda_{rj} \phi_{jk} = \delta_{rk}, \tag{7}$$

where
$$\phi_{jk} = \Sigma w_i x_i^{j+k}$$

as in the least-squares theory. Hence λ_{rj} is the element χ_{rj} of the inverse matrix $\boldsymbol{\phi}^{-1}$, and so from (1) the minimum variance estimate is

$$b_{pr} = \sum_{j=0}^{p} \chi_{rj} \sum_i w_i x_i^j y_i = \sum_{j=0}^{p} \chi_{rj} M_j, \tag{8}$$

which is identical with the least-squares estimate.

8.3.1 *The normal equations for correlated variables*

When the deviations of the observations are correlated, so that

$$\left.\begin{aligned} E(y_i - Y_i)(y_h - Y_h) &= \rho_{ih}\sigma_i\sigma_h = \sigma_{ih}, \\ E(y_i - Y_i)^2 &= \sigma_i^2 = \sigma_{ii}, \end{aligned}\right\} \tag{1}$$

the values σ_{ih} defined by (1) form an $n \times n$ matrix $\boldsymbol{\sigma}$. The inverse of $\boldsymbol{\sigma}$ will be noted by $\mathbf{W}$. A natural extension of the least-squares principle, which for uncorrelated observations states that $\Sigma w_i v_i^2$ is to be minimized, is that in the present case the quantity

$$\mathbf{v}^T \mathbf{W} \mathbf{v} \tag{2}$$

should be minimized. Then, as in § 8.1.3.1, the normal equations are

$$\boldsymbol{\phi}\mathbf{b} = \mathbf{M}, \tag{3a}$$

where

$$\boldsymbol{\phi} = \mathbf{X}\mathbf{W}\mathbf{X}^T \tag{3b}$$

and

$$\mathbf{M} = \mathbf{X}\mathbf{W}\mathbf{y}. \tag{3c}$$

The elements of (3*b*) and (3*c*) are

$$\phi_{rs} = \sum_i \sum_h w_{ih}\, x_i^r x_h^s \tag{3d}$$

and

$$M_r = \sum_i \sum_h w_{ih}\, x_i^r y_h. \tag{3e}$$

These quantities are considerably more complicated than the corresponding quantities for the uncorrelated case.

As in the uncorrelated case, it can be shown that the estimates b_{pj} given by the least-squares principle are identical with the minimum variance estimates. This will be established in the next section.

8.3.2 *The generalized Gauss–Markoff theorem in matrix notation*

The elements z_{ri} in (8.3,1) form a $(p+1)\times n$ matrix $\mathbf{Z}$, such that

$$\mathbf{b} = \mathbf{Z}\mathbf{y}. \tag{1}$$

If $\mathbf{b}$ is to be an unbiased estimate of $\mathbf{B}$,

$$E(\mathbf{b}) = \mathbf{Z}E(\mathbf{y}) = \mathbf{Z}\mathbf{X}^T\mathbf{B},$$

and so

$$\mathbf{Z}\mathbf{X}^T = \mathbf{I}. \tag{2}$$

The covariance matrix for the b_{pj} is

$$E(\mathbf{b}-\mathbf{B})(\mathbf{b}^T-\mathbf{B}^T) = \mathbf{Z}E(\mathbf{y}-\mathbf{Y})(\mathbf{y}^T-\mathbf{Y}^T)\mathbf{Z}^T = \mathbf{Z}\boldsymbol{\sigma}\mathbf{Z}^T, \tag{3}$$

where $\boldsymbol{\sigma}$ is the matrix defined in (8.3.1,1).

The variances are the diagonal elements

$$\sum_i \sum_h z_{ri}\,\sigma_{ih}\,z_{rh},$$

and so the condition for minimum variance of b_{pr} is

$$\sum_i \left\{\sum_h z_{rh}\,\sigma_{ih}\right\} \Delta z_{ri} = 0$$

for arbitrary small variations Δz_{ri}. Hence the minimum variance conditions are

$$\sum_i \{\mathbf{Z\sigma}\}_{ri}\,\Delta z_{ri} = 0 \tag{4a}$$

with the conditions

$$\sum_i \{\mathbf{X}\}_{si}\,\Delta z_{ri} = 0 \tag{4b}$$

imposed by the requirement (2) for unbiased estimates.

A Lagrangian multiplier matrix $\boldsymbol{\lambda}$, of order $(p+1)\times(p+1)$, is used to combine the conditions (4*a*) and (4*b*) into a single equation

$$\sum_i (\mathbf{Z\sigma}-\boldsymbol{\lambda}\mathbf{X})_{ri}\,\Delta z_{ri} = 0,$$

and, as the variations Δz_{ri} can now be considered independent,

$$\mathbf{Z\sigma} = \boldsymbol{\lambda}\mathbf{X}. \tag{5}$$

Hence, on using (2), $\boldsymbol{\lambda}\mathbf{X}\boldsymbol{\sigma}^{-1}\mathbf{X}^T = \mathbf{I}$,

or

$$\boldsymbol{\lambda} = (\mathbf{XWX}^T)^{-1} = \boldsymbol{\phi}^{-1}, \tag{6}$$

where $\mathbf{W} = \boldsymbol{\sigma}^{-1}$ is the weight matrix. Thus

$$\mathbf{b} = \mathbf{Zy} = \boldsymbol{\lambda}\mathbf{XWy} = \boldsymbol{\phi}^{-1}\mathbf{M}, \tag{7}$$

where $\boldsymbol{\phi}$ and $\mathbf{M}$ are identical with the corresponding quantities derived by the generalized least-squares method of § 8.3.1. The Gauss–Markoff theorem on the equality of the least-squares and minimum variance estimates is therefore established.

When the deviations $y_i - Y_i$ are independent, $\boldsymbol{\sigma}$ is the diagonal matrix whose elements are σ_i^2 and $\mathbf{W}$ the diagonal matrix whose elements are $w_i = 1/\sigma_i^2$. The quantities ϕ_{jk} and M_k then have the simpler forms (7.1,2).

8.4 TABLES OF STANDARD DEVIATIONS FOR THE EQUALLY-SPACED CASE

Formulae and tables giving the standard deviations of the fitted values have been prepared for the case when the observations are equally-spaced and of unit weight. The standard deviation is, from (8.1,6),

$$\sigma[u_p(\epsilon)] = \left[\sum_{j=0}^{p}\{T_j^2(\epsilon)/\Sigma T_j^2(\epsilon_i)\}\right]^{\frac{1}{2}}\sigma. \tag{1}$$

On changing the variable to

$$k = 2\epsilon/n = 2(x-\bar{x})/n\Delta x, \tag{2}$$

the standard deviation can be put in the form

$$\sigma[u_p(k)] = n^{-\frac{1}{2}}\,\rho_{p0}(k,n)\,\sigma, \tag{3}$$

where $\rho_{p0}(k,n)$ is a function of k and n which can be evaluated by means of the standard expressions for $T_j(\epsilon)$ and $\Sigma T_j^2(\epsilon_i)$. It is found that $\rho_{p0}(k, n)$ only varies slowly with n.

The range of the variable ϵ may be divided into two parts: the region of interpolation, comprising the values $|\epsilon| < \frac{1}{2}n$, i.e. $|k| < 1$; and the region of extrapolation, comprising the values $|\epsilon| > \frac{1}{2}n$, i.e. $|k| > 1$.

In the region of interpolation the variation in $\rho_{p0}(k, n)$ is comparatively small. Table 8.8*a* gives the values $\rho_{p0}(k, n)$ for $|k| \leqslant 1$ and for various selected values of n. Intermediate values may be obtained by linear interpolation between the tabulated values. The error arising from interpolation is generally less than 1 per cent, and never exceeds 2 per cent.

In the region of extrapolation $\rho_{p0}(k, n)$ may be split up into two parts,

$$\rho_{p0}(k, n) = \rho_{p0}(k)\,\phi_p(n), \tag{4a}$$

where

$$\rho_{p0}(k) = \rho_{p0}(k, \infty) \tag{4b}$$

and

$$\{\phi_p(n)\}^{-2} = (1 - n^{-2})(1 - 4n^{-2})\ldots(1 - p^2 n^{-2}). \tag{4c}$$

These functions are listed in Table 8.8*b*. The error introduced by splitting $\rho_{p0}(k, n)$ into two factors is less than 1 per cent for n greater than 12. When n is less than 12, the error is somewhat greater for values of k near 1; it is always less than 5 per cent, except in the single case $n = 7, p = 5$.

Table 8.8*b* extends to $k = 3$. Beyond this value the function $\rho_{p0}(k)$ can be written to a reasonable approximation as $\alpha_p k^p$, where α_p is a constant.

8.4.1 *Variation of the standard deviation with the location of the point*

Examination of Table 8.8*a* shows that, for values of $|k|$ less than 0·9, $\rho_{p0}(k, n)$ differs from $\rho_{p0}(k) \equiv \rho_{p0}(k, \infty)$ by at most 2 per cent for n greater than 18, and by at most 1 per cent for n greater than 25. Hence for reasonable values of n the curves giving the variation in standard deviation are practically identical with the curves for $n = \infty$.

The curves of $\rho_{p0}(k)$ in the region of interpolation are drawn in Figs. 8.4.1*a* and 8.4.1*b*. The curves have p minima and $p - 1$ maxima, all symmetrically located about $\epsilon = 0$. The maxima and minima are shallow, and the general trend is for the standard deviation to increase slowly as $|\epsilon|$ increases; that is, successive maxima and minima are somewhat higher than the preceding ones. Beyond the last minimum the standard deviation increases quite rapidly. There is a transition region between the region of interpolation and the region of extrapolation.

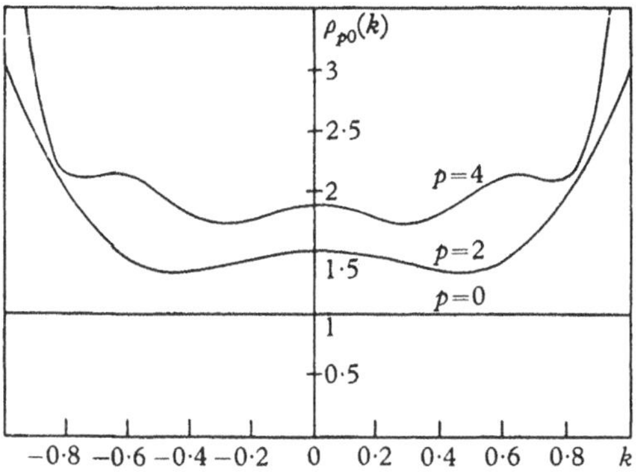

Fig. 8.4.1*a*. Variation of standard deviation in the region of interpolation—p even.

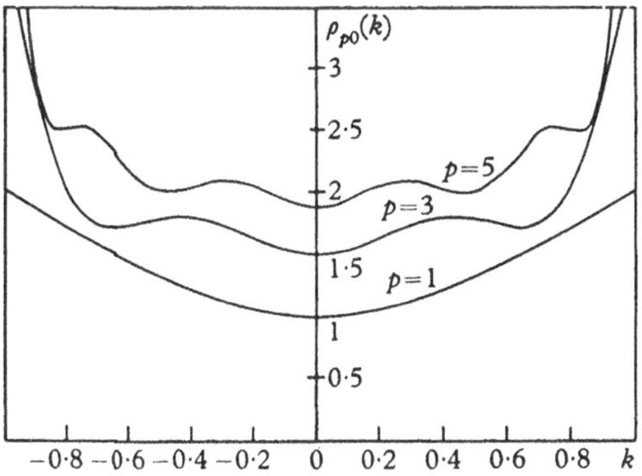

Fig. 8.4.1*b*. Variation of standard deviation in the region of interpolation—p odd.

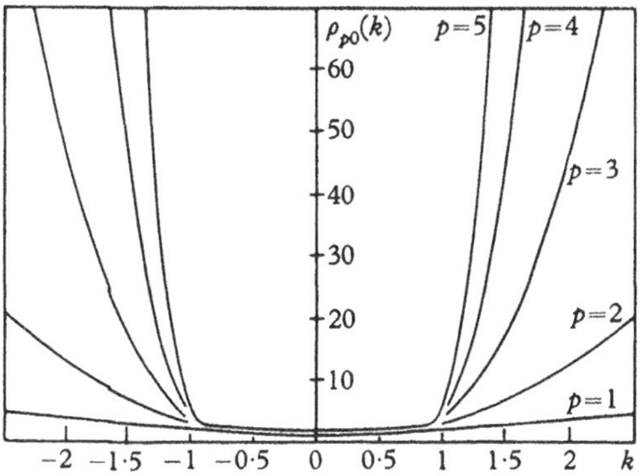

Fig. 8.4.1*c*. Variation of standard deviation in the region of extrapolation.

The curves of $\rho_{p0}(k)$ in the region of extrapolation are drawn in Fig. 8.4.1c. It will be observed that the increase is very rapid beyond $|k| = 1$, especially for the higher values of p. Considerable caution should be exercised in extrapolating polynomials, especially if the degree is not definitely known.

8.4.2 *The use of the tables*

In a practical problem, σ is usually unknown, and an estimated value s has to be used in (8.4,3). The procedure for finding the estimated standard deviation $s[u_p(\epsilon)]$ of the fitted value at the point ϵ may be summarized as follows:

(i) Evaluate $|k| = 2|\epsilon|/n$ to three decimal places.

(ii) (*a*) If $|k| < 1$, determine the value of $\rho_{p0}(k, n')$ by interpolating between the listed values of k in Table 8.8*a*, using the listed value n' closest to n. Estimate $\rho_{p0}(k, n)$ by examining the variation with n at this point in the table.

(*b*) If $|k| > 1$, determine $\rho_{p0}(k)$ by interpolation in Table 8.8*b*, and $\phi_p(n)$ from the lower section of this table.

(iii) (*a*) If $|k| < 1$, $s[u_p(\epsilon)] = \rho_{p0}(k, n)\, s/\sqrt{n}$. (1)

(*b*) If $|k| > 1$, $s[u_p(\epsilon)] = \rho_{p0}(k)\, \phi_p(n)\, s/\sqrt{n}$. (2)

8.4.2.1 *Example*

The calculation of the standard deviations of the fitted values at the points of observation by formula (8.4.2,1) is shown in Table 8.4.2 for the fourth-degree polynomial of Tables 7.6.2.2, 7.6.2.2*a*, and 7.6.2.3*b*. In this example n is 25 and s_4 0·0328.

The standard deviations at points beyond the range of the observations are calculated by formula (8.4.2,2) in the lower portion of Table 8.4.2.

8.4.3 *Rough approximations to the standard deviations*

In many cases only a rough approximation to the standard deviation of the fitted value is required. Table 8.4.3 gives expressions for the standard deviation which are accurate to within 20 per cent over the ranges of ϵ and n shown.

In Example 8.4.2.1, $s_4 = 0{\cdot}0328$ and $n = 25$, and so the rough value of $s[u_4(\epsilon)]$ is 0·013. This is close to the accurate values calculated in Table 8.4.2 when $|\epsilon|$ is less than 10.

8.4.4 *Variance of the residuals*

From (8.1.1,1),

$$\operatorname{var} v_i = \sigma_i^2 - \operatorname{var} u_p(x_i), \tag{1}$$

and so in the equally-spaced case

$$\operatorname{var} v_i = \sigma^2[1 - n^{-1}\, \rho_{p0}^2(k_i, n)]. \tag{2}$$

It is apparent that the variance of the residuals at the extremes of the range is a little less than at the centre of the range, although the difference is small if n is large. This is equivalent to the statement that the curve 'fits' the extreme observations somewhat better than it does the central observations.

TABLE 8.4.2

Standard deviations of fitted values (Example 8.4.2.1)

$s_4 = 0{\cdot}0328 \quad n = 25$

(a) Region of interpolation

$s_4/\sqrt{n} = 0{\cdot}00656$

$\lvert\epsilon\rvert$	$\lvert k\rvert = 2\lvert\epsilon\rvert/n$	$\rho_{40}(k, 25)$	$s[u_4(\epsilon)]$
0	0	1·88	0·0123
1	0·08	1·86	0·0122
2	0·16	1·80	0·0118
3	0·24	1·75	0·0115
4	0·32	1·75	0·0115
5	0·40	1·82	0·0119
6	0·48	1·95	0·0128
7	0·56	2·08	0·0136
8	0·64	2·15	0·0141
9	0·72	2·12	0·0139
10	0·80	2·12	0·0139
11	0·88	2·60	0·0171
12	0·96	4·02	0·026

(b) Region of extrapolation

$\phi_4(25) = 1{\cdot}02 \qquad \phi_4(25)\, s_4/\sqrt{n} = 0{\cdot}00669$

$\lvert\epsilon\rvert$	$\lvert k\rvert$	$\rho_{40}(k)$	$s[u_4(\epsilon)]$
14	1·12	9·83	0·066
16	1·28	20·6	0·138
18	1·44	37·6	0·25
20	1·60	62·4	0·42
22	1·76	97	0·65

TABLE 8.4.3

Rough approximations to $s[u_p(\epsilon)]$

Degree p	$s[u_p(\epsilon)]$	Range of $\lvert\epsilon\rvert$	Range of n
1	$1{\cdot}2s_1/\sqrt{n}$	$0-0{\cdot}32n$	∞ to 7
2	$1{\cdot}5s_2/\sqrt{n}$	$0-0{\cdot}38n$	∞ to 7
3	$1{\cdot}8s_3/\sqrt{n}$	$0-0{\cdot}41n$	∞ to 7
4	$2{\cdot}0s_4/\sqrt{n}$	$0-0{\cdot}42n$	∞ to 7
5	$2{\cdot}2s_5/\sqrt{n}$	$0-0{\cdot}44n$	∞ to 10

The residuals v_h and the fitted values u_i are statistically independent. For from (8.1,2*b*),

$$\operatorname{cov}(y_h, u_i) = \sum_j T_j(x_i)\operatorname{cov}(y_h, a_j) = \sum_j T_j(x_h)\,T_j(x_i)\,\sigma^2/S_{jj},$$

while

$$\operatorname{cov}(u_h, u_i) = \sum_{j,k} T_j(x_h)\,T_k(x_i)\operatorname{cov}(a_j, a_k) = \sum_j T_j(x_h)\,T_j(x_i)\,\sigma^2/S_{jj}.$$

Hence

$$\operatorname{cov}(y_h, u_i) = \operatorname{cov}(u_h, u_i) = \sum_j T_j(x_h)\,T_j(x_i)\,\sigma^2/S_{jj} \tag{3a}$$

and

$$\operatorname{cov}(v_h, u_i) = \operatorname{cov}(y_h - u_h, u_i) = 0. \tag{3b}$$

8.4.5 *The polynomial coefficients*

The standard deviations of the polynomial coefficients may also be calculated from tabulated functions $\rho_{pj}(k, n)$. If the curve is expressed in the form

$$u_p(z) = \Sigma c_{pj}\, z^j \tag{1a}$$

with

$$z = \epsilon - g, \tag{1b}$$

then from (8.1,5),

$$\sigma(c_{pj}) = \left\{\sum_{k=j}^{p} \gamma_{jk}^2/S_{kk}\right\}^{\frac{1}{2}} \sigma. \tag{1c}$$

Since γ_{jk} is a known function of β_{jk} and g, (1*c*) can be written as a function of n and g in the form

$$\sigma[c_{pj}(k)] = \rho_{pj}(k, n)\,\sigma/n^{j+\frac{1}{2}}, \tag{2a}$$

where

$$k = 2g/n. \tag{2b}$$

The values $\rho_{pj}(k, n)$ have been tabulated by Guest (1950*c*) for polynomials up to the fifth degree. Since these tables will only be required occasionally, they will not be reproduced here.

TABLE 8.4.5

Standard deviations of power-series coefficients (*Example* 8.4.5.1)

	$s_4 = 0{\cdot}0328$ $n = 25$ $\lvert k \rvert = 0{\cdot}96$		
j	$\rho_{4j}(0{\cdot}96, 25)$	$n^{j+\frac{1}{2}}$	$s(c_{pj})$
0	4·02	5	$2{\cdot}64 \times 10^{-2}$
1	60·5	125	$1{\cdot}59 \times 10^{-2}$
2	263	3125	$2{\cdot}76 \times 10^{-3}$
3	416	781×10^2	$1{\cdot}75 \times 10^{-4}$
4	215	1953×10^3	$3{\cdot}61 \times 10^{-6}$

8.4.5.1 *Example*

For the fourth-degree polynomial of Table 7.6.2.2 n is 25, and so if the origin is chosen at $\epsilon = -12$, $|k| = 24/25 = 0{\cdot}96$. From the tables (Guest 1950*c*), the values $\rho_{pj}(k, n)$ are as listed in Table 8.4.5. The estimated standard deviations of the coefficients are then given by

$$s(c_{4j}) = \rho_{4j}(0{\cdot}96, 25)\, s_4/n^{j+\frac{1}{2}}.$$

These values are listed in the last column of the table.

8.5 STANDARD DEVIATIONS IN THE UNEQUALLY-SPACED CASE

In the present section formulae will be obtained which will give, at least approximately, the standard deviations of the polynomial coefficients and fitted values when the observations are unequally-spaced. The procedure adopted is to characterize any particular set of observations by two parameters, denoted by κ_2 and κ_3. The parameter κ_2 is a measure of the departure of the independent variable x from symmetry about the central value, while the parameter κ_3 is a measure of the relative concentration of the observations towards the central values of x as opposed to the extreme values. Tables of the standard deviations in terms of these parameters will be given.

Although the tables were calculated principally for use in theoretical discussions, they may also be of use in practical examples, either for the rough calculation of the standard deviations or for the checking of the values obtained by the more usual methods.

Unfortunately, the treatment in terms of the parameters κ_2 and κ_3 is not adequate for all possible sets of data. In certain cases it would be desirable to take into account higher-order parameters κ_4 and κ_5. However, it is found that the treatment given here is adequate for practically all cases in which the curve is of the first or second degree, and for a large proportion of the cases in which the curve is of the third degree.

8.5.1 *The smoothing of the points of observation*

When the values of the independent variable x_i at the n points of observation are arranged in order of magnitude, each observation may be identified by a number ϵ_i giving its position in the sequence, ϵ_i taking the integral or half-integral values from $-\frac{1}{2}(n-1)$ to $+\frac{1}{2}(n-1)$. In the present discussion the system of points x_i will be replaced by a smoothed-out system X_i obtained by fitting a curve of the third degree in ϵ to the values x_i. The smoothed-out system of points is given by the equation

$$X_i = k_0 + k_1 T_1^e(\epsilon_i) + k_2 T_2^e(\epsilon_i) + k_3 T_3^e(\epsilon_i), \tag{1a}$$

where

$$k_j = \sum_i T_j^e(\epsilon_i)\, x_i \Big/ \sum_i \left\{T_j^e(\epsilon_i)\right\}^2 \tag{1b}$$

and $T_j^e(\epsilon)$ is the orthogonal polynomial of degree j in ϵ for the equally-spaced case, whose properties have been discussed in § 7.7. The superscript e is added to distinguish the polynomial from the orthogonal polynomials in the variables x or X.

By a change of origin the term k_0 can be made to vanish, and X_i can then be written in the form

$$X_i = \phi^{-1}[\kappa_1 T_1^e(\epsilon_i) + n^{-1}\kappa_2 T_2^e(\epsilon_i) + 2n^{-2}\kappa_3 T_3^e(\epsilon_i)], \tag{2a}$$

where

$$\kappa_1 = \phi k_1, \quad \kappa_2 = n\phi k_2, \quad \kappa_3 = \tfrac{1}{2}n^2\phi k_3. \tag{2b}$$

The values κ_j are usually of the same order of magnitude. The scale factor ϕ will be chosen so that the range of the values ϕX_i is approximately equal to $(n-1)$, as in the equally-spaced case. It is found that the form for ϕ which is the most convenient in the arithmetical manipulations is

$$\phi = [k_1 + \tfrac{1}{10}(n^2+1)\,k_3]^{-1}. \tag{3}$$

Then

$$\phi k_1 + \tfrac{1}{10}(n^2+1)\,\phi k_3 = 1 = \kappa_1 + \tfrac{1}{5}(1+n^{-2})\,\kappa_3,$$

or

$$\kappa_1 = 1 - \tfrac{1}{5}\kappa_3 - \tfrac{1}{5}n^{-2}\kappa_3. \tag{4}$$

The original set of points x_i is thus replaced by a smoothed-out set characterized by the three parameters ϕ, κ_2, and κ_3.

The range of ϕX is, by use of the standard formulae for $T_j^e(\epsilon)$,

$$\kappa_1(n-1) + 2n^{-2}\kappa_3[(n-1)^3/4 - \{(3n^2-7)/20\}(n-1)]$$
$$= (n-1)[\kappa_1 + \tfrac{1}{5}(1 - 5n^{-1} + 6n^{-2})\,\kappa_3],$$

and, on substituting for κ_1 from (4),

$$Ra[\phi X] = (n-1) - (1-n^{-1})^2\kappa_3. \tag{5}$$

8.5.2 *The parameters κ_2 and κ_3*

Values of k_1, k_2, and k_3 could, if necessary, be found by the conventional least-squares procedure for equally-spaced data, x being treated as the dependent variable and ϵ as the independent variable. However, very often rough approximations will suffice. If $X_{+\frac{1}{2}}$, $X_{+\frac{1}{4}}$, X_0, $X_{-\frac{1}{4}}$, $X_{-\frac{1}{2}}$ denote the values of X for which ϵ takes the values $+\tfrac{1}{2}(n-1)$, $+\tfrac{1}{4}(n-1)$, 0, $-\tfrac{1}{4}(n-1)$, $-\tfrac{1}{2}(n-1)$, then, from (8.5.1, 1a), and the standard expressions for the orthogonal polynomials,

$$X_{+\frac{1}{2}} + X_{-\frac{1}{2}} - 2X_0 = \tfrac{1}{2}k_2(n-1)^2, \tag{1a}$$

$$X_{+\frac{1}{2}} - X_{-\frac{1}{2}} - 2(X_{+\frac{1}{4}} - X_{-\frac{1}{4}}) = \tfrac{3}{16}k_3(n-1)^3, \tag{1b}$$

and, from (8.5.1,5),

$$X_{+\frac{1}{2}} - X_{-\frac{1}{2}} = \{\phi^{-1} - \tfrac{1}{2}(n-1)\,k_3\}\,(n-1). \tag{1c}$$

Hence, if $x_{+\frac{1}{2}}$, etc., denote the observed values corresponding to the same five values of ϵ, the values of the parameters can be estimated from the formulae

$$k_2 = \frac{2(x_{+\frac{1}{2}} + x_{-\frac{1}{2}} - 2x_0)}{(n-1)^2}, \tag{2a}$$

$$k_3 = \frac{16\{(x_{+\frac{1}{2}} - x_{-\frac{1}{2}}) - 2(x_{+\frac{1}{4}} - x_{-\frac{1}{4}})\}}{3(n-1)^3}, \tag{2b}$$

$$\phi^{-1} = \frac{x_{+\frac{1}{2}} - x_{-\frac{1}{2}}}{n-1} + \tfrac{1}{2}(n-1)\,k_3, \tag{3a}$$

$$\kappa_2 = n\phi k_2, \quad \kappa_3 = \tfrac{1}{2}n^2\phi k_3. \tag{3b}$$

Approximations which can be calculated even more rapidly are obtained by using the value n for $n-1$, and neglecting the second term in (3a) in calculating κ_2 and κ_3. Then

$$\kappa_2 = \frac{2(x_{+\frac{1}{2}} + x_{-\frac{1}{2}} - 2x_0)}{x_{+\frac{1}{2}} - x_{-\frac{1}{2}}}, \tag{4a}$$

$$\kappa_3 = \frac{8}{3}\,\frac{(x_{+\frac{1}{2}} - x_{-\frac{1}{2}}) - 2(x_{+\frac{1}{4}} - x_{-\frac{1}{4}})}{x_{+\frac{1}{2}} - x_{-\frac{1}{2}}}, \tag{4b}$$

and

$$\phi = \frac{n-1-\kappa_3}{x_{+\frac{1}{2}} - x_{-\frac{1}{2}}}. \tag{4c}$$

The significance of the parameters κ_2 and κ_3 can be brought out by rewriting (4a) and (4b) in the forms

$$(x_{+\frac{1}{2}} - x_0)/(x_{+\frac{1}{2}} - x_{-\frac{1}{2}}) = (2 + \kappa_2)/4, \tag{5a}$$

$$(x_{+\frac{1}{4}} - x_{-\frac{1}{4}})/(x_{+\frac{1}{2}} - x_{-\frac{1}{2}}) = (8 - 3\kappa_3)/16. \tag{5b}$$

Thus κ_2 is a measure of the departure from symmetry about the central value x_0 ($\epsilon = 0$). κ_3 is a measure of the relative concentration of the observations towards the centre of the range. For the equally-spaced case, $\kappa_2 = 0 = \kappa_3$. When κ_2 is $+1$, the first half of the observations (for which ϵ is positive) is spread over three-quarters of the range of x. When κ_3 is $+4/3$, the central half of the observations (for which $|\epsilon|$ is less than $\frac{1}{4}(n-1)$) is confined to a quarter of the range of x.

8.5.2.1 *Example*

To find the values of κ_2 and κ_3 for the 67 observations listed in Table 7.1.1, the values of x are required for the observations numbered 1, $17\frac{1}{2}$, 34, $50\frac{1}{2}$, and 67. These numbers are $1 + j(n-1)/4$. The value of x corresponding to

$17\frac{1}{2}$ is taken as the mean of the values for 17 and 18. Then

$$x_{+\frac{1}{2}}\ 29{\cdot}6, \qquad x_{+\frac{1}{4}}\ 11{\cdot}0, \qquad x_0\ 1{\cdot}1,$$
$$x_{-\frac{1}{2}}\ -15{\cdot}2, \quad x_{-\frac{1}{4}}\ -6{\cdot}5.$$

The values obtained from (8.5.2,4) are

$$\kappa_2 = 2(29{\cdot}6 - 15{\cdot}2 - 2\times 1{\cdot}1)/(29{\cdot}6 + 15{\cdot}2) = +0{\cdot}54,$$
$$\kappa_3 = 2{\cdot}67(29{\cdot}6 + 15{\cdot}2 - 2\times 11{\cdot}0 - 2\times 6{\cdot}5)/(29{\cdot}6 + 15{\cdot}2) = +0{\cdot}58,$$
$$\phi = (67 - 1 - 0{\cdot}58)/(29{\cdot}6 + 15{\cdot}2) = 1{\cdot}460.$$

8.5.2.2 *Range of the parameters* κ_2 *and* κ_3. There does not appear to be any very simple criterion which fixes the ranges of the values κ_2 and κ_3 likely to be encountered in practical examples. However, it seems that neither parameter commonly exceeds 1 in magnitude, and so tables will only be given for values between $+1$ and -1. If $|\kappa_2|$ and $|\kappa_3|$ are much greater than 1, it is not likely that the approximation (8.5.1,2a) would be an adequate representation of the points x_i.

8.5.3 *The orthogonal polynomials* $T_j(X)$

The orthogonal polynomials $T_j(X)$ are written in the form

$$T_j(X) = \sum_{k=0}^{j} \beta_{kj} X^k \tag{1a}$$

or

$$T_j(X) = X^j - \sum_{k=0}^{j-1} \alpha_{kj} T_k(X), \tag{1b}$$

with

$$\beta_{kj} = -\sum_{m=k}^{j-1} \beta_{km} \alpha_{mj} \tag{1c}$$

and

$$\alpha_{kj} = \sum_i X_i^j T_k(X_i) / \sum_i T_k^2(X_i). \tag{1d}$$

These polynomials can be calculated in turn from the polynomials of lower degree.

General formulae can be derived by using the known expressions for $T_j^e(\epsilon_i)$ and $\Sigma\{T_j^e(\epsilon_i)\}^2$. It is found that n occurs in these expressions as powers of n^{-2}. For, from Table 7.10a,

$$\Sigma\{T_j^e(\epsilon_i)\}^2 = (n^{2j+1}/R_j)(1 + r_j n^{-2} + s_j n^{-4} + \ldots)$$

and

$$\beta_{kj}^e = (n^{j-k}/Q_j)(1 + q_j n^{-2} + \ldots).$$

If powers of n^{-2} above the first are neglected

$$\Sigma T_j^2(X_i) = \phi^{-2j}(n^{2j+1}/R_j) f_j (1 - n^{-2} g_j), \tag{2a}$$

and

$$\alpha_{kj} = (n\phi^{-1})^{j-k} \theta_{kj} (1 - n^{-2} \omega_{kj}), \tag{2b}$$

where f_j, g_j, θ_{kj}, and ω_{kj} are functions of κ_2 and κ_3. These functions are too complicated to be written out explicitly, but they have been calculated for selected values of κ_2 and κ_3. f_j and g_j are tabulated in Table 8.8*c*. R_j is the numerical factor occurring in the expression for $\Sigma\{T_j^e(\epsilon_i)\}^2$, and it is tabulated in Table 7.10*a*. The first three values are $R_1 = 12$, $R_2 = 180$, $R_3 = 2800$.

8.5.4 *Standard deviations of the orthogonal coefficients*

The standard deviation of the coefficient a_j is given by the equation

$$\operatorname{var} a_j = \sigma^2[a_j] = \sigma^2/\Sigma T_j^2(x_i).$$

If the approximation $\Sigma T_j^2(X_i)$ is used for $\Sigma T_j^2(x_i)$, then

$$\operatorname{var} a_j = \phi^{2j}\sigma^2(R_j/n^{2j+1})/f_j(1-n^{-2}g_j). \tag{1}$$

The term $n^{-2}g_j$ can usually be neglected. Since

$$\Sigma\{T_j^e(\epsilon_i)\}^2 \doteqdot n^{2j+1}/R_j,$$

f_j gives the ratio of the variance in the equally-spaced case to the variance in the general case for the same value of ϕ.

It will be seen from Table 8.8*c* that, for all three values of j, negative values of κ_3 yield values for f_j greater than unity and positive values of κ_3 yield values for f_j less than unity. That is, the standard deviations are reduced if the observations are crowded towards the extremes of the range and increased if the observations are crowded towards the centre of the range.

8.5.5 *Standard deviations of the fitted values*

The standard deviation of the fitted value at the point x is given by

$$\operatorname{var} u_p(x) = \sigma^2[u_p(x)] = \sigma^2\sum_{j=0}^{p}\left\{T_j^2(x)\Big/\sum_i T_j^2(x_i)\right\}.$$

If the smoothed values X_i are used in place of the observed values x_i, this becomes

$$\sigma^2[u_p(x)]/\sigma^2 = \sum_{j=0}^{p}\phi^{2j}(\Sigma\beta_{kj}X^k)^2/(n^{2j+1}f_j/R_j)$$

or

$$\sigma^2[u_p(x)]/\sigma^2 = n^{-1}\sum_{j=0}^{p}\{\Sigma\beta_{kj}(n\phi^{-1})^{k-j}(X\phi/n)^k\}^2 R_j/f_j. \tag{1}$$

The coefficients β_{kj} can be evaluated from the expressions for α_{kj} given in (8.5.3,2*b*).

The first-degree polynomial has been discussed in § 6.3.1. For the second- and third-degree polynomials the curves are found to be roughly symmetrical about the value of x given by

$$\phi X/n = \kappa_2/10.$$

Hence for purposes of tabulation the variable

$$k = (2\phi/n)(x-\bar{x}) - \kappa_2/5 \tag{2}$$

is convenient. Then

$$\sigma[u_p(x)] = n^{-\frac{1}{2}}\, \rho_{p0}(k, \kappa_2, \kappa_3)\, \sigma. \tag{3}$$

The functions $\rho_{20}(k)$ and $\rho_{30}(k)$ are analogous to the corresponding functions $\rho_{p0}(k)$ in the equally-spaced case. They are given in Table 8.8*d* for the range $k - 1{\cdot}4(0{\cdot}2) + 1{\cdot}4$, $\kappa_2 - 1{\cdot}0(0{\cdot}5) + 1{\cdot}0$, $\kappa_3 - 1{\cdot}0(0{\cdot}25) + 1{\cdot}0$.

When $|\kappa_2|$ is large the values of ρ_{20} and ρ_{30} near $|k| = 0{\cdot}5$ are increased for points with $k\kappa_2$ positive and decreased for points with $k\kappa_2$ negative. The parameter κ_3 has a much less marked effect. In general, the values of ρ_{20} and ρ_{30} are increased when κ_3 is positive and decreased when κ_3 is negative.

8.5.6 *Estimation of standard deviations of fitted values from the tables in practical examples*

The steps in the estimation of $s[u_p(x)]$ from the tables when the polynomial is of the second or third degree may be summarized as follows:

(i) The observations are supposed numbered in decreasing order of x from 1 to n. Write down the values $x_{+\frac{1}{2}}, x_{+\frac{1}{4}}, x_0, x_{-\frac{1}{4}}, x_{-\frac{1}{2}}$ of the variable x_i for the observations numbered 1, $1+\frac{1}{4}(n-1)$, $1+\frac{2}{4}(n-1)$, $1+\frac{3}{4}(n-1)$, n, interpolating where necessary.

(ii) Calculate

$$\kappa_2 = 2(x_{+\frac{1}{2}} + x_{-\frac{1}{2}} - 2x_0)/(x_{+\frac{1}{2}} - x_{-\frac{1}{2}}),$$

$$\kappa_3 = 2{\cdot}67\{x_{+\frac{1}{2}} - x_{-\frac{1}{2}} - 2(x_{+\frac{1}{4}} - x_{-\frac{1}{4}})\}/(x_{+\frac{1}{2}} - x_{-\frac{1}{2}}),$$

$$\phi = (n-1-\kappa_3)/(x_{+\frac{1}{2}} - x_{-\frac{1}{2}}).$$

(iii) Calculate

$$k = \frac{2\phi}{n}x - \left(\frac{2\phi}{n}\frac{\Sigma x_i}{n} + \frac{\kappa_2}{5}\right)$$

for each value x at which the standard deviation is required.

(iv) Find $\rho_{p0}(k, \kappa_2', \kappa_3')$ by interpolation in Table 8.8*d* using the values κ_2', κ_3' nearest to κ_2 and κ_3.

(v) Then

$$s[u_p(x)] \doteqdot \rho_{p0}(k, \kappa_2', \kappa_3')\, s_p/\sqrt{n},$$

where s_p is an estimate of σ.

8.5.6.1 *Example*

The values

$$\kappa_2 = +0{\cdot}54, \quad \kappa_3 = +0{\cdot}58, \quad \phi = 1{\cdot}460,$$

were derived in § 8.5.2.1 for the observations listed in Table 7.1.1. Now $2\phi/n = 0{\cdot}0436$ and $\Sigma x_i/n = 3{\cdot}01$, and so

$$k = \frac{2\phi}{n}x - \left(\frac{2\phi}{n}\frac{\Sigma x_i}{n} + \frac{\kappa_2}{5}\right) = 0{\cdot}0436x - 0{\cdot}239.$$

The values k for various values of x are calculated in Table 8.5.6. The values ρ_{20}, ρ_{30} are entered from Table 8.8*d*, taking $\kappa_2 = 0{\cdot}5$, $\kappa_3 = 0{\cdot}5$.

For the second-degree polynomial the values ρ_{20} are to be multiplied by $s_2/\sqrt{n}$. From Table 7.1.8, s_2 is $0{\cdot}0273 \times 10^r\sqrt{10^s}$, or 27·3. For the third-degree polynomial the values ρ_{30} are multiplied by $s_3/\sqrt{n}$, where s_3 is 24·3.

In each case the values of standard deviations calculated from the inverse matrix (cf. Table 8.1.2) are shown for comparison. The difference between the approximate value and the accurately calculated value is small except near the ends of the range. The method of the present section is much quicker than that of Table 8.1.2, and it does not require the values χ_{jk} of the elements of the inverse matrix.

TABLE 8.5.6

Estimation of standard deviations of fitted values from the tables

s_3 24·3, $s_3/\sqrt{n}$ 2·97; s_2 27·3, $s_2/\sqrt{n}$ 3·34.
$k = 0{\cdot}0436x - 0{\cdot}239$

		3rd degree polynomial			2nd degree polynomial		
x	k	ρ_{30}	$s[u_3]$	Using inverse matrix	ρ_{20}	$s[u_2]$	Using inverse matrix
−20	−1·11	7·5	22	23·3	4·2	14·0	14·4
−15	−0·89	3·5	10·4	10·7	2·66	8·9	9·3
−10	−0·68	1·84	5·5	5·0	1·70	5·7	5·6
− 5	−0·46	1·56	4·6	4·5	1·28	4·3	4·1
0	−0·24	1·51	4·5	4·4	1·33	4·4	4·3
+ 5	−0·02	1·46	4·3	4·2	1·45	4·8	4·8
+10	+0·20	1·70	5·0	5·1	1·47	4·9	5·0
+15	+0·42	1·98	5·9	6·2	1·49	5·0	5·1
+20	+0·63	2·08	6·2	6·5	1·77	5·9	5·9
+25	+0·85	2·77	8·2	7·5	2·58	8·6	8·4
+30	+1·07	5·6	16·6	14·6	3·9	13·0	12·6

8.5.7 *Variation of standard deviations with range and number of observations*

The standard deviation of the coefficient a_j is from (8.5.4,1) approximately equal to

$$\phi^j (R_j/f_j)^{\frac{1}{2}} \sigma/n^{j+\frac{1}{2}}, \tag{1}$$

while from (8.5.1,5),

$$\phi \doteqdot (n-1)/(\text{Ra}\, x). \tag{2}$$

Hence, as $f_j^{\frac{1}{2}}$ is close to unity,

$$\sigma[a_j] \sim \sigma \sqrt{R_j}/(\text{Ra}\, x)^j \sqrt{n}. \tag{3}$$

The standard deviation of a_j is then inversely proportional to the jth power of the range and to the square root of the number of observations.

The standard deviation of the fitted value in the region of interpolation is, from (8.5.5,3) and Table 8.8*d*, of the order of $2\sigma/\sqrt{n}$ for the second- and third-degree polynomials, and so is inversely proportional to the square root of the number of observations. In the region of extrapolation the coefficient of highest degree $b_{pp} = a_p$ becomes dominant and the standard deviation varies as this coefficient.

In general, the coefficient b_{pj} varies approximately as a_j when the origin is in the region of interpolation and as a_p when the origin is in the region of extrapolation. The transition region between these two regions may start at fairly low values of $|k|$, and in fact $b_{p,p-1}$ varies as a_p even when $|k|$ is quite small. For high values of j even a small increase in the range of the observations leads to a considerable reduction in the standard deviations.

8.6 OPTIMUM SPACING OF OBSERVATIONS

When the degree of the polynomial is not known with certainty, or when it is desired to check whether the observations can in fact be adequately fitted by a polynomial of given degree, it is probably best to space the observations more or less uniformly throughout the range. If, however, the degree of the polynomial is known, other spacing may be preferable. de la Garza (1954) has discussed how the spacing may be chosen so that the maximum variance of the fitted value in the region of interpolation is as small as possible.

It will be supposed that the experiment is to consist of n observations of equal weight in a given range of x. By a suitable choice of origin and scale the range may be taken as $+1$ to -1. From (8.1.3.2,3),

$$\{\text{var}\, u_p(x)\}/\sigma^2 = \mathbf{x}^T \boldsymbol{\phi}^{-1} \mathbf{x} = \mathbf{x}^T(\mathbf{XWX}^T)^{-1}\mathbf{x}. \tag{1}$$

de la Garza, in the reference quoted, establishes the important result that it is possible to find $p+1$ values ξ_i, with $|\xi_i| \leqslant 1$, and $p+1$ positive values ω_i, with $\Sigma\omega_i = n = \Sigma w_i$, such that

$$\boldsymbol{\xi\omega\xi}^T = \mathbf{XWX}^T, \tag{2}$$

where the elements of $\boldsymbol{\xi}$ are ξ_i^j and $\boldsymbol{\omega}$ is a diagonal matrix. It follows that, as far as the variances of the fitted values are concerned, any experiment gives the same variances as the corresponding experiment with the $p+1$ coordinates ξ_i and weights ω_i obtained by solving (2).

Thus to find the optimum design it is only necessary to consider the simple cases where the n observations are divided among the $p+1$ points ξ_i, with $n_i(\equiv \omega_i)$ observations y_i at each point. In these cases the least-squares polynomial becomes the polynomial which passes through the $p+1$ points $\bar{y}_i$, ξ_i, where $\bar{y}_i$ is the mean of the n_i values y_i. The polynomial can be written as

$$u_p(x) = \frac{(x-\xi_1)(x-\xi_2)\dots(x-\xi_p)}{(\xi_0-\xi_1)(\xi_0-\xi_2)\dots(\xi_0-\xi_p)}\bar{y}_0 + \frac{(x-\xi_0)(x-\xi_2)\dots(x-\xi_p)}{(\xi_1-\xi_0)(\xi_1-\xi_2)\dots(\xi_1-\xi_p)}\bar{y}_1 + \dots,$$

or

$$u_p(x) = \sum_{i=0}^{p}\left\{\left[\prod_{j\neq i}(x-\xi_j)\right]\Big/\left[\prod_{j\neq i}(\xi_i-\xi_j)\right]\right\}\bar{y}_i. \tag{3}$$

This is known as the Lagrange interpolation formula. At $x=\xi_i$ the coefficients of all the $\bar{y}$ vanish, except the coefficient of $\bar{y}_i$, and as this is unity $u_p(\xi_i)$ is $\bar{y}_i$.

The variance of the fitted value is given by

$$\{\operatorname{var} u_p(x)\}/\sigma^2 = \sum_{i=0}^{p}\left\{\left[\prod_{j\neq i}(x-\xi_j)\right]\Big/\left[\prod_{j\neq i}(\xi_i-\xi_j)\right]\right\}^2\Big/n_i. \tag{4a}$$

For any of the points ξ_i,

$$\{\operatorname{var} u_p(\xi_i)\}/\sigma^2 = 1/n_i, \tag{4b}$$

and so at best the smallest maximum variance (conveniently described as the minimax variance) is the largest of the values σ^2/n_i. This is as small as possible when all the n_i are equal, so that

$$\text{minimax}\operatorname{var} u_p(\xi_i) = (p+1)\,\sigma^2/n. \tag{4c}$$

Clearly, if $\operatorname{var} u_p(x)$ is less than this value at other points in the region $-1 \leqslant x \leqslant +1$ the minimax conditions will have been obtained. Now if ξ_0 and ξ_p are the greatest and least of the ξ_i, all the coefficients in (4a) increase as x becomes greater than ξ_0, and also as x becomes less than ξ_p, so that ξ_0 and ξ_p must be at the ends of the range. The function (4a) has $p-1$ maxima in the region of interpolation, and so if these maxima occur at the $p-1$ points ξ_i, then the minimax condition is satisfied. Differentiation of (4a) leads to the equations

$$\sum_{i \neq m} (\xi_m - \xi_i)^{-1} = 0, \quad m = 1 \text{ to } p-1. \tag{5}$$

These equations can then be solved to give the optimum spacing.

The solutions are given in Table 8.6. For the curve of degree p, the observations should be divided into $p+1$ groups taken at values of x equal to $\frac{1}{2}\xi_i(x_0 - x_p) + \frac{1}{2}(x_0 + x_p)$, where x_0 and x_p are the limits of the range and ξ_i takes the values listed in Table 8.6.

The maximum standard deviations occur at the points ξ_i, and from (4c) have the value $(p+1)^{\frac{1}{2}}\sigma/\sqrt{n}$. Compared to the equally-spaced case, where the standard deviation is $\rho_{p0}(k)\,\sigma/\sqrt{n}$, the standard deviation for the minimax case is greater near the centre of the range but less near the ends of the range. For the straight line ($p = 1$) the standard deviation is decreased at all points except $\xi = 0$.

TABLE 8.6

Location of observations for minimax variance of the fitted value

	$\xi_i = 2\{x_i - \frac{1}{2}(x_0+x_p)\}/(x_0-x_p)$				
Degree p	1	2	3	4	5
No. per point	$n/2$	$n/3$	$n/4$	$n/5$	$n/6$
ξ_0	+1	+1	+1	+1	+1
ξ_1	−1	0	+0·447	+0·655	+0·765
ξ_2		−1	−0·447	0	+0·285
ξ_3			−1	−0·655	−0·285
ξ_4				−1	−0·765
ξ_5					−1

[0·447 = $\sqrt{(1/5)}$; 0·655 = $\sqrt{(3/7)}$; 0·765, 0·285 = $\sqrt{\{(1 \pm 2/\sqrt{7})/3\}}$.]

8.6.1 *Calculation of the polynomial*

The fitted polynomial can be obtained in power-series form by the expansion of (8.6,3), or by the use of divided differences (Birge, 1949). Table 8.6.1 lists explicit formulae for the coefficients b_{pj} when the degree p is 1, 2, or 3, and the range of the observations is +1 to −1.

8.6.1.1 *Example*

Table 8.6.1.1 shows the calculation of the power-series coefficients for a third-degree polynomial, using the formulae of Table 8.6.1. The coefficients are checked by computing the fitted values at the points of observation.

TABLE 8.6.1

Coefficients b_{pj} as explicit functions of the observed means $\bar{y}_i$

(*a*) 1st degree

$$b_{11} = \tfrac{1}{2}(\bar{y}_0 - \bar{y}_1)$$
$$b_{10} = \tfrac{1}{2}(\bar{y}_0 + \bar{y}_1)$$

(*b*) 2nd degree

$$b_{22} = \tfrac{1}{2}(\bar{y}_0 + \bar{y}_2) - \bar{y}_1$$
$$b_{21} = \tfrac{1}{2}(\bar{y}_0 - \bar{y}_2)$$
$$b_{20} = \bar{y}_1$$

(*c*) 3rd degree $\xi \doteq 1/\sqrt{5} = 0{\cdot}477$

$$b_{33} = \{\xi(\bar{y}_0 - \bar{y}_3) - (\bar{y}_1 - \bar{y}_2)\}/2\xi(1-\xi^2)$$
$$b_{31} = -\{\xi^3(\bar{y}_0 - \bar{y}_3) - (\bar{y}_1 - \bar{y}_2)\}/2\xi(1-\xi^2)$$
$$b_{32} = \{(\bar{y}_0 + \bar{y}_3) - (\bar{y}_1 + \bar{y}_2)\}/2(1-\xi^2)$$
$$b_{30} = -\{\xi^2(\bar{y}_0 + \bar{y}_3) - (\bar{y}_1 + \bar{y}_2)\}/2(1-\xi^2)$$

TABLE 8.6.1.1

Calculation of the polynomial coefficients for a cubic curve

(*a*) Observed means

$\bar{y}_i$	2·07	0·43	2·02	2·01
ξ_i	1	0·45	−0·45	−1

(*b*) Sums and differences

	2·07	0·43
	2·01	2·02
Sum	4·08	2·45
Diff.	0·06	−1·59

(*c*) Powers of ξ

$\xi = 0{\cdot}45 \quad \xi^2 = 0{\cdot}2025 \quad \xi^3 = 0{\cdot}091125$

$2(1-\xi^2) = 1{\cdot}5950 \quad 2\xi(1-\xi^2) = 0{\cdot}717750$

(*d*) Coefficients

$$b_{33} = (0{\cdot}45 \times 0{\cdot}06 + 1{\cdot}59)/0{\cdot}71775 = 2{\cdot}252874$$
$$b_{31} = -(0{\cdot}091125 \times 0{\cdot}06 + 1{\cdot}59)/0{\cdot}71775 = -2{\cdot}222874$$
$$b_{32} = (4{\cdot}08 - 2{\cdot}45)/1{\cdot}595 = 1{\cdot}021944$$
$$b_{30} = -(0{\cdot}2025 \times 4{\cdot}08 - 2{\cdot}45)/1{\cdot}595 = 1{\cdot}018056$$

(*e*) Check

ξ_i	$\Sigma b_{3j}\,\xi^i$
+1	2·0700000
+0·45	0·4299995
−0·45	2·0199998
−1	2·0100000

8.7 NOTES AND REFERENCES

(8.1) The matrix treatment of the least-squares theory is given by Hayes and Vickers (1951).

(8.2) The choice of the degree of the polynomial is often a matter of individual judgment, and the tests given here should not be regarded as binding. Thus on physical grounds the cubic curve in § 8.2.2.2 might be rejected as a representation of the variation of J, since a negative value of a_3 implies a fall in J at higher temperatures in the region of extrapolation. The fact that the cubic curve gives a better representation in the region of interpolation does not necessarily mean that the 'true' curve is a cubic.

Another difficulty is that one or two very large residuals may dominate the sum Σv^2, and then only a small change in Σv^2 is produced by increasing the degree of the polynomial (Guest, 1950*a*).

If the aim is merely to smooth the observations, the techniques described in § 10.4 may be preferable.

(8.3) References on the Gauss–Markoff theorem are: Aitken (1933*c*, 1935), David and Neyman (1938), Kavanagh (1941), and Cohen (1953). A historical discussion is given by Plackett (1949).

(8.4) The details of the calculations of the tables are given in two papers by Guest (1950*b*, 1950*c*).

(8.5) A more detailed account of the representation by two parameters κ_2 and κ_3 is given by Guest (1953*a*); see also Guest (1956).

(8.6) This section is based on two papers by de la Garza (1954, 1955); for the straight line see also Daniel and Heerema (1950). K. Smith (*Biometrika*, **12** (1918), 1) earlier gave a discussion of both uniform spacing and minimax variance methods; see also Guest, *Ann. Math. Statist.*, March 1958.

TABLE 8.8*a*

Values of $\rho_{p0}(k, n)$

k \ n	$p = 1$ $\infty - 7$	$p = 2$ ∞	14	11	8	7
0·00	1·00	1·50	1·51	1·51	1·52	1·53
0·05	1·00	1·50	1·50	1·51	1·52	1·52
0·10	1·01	1·49	1·49	1·50	1·50	1·51
0·15	1·03	1·47	1·47	1·48	1·49	1·49
0·20	1·06	1·44	1·45	1·45	1·46	1·47
0·25	1·09	1·42	1·42	1·43	1·43	1·44
0·30	1·13	1·39	1·39	1·40	1·40	1·41
0·35	1·17	1·37	1·37	1·37	1·37	1·38
0·40	1·22	1·35	1·35	1·35	1·35	1·36
0·45	1·27	1·34	1·34	1·34	1·35	1·35
0·50	1·32	1·35	1·35	1·35	1·36	1·36
0·55	1·38	1·38	1·39	1·39	1·39	1·39
0·60	1·44	1·44	1·45	1·45	1·45	1·45
0·65	1·51	1·54	1·54	1·54	1·55	1·55
0·70	1·57	1·66	1·66	1·67	1·68	1·68
0·75	1·64	1·81	1·82	1·83	1·84	1·85
0·80	1·71	1·99	2·01	2·01	2·03	2·05
0·85	1·78	2·21	2·22	2·23	2·26	2·27
0·90	1·85	2·45	2·47	2·48	2·51	2·53
0·95	1·93	2·71	2·74	2·75	2·79	2·81
1·00	2·00	3·00	3·03	3·05	3·09	3·12

k \ n	$p = 3$ ∞	25	18	14	11	8	7
0·00	1·50	1·50	1·50	1·51	1·51	1·52	1·53
0·05	1·51	1·51	1·51	1·52	1·52	1·53	1·54
0·10	1·54	1·54	1·54	1·54	1·55	1·56	1·57
0·15	1·58	1·58	1·58	1·59	1·59	1·61	1·62
0·20	1·62	1·63	1·63	1·64	1·64	1·67	1·68
0·25	1·67	1·68	1·68	1·69	1·70	1·72	1·74
0·30	1·72	1·73	1·73	1·74	1·75	1·77	1·79
0·35	1·76	1·76	1·77	1·77	1·79	1·81	1·83
0·40	1·78	1·79	1·79	1·80	1·81	1·84	1·86
0·45	1·79	1·79	1·80	1·80	1·81	1·84	1·86
0·50	1·78	1·78	1·79	1·79	1·80	1·82	1·84
0·55	1·76	1·76	1·76	1·77	1·77	1·79	1·80
0·60	1·73	1·73	1·73	1·74	1·74	1·75	1·75
0·65	1·71	1·72	1·72	1·72	1·72	1·72	1·72
0·70	1·73	1·73	1·73	1·74	1·74	1·74	1·74
0·75	1·82	1·82	1·82	1·83	1·83	1·84	1·85
0·80	2·01	2·01	2·02	2·02	2·04	2·07	2·09
0·85	2·31	2·32	2·33	2·35	2·37	2·43	2·48
0·90	2·75	2·77	2·78	2·81	2·85	2·95	3·02
0·95	3·31	3·34	3·36	3·40	3·46	3·61	3·71
1·00	4·00	4·04	4·07	4·12	4·20	4·40	4·55

TABLE 8.8*a* (cont.)

k \ n	$p = 4$ ∞	40	25	18	14	11	9	8	7
0·00	1·88	1·88	1·88	1·89	1·90	1·91	1·94	1·96	1·99
0·05	1·87	1·87	1·87	1·88	1·89	1·90	1·93	1·95	1·98
0·10	1·84	1·84	1·85	1·85	1·86	1·87	1·89	1·91	1·94
0·15	1·80	1·81	1·81	1·81	1·82	1·83	1·85	1·86	1·88
0·20	1·77	1·77	1·77	1·77	1·78	1·79	1·80	1·81	1·83
0·25	1·74	1·74	1·74	1·75	1·75	1·76	1·77	1·78	1·79
0·30	1·73	1·74	1·74	1·74	1·75	1·76	1·77	1·78	1·80
0·35	1·76	1·76	1·76	1·77	1·78	1·79	1·81	1·82	1·85
0·40	1·81	1·82	1·82	1·83	1·84	1·86	1·88	1·91	1·94
0·45	1·89	1·90	1·90	1·91	1·93	1·95	1·98	2·02	2·07
0·50	1·98	1·98	1·99	2·00	2·02	2·05	2·09	2·13	2·19
0·55	2·06	2·07	2·07	2·09	2·11	2·14	2·18	2·22	2·29
0·60	2·12	2·12	2·13	2·14	2·16	2·19	2·24	2·27	2·34
0·65	2·14	2·15	2·15	2·16	2·18	2·20	2·23	2·27	2·32
0·70	2·13	2·13	2·14	2·14	2·15	2·16	2·18	2·20	2·22
0·75	2·10	2·10	2·10	2·10	2·11	2·11	2·11	2·11	2·11
0·800	2·12	2·12	2·12	2·12	2·12	2·12	2·12	2·13	2·13
0·825	2·19	2·19	2·19	2·20	2·20	2·21	2·22	2·24	2·27
0·850	2·32	2·32	2·33	2·33	2·35	2·37	2·41	2·45	2·53
0·875	2·52	2·53	2·54	2·56	2·59	2·64	2·71	2·79	2·93
0·900	2·82	2·83	2·85	2·88	2·93	3·01	3·14	3·26	3·48
0·925	3·21	3·23	3·26	3·31	3·38	3·50	3·68	3·86	4·16
0·950	3·71	3·73	3·77	3·84	3·93	4·10	4·35	4·58	4·98
0·975	4·30	4·34	4·39	4·48	4·60	4·82	5·13	5·43	5·94
1·000	5·00	5·04	5·11	5·22	5·38	5·64	6·04	6·40	7·03

k \ n	$p = 5$ ∞	40	25	18	14	11	9	8	7
0·00	1·88	1·88	1·88	1·89	1·90	1·91	1·94	1·96	1·99
0·05	1·89	1·89	1·90	1·91	1·92	1·94	1·96	1·98	2·03
0·10	1·93	1·94	1·94	1·95	1·97	1·99	2·02	2·05	2·11
0·15	1·99	1·99	2·00	2·01	2·03	2·05	2·10	2·14	2·21
0·20	2·04	2·04	2·05	2·06	2·08	2·11	2·16	2·21	2·29
0·25	2·07	2·08	2·09	2·10	2·11	2·15	2·19	2·24	2·32
0·30	2·08	2·08	2·09	2·10	2·12	2·14	2·18	2·22	2·29
0·35	2·06	2·06	2·07	2·08	2·09	2·11	2·14	2·16	2·20
0·40	2·02	2·03	2·03	2·04	2·04	2·06	2·08	2·09	2·12
0·45	2·00	2·00	2·00	2·01	2·02	2·03	2·05	2·07	2·10
0·50	2·00	2·01	2·01	2·02	2·03	2·06	2·09	2·13	2·21
0·55	2·06	2·07	2·08	2·09	2·12	2·16	2·23	2·30	2·46
0·60	2·18	2·19	2·20	2·23	2·26	2·33	2·43	2·55	2·78
0·65	2·32	2·33	2·35	2·38	2·43	2·51	2·64	2·78	3·05
0·70	2·45	2·46	2·48	2·51	2·56	2·64	2·77	2·90	3·16
0·75	2·51	2·52	2·54	2·56	2·59	2·65	2·74	2·82	2·99
0·800	2·50	2·51	2·51	2·52	2·53	2·54	2·55	2·56	2·56
0·825	2·49	2·49	2·49	2·49	2·49	2·48	2·46	2·45	2·41
0·850	2·50	2·50	2·50	2·50	2·49	2·49	2·48	2·48	2·53
0·875	2·59	2·59	2·60	2·60	2·62	2·64	2·72	2·82	3·12
0·900	2·82	2·83	2·85	2·88	2·94	3·05	3·28	3·57	4·24
0·925	3·25	3·27	3·32	3·39	3·52	3·76	4·21	4·72	5·84
0·950	3·91	3·95	4·04	4·17	4·38	4·79	5·50	6·27	7·89
0·975	4·82	4·90	5·02	5·23	5·54	6·14	7·14	8·21	10·38
1·000	6·00	6·11	6·28	6·57	7·00	7·81	9·15	10·54	13·35

TABLE 8.8b

$\rho_{p0}(k)$ *and* $\phi_p(n)$ *in the region of extrapolation*

k \ p	1	2	$\rho_{p0}(k)$ 3	4	5
1·000	2·00	3·00	4·00	5·00	6·00
1·025	2·04	3·15	4·39	5·80	7·45
1·050	2·08	3·31	4·81	6·71	9·18
1·075	2·11	3·47	5·26	7·72	11·21
1·100	2·15	3·64	5·74	8·85	13·56
1·125	2·19	3·82	6·25	10·08	16·2
1·150	2·23	4·00	6·80	11·44	19·3
1·175	2·27	4·18	7·37	12·92	22·8
1·20	2·31	4·37	7·97	14·5	26·6
1·25	2·38	4·76	9·27	18·1	35·7
1·30	2·46	5·17	10·71	22·3	47·0
1·35	2·54	5·60	12·27	27·1	60·5
1·40	2·62	6·05	14·0	32·6	76·8
1·45	2·70	6·52	15·8	38·9	96·2
1·50	2·78	7·01	17·8	45·9	119
1·55	2·86	7·51	19·9	53·7	146
1·60	2·95	8·03	22·2	62·4	176
1·65	3·03	8·57	24·7	72·1	212
1·70	3·11	9·12	27·3	82·9	253
1·75	3·19	9·69	30·1	94·7	299
1·8	3·27	10·3	33·1	108	352
1·9	3·44	11·5	39·5	137	480
2·0	3·61	12·8	46·8	173	640
2·1	3·77	14·2	54·8	214	839
2·2	3·94	15·6	63·6	262	1083
2·3	4·11	17·1	73·4	317	1379
2·4	4·28	18·7	84·0	381	1740
2·5	4·44	20·3	95·6	454	2160
2·6	4·61	22·0	108·2	536	2660
2·7	4·78	23·8	121·8	629	3250
2·8	4·95	25·7	136·5	732	3940
2·9	5·12	27·6	152·3	849	4740
3·0	5·29	29·5	169·3	978	5670
$k > 3$	$1{\cdot}73k$	$3{\cdot}35k^2$	$6{\cdot}61k^3$	$13{\cdot}1k^4$	$26{\cdot}1k^5$

n \ p	1	2	$\phi_p(n)$ 3	4	5
∞	1·00	1·00	1·00	1·00	1·00
100	1·00	1·00	1·00	1·00	1·00
50	1·00	1·00	1·00	1·01	1·01
40	1·00	1·00	1·00	1·01	1·02
30	1·00	1·00	1·01	1·02	1·03
25	1·00	1·00	1·01	1·02	1·05
20	1·00	1·01	1·02	1·04	1·07
18	1·00	1·01	1·02	1·05	1·09
16	1·00	1·01	1·03	1·06	1·12
15	1·00	1·01	1·03	1·07	1·14
14	1·00	1·01	1·04	1·08	1·16
13	1·00	1·02	1·04	1·10	1·19
12	1·00	1·02	1·05	1·11	1·23
11	1·00	1·02	1·06	1·14	1·28
10	1·01	1·03	1·08	1·17	1·35
9	1·01	1·03	1·09	1·22	1·47
8	1·01	1·04	1·12	1·30	1·66
7	1·01	1·05	1·17	1·42	2·03

TABLE 8.8c

The quantities f_j and g_j as functions of the parameters κ_2 and κ_3

κ_3 \ κ_2^2	0	0·5	1	0	0·5	1
		f_1			g_1	
−1·0	1·457	1·490	1·524	0·824	0·917	1·006
−0·5	1·214	1·248	1·281	0·865	0·975	1·080
0	1	1·033	1·067	1	1·129	1·25
0·5	0·814	0·848	0·881	1·289	1·435	1·570
1·0	0·657	0·690	0·724	1·826	1·979	2·118
		f_2			g_2	
−1·0	1·394	1·478	1·570	1·811	2·877	3·834
−0·5	1·167	1·217	1·278	2·871	4·077	5·200
0	1	1·016	1·045	5	6·260	7·5
0·5	0·876	0·858	0·857	8·325	9·612	10·970
1·0	0·781	0·731	0·702	12·888	14·322	15·924
		f_3			g_3	
−1·0	1·361	1·505	1·685	2·807	6·589	9·683
−0·5	1·149	1·206	1·294	6·493	10·771	14·708
0	1	1·009	1·033	14	18·387	22·750
0·5	0·872	0·873	0·864	25·508	30·076	34·623
1·0	0·745	0·766	0·752	41·610	46·319	50·850

TABLE 8.8*d*

Values of ρ_{20} and ρ_{30} for the unequally-spaced case

		ρ_{20}														
		$k\kappa_2$ negative								$k\kappa_2$ positive						
	$\lvert k \rvert$	1·4	1·2	1·0	0·8	0·6	0·4	0·2	0	0·2	0·4	0·6	0·8	1·0	1·2	1·4
$\kappa_3 = +1{\cdot}00$	$\lvert \kappa_2 \rvert = 1{\cdot}0$	7·59	5·46	3·68	2·27	1·36	1·16	1·39	1·61	1·69	1·72	1·91	2·52	3·62	5·15	7·07
	0·5	7·54	5·51	3·81	2·48	1·57	1·20	1·26	1·38	1·44	1·51	1·81	2·55	3·71	5·26	7·16
	0	7·32	5·39	3·79	2·54	1·71	1·33	1·28	1·30	1·28	1·33	1·71	2·54	3·79	5·39	7·32
$\kappa_3 = +0{\cdot}75$	$\lvert \kappa_2 \rvert = 1{\cdot}0$	7·02	5·03	3·36	2·07	1·30	1·19	1·42	1·62	1·68	1·68	1·85	2·43	3·48	4·96	6·79
	0·5	7·11	5·17	3·56	2·30	1·48	1·21	1·30	1·42	1·45	1·48	1·74	2·43	3·55	5·06	6·90
	0	7·01	5·14	3·59	2·40	1·63	1·32	1·32	1·35	1·32	1·32	1·63	2·40	3·59	5·14	7·01
$\kappa_3 = +0{\cdot}50$	$\lvert \kappa_2 \rvert = 1{\cdot}0$	6·49	4·62	3·07	1·90	1·25	1·23	1·46	1·63	1·66	1·64	1·78	2·33	3·34	4·75	6·52
	0·5	6·70	4·85	3·31	2·13	1·41	1·23	1·35	1·46	1·47	1·47	1·68	2·31	3·39	4·84	6·63
	0	6·70	4·88	3·39	2·26	1·56	1·32	1·35	1·39	1·35	1·32	1·56	2·26	3·39	4·88	6·70
$\kappa_3 = +0{\cdot}25$	$\lvert \kappa_2 \rvert = 1{\cdot}0$	6·00	4·24	2·80	1·74	1·22	1·27	1·50	1·64	1·66	1·62	1·73	2·23	3·19	4·55	6·24
	0·5	6·30	4·53	3·07	1·98	1·36	1·26	1·40	1·50	1·50	1·46	1·62	2·20	3·22	4·62	6·35
	0	6·38	4·63	3·19	2·12	1·50	1·33	1·40	1·45	1·40	1·33	1·50	2·12	3·19	4·63	6·38
$\kappa_3 = 0$	$\lvert \kappa_2 \rvert = 1{\cdot}0$	5·54	3·89	2·56	1·61	1·21	1·32	1·54	1·66	1·66	1·60	1·67	2·14	3·05	4·34	5·97
	0·5	5·91	4·22	2·85	1·84	1·32	1·29	1·45	1·55	1·53	1·46	1·56	2·09	3·06	4·40	6·07
	0	6·05	4·37	3·00	1·99	1·44	1·35	1·44	1·50	1·44	1·35	1·44	1·99	3·00	4·37	6·05
$\kappa_3 = -0{\cdot}25$	$\lvert \kappa_2 \rvert = 1{\cdot}0$	5·11	3·57	2·34	1·50	1·22	1·36	1·58	1·68	1·66	1·58	1·63	2·04	2·90	4·14	5·69
	0·5	5·52	3·92	2·63	1·71	1·30	1·33	1·50	1·59	1·56	1·47	1·52	1·98	2·89	4·18	5·77
	0	5·73	4·11	2·81	1·87	1·41	1·37	1·49	1·55	1·49	1·37	1·41	1·87	2·81	4·11	5·73
$\kappa_3 = -0{\cdot}50$	$\lvert \kappa_2 \rvert = 1{\cdot}0$	4·71	3·27	2·14	1·42	1·24	1·41	1·61	1·71	1·67	1·58	1·59	1·95	2·76	3·93	5·41
	0·5	5·15	3·64	2·43	1·61	1·29	1·38	1·55	1·63	1·59	1·48	1·49	1·88	2·73	3·95	5·48
	0	5·40	3·86	2·62	1·76	1·38	1·40	1·54	1·61	1·54	1·40	1·38	1·76	2·62	3·86	5·40
$\kappa_3 = -0{\cdot}75$	$\lvert \kappa_2 \rvert = 1{\cdot}0$	4·33	2·99	1·96	1·35	1·26	1·46	1·65	1·73	1·69	1·58	1·56	1·87	2·62	3·72	5·13
	0·5	4·80	3·37	2·24	1·52	1·29	1·42	1·60	1·68	1·63	1·50	1·46	1·79	2·57	3·72	5·18
	0	5·07	3·61	2·45	1·67	1·37	1·44	1·59	1·66	1·59	1·44	1·37	1·67	2·45	3·61	5·07
$\kappa_3 = -1{\cdot}00$	$\lvert \kappa_2 \rvert = 1{\cdot}0$	3·99	2·74	1·81	1·31	1·30	1·51	1·69	1·76	1·70	1·58	1·53	1·79	2·48	3·52	4·86
	0·5	4·45	3·11	2·07	1·45	1·31	1·47	1·65	1·72	1·66	1·52	1·45	1·70	2·41	3·50	4·88
	0	4·75	3·36	2·28	1·58	1·37	1·48	1·64	1·70	1·64	1·48	1·37	1·58	2·28	3·36	4·75

TABLE 8.8d (cont.)

		ρ_{30}														
		$k\kappa_2$ negative								$k\kappa_2$ positive						
	$\lvert k \rvert$	1·4	1·2	1·0	0·8	0·6	0·4	0·2	0	0·2	0·4	0·6	0·8	1·0	1·2	1·4
$\kappa_3 = +1{\cdot}00$	$\lvert\kappa_2\rvert = 1{\cdot}0$	19·7	11·8	6·27	2·81	1·38	1·41	1·50	1·61	1·95	2·32	2·45	2·58	4·06	7·83	14·0
	0·5	18·1	10·8	5·74	2·75	1·62	1·48	1·37	1·39	1·73	2·07	2·18	2·55	4·61	8·90	15·5
	0	16·9	9·91	5·20	2·62	1·89	1·76	1·49	1·30	1·49	1·76	1·89	2·62	5·20	9·91	16·9
$\kappa_3 = +0{\cdot}75$	$\lvert\kappa_2\rvert = 1{\cdot}0$	18·5	11·0	5·71	2·49	1·35	1·50	1·56	1·62	1·91	2·26	2·38	2·48	3·89	7·54	13·5
	0·5	17·3	10·2	5·37	2·52	1·56	1·52	1·43	1·42	1·71	2·01	2·10	2·43	4·39	8·51	14·9
	0	16·2	9·42	4·91	2·46	1·82	1·74	1·51	1·35	1·51	1·74	1·82	2·46	4·91	9·42	16·2
$\kappa_3 = +0{\cdot}50$	$\lvert\kappa_2\rvert = 1{\cdot}0$	17·3	10·1	5·14	2·17	1·36	1·59	1·62	1·63	1·88	2·20	2·31	2·39	3·72	7·23	13·0
	0·5	16·4	9·63	4·98	2·30	1·53	1·57	1·49	1·46	1·70	1·97	2·04	2·31	4·17	8·12	14·2
	0	15·4	8·94	4·61	2·31	1·77	1·74	1·54	1·39	1·54	1·74	1·77	2·31	4·61	8·94	15·4
$\kappa_3 = +0{\cdot}25$	$\lvert\kappa_2\rvert = 1{\cdot}0$	15·9	9·19	4·55	1·89	1·42	1·69	1·68	1·64	1·85	2·15	2·25	2·30	3·55	6·92	12·4
	0·5	15·5	9·02	4·58	2·09	1·53	1·63	1·56	1·50	1·70	1·95	1·99	2·20	3·94	7·74	13·6
	0	14·7	8·46	4·31	2·15	1·74	1·75	1·58	1·45	1·58	1·75	1·74	2·15	4·31	8·46	14·7
$\kappa_3 = 0$	$\lvert\kappa_2\rvert = 1{\cdot}0$	14·6	8·26	3·97	1·66	1·50	1·77	1·73	1·66	1·84	2·12	2·20	2·21	3·37	6·60	11·9
	0·5	14·6	8·37	4·15	1·89	1·56	1·71	1·62	1·55	1·71	1·94	1·96	2·09	3·70	7·34	13·0
	0	14·0	7·97	4·00	2·01	1·73	1·78	1·62	1·50	1·62	1·78	1·73	2·01	4·00	7·97	14·0
$\kappa_3 = -0{\cdot}25$	$\lvert\kappa_2\rvert = 1{\cdot}0$	13·2	7·32	3·41	1·51	1·60	1·85	1·78	1·68	1·84	2·09	2·15	2·13	3·19	6·26	11·3
	0·5	13·6	7·70	3·73	1·73	1·61	1·78	1·69	1·59	1·74	1·94	1·94	1·99	3·46	6·94	12·4
	0	13·2	7·47	3·68	1·88	1·74	1·82	1·67	1·55	1·67	1·82	1·74	1·88	3·68	7·47	13·2
$\kappa_3 = -0{\cdot}50$	$\lvert\kappa_2\rvert = 1{\cdot}0$	11·8	6·41	2·89	1·43	1·71	1·92	1·81	1·71	1·85	2·08	2·12	2·06	3·01	5·90	10·7
	0·5	12·6	7·00	3·30	1·61	1·69	1·86	1·74	1·64	1·76	1·96	1·94	1·91	3·22	6·52	11·8
	0	12·5	6·94	3·36	1·77	1·78	1·87	1·72	1·61	1·72	1·87	1·78	1·77	3·36	6·94	12·5
$\kappa_3 = -0{\cdot}75$	$\lvert\kappa_2\rvert = 1{\cdot}0$	10·4	5·55	2·44	1·43	1·80	1·97	1·84	1·73	1·86	2·07	2·09	1·99	2·82	5·54	10·1
	0·5	11·5	6·29	2·89	1·54	1·77	1·93	1·80	1·68	1·80	1·98	1·95	1·84	2·97	6·09	11·1
	0	11·6	6·39	3·03	1·69	1·82	1·93	1·77	1·66	1·77	1·93	1·82	1·69	3·03	6·39	11·6
$\kappa_3 = -1{\cdot}00$	$\lvert\kappa_2\rvert = 1{\cdot}0$	9·15	4·75	2·06	1·48	1·88	2·00	1·86	1·76	1·88	2·07	2·08	1·94	2·64	5·17	9·46
	0·5	10·5	5·57	2·50	1·53	1·86	1·99	1·84	1·72	1·83	2·01	1·97	1·79	2·72	5·63	10·4
	0	10·8	5·83	2·71	1·65	1·88	1·98	1·82	1·70	1·82	1·98	1·88	1·65	2·71	5·83	10·8

CHAPTER 9

THE GROUPING OF OBSERVATIONS

When polynomial curves are to be fitted to observational data, the time required for the arithmetical calculations increases very rapidly with the number of observations n. Hence if n is at all large it is usually desirable to reduce the computing labour by combining the observations into a number of groups, and treating the mean value in each group as a single observation. The estimates of the polynomial coefficients obtained from the grouped sets will be less efficient than those obtained from the original observations, and they may also be biased. In this chapter these questions of efficiency and bias will be investigated, firstly for the case of a series of equally-spaced observations of equal weight, and secondly for the general case when the spacing between successive observations is non-uniform.

9.1 EQUALLY-SPACED OBSERVATIONS

The n observations $y(\epsilon)$ are first converted into N groups each containing r observations. The sum of the values in a particular group will be represented by the symbol $y_N(\epsilon)$. The values of the independent variable ϵ corresponding to the observations are spaced at unit intervals in the range $-\frac{1}{2}(n-1)$ to $+\frac{1}{2}(n-1)$ for the original observations, and in the range $-\frac{1}{2}(N-1)$ to $+\frac{1}{2}(N-1)$ for the grouped values. Then

$$y_N(\epsilon) = \sum_{z=-\frac{1}{2}(r-1)}^{+\frac{1}{2}(r-1)} y(r\epsilon+z). \tag{1}$$

A least-squares polynomial

$$\mathscr{U}_p(\epsilon) = \sum_{j=0}^{p} \mathscr{A}_{jN}\, T_{jN}(\epsilon) \tag{2}$$

is fitted to the N grouped values $y_N(\epsilon)$. The coefficients $\mathscr{A}_{jN}$ are given by the equations

$$\mathscr{A}_{jN} = \sum_{\epsilon} T_{jN}(\epsilon)\, y_N(\epsilon) \Big/ \sum_{\epsilon} T^2_{jN}(\epsilon). \tag{3}$$

Now
$$E(\mathscr{A}_{jN}) = \sum_{\epsilon} \left\{ T_{jN}(\epsilon) \Big/ \sum_{\epsilon} T^2_{jN}(\epsilon) \right\} E\left\{ \sum_{z} y(r\epsilon+z) \right\}$$

$$= \sum_{\epsilon} \left\{ T_{jN}(\epsilon) \Big/ \sum_{\epsilon} T^2_{jN}(\epsilon) \right\} \sum_{z} \sum_{k} A_{kn}\, T_{kn}(r\epsilon+z), \tag{4}$$

where A_{kn} is the coefficient of the orthogonal polynomial of order n for the true curve.

To evaluate this expression it is necessary to expand

$$\sum_z T_{kn}(r\epsilon+z)$$

in terms of the polynomials $T_{jN}(\epsilon)$ of order N. Taking as an example the case $k=3$,

$$\begin{aligned}\sum_z T_{3n}(r\epsilon+z) &= \sum_z [(r\epsilon+z)^3-\{(3n^2-7)/20\}(r\epsilon+z)]\\ &= \sum_z [r^3\epsilon^3+3r\epsilon z^2-\{(3n^2-7)/20\}r\epsilon]\\ &= r^4[\epsilon^3-\{(3N^2-7)/20\}\epsilon]-\{r^2(r^2-1)/10\}\epsilon\\ &= r^4 T_{3N}(\epsilon)-\{r^2(r^2-1)/10\}T_{1N}(\epsilon).\end{aligned}$$

When these expansions are substituted in (4), $E(\mathscr{A}_{jN})$ is obtained as a linear function of A_{kN}. This set of linear equations can be solved to give A_{kN} as a linear function of $E(\mathscr{A}_{jN})$. Thus

$$A_{2n} = r^{-3}E[\mathscr{A}_{2N}+\tfrac{3}{7}(1-r^{-2})\mathscr{A}_{4N}], \text{ etc.}$$

Hence the value

$$a_{2n} = r^{-3}[\mathscr{A}_{2N}+\tfrac{3}{7}(1-r^{-2})\mathscr{A}_{4N}]$$

will provide an unbiased estimate of A_{2n}, and so from the $\mathscr{A}_{jN}$ a set of unbiased estimates a_{kn} of the A_{kn} can be obtained. The estimated curve will be

$$u_{pn}(\epsilon) = \sum_{j=0}^{p} a_{jn}T_{jn}(\epsilon), \tag{5}$$

the coefficients a_{jn} being given by the formulae listed in Table 9.8*a*.

In general,

$$a_{jn} = r^{-(j+1)}\sum_{k=j}^{p}\gamma_{jk}\mathscr{A}_{kN}, \tag{6}$$

where γ_{jj} is unity and γ_{jk} is zero for odd values of $j+k$. For polynomials up to the fourth degree the only non-zero coefficients other than γ_{jj} are γ_{13} and γ_{24}. These are listed in Table 9.8*b* for values of r from 2 to 15.

If the fitting is done by means of tables of the orthogonal polynomials $T'_j(\epsilon)$ (§ 7.6.3), then (5) and (6) become

$$u_{pn}(\epsilon) = \sum_{j=0}^{p} a'_{jn}T'_{jn}(\epsilon) \tag{7a}$$

and $$a'_{jn} = r^{-(j+1)}\sum_{k=j}^{p}\gamma_{jk}(\beta'_{kkN}/\beta'_{jjn})\mathscr{A}'_{kN}, \tag{7b}$$

where $$\mathscr{A}'_{kN} = \sum_\epsilon T'_{kN}(\epsilon)\,y_N(\epsilon)\Big/\sum_\epsilon T'^2_{kN}(\epsilon). \tag{7c}$$

9.1.1 *Example*

In Table 9.1.1 the 62 observations of Table 7.6.2.1 have been grouped in pairs ($r = 2$). The calculations are performed in the same way as in Tables 7.6.2.2 and 7.6.2.1*a*. The quantities a_j obtained from the calculating scheme correspond to the quantities $\mathscr{A}_{jN}$ in the notation of this chapter.

The calculation of the unbiased estimates a_{jn} from these values $\mathscr{A}_{jN}$ is shown in Table 9.1.1*a*. The estimates agree very well with those found in Table 7.6.2.1*a*. From the values a_{jn} the power-series coefficients b_{3j} are calculated by means of the quantities β_{kjn}.

When the orthogonal polynomials for $n = 31$ are used with the grouped values, the coefficients $\mathscr{A}'_{kN}$ listed in Table 9.1.1*b* are obtained. The values for a'_{jn} are calculated in the same table.

TABLE 9.1.1

The fitting of a cubic to 31 grouped observations

$M_j = \Sigma \epsilon^j (y_+ + (-)^j y_-)$

ϵ^3	ϵ	Diff.	y_+	y_-	Sum	ϵ^2
0	0	—	18		18	0
1	1	2	16	14	30	1
8	2	3	18	15	33	4
27	3	1	23	22	45	9
64	4	−5	10	15	25	16
125	5	38	45	7	52	25
216	6	66	75	9	84	36
343	7	115	124	9	133	49
512	8	−17	41	58	99	64
729	9	29	79	50	129	81
1000	10	−15	30	45	75	100
1331	11	−40	33	73	106	121
1728	12	−93	16	109	125	144
2197	13	−100	4	104	108	169
2744	14	−124	4	128	132	196
3375	15	−122	10	132	142	225
Sum:		−262	528	790	1336	
			y_0 18			

S_{00}	31		
S_{11}	2480		
S_{22}	158224		
S_{33}	9,683308·8		
β_{02}	−80		
β_{13}	−143·8		
$a_j = \mathscr{M}_j/S_{jj}$		Σy^2	105922
$M_0 = \mathscr{M}_0$	1336	$a_0 \mathscr{M}_0$	57577
a_0	43·096774	Σv_0^2	48345
$M_1 = \mathscr{M}_1$	−5065	$a_1 \mathscr{M}_1$	10344
a_1	−2·0423387	Σv_1^2	38001
M_2	142993	$a_2 \mathscr{M}_2$	8242
$+\beta_{02} M_0$	−106880	Σv_2^2	29759
$= \mathscr{M}_2$	36113		
a_2	+0·22823971		
M_3	−1130029	$a_3 \mathscr{M}_3$	16663
$+\beta_{13} M_1$	+728347	Σv_3^2	13096
$= \mathscr{M}_3$	−401682		
a_3	−0·041481895		

TABLE 9.1.1*a*

Calculation of unbiased estimates from the values $\mathscr{A}_{jN}$

$\mathscr{A}_{3N}$	−0·041481895	$\mathscr{A}_{2N}$	+0·22823971	
$\mathscr{A}_{1N}$	−2·0423387	$\mathscr{A}_{0N}$	43·096774	
r	$= 2$			
a_{3n}	$= r^{-4} \mathscr{A}_{3N}$		−0·002592618	
a_{1n}	$= r^{-2}\{\mathscr{A}_{1N} + \frac{1}{10}(1-r^{-2}) \mathscr{A}_{3N}\}$		−0·5113625	
a_{2n}	$= r^{-3} \mathscr{A}_{2N}$		+0·02852996	
a_{0n}	$= r^{-1} \mathscr{A}_{0N}$		+21·548387	
β_{02n}	−320·25	β_{13n}	−576·25	
b_{33}	$= a_{3n} - 0·002592618$	b_{31}	$= a_{1n} + \beta_{13n} a_{3n}$	+0·9826336
b_{32}	$= a_{2n} + 0·02852996$	b_{30}	$= a_{0n} + \beta_{02n} a_{2n}$	+12·411667

TABLE 9.1.1*b*

Calculation of unbiased estimates from the values $\mathscr{A}'_{kN}$

$\mathscr{A}'_{0N}$ 1336/31 $= 43{\cdot}096774$		β'_{00N} 1	β'_{00n} 1
$\mathscr{A}'_{1N}$ $-5065/2480$ $= -2{\cdot}0423387$		β'_{11N} 1	β'_{11n} 2
$\mathscr{A}'_{2N}$ 36113/158224 $= +0{\cdot}22823971$		β'_{22N} 1	β'_{22n} 0·5
$\mathscr{A}'_{3N}$ $-334735/6724520 = -0{\cdot}049778274$		β'_{33N} 0·83r	β'_{33n} 0·3r

$a'_{jn} = r^{-(j+1)} \sum_{k=j}^{p} \gamma_{jk}(\beta'_{kkN}/\beta'_{jjn})\,\mathscr{A}'_{kN} \quad r = 2 \quad \gamma_{13} = 3/40$

$a'_{3n} = 2^{-4}(2{\cdot}5\mathscr{A}'_{3N}) = -0{\cdot}007777855$

$a'_{1n} = 2^{-2}(\mathscr{A}'_{1N}/2 + \mathscr{A}'_{3N}/32) = -0{\cdot}2556812$

$a'_{2n} = 2^{-3}(2\mathscr{A}'_{2N}) = +0{\cdot}05705993$

$a'_{0n} = 2^{-1}(\mathscr{A}'_{0N}) = 21{\cdot}548387$

9.1.2 *Standard deviations of the orthogonal coefficients*

Since y_N is the sum of r observations y,

$$\sigma^2(y_N) = r\sigma^2(y), \tag{1}$$

while, from (9.1,3),

$$\sigma^2(\mathscr{A}_{jN}) = \sigma^2(y_N)/\Sigma T^2_{jN}(\epsilon). \tag{2}$$

It is apparent from (9.1,6) that $\sigma^2(a_{jn})$ will depend on p as well as on j. However, provided N is not less than 10 or so, the variation with p turns out to be insignificant, the contributions to $\sigma^2(a_{jn})$ of the terms $\mathscr{A}_{kN}$, $k > j$, being negligible. Thus

$$\sigma^2(a_{jn}) = r^{-2(j+1)}\,\sigma^2(\mathscr{A}_{jN}). \tag{3}$$

9.1.3 *Efficiencies of the estimated coefficients* a_{jn}

Since the standard deviation of the coefficient obtained by fitting a curve to the n original observations is $\sigma^2(y)/\Sigma T^2_{jn}(\epsilon)$, the efficiency η_j of the estimate a_{jn}, from (9.1.2,2), (9.1.2,3), and the expressions for $\Sigma T^2_{jn}(\epsilon)$ and $\Sigma T^2_{jN}(\epsilon)$, is given by

$$\eta_j = \frac{(n^2-r^2)(n^2-4r^2)\dots(n^2-j^2r^2)}{(n^2-1)(n^2-4)\dots(n^2-j^2)}, \tag{1}$$

or

$$\eta_j = 1 - \frac{j(j+1)(2j+1)}{6N^2}\{1-r^{-2}\}$$
$$+\frac{j(j^2-1)(2j-1)(10j^2+17j+6)}{360N^4}\{1-r^{-4}\}$$
$$-\frac{\{j(j+1)(2j+1)\}^2}{36N^4}\{r^{-2}-r^{-4}\}+\dots. \tag{2}$$

From this expression the minimum number N of groups which must be retained for the efficiency to exceed a specified value can be determined. Table 9.1.3 gives the minimum number of groups required to ensure that the efficiencies do not fall below 0·98, 0·95, 0·90, 0·80, and 0·70.

TABLE 9.1.3

Minimum values of N, the number of groups, to obtain stated efficiencies

Degree j	1	2	3	4	5
Efficiency η_j					
0·98	8	16	27	39	53
0·95	5	10	17	25	33
0·90	4	8	12	18	24
0·80	3	5	9	12	16
0·70	—	4	7	10	13

The values in Table 9.1.3 have been calculated from (2) by neglecting terms r^{-k}. If r is small, a slightly smaller number of groups would be permissible. As an example, for a_3 with $r = 2$ the value of N for an efficiency of 0·95 drops from 17 to 15.

For the example of § 9.1.1,

$$\eta_3 = 2^7 S_{33N}/S_{33n} = 2^7 \times 9{\cdot}683 \times 10^6/1253{\cdot}1 \times 10^6 = 0{\cdot}989.$$

9.1.4 *Standard deviations of the fitted values*

The standard deviations of the fitted values

$$u_p(\epsilon) = \Sigma a_{jn} T_{jn}(\epsilon)$$

may be evaluated by expanding the coefficients a_{jn} using (9.1,6), collecting terms in $\mathscr{A}_{jN}$, and using (9.1.2,2). However, it is found that, to a reasonable approximation,

$$\sigma^2[u_p(\epsilon)] \doteqdot \sum_{j=0}^{p} T_{jn}^2(\epsilon)\,\sigma^2(a_{jn}) \doteqdot \sigma^2 \sum_{j=0}^{p} \eta_j\, T_{jn}^2(\epsilon)/S_{jjn}. \qquad (1)$$

In the region of extrapolation the term $\sigma^2(a_{pn})$ becomes dominant, and the efficiency of the fitted value approaches the efficiency of the coefficient of highest degree. In the region of interpolation the efficiency will be somewhat greater than this value.

It will usually be sufficient to calculate $\sigma[u_p(\epsilon)]$ on the assumption that η_j in (1) is unity. The formula then reduces to the standard form, and the tables of $\rho_{p0}(k, n)$ can be used to give a quick estimate. The proportional inaccuracy should not be greater than η_p.

9.1.5 *Estimation of the standard deviation of an observation*

The standard deviation of an observation may be estimated from the residuals

$$v_{pN}(\epsilon) = y_N(\epsilon) - \sum_{j=0}^{p} \mathscr{A}_{jN} T_{jN}(\epsilon).$$

From § 8.1.1 it follows that

$$E\Sigma v_{pN}^2(\epsilon) = (N-p-1)\,\sigma^2(y_N) = r(N-p-1)\,\sigma^2(y), \qquad (1)$$

and so the expression

$$s_{pN} = \left\{\frac{\Sigma v_{pN}^2(\epsilon)}{r(N-p-1)}\right\}^{\frac{1}{2}} \qquad (2)$$

will provide an estimate of $\sigma(y)$. The quantity $\Sigma v_{pN}^2(\epsilon)$ may be calculated from the formula

$$\Sigma v_{pN}^2(\epsilon) = \Sigma y_N^2(\epsilon) - \sum_j \mathscr{A}_{jN}^2 \Sigma T_{jN}^2(\epsilon). \qquad (3)$$

Another estimate of $\sigma(y)$ may be obtained from the residuals

$$v_{pn}(\epsilon) = y(\epsilon) - \sum_{j=0}^{p} a_{jn} T_{jn}(\epsilon).$$

It will be shown in § 9.1.5.1 below that

$$E\Sigma v_{pn}^2(\epsilon) = \left[n-p-1+\sum_j \{(1/\eta_j)-1\}\right]\sigma^2. \qquad (4)$$

In the present case, from (9.1.3,2), the last term in the square brackets is found to be approximately $p^4/12N^2$, and so it can be neglected. Hence

$$s_{pn} = \left\{\frac{\Sigma v_{pn}^2(\epsilon)}{n-p-1}\right\}^{\frac{1}{2}} \qquad (5)$$

will provide an estimate of $\sigma(y)$.

Since s_{pn} is very close to the least-squares estimate, $\nu s_{pn}^2/\sigma^2$ should be distributed at least approximately as χ^2 with $\nu = n-p-1$ d.f. The distribution of s_{pN} will be similarly related to the χ^2 distribution with $\nu = N-p-1$ d.f., and so the relative efficiency of this estimate will be, from (2.4.1,3),

$$(N-p-1)/(n-p-1) \doteqdot 1/r.$$

The estimate s_{pN} is then rather inefficient, but it can easily be calculated by means of (3), while there is no comparable formula by means of which s_{pn} could be obtained.

For the example of § 9.1.1, from Table 9.1.1, $\Sigma v_{3N}^2(\epsilon)$ is 13096, and so (2) gives

$$s_{3N} = \{13096/27 \times 2\}^{\frac{1}{2}} = 15{\cdot}6.$$

The value found by fitting a curve to the original observations is 11·4 (Table 7.6.2.1a).

9.1.5.1 *Relation of certain least-squares estimates to estimates of lower efficiency.* First it will be shown that, if a_j^* is the least-squares estimate of the orthogonal polynomial coefficient A_j and a_j is any other unbiased estimate obtained from a linear combination of the same observations y_i, then for any linear function f of the a_j,

$$E(f-f^*)^2 = E(f-F)^2 - E(f^*-F)^2, \tag{1a}$$

where

$$f = \Sigma\lambda_j a_j, \quad f^* = \Sigma\lambda_j a_j^*, \quad F = \Sigma\lambda_j A_j. \tag{1b}$$

For

$$E(f-f^*)^2 = E(f-F)^2 + E(f^*-F)^2 - 2E(f-F)(f^*-F), \tag{2}$$

and

$$E(f-F)(f^*-F) = E\{\Sigma\lambda_j(a_j-A_j)\}\{\Sigma\lambda_k(a_k^*-A_k)\}. \tag{3}$$

Now, a_j is of the form

$$a_j = \sum_i \psi_j(x_i)\, y_i$$

and, since a_j is unbiased,

$$E(a_j) = A_j = \sum_i \psi_j(x_i)\, E(y_i) = \sum_i \psi_j(x_i) \sum_k A_k T_k(x_i).$$

Hence it follows that

$$\sum_i \psi_j(x_i)\, T_k(x_i) = \delta_{jk},$$

where δ_{jk} is the Kronecker delta. From (3),

$$\begin{aligned} E(f-F)(f^*-F) &= \sum_{j,\,k} \lambda_j \lambda_k \sum_i \psi_j(x_i)\{w_i T_k(x_i)/\Sigma w_i T_k^2(x_i)\}\,\sigma_i^2 \\ &= \sum_j \lambda_j^2\, \sigma^2/\Sigma w_i T_j^2(x_i) \\ &= E(f^*-F)^2, \end{aligned}$$

and on substituting this in (2), (1a) is established.

In particular,

$$E(a_j-a_j^*)^2 = \sigma^2(a_j) - \sigma^2(a_j^*) = \{(1/\eta_j)-1\}\,\sigma^2(a_j^*). \tag{4}$$

For the residuals in the two cases

$$\begin{aligned} \Sigma w_i v_i^2 - \Sigma w_i v_i^{*2} &= \Sigma w_i\{y_i - \Sigma a_j T_j(x_i)\}^2 - \Sigma w_i\{y_i - \Sigma a_j^* T_j(x_i)\}^2 \\ &= \sum_j [-2(a_j-a_j^*)\,\Sigma w_i y_i T_j(x_i) \\ &\qquad + (a_j^2 - a_j^{*2})\,\Sigma w_i T_j^2(x_i)], \end{aligned}$$

or

$$\Sigma w_i v_i^2 - \Sigma w_i v_i^{*2} = \sum_j (a_j-a_j^*)^2\, \Sigma w_i T_j^2(x_i). \tag{5}$$

On combining (4) and (5),

$$E[\Sigma w_i v_i^2 - \Sigma w_i v_i^{*2}] = \sum_j \{(1/\eta_j)-1\}\,\sigma^2,$$

and so, since
$$E\Sigma w_i v_i^{*2} = (n-p-1)\,\sigma^2,$$
$$E\Sigma w_i v_i^2 = \left[(n-p-1)+\sum_j \{(1/\eta_j)-1\}\right]\sigma^2. \tag{6}$$

This is the result (9.1.5,4) used in § 9.1.5 above.

9.1.6 *The dropping of observations before grouping*

If the number of observations n is prime, it will be necessary to drop one or more observations at the ends of the range before grouping. The standard deviation of a_j is proportional to $n^{-(j+\frac{1}{2})}$. Hence if the number of observations is reduced to n' by dropping ν observations before grouping, the efficiency of the estimate a_j will be reduced by the factor $(n'/n)^{2j+1}$. This factor may cause a considerable drop in efficiency. For example, if one observation is omitted when n is 50, the efficiency of a_3 is reduced by the factor 0·868. Table 9.1.6 shows the relative efficiencies of the coefficients a_j for various values of n when 1, 2, and 3 observations are omitted.

The efficiency of the grouped estimate a_{jN} is then the expression (9.1.3,2) multiplied by the factor $(n'/n)^{2j+1}$.

When the 62 observations of Table 7.6.2.1 are reduced to 15 groups, the two end observations must be omitted. The efficiency of the estimate a_3 will be $S_{33N}r^7/S_{33n}$, which is, from Table 7.10*b*, 0·749.

The effect on the fitted values of dropping observations can be found by means of the tables of the functions $\rho_{p0}(k)$ (Table 8.8*a*). The efficiency of the fitted value at the point ϵ will be given by
$$\eta[u_p(\epsilon)] = \frac{n'}{n}\left(\frac{\rho_{p0}(k)}{\rho_{p0}(k')}\right)^2, \tag{1}$$
where (assuming the observations are omitted symmetrically from the two ends of the range)
$$k' = 2\epsilon/n', \qquad \nu \text{ even};$$
$$k' = (2\epsilon \pm 1)/n', \quad \nu \text{ odd}.$$

$\rho_{p0}(k)$ varies only slowly with k for a considerable part of the region of interpolation, and so in this part the efficiency of the fitted value is close to n'/n, the efficiency of the coefficient a_0. Beyond a value of $|k|$ given in Table 9.1.6.1 the variation of $\rho_{p0}(k)$ with k becomes very rapid, and the efficiency drops sharply. For large values of $|k|$ the efficiency approaches that for the coefficient a_p.

TABLE 9.1.6

Relative efficiencies of a_j when ν observations are dropped from the original set of n observations

	a_0			a_1			a_2		
ν	1	2	3	1	2	3	1	2	3
n									
20	0·950	0·900	0·850	0·857	0·729	0·614	0·774	0·590	0·444
30	0·967	0·933	0·900	0·903	0·813	0·729	0·844	0·708	0·590
40	0·975	0·950	0·925	0·927	0·857	0·791	0·881	0·774	0·677
50	0·980	0·960	0·940	0·941	0·885	0·831	0·904	0·815	0·734
75	0·987	0·973	0·960	0·961	0·922	0·885	0·935	0·874	0·815
100	0·990	0·980	0·970	0·970	0·941	0·913	0·951	0·904	0·859
150	0·993	0·987	0·980	0·980	0·961	0·941	0·967	0·935	0·904
250	0·996	0·992	0·988	0·988	0·976	0·964	0·980	0·961	0·941
500	0·998	0·996	0·994	0·994	0·988	0·982	0·990	0·980	0·970

	a_3			a_4			a_5		
ν	1	2	3	1	2	3	1	2	3
n									
20	0·698	0·478	0·321	0·630	0·387	0·232	0·569	0·314	0·167
30	0·789	0·617	0·478	0·737	0·537	0·387	0·689	0·468	0·314
40	0·838	0·698	0·579	0·796	0·630	0·496	0·757	0·569	0·424
50	0·868	0·751	0·648	0·834	0·693	0·573	0·801	0·638	0·506
75	0·910	0·828	0·751	0·886	0·784	0·693	0·863	0·743	0·638
100	0·932	0·868	0·808	0·914	0·834	0·760	0·895	0·801	0·715
150	0·954	0·910	0·868	0·942	0·886	0·834	0·929	0·863	0·801
250	0·972	0·945	0·919	0·965	0·930	0·897	0·957	0·915	0·876
500	0·986	0·972	0·959	0·982	0·965	0·947	0·978	0·957	0·936

TABLE 9.1.6.1

Range of $|k|$ within which the efficiency of the fitted value approximates to n'/n

Degree p	Range of $\|k\|$
2	0 to 0·50
3	0 to 0·70
4	0 to 0·80
5	0 to 0·85

9.2 STEP FUNCTION METHODS

The use of step functions for the estimation of the slope of a straight line was discussed in § 6.4. Similar methods can be used to estimate the coefficients for higher degree polynomials.

9.2.1 *Second-degree polynomials*

To obtain an estimate b_{22} of the second-degree coefficient, a function $W_2(\epsilon)$ is required such that

$$\Sigma W_2(\epsilon) = 0 \tag{1a}$$

and

$$\Sigma W_2(\epsilon)\,\epsilon = 0. \tag{1b}$$

The estimate is then

$$b_{22} = \Sigma W_2(\epsilon)\, y(\epsilon)/\Sigma W_2(\epsilon)\,\epsilon^2, \tag{2}$$

and its standard deviation is given by

$$\frac{\sigma^2(y)}{\sigma^2(b_{22})} = \frac{\{\Sigma W_2(\epsilon)\,\epsilon^2\}^2}{\Sigma W_2^2(\epsilon)}. \tag{3}$$

Condition (1b) is satisfied if $W_2(\epsilon)$ is an even function of ϵ. Condition (1a) requires that the values $W_2(\epsilon)$ should fall into two groups, one with $W_2(\epsilon)$ positive and the other with $W_2(\epsilon)$ negative. The appropriate step functions will be such that, for any arbitrary function $f(\epsilon)$, $\Sigma W_2(\epsilon) f(\epsilon)$ is of the form

$$\left[\left\{q_1\left(\sum_{-\frac{1}{2}(n-1)}^{\frac{1}{2}(n-1)} - \sum_{-\frac{1}{2}(a_1-1)}^{\frac{1}{2}(a_1-1)}\right) + q_2\left(\sum_{-\frac{1}{2}(a_1-1)}^{\frac{1}{2}(a_1-1)} - \sum_{-\frac{1}{2}(a_2-1)}^{\frac{1}{2}(a_2-1)}\right) + \ldots\right\}\right.$$
$$\left. - \left\{r_1\left(\sum_{-\frac{1}{2}(b_1-1)}^{\frac{1}{2}(b_1-1)} - \sum_{-\frac{1}{2}(b_2-1)}^{\frac{1}{2}(b_2-1)}\right) + r_2\left(\sum_{-\frac{1}{2}(b_2-1)}^{\frac{1}{2}(b_2-1)} - \sum_{-\frac{1}{2}(b_3-1)}^{\frac{1}{2}(b_3-1)}\right) + \ldots\right\}\right][f(\epsilon)]. \tag{4}$$

$\frac{1}{2}(a_j-1)$ and $\frac{1}{2}(b_j-1)$ are the values of ϵ at the boundaries of the steps. The functions will be called single-step functions if there is one step in each group, double-step functions if there are two steps.

The expression $\Sigma W_2(\epsilon)\,\epsilon^2$ will contain terms

$$\sum_{-\frac{1}{2}(a_j-1)}^{\frac{1}{2}(a_j-1)} = a_j(a_j^2-1)/12$$

for which the approximations $n^3\alpha_j^3/12$ will be used, where $\alpha_j = a_j/n$ is the parameter specifying the boundary of the step. Since for the least-squares estimate

$$\sigma^2(y)/\sigma^2(b_{22}^*) = n(n^2-1)(n^2-4)/180 \doteqdot n^5/180,$$

the efficiency of the estimate b_{22} is, from (3) and (4),

$$\eta(b_{22}) = \frac{5}{4}\,\frac{[\{q_1(1-\alpha_1^3)+\ldots\}-\{r_1(\beta_1^3-\beta_2^3)+\ldots\}]^2}{\{q_1^2(1-\alpha_1)+\ldots\}+\{r_1^2(\beta_1-\beta_2)+\ldots\}}. \tag{5a}$$

The values q, r, α, β, for maximum efficiency can be determined, subject to the condition (1a)

$$\{q_1(1-\alpha_1)+\ldots\}-\{r_1(\beta_1-\beta_2)+\ldots\} = 0. \tag{5b}$$

The optimum values are listed in Table 9.2.1 for single-step and double-step functions. The corresponding efficiencies are 0·8958 for single-step and 0·9630 for double-step functions.

TABLE 9.2.1

Values of parameters for maximum efficiencies

Second-degree coefficient				Third-degree coefficient			
Single-step functions		Double-step functions		Single-step functions		Double-step functions	
α	0·7363	α_1	0·8482	α	0·8621	α_1	0·9222
		α_2	0·6757			α_2	0·8309
β	0·4414	β_1	0·5007	β_1	0·7024	β_1	0·7399
		β_2	0·3207	β_2	0·1205	β_2	0·6536
						β_3	0·1265
q	1·3042	q_1	1·5774	q	1·0393	q_1	1·2465
		q_2	0·7590			q_2	0·5940
r	0·7793	r_1	0·4762	r	0·5573	r_1	0·3167
		r_2	0·8875			r_2	0·5919
η_2	0·8958	η_2	0·9630	η_3	0·9014	η_3	0·9473

9.2.2 *Third-degree polynomials*

To obtain an estimate b_{33} of the third-degree coefficient a function $W_3(\epsilon)$ is required such that

$$\Sigma W_3(\epsilon) = 0 = \Sigma W_3(\epsilon)\,\epsilon^2, \tag{1a}$$

$$\Sigma W_3(\epsilon)\,\epsilon = 0. \tag{1b}$$

The estimate is then

$$b_{33} = \Sigma W_3(\epsilon)\,y(\epsilon)/\Sigma W_3(\epsilon)\,\epsilon^3, \tag{2}$$

and its standard deviation is given by

$$\frac{\sigma^2(y)}{\sigma^2(b_{33})} = \frac{\{\Sigma W_3(\epsilon)\,\epsilon^3\}^2}{\Sigma W_3^2(\epsilon)}. \tag{3}$$

Condition (1a) requires that $W_3(\epsilon)$ be an odd function of ϵ. Proceeding as in § 9.2.1, the efficiency of the estimate is found to be given by the expression

$$\eta(b_{33}) = \frac{175}{64}\frac{[\{q_1(1-\alpha_1^4)+\ldots\}-\{r_1(\beta_1^4-\beta_2^4)+\ldots\}]^2}{\{q_1^2(1-\alpha_1)+\ldots\}+\{r_1^2(\beta_1-\beta_2)+\ldots\}}. \tag{4a}$$

The values of the parameters which make this expression a maximum, subject to the condition (1*b*)

$$\{q_1(1-\alpha_1^2)+\ldots\}-\{r_1(\beta_1^2-\beta_2^2)+\ldots\} = 0, \tag{4b}$$

are listed in Table 9.2.1. The maximum efficiency for the estimate b_{33} is 0·9014 for single-step functions and 0·9473 for double-step functions.

9.2.3 *The polynomial coefficients*

Estimates b_{pk} of the polynomial coefficients can be obtained from linear functions of the quantities

$$b_{kk} = \Sigma W_k(\epsilon)\, y(\epsilon)/\Sigma W_k(\epsilon)\, \epsilon^k, \tag{1a}$$

where $$\Sigma W_k(\epsilon)\, \epsilon^m = 0, \quad m < k. \tag{1b}$$

For if the estimates b_{pk} are to be unbiased (cf. § 9.5),

$$\Sigma W_j(\epsilon)\, y(\epsilon) = \Sigma W_j(\epsilon) \sum_{k=j}^{p} b_{pk}\, \epsilon^k. \tag{2}$$

On dividing this equation by $\Sigma W_j(\epsilon)\, \epsilon^j$, it can be put in the form

$$b_{jj} = b_{pj} + \sum_{j+1}^{p} \alpha_{jk}\, b_{pk}, \tag{3}$$

or $$b_{pj} = b_{jj} - \sum_{j+1}^{p} \alpha_{jk}\, b_{pk}.$$

By continual expansion of the terms b_{pk}, this equation can be put in the form

$$b_{pj} = \sum_{k=j}^{p} \beta_{jk}\, b_{kk}. \tag{4}$$

β_{jj} is unity, and β_{jk} vanishes if $j+k$ is odd. For the second- and third-degree polynomials,

$$b_{20}, b_{30} = b_{00}+\beta_{02}\, b_{22}, \quad b_{32} = b_{22},$$

$$b_{31} = b_{11}+\beta_{13}\, b_{33}, \quad b_{21} = b_{11},$$

where

$$\beta_{02} = -\Sigma\epsilon^2/n = -(n^2-1)/12, \quad \beta_{13} = -\Sigma W_1(\epsilon)\, \epsilon^3/\Sigma W_1(\epsilon)\, \epsilon.$$

9.2.4 *The fitted values*

The variance of the fitted value

$$u_p(\epsilon) = \sum_{j=0}^{p} b_{pj}\, \epsilon^j$$

can be found in terms of the variances and covariances of the b_{pj}, and the efficiency by dividing this value by the corresponding least-squares variance. The efficiencies are very nearly equal to the efficiency of the quantity b_{pp} in the region of extrapolation, and are somewhat greater than this in the region of interpolation.

9.2.5 *Tables of step functions*

From the values α and β given in Table 9.2.1 the numbers of observations in each step can be found. Thus for the step (α_j, α_{j+1}) the number is

$$\tfrac{1}{2}(a_j - 1) - \tfrac{1}{2}(a_{j+1} - 1) = \tfrac{1}{2}n(\alpha_j - \alpha_{j+1}).$$

The weights can then be found from the conditions that $\Sigma W_j(\epsilon)\,\epsilon^k = 0$ for $k < j$. The functions $W_j(\epsilon)$ for $j = 4$ and $j = 5$ can be determined in a similar way.

The optimum single-step functions are given in Table 9.8*c* for polynomials of degree $p \leqslant 5$, and for values $n(1)75$. The quantities β_{kj} and $\Sigma W_j(\epsilon)\,\epsilon^j$ are also listed. For the tabulation of the function $W_j(\epsilon)$, the observations are supposed numbered by the value of $|\epsilon|$ if n is odd, and by the value of $|\epsilon| + \frac{1}{2}$ if n is even. Thus for 62 observations the numbers are 1 to 31, for 63 observations 0 to 31. For coefficients of even degree observations of equal $|\epsilon|$ are added, for coefficients of odd degree they are subtracted. This is indicated by the suffix $+$ or $-$ under the summation sign. The expression Σa means the sum of all observations numbered 0 to a.

9.2.6 *Example*

Table 9.2.6 shows the calculations for the cubic curve fitted to the observations of Table 7.6.2.1. The quantities W_j, $\Sigma W_j\,\epsilon^j$, and β_{kj} are first entered on the right of the calculating scheme from the tables. The observations are entered at the left, the observation for the largest value of ϵ being at the bottom of the y_+ column.

Lines are drawn to indicate the sums Σa required. The sums of the corresponding observations are added starting from the top, the progressive total being entered in the $\sum\limits_{+}$ column wherever a line is drawn. The differences are then added, the progressive totals being entered in the $\sum\limits_{-}$ column. As a check, the final $\sum\limits_{+}$ total should be $\Sigma y_+ + \Sigma y_- + y_0$ and the final $\sum\limits_{-}$ total $\Sigma y_+ - \Sigma y_-$. The calculations of the polynomial coefficients are then carried out at the right of the scheme.

The efficiency of the estimate b_{33} is, from (9.2.2,3),

$$(\Sigma W_3\,\epsilon^3)^2/\Sigma W_3^2 S_{33} = \frac{(16151 \times 10^3)^2}{(8 \times 117^2 + 36 \times 58^2) \times 1253{\cdot}1 \times 10^6} = 0{\cdot}903.$$

The value given in Table 9.2.1 is 0·901.

TABLE 9.2.6

Calculating scheme using single-step functions

No.	$\sum_{-}$	y_+	y_-	$\sum_{+}$
0				
1		5	13	
2		6	8	
3		10	6	
4	−3	8	5	
5		10	10	
6		13	13	
7		10	9	
8		3	10	
9		7	5	
10	+4	16	5	
11		29	2	
12		37	1	
13		38	8	
14		50	3	340
15		74	6	
16		22	22	
17		19	36	
18		44	30	
19		35	20	
20		15	21	
21		15	24	
22	+199	18	28	
23		15	45	829
24		10	52	
25		6	57	
26		4	56	
27	−24	0	48	
28		3	55	
29		1	73	
30		3	65	
31	−270	7	67	1336
32		533	803	
33	Check			
34				
35				
36				
37				

$\sum_{+} 31/62 = b_{00}$ 21·548387

β_{02} −320·25 β_{04}

$b_{00}+\beta_{02}\,b_{22}+\beta_{04}\,b_{44} = b_{p0}$ 12·192375

W_2: $7\left(\sum_{+} 31 - \sum_{+} 23\right) - 4\sum_{+} 14$

$\sum_{+} W_2\,\epsilon^2$ 74928

b_{22} 0·029214713

β_{24}

$b_{22}+\beta_{24}\,b_{44} = b_{p2}$ 0·029214713

W_4:

$\sum_{+} W_4\,\epsilon^4$

b_{44}

$\left(\sum_{-} 31 - \sum_{-} 10\right)\Big/861 = b_{11}$ −0·31823461

β_{13} −530·25 β_{15}

$b_{11}+\beta_{13}\,b_{33}+\beta_{15}\,b_{55} = b_{p1}$ +1·0113728

W_3: $117\left(\sum_{-} 31 - \sum_{-} 27\right) - 58\left(\sum_{-} 22 - \sum_{-} 4\right)$

$\sum_{-} W_3\,\epsilon^3$ 16,150680

b_{33} −0·0025075105

β_{35}

$b_{33}+\beta_{35}\,b_{55} = b_{p3}$ −0·0025075105

W_5:

$\sum_{-} W_5\,\epsilon^5$

b_{55}

9.3 GENERAL SUMMARY FOR THE EQUALLY-SPACED CASE

Table 9.3 gives a summary of the solutions by different methods of Example 7.6.2.1 on coded sugar prices. The time taken for each method by an experienced computer is also shown.

TABLE 9.3

Solutions of Example 7.6.2.1 *obtained by different methods*

	Least-squares methods			Step function methods	
	62 observations	31 groups	15 groups	Single-step	Double-step
$-b_{33}\times 10^3$	$2{\cdot}59\pm 0{\cdot}32$	2·59	2·95	2·51	2·44
$b_{32}\times 10^2$	$2{\cdot}86\pm 0{\cdot}50$	2·85	2·92	2·92	2·84
$b_{31}\times 10$	$9{\cdot}8\ \pm 2{\cdot}0$	9·8	12·4	10·1	8·7
b_{30}	$12{\cdot}4\ \pm 2{\cdot}2$	12·4	11·7	12·2	12·4
Minutes required:					
by machine	43	31	29	20	28
by logarithms	—	—	—	41	62
Efficiency of b_{33}	1·000	0·989	0·749	0·903	0·949

An examination of Table 9.3 shows that the reduction in time brought about by the use of grouping methods is not as large as might have been expected. More significant is the reduction of strain on the computer, since the multiplying factors are smaller in magnitude and fewer in number and the chance of making a mistake is much reduced. In fact, the step function method is almost foolproof, except for the possibility of copying errors. Step function methods can be used even when no calculating machine is available, the quantities $\Sigma W_j(\epsilon)\,y(\epsilon)$ being obtained by simple addition and the coefficients b_{pj} by logarithms.

The coefficients obtained for $N = 31$ are close to the values for $n = 62$, while the coefficients for $N = 15$ are somewhat different. Since in the latter case the two end observations have been omitted, this would indicate that the deviations from a smooth curve are not entirely random in this example. This, of course, would be expected from the nature of the data. However, the differences between the various estimates are usually less than the standard deviations.

9.4 UNEQUALLY-SPACED OBSERVATIONS

Following the treatment of Ch. 8, the values x_i will be replaced by the smoothed-out set

$$X_{ni} = k_{1n}\,T^e_{1n}(\epsilon_i) + k_{2n}\,T^e_{2n}(\epsilon_i) + k_{3n}\,T^e_{3n}(\epsilon_i), \qquad (1a)$$

where
$$k_{jn} = \sum_i T^e_{jn}(\epsilon_i)\, x_i \Big/ \sum_i \{T^e_{jn}(\epsilon_i)\}^2. \tag{1b}$$

If the n observations are converted by grouping into $N = n/r$ values
$$y_N(\epsilon_i) = \sum_{z=-\frac{1}{2}(r-1)}^{\frac{1}{2}(r-1)} y_n(r\epsilon_i + z)$$
at points
$$x_{Ni} = \sum_z x_n(r\epsilon_i + z),$$
the smoothed-out system for the grouped observations may be written
$$X_{Ni} = k_{1N}\, T^e_{1N}(\epsilon_i) + k_{2N}\, T^e_{2N}(\epsilon_i) + k_{3N}\, T^e_{3N}(\epsilon_i). \tag{2}$$
The values k_{jN} may be found in terms of the k_{jn} by using the expansion of $\sum_z T^e_{jn}(r\epsilon_i + z)$ as a series in $T^e_{kN}(\epsilon_i)$ (Guest, 1954). In fact,
$$\left.\begin{aligned} k_{1N} &= r^2[k_{1n} - \tfrac{1}{10}(r^2-1)\, k_{3n}] \\ k_{2N} &= r^3\, k_{2n} \\ k_{3N} &= r^4\, k_{3n} \end{aligned}\right\}. \tag{3}$$
The scale factor
$$\phi = \{k_{1n} + \tfrac{1}{10}(n^2+1)\, k_{3n}\}^{-1} \tag{4}$$
is now removed from (1a) and (2), giving
$$X_{ni} = \phi^{-1}\{\kappa_{1n}\, T^e_{1n}(\epsilon_i) + n^{-1}\,\kappa_{2n}\, T^e_{2n}(\epsilon_i) + 2n^{-2}\,\kappa_{3n}\, T^e_{3n}(\epsilon_i)\}, \tag{5a}$$
$$X_{Ni} = r^2\phi^{-1}\{\kappa_{1N}\, T^e_{1N}(\epsilon_i) + N^{-1}\,\kappa_{2N}\, T^e_{2N}(\epsilon_i) + 2N^{-2}\,\kappa_{3N}\, T^e_{3N}(\epsilon_i)\}. \tag{5b}$$
Then, by comparing (1a) and (5a), (2) and (5b), and using (3) and (4),
$$\kappa_{3N} = \kappa_{3n}, \quad \kappa_{2N} = \kappa_{2n}, \quad \kappa_{1N} = 1 - \tfrac{1}{5}\kappa_{3N} - \tfrac{1}{5}N^{-2}\,\kappa_{3N}. \tag{6}$$
Hence the grouped variable (5b) is described by the same two parameters $\kappa_2 \equiv \kappa_{2n}$ and $\kappa_3 \equiv \kappa_{3n}$ as the ungrouped variable, n being replaced by N in the various equations.

9.4.1 *Least-squares curve fitted to the grouped estimates*

When a least-squares curve is fitted to the N observations y_{Ni}, x_{Ni}, the standard deviation of the orthogonal coefficient a_{jN} is given by
$$\operatorname{var} a_{jN} = \operatorname{var} y_N / \Sigma T^2_j(x_{Ni}). \tag{1}$$
An approximate value for this expression can be found by replacing the x_{Ni} by the smoothed set X_{Ni}, and using the formulae of § 8.5.4. Thus
$$\operatorname{var} a_{jN} \doteqdot (\phi/r^2)^{2j}\{(N^{2j+1}/R_j)\, f_j(1 - N^{-2}g_j)\}^{-1} \operatorname{var} y_N. \tag{2}$$

A similar expression, with n replacing N, holds for $\text{var}\, a_{jn}$. It will be assumed throughout this chapter that n is sufficiently large to permit the neglecting of terms of order n^{-2}. Then

$$(\text{var}\, a_{jn})/(\text{var}\, a_{jN}) \doteqdot r^{2(j-1)}(1-N^{-2}g_j), \tag{3}$$

and the efficiency of the estimate a_{jN} is given by

$$\eta(a_{jN}) = 1-N^{-2}g_j. \tag{4}$$

g_j has been tabulated in Table 8.8c. From these quantities the efficiencies for polynomials of the first, second, and third degrees

TABLE 9.4.1

Percentage efficiencies for suggested values of N

κ_2^2 / κ_3			0		
	−1·0	−0·5	0	0·5	1·0
			1st-degree polynomial		
$\eta(a_{1N})$ $N = 5$	96·7	96·6	96·0	94·8	92·7
			2nd-degree polynomial		
$\eta(a_{2N})$ $N = 9$	97·8	96·5	93·8	89·7	84·1
$N = 12$	98·7	98·0	96·5	94·2	91·0
			3rd-degree polynomial		
$\eta(a_{3N})$ $N = 16$	98·9	97·5	94·5	90·0	83·7
$N = 21$	99·4	98·5	96·8	94·2	90·6
κ_2^2 / κ_3			0·5		
	−1·0	−0·5	0	0·5	1·0
			1st-degree polynomial		
$\eta(a_{1N})$ $N = 5$	96·3	96·1	95·5	94·2	92·1
			2nd-degree polynomial		
$\eta(a_{2N})$ $N = 9$	96·4	95·0	92·3	88·1	82·3
$N = 12$	98·0	97·2	95·7	93·3	90·1
			3rd-degree polynomial		
$\eta(a_{3N})$ $N = 16$	97·4	95·8	92·8	88·3	81·9
$N = 21$	98·5	97·6	95·8	93·2	89·5
κ_2^2 / κ_3			1·0		
	−1·0	−0·5	0	0·5	1·0
			1st-degree polynomial		
$\eta(a_{1N})$ $N = 5$	96·0	95·7	95·0	93·7	91·5
			2nd-degree polynomial		
$\eta(a_{2N})$ $N = 9$	95·3	93·6	90·7	86·5	80·3
$N = 12$	97·3	96·4	94·8	92·4	88·9
			3rd-degree polynomial		
$\eta(a_{3N})$ $N = 16$	96·2	94·3	91·1	86·5	80·1
$N = 21$	97·8	96·7	94·8	92·1	88·5

can be determined for various values of N. Clearly the larger the value of N the higher will be the efficiencies, but also the longer will be the time taken to calculate the fitted polynomial. Some compromise is necessary, and it appears that suitable values of N lie in the range 9 to 12 for a second-degree curve, and in the range 16 to 21 for a third-degree curve. For a first-degree curve, 5 groups are sufficient. The efficiencies for these suggested values of N are shown in Table 9.4.1.

9.4.2 *Bias of the estimates*

If the polynomial is of the second or third degree, grouping will usually lead to bias in the estimates. The origin of the bias may be seen by considering the power-series representation. Thus if the 'true' curve is $\Sigma B_{pj} x^j$, the expectation of the observation y_i corresponding to a given value x_i is

$$E(y_i) = \sum_j B_{pj} x_i^j.$$

Hence

$$E(y_{Ni}) = \sum_j B_{pj} \sum_z x_i^j,$$

where the suffix z indicates summation over the r values x_i in a particular group. Now

$$\sum_z x_i^j \neq r^{-j+1}\left(\sum_z x_i\right)^j$$

unless j is 0 or 1, and so

$$E(y_{Ni}) \neq \sum_j (r^{-j+1} B_{pj})\, x_{Ni}^j. \tag{1a}$$

The true value for the grouped curve is

$$B_{pjN} = r^{-j+1} B_{pj},$$

and so (1*a*) becomes

$$E\left(y_{Ni} - \sum_j B_{pjN} x_{Ni}^j\right) \neq 0. \tag{1b}$$

Because of this inequality the estimates obtained from the normal equations

$$\Sigma(y_{Ni} - \Sigma b_{pjN} x_{Ni}^j)\, x_{Ni}^k = 0$$

will be biased estimates.

It is more convenient to calculate expressions for the bias in the orthogonal coefficients a_{kN} rather than the power-series coefficients b_{pk}, though the arithmetical procedure is still very complicated even for the orthogonal coefficients. The bias in a_{kN} is defined to be the difference between the expectation and the true value, $E(a_{kN}) - A_{kN}$. The detailed calculations will not be given here. The bias is found to be given, to order N^{-2}, by an expression of the form

$$N^{-2} \sum_{j=0}^{p} g_{jk}(\phi^{-1} N r^2)^{j-k} A_{jN}, \tag{2a}$$

where the g_{jk} are functions of κ_2 and κ_3.

From (8.5.1,5), $n\phi^{-1}$ is approximately the range of x_{ni}. Hence $\phi^{-1} N r^2$ is approximately the range of x_{Ni}. Thus the estimate of bias can be written as

$$N^{-2} \sum_{j=0}^{p} g_{jk}(\mathrm{Ra}\, x_{Ni})^{j-k} A_{jN}. \tag{2b}$$

The ratio of bias to standard deviation will be more useful than the actual bias in deciding whether the grouping will be satisfactory. From (9.4.1,2), this ratio is given by

$$\frac{\text{Bias}\, a_{kN}}{\text{S.D.}\, a_{kN}} = \frac{1}{\sigma_N} N^{-\frac{3}{2}} \sum_{j} G_{jk} A_{jN} (\mathrm{Ra}\, x_{Ni})^j, \tag{3}$$

where σ_N is the standard deviation of a grouped observation y_{Ni}, and G_{jk} is a function of the parameters κ_2 and κ_3. G_{0j} and G_{1j} vanish, and so for the third-degree polynomial the bias depends only on A_2 and A_3, while for the second-degree polynomial it depends only on A_2. Selected values of G_{jk} are given in Table 9.8*d*.

9.4.2.1 *Checking for bias before grouping.* It will often be desirable to ascertain, before proceeding with the calculations, whether the grouping will give rise to a significant bias. In terms of the observations before grouping, (9.4.2,3) becomes

$$\frac{\text{Bias}\, a_{kN}}{\text{S.D.}\, a_{kN}} \doteqdot \frac{\sqrt{n}}{\sigma} N^{-2} \Sigma G_{jk} A'_j, \tag{1}$$

where σ is the standard deviation of an observation y_i and A'_j is written for $A_{jn}(\mathrm{Ra}\, x_{ni})^j$.

The values A'_j can be estimated from the five values of y_i spaced at intervals of one-quarter of the range of x_i. If these values are denoted by $y_{+\frac{1}{2}}, y_{+\frac{1}{4}}, y_0, y_{-\frac{1}{4}}, y_{-\frac{1}{2}}$, then

$$A'_2 \doteqdot 2(y_{+\frac{1}{2}} + y_{-\frac{1}{2}} - 2y_0), \tag{2a}$$

$$A'_3 \doteqdot \tfrac{16}{3}\{(y_{+\frac{1}{2}} - y_{-\frac{1}{2}}) - 2(y_{+\frac{1}{4}} - y_{-\frac{1}{4}})\}. \tag{2b}$$

These are actually the estimates for an equally-spaced curve passing through these five points, as derived in § 8.5.2. If the scatter of the observations is large an average value of y in the particular region should be taken. σ can be estimated from the scatter of the observations and κ_2 and κ_3 from (8.5.2,4*a*–*c*).

9.4.3 *Example*

Table 7.1.1 contains 67 observations. If it is desired to fit a cubic curve using grouped values, these observations can be grouped to give the 17 values shown in Table 9.4.3. The fitted coefficients can then be calculated using the Doolittle technique, the values obtained being listed in section (*b*) of the table. It is seen that the results agree very well with the values a^*_{jn} and b^*_{3j} obtained in Ch. 7 by fitting a curve to the original observations. The efficiencies are listed in section (*c*) of the table, and are compared with the values $1-N^{-2}g_j$ obtained from Table 8.8*c* for $\kappa_2^2 = 0{\cdot}3$, $\kappa_3 = 0{\cdot}6$.

TABLE 9.4.3

Grouping methods applied to the example of Table 7.1.1

(*a*) Grouped observations ($N = 17$)

x	y	x	y	x	y	x	y
+110·71	387	+40·43	282	−2·65	446	−29·88	622
+92·40	298	+25·88	228	−7·20	438	−35·59	706
+65·37	376	+18·25	272	−14·98	475	−46·32	891
+55·23	295	+8·98	314	−24·12	566	−57·07	1110
		+3·05	379				

(*b*) Coefficients a_j and b_{3j}

j	a^*_{jn}	$r^{j-1}\,a_{jN}$	b^*_{3jn}	$r^{j-1}\,b_{3jN}$
0			92·4 ± 4·4	93·0
1	−3·59 ± 0·25	−3·59	−5·57 ± 0·48	−5·60
2	+0·261 ± 0·020	+0·264	+0·407 ± 0·040	+0·401
3	−0·0074 ± 0·0018	−0·0072	$b_{33} = a_3$	

(*c*) Efficiencies of the estimates a_{jN}

j	ΣT^2_{jn}	ΣT^2_{jN}	$\Sigma T^2_{jN}/r^{2j-1}\,\Sigma T^2_{jn}$	$1-g_j/N^2$ (Table 8.8*c*)
1	9265	36861	0·9946	1 − 1·46/289 = 0·9950
2	143·66 × 10⁴	8922 × 10⁴	0·970	1 − 9·9 /289 = 0·966
3	1·913 × 10⁸	1780 × 10⁸	0·909	1 − 31·1 /289 = 0·892

The order of the bias in the estimates a_{jN} can be found from equation (9.4.2,3). The value of σ_N obtained from the residuals is 37. In § 8.5.2.1 it was found that $\kappa_2 = 0{\cdot}54$, $\kappa_3 = 0{\cdot}58$. From Table 9.8*d*, with $\kappa_2^2 = 0{\cdot}5$ and $\kappa_3 = 0{\cdot}5$,

$$G_{21} \sim 0{\cdot}09, \quad G_{22} \sim 0{\cdot}04, \quad G_{23} \sim 0{\cdot}00,$$

$$G_{31} \sim 0{\cdot}13, \quad G_{32} \sim 0{\cdot}07, \quad G_{33} \sim 0{\cdot}03.$$

The range of x_N is 168, and

$$a_{2N}\,(\mathrm{Ra}\,x_N)^2 = 1{\cdot}87 \times 10^3, \quad a_{3N}\,(\mathrm{Ra}\,x_N)^3 = -2{\cdot}14 \times 10^3.$$

Thus

$$\text{Bias } a_{1N}/\text{S.D. } a_{1N} = (0{\cdot}09 \times 1{\cdot}87 - 0{\cdot}13 \times 2{\cdot}14) \times 10^3/37 \times 17\sqrt{17} = -0{\cdot}04,$$

$$\text{Bias } a_{2N}/\text{S.D. } a_{2N} = (0{\cdot}04 \times 1{\cdot}87 - 0{\cdot}07 \times 2{\cdot}14) \times 10^3/37 \times 17\sqrt{17} = -0{\cdot}03,$$

$$\text{Bias } a_{3N}/\text{S.D. } a_{3N} = (-0{\cdot}03 \times 2{\cdot}14) \times 10^3/37 \times 17\sqrt{17} = -0{\cdot}02.$$

The bias in each coefficient is negligible.

The bias before grouping could have been checked by means of equation (9.4.2.1,1). Since the range of x is 45, rough values of y at $x = 30, 19, 8, -4, -15$ are needed. These values are

$$y_{+\frac{1}{2}} = 110, \quad y_{+\frac{1}{4}} = 80, \quad y_0 = 50, \quad y_{-\frac{1}{4}} = 120, \quad y_{-\frac{1}{2}} = 290,$$

and so from (9.4.2.1,2a, b)

$$A_2' = 2(110 + 290 - 2 \times 50) = +600,$$

$$A_3' = \tfrac{16}{3}(110 - 290 - 2 \times 80 + 2 \times 120) = -550.$$

Then, if σ is taken as 20,

$$\text{Bias } a_{1N}/\text{S.D. } a_{1N} = \sqrt{67}(0{\cdot}09 \times 600 - 0{\cdot}13 \times 550)/20 \times 289 = -0{\cdot}02,$$

which is of the same order as the value obtained from the fitted coefficients a_{jN}.

9.4.4 *Grouped observations of different weight*

Often it is not possible to find a suitable pair of values such that $n = Nr$. More usually,

$$n = Nr + \nu, \tag{1}$$

where ν is small. The ν additional observations should be included in ν groups near the centre of the range of x, and in forming these groups the sums of the x and y values must be multiplied by the factor $r/(r+1)$ to bring them to the same scale as the other groups. If ν is negative the factor is $r/(r-1)$. For example, the y values in the central group of Table 7.1.1 are 99, 96, 89, and their sum 284 is multiplied by 4/3 to give the value 379 in Table 9.4.3.

These ν groups should strictly be given a weight $(r \pm 1)/r$ in forming the moments and the sums of the powers. This weighting greatly increases the time required to evaluate these quantities, and since the reduction in efficiency due to the omission of these weights is negligible, it is recommended that the grouped observations be treated as if they were all of equal weight.

If the original observations have different weights w_i, Σw_i replaces n in (1), each observation of weight w being regarded as equivalent to w observations of unit weight. The observations are divided into N groups each having the same value of Σw_i.

9.5 STEP FUNCTION METHODS FOR THE UNEQUALLY-SPACED CASE

If the symbols $W_j(x)$ represent a set of $p+1$ functions, then

$$E\left\{\sum_i W_j(x_i)\,y_i\right\} = \sum_i W_j(x_i) \sum_k B_{pk}\,x_i^k, \tag{1a}$$

where the B_{pk} are the coefficients of the 'true' curve. In matrix notation this equation is

$$E\{\mathbf{Wy}\} = \mathbf{WX}^T\mathbf{B}, \tag{1b}$$

where $\mathbf{W}$ is the $(p+1)\times n$ matrix whose elements are $W_j(x_i)$ and $\mathbf{X}$ the $(p+1)\times n$ matrix whose elements are x_i^j. Then

$$E\{(\mathbf{WX}^T)^{-1}\,\mathbf{Wy}\} = \mathbf{B},$$

and so the quantities b_{pk} satisfying

$$(\mathbf{WX}^T)^{-1}\,\mathbf{Wy} = \mathbf{b} \tag{2a}$$

will be unbiased estimates of B_{pk}. Equation (2a) is equivalent to

$$\mathbf{Wy} = \mathbf{WX}^T\mathbf{b}, \tag{2b}$$

which when written out is

$$\sum_k b_{pk} \sum_i W_j(x_i)\,x_i^k = \sum_i W_j(x_i)\,y_i. \tag{2c}$$

These equations correspond to the normal equations in the least-squares case, and their solutions will provide unbiased estimates of the power-series coefficients.

The functions $W_j(x_i)$ to be considered in this section will be step functions. There are very many types of step function which could be used, but only the simplest ones will be discussed here. Attention will be confined to functions $W_j(x)$ which are independent of the degree p of the polynomial, and $W_p(x)$ will be chosen to maximize the efficiency of b_{pp}. Only single-step functions of the simplest type will be considered.

The form of the optimum function $W_p(x)$ will clearly depend on the distribution of the values x_i. However, the determination of the optimum function for the particular set of values in each example would complicate the method intolerably. The optimum form will be determined for the equally-spaced case, and the same form will be used in the general case where the spacing is non-uniform.

9.5.1 *The second-degree polynomial*

The function $W_0(x)$ will be taken equal to unity everywhere. For the discussion of the equally-spaced case the function $W_2(\epsilon)$ will be chosen, by analogy with the orthogonal polynomial $T_2(\epsilon)$, so that for any arbitrary function $f(\epsilon)$,

$$\sum_i W_2(\epsilon_i) f(\epsilon_i)$$

is of the form

$$\left[q\left\{\sum_{-\frac{1}{2}(n-1)}^{\frac{1}{2}(n-1)} - \sum_{-\frac{1}{2}(a-1)}^{\frac{1}{2}(a-1)}\right\} - r\left\{\sum_{-\frac{1}{2}(b-1)}^{\frac{1}{2}(b-1)}\right\}\right] f(\epsilon). \tag{1}$$

The numbers $\frac{1}{2}(a-1)$ and $\frac{1}{2}(b-1)$ are the values of ϵ at the end of each step. The parameters $\alpha = a/n$ and $\beta = b/n$ will be introduced, and the approximations

$$\sum_{-\frac{1}{2}(a-1)}^{\frac{1}{2}(a-1)} \epsilon^0 = a = \alpha n, \qquad \sum_{-\frac{1}{2}(a-1)}^{\frac{1}{2}(a-1)} \epsilon^2 = \tfrac{1}{12} a(a^2-1) \doteqdot \tfrac{1}{12}\alpha^3 n^3,$$

$$2\sum_{0,\frac{1}{2}}^{\frac{1}{2}(a-1)} \epsilon = \tfrac{1}{4}(a^2-1) \doteqdot \tfrac{1}{4}\alpha^2 n^2, \qquad 2\sum_{0,\frac{1}{2}}^{\frac{1}{2}(a-1)} \epsilon^3 = \tfrac{1}{32}(a^2-1)^2 \doteqdot \tfrac{1}{32}\alpha^4 n^4,$$

will be used. Thus

$$\Sigma W_2 \epsilon^2 = \tfrac{1}{12} n^3\{q(1-\alpha^3) - r\beta^3\}, \tag{2a}$$

$$\Sigma W_2 = n\{q(1-\alpha) - r\beta\}, \tag{2b}$$

$$\Sigma W_2^2 = n\{q^2(1-\alpha) + r^2\beta\}, \tag{2c}$$

while

$$\Sigma W_0 \epsilon^2 = \tfrac{1}{12} n^3, \tag{3a}$$

$$\Sigma W_0 = n = \Sigma W_0^2. \tag{3b}$$

W_0 and W_2 are even functions of ϵ, while W_1 as defined in §§ 6.4.3 and 6.4.1.1 is an odd function of ϵ. Hence the 'normal' equations (9.5,2c) for even values of k are

$$b_{20}\Sigma W_0 + b_{22}\Sigma W_0\epsilon^2 = \Sigma W_0 y, \tag{4a}$$

$$b_{20}\Sigma W_2 + b_{22}\Sigma W_2\epsilon^2 = \Sigma W_2 y. \tag{4b}$$

These equations will give, on substituting from (2a–c) and (3a, b), an expression for b_{22} in terms of the parameters q, r, α, β. The efficiency of the estimate b_{22} may be evaluated by comparing its variance with the value $n^5/180$ for the least-squares estimate. It is found that

$$\eta(b_{22}) = \frac{5}{4}\,\frac{[\{q(1-\alpha^3) - r\beta^3\} - \{q(1-\alpha) - r\beta\}]^2}{\{q^2(1-\alpha) + r^2\beta\} - \{q(1-\alpha) - r\beta\}^2}. \tag{5}$$

The values of the parameters which make this expression a maximum are

$$\alpha = 0{\cdot}776, \quad \beta = 0{\cdot}497, \quad q = 1{\cdot}143, \quad r = 0{\cdot}988,$$

the efficiency being then 0·903. In practice, it is more convenient to use the values $q = r = 1$, and this reduces the efficiency to 0·902. The groups, then, will be given equal weights, and there will be

$$\tfrac{1}{2}(n-1)-\tfrac{1}{2}(a-1) = \tfrac{1}{2}n(1-\alpha) = 0{\cdot}11n$$

observations in the extreme groups and $0{\cdot}50n$ observations in the central group.

9.5.2 *The third-degree polynomial*

The function $W_3(\epsilon)$ will be chosen, by analogy with $T_3(\epsilon)$, so that for any arbitrary function

$$\Sigma W_3(\epsilon)f(\epsilon) = \left[q\left(\sum_{0,\frac{1}{2}}^{\frac{1}{2}(n-1)} - \sum_{0,\frac{1}{2}}^{\frac{1}{2}(a-1)}\right) - r\left(\sum_{0,\frac{1}{2}}^{\frac{1}{2}(b_1-1)} - \sum_{0,\frac{1}{2}}^{\frac{1}{2}(b_2-1)}\right)\right][f(\epsilon)-f(-\epsilon)]. \tag{1}$$

As in the previous section, parameters $\alpha = a/n$, $\beta_1 = b_1/n$, $\beta_2 = b_2/n$ are introduced, and these are adjusted to maximize $\eta(b_{33})$. The values obtained are

$$\alpha = 0{\cdot}859, \quad \beta_1 = 0{\cdot}697, \quad \beta_2 = 0{\cdot}117, \quad q = 0{\cdot}297, \quad r = 0{\cdot}153,$$

the resulting efficiency being 0·902. It is more convenient to take $q = 2$, $r = 1$, the reduction in efficiency by reason of this choice being negligible. Of course, only the ratio of q to r and not the individual magnitudes is significant. The adopted step functions will then contain $0{\cdot}07n$ observations in the outside groups and $0{\cdot}29n$ observations in the central groups, the two groups being weighted in the ratio 2 : 1. $0{\cdot}08n$ observations between each extreme and central group, and the $0{\cdot}12n$ central observations, are omitted.

9.5.3 *Tables of step functions*

To specify the step functions for a given value of n, the observations are supposed numbered in decreasing order of x from 1 to n. The symbol Σn_{jk} represents the sum (of x_i^k or y_i) for all observations from 1 to n_{jk}. The number $n-n_{jk}$ is denoted by the symbol n'_{jk}. Table 9.5.3 gives the recommended step functions in terms of the n_{jk}. The n_{jk} are given as functions of n in the lower half of the table. The integers nearest those shown in the table would be used in a practical example.

Table 9.8*e* gives directly the step functions for each value of n between 10 and 100, without the necessity of calculating n_{jk} from Table 9.5.3.

TABLE 9.5.3

Recommended step functions for the unequally-spaced case

Zero degree : Σn;
First degree : $\Sigma n_{11} + \Sigma n'_{11} - \Sigma n$;
Second degree : $\Sigma n_{21} + \Sigma n_{22} - \Sigma n'_{22} - \Sigma n'_{21} + \Sigma n$;
Third degree : $2\Sigma n_{31} + \Sigma n_{32} - \Sigma n_{33} - \Sigma n'_{33} + \Sigma n'_{32} + 2\Sigma n'_{31} - 2\Sigma n$.

$n_{11} : 0{\cdot}33n$; $n_{21} : 0{\cdot}11n$; $n_{22} : 0{\cdot}25n$;
$n_{31} : 0{\cdot}07n$; $n_{32} : 0{\cdot}15n$; $n_{33} : 0{\cdot}44n$;
$n'_{jk} : n - n_{jk}$.

9.5.4 *Example*

The method of obtaining the 'normal' equations is shown in Table 9.5.4 for the observations listed in Table 7.1.1.

The sums to be calculated are first entered from Table 9.8*e*. A separate table of values x^j, y is formed, the observations being listed in decreasing order of x and suitable powers of 10 removed to bring the values to the

TABLE 9.5.4

Solution of Example 7.1.1 *by use of step functions*

Degree

Zero : $\Sigma(67)$;
First : $\Sigma(22) + \Sigma(45) - \Sigma(67)$;
Second : $\Sigma(7) + \Sigma(16) - \Sigma(51) - \Sigma(60) + \Sigma(67)$;
Third : $2\Sigma(5) + \Sigma(10) - \Sigma(30) - \Sigma(37) + \Sigma(57) + 2\Sigma(62) - 2\Sigma(67)$.

Factors removed : x, 10^q, $q = 1$; y, 10^r, $r = 2$.

(*a*) Partial sums

	Σx^0	Σx	Σx^2	Σx^3	Σy	Check
	22	37·823	75·755047	169·357190	17·45	322·385237
	45	40·341	77·912973	170·132765	38·47	371·856738
	7	18·370	48·513126	128·922157	6·19	208·995283
	16	32·371	70·626595	164·386125	13·56	296·943720
	51	37·059	79·771697	169·048687	46·61	383·489384
	60	29·407	86·410847	163·162986	61·96	400·940833
	5	13·627	37·257289	102·191885	4·12	162·196174
	10	23·694	58·004464	145·921859	8·93	246·550323
	30	41·404	77·466790	170·218818	22·89	341·979608
	37	41·924	77·545716	170·230401	29·61	356·310117
	57	32·377	83·442129	166·157945	56·11	395·087074
	62	27·139	88·983209	160·244907	66·26	404·627116
	67	20·173	98·726409	146·564412	79·90	412·363821

(*b*) 'Normal' equations

Zero	67	20·173	98·726409	146·564412	79·90	412·363821
First	0	57·991	54·941611	192·925543	−23·98	281·878154
Second	−21	4·448	51·683586	107·661021	−8·92	133·872607
Third	0	13·929	41·462265	203·375345	−6·50	252·266010

order of unity. Lines are drawn across the table to indicate the partial sums required. The columns are summed and the partial sums recorded in part (*a*) of Table 9.5.4. These sums can be checked by summing across the rows, allowing the entries to accumulate till a sub-total to be checked is reached.

The 'normal' equations are then formed by combining these partial sums. For example, the first-degree entry for Σx is

$$(22)+(45)-(67) = 37{\cdot}823+40{\cdot}341-20{\cdot}173 = 57{\cdot}991.$$

The entries for each degree are checked by comparing their sum with the check column entry.

9.5.5 *Solution of non-symmetric equations*

The normal equations obtained in the step function method are non-symmetric, and so they cannot be solved by the Doolittle technique. However, the method of single division can still be used. Probably the safest scheme is that given in Table 7.1.1, with the addition of a check column.

TABLE 9.5.5

Compact method of single division

ϕ_{00}	ϕ_{01}	ϕ_{02}	M_0	C_0
ϕ_{10}	ϕ_{11}	ϕ_{12}	M_1	C_1
ϕ_{20}	ϕ_{21}	ϕ_{22}	M_2	C_2
C'_0	C'_1	C'_2		
$S_{00}/1$	α_{01}	α_{02}	a_0	c_0
S_{10}	$S_{11}/1$	α_{12}	a_1	c_1
S_{20}	S_{21}	$S_{22}/1$	a_2	c_2
c'_0	c'_1	c'_2		
b_{20}	b_{21}	b_{22}		
d_{20}	d_{21}	d_{22}		

For more experienced computors, an abbreviated scheme is available, based on (7.1.2,6*a*–*c*,7*a*,*b*,8). This scheme, called by Dwyer the compact method of single division, is shown in Table 9.5.5. The first column of the lower matrix, identical with the first column of the upper matrix, is written down. The first row of the lower matrix is then calculated by division of the first row of the upper matrix by S_{00}. The second column is then calculated, using (7.1.2,6*a*), and the second row, using (7.1.2,6*b*) and (7.1.2,6*c*). The remaining columns and rows are then calculated in order. The values c'_j and c_j check the formation of the columns and rows respectively.

In calculating the element in row j and column k, the products of corresponding elements in row j and column k of the lower matrix are subtracted from ϕ_{jk}, and divided by S_{jj} if $j<k$. Table

9.5.5.1 illustrates the computation for a_2. The appropriate elements can be selected by a pair of cards or a right-angled template placed as shown. After some practice the correct elements will be selected automatically.

TABLE 9.5.5.1

Selection of elements in the compact method of single division

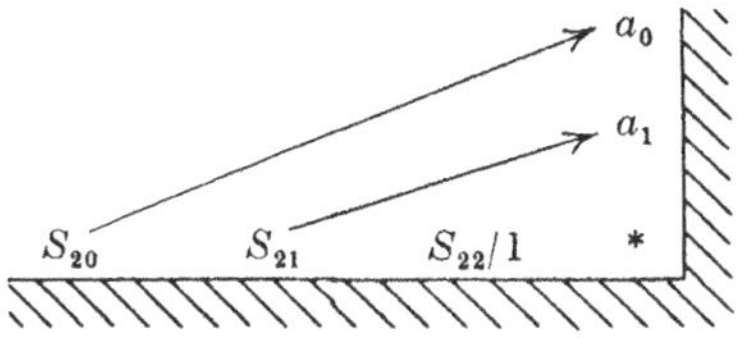

$$a_2 = (M_2 - S_{20}a_0 - S_{21}a_1)/S_{22}$$

9.5.5.1 *Example*

Table 9.5.5.2 shows the solution by the compact method of single division of the 'normal' equations obtained in Example 9.5.4.

TABLE 9.5.5.2

Solution of 'normal' equations obtained by step function methods
Factors removed: x, 10^q, $q = 1$: y, 10^r, $r = 2$;
elements divided by 10^s, $s = 2$

0·670000	0·201730	0·987264	1·465644	0·799000	4·123638
0	0·579910	0·549416	1·929256	−0·239800	2·818782
−0·210000	0·044480	0·516836	1·076610	−0·089200	1·338726
0	0·139290	0·414623	2·033754	−0·065000	2·522667
0·460000	0·965410	2·468139	6·505264		
0·670000 | 1	0·301090	1·473528	2·187528	1·192537	6·154684
0	0·579910 | 1	0·947416	3·326820	−0·413512	4·860723
−0·210000	0·107709	0·724232 | 1	1·626084	+0·284124	2·910208
0	0·139290	0·282657	1·110737 | 1	−0·078967	0·921034
0·460000	0·826909	1·006889	1·110737		
0·920486	−0·541641	0·412531	−0·078967		
1·920489	0·458356	1·412529	0·921034		

The fitted curve is

$$u_3(x) = 92{\cdot}049 - 5{\cdot}4164x + 0{\cdot}41253x^2 - 0{\cdot}0078967x^3.$$

9.5.6 *Efficiencies for non-uniform spacing*

The step functions are intended for cases where the spacing is non-uniform, but the steps were in fact chosen to give maximum efficiency for uniform spacing. When the spacing is non-uniform the efficiencies are usually lower. The standard deviations can be evaluated by replacing $\Sigma W_j x^k$ by $\Sigma W_j X^k$, where X is the smoothed variable of (8.5.1,2a) specified by the two parameters κ_2 and κ_3. The efficiencies can then be obtained in terms of these two parameters. Table 9.5.6 shows the efficiencies for b_{22} and b_{33}.

TABLE 9.5.6

Efficiencies of the coefficients b_{pp}

κ_3 \ κ_2^2	b_{22} 0	0·5	1·0	b_{33} 0	0·25	0·5
−1·0	0·853	0·751	0·660	0·774	0·612	0·477
−0·5	0·892	0·789	0·692	0·872	0·723	0·590
0	0·902	0·804	0·707	0·902	0·778	0·662
+0·5	0·881	0·796	0·702	0·868	0·768	0·676
+1·0	0·833	0·760	0·673	0·778	0·698	0·627

The effect of departures from uniform spacing is summarized in Table 9.5.6.1. Since the efficiency of the fitted value is at worst only slightly less than the efficiency of b_{pp}, this table will also give the limiting efficiencies for the fitted values.

The loss in efficiency will not usually be important for first- and second-degree polynomials, but it may be serious for polynomials of the third degree. However, the efficiency may be too severe a criterion for judging the value of grouping methods. Perhaps a more suitable criterion would be the increase in range of the observations necessary to offset the drop in efficiency. Calculations based on Table 9.5.6.1 show that a 10% increase in range is required when the departure from uniformity is pronounced. It is clear that it might often be more convenient to take a few extra readings and use the simpler method of calculation employing step functions than to perform the full least-squares computation for the original observations.

For the example of §§ 9.5.4 and 9.5.5.1, the efficiency of the estimate b_{33} can be shown to be 0·724. This is in agreement with the value 0·768 given in Table 9.5.6 for the case $\kappa_2^2 = 0{\cdot}25$, $\kappa_3 = 0{\cdot}5$.

TABLE 9.5.6.1

Effect of departures from uniform spacing

Departure from uniformity		Efficiency b_{11}	b_{22}	b_{33}
Slight	$\|\kappa_2\|, \|\kappa_3\| < 0{\cdot}25$	$>0{\cdot}875$	$>0{\cdot}875$	$>0{\cdot}870$
Moderate	$\|\kappa_2\|, \|\kappa_3\| < 0{\cdot}50$	$>0{\cdot}850$	$>0{\cdot}840$	$>0{\cdot}720$
Pronounced	$\|\kappa_2\|, \|\kappa_3\| < 0{\cdot}75$	$>0{\cdot}800$	$>0{\cdot}750$	$>0{\cdot}520$

9.6 GENERAL SUMMARY FOR THE UNEQUALLY-SPACED CASE

Table 9.6 gives the times required to fit polynomials of the first, second, and third degrees to the example used in this chapter. The efficiencies and the estimates $a_p = b_{pp}$ are also given. It will be seen that for the first-degree curve the step function method is by far the most rapid. For the second-degree curve the step function and least-squares grouped methods require about the same time, while for the third-degree curve the least-squares grouped method is the most rapid.

TABLE 9.6

Comparison of different methods of calculating the polynomials

	First degree		Second degree			Third degree		
	Time (min.)	Effic. $\eta(a_1)$	Time (min.)	Effic. $\eta(a_2)$	a_2 ($\times 10^2$)	Time (min.)	Effic. $\eta(a_3)$	a_3 ($\times 10^3$)
Least-squares	58	1·000	158	1.000	$26{\cdot}1 \pm 2{\cdot}3$	246	1·000	$-7{\cdot}4 \pm 1{\cdot}8$
Least-squares grouped	29	0·936	61	0·925	25·8	100	0·909	−7·2
Single-step functions	16	0·825	60	0·831	28·4	128	0·725	−7·9
Double-step functions	17	0·898	67	0·835	28·9	145	0·763	−7·1

Each method has its advantages and disadvantages, and each will be useful in some examples and not in others. It is probably true to say that the least-squares grouped method is the most satisfactory in cases where it can be used. For the bias to be small with this method the second- and third-degree coefficients must not be very large. When these coefficients are large, cases may arise in which it is convenient to remove the greater part of their contribution to y_i by subtracting $b_2 x_i^2 + b_3 x_i^3$ from y_i before grouping, b_2 and b_3 being approximate values of the coefficients.

9.7 NOTES AND REFERENCES

(9.1) The least-squares grouping method was discussed in a paper by Guest (1954).

(9.4) This treatment is based on a paper by Guest (1956).

(9.5) A less efficient method of grouping is described by Nair and Shrivastava (1942).

9.8 TABLES

TABLE 9.8*a*

Formulae connecting the coefficients a_{jn} and $\mathscr{A}_{jN}$

$$
\begin{aligned}
a_{5n} &= r^{-6}\,\mathscr{A}_{5N}\\
a_{4n} &= r^{-5}\,\mathscr{A}_{4N}\\
a_{3n} &= r^{-4}\{\mathscr{A}_{3N} + \tfrac{10}{9}(1 - r^{-2})\,\mathscr{A}_{5N}\}\\
a_{2n} &= r^{-3}\{\mathscr{A}_{2N} + \tfrac{3}{7}(1 - r^{-2})\,\mathscr{A}_{4N}\}\\
a_{1n} &= r^{-2}\{\mathscr{A}_{1N} + \tfrac{1}{10}(1 - r^{-2})\,\mathscr{A}_{3N} + \tfrac{1}{126}(1 - r^{-2})\,(N^2 + 8 - 6r^{-2})\,\mathscr{A}_{5N}\}\\
a_{0n} &= r^{-1}\,\mathscr{A}_{0N}
\end{aligned}
$$

TABLE 9.8*b*

Values of γ_{13} *and* γ_{24} (9.1,6)

r	γ_{13}	γ_{24}	r	γ_{13}	γ_{24}
2	3/40	9/28	9	8/81	80/189
3	4/45	8/21	10	99/1000	297/700
4	3/32	45/112	11	12/121	360/847
5	12/125	72/175	12	143/1440	143/336
6	7/72	5/12	13	84/845	72/169
7	24/245	144/343	14	39/392	585/1372
8	63/640	27/64	15	112/1125	32/75

TABLE 9.8*c*

Single-step functions for the equally-spaced case

	Single-step functions					
	Zero degree $b_{00} = \sum_{+} y/n$		First degree $b_{11} = \sum_{-} W_1 y \big/ \sum_{-} W_1 \epsilon$			
n	β_{02}	β_{04}	W_1	$\sum_{-} W_1 \epsilon$	β_{13}	β_{15}
7	−4	10·56	Σ3 − Σ1	10	−7	26·6666667
8	−5·25	21·8352273	Σ4 − Σ1	15	−8·25	46·5625
9	−6·6r	32·3478261	Σ4 − Σ1	18	−11	86·6666667
10	−8·25	56·0782895	Σ5 − Σ2	21	−14·25	152·0625
11	−10	56·4324324	Σ5 − Σ1	28	−16	171·578947
12	−11·916r	92·4430970	Σ6 − Σ2	32	−19·75	277·198864
13	−14	118·484211	Σ6 − Σ2	36	−24	400·307692
14	−16·25	178·9125	Σ7 − Σ2	45	−26·25	555·757415
15	−18·6r	259·2	Σ7 − Σ2	50	−31	761·176471
16	−21·25	363·085227	Σ8 − Σ3	55	−36·25	1054·98355
17	−24	440·228571	Σ8 − Σ2	66	−39	1200·25
18	−26·916r	591·596398	Σ9 − Σ3	72	−44·75	1550·67361
19	−30	596·689655	Σ9 − Σ3	78	−51	2053·08475
20	−33·25	700·141810	Σ10 − Σ3	91	−54·25	2471·49761
21	−36·6r	915·2	Σ10 − Σ3	98	−61	2677·97849
22	−40·25	1174·97540	Σ11 − Σ4	105	−68·25	3303·74133
23	−44	1485	Σ11 − Σ3	120	−72	4017·72973
24	−47·916r	1699·51705	Σ12 − Σ4	128	−79·75	4867·71991
25	−52	2103·90448	Σ12 − Σ4	136	−88	5462
26	−56·25	2575·15754	Σ13 − Σ4	153	−92·25	6518·0625
27	−60·6r	2593·69038	Σ13 − Σ4	162	−101	7717·36937
28	−65·25	2902·9725	Σ14 − Σ5	171	−110·25	9321·53708
29	−70	3505·46341	Σ14 − Σ4	190	−115	10591·5486
30	−74·916r	4194·79821	Σ15 − Σ5	200	−124·75	12598·6870
31	−80	4653·87611	Σ15 − Σ5	210	−135	13542·5806
32	−85·25	5500·39123	Σ16 − Σ5	231	−140·25	15553·9937
33	−90·6r	6455·21590	Σ16 − Σ5	242	−151	17882·4828
34	−96·25	7105·3125	Σ17 − Σ6	253	−162·25	20818·9764
35	−102	7099·2	Σ17 − Σ5	276	−168	23577

TABLE 9.8c (cont.)

	Single-step functions					
	Zero degree $b_{00} = \sum_{+} y/n$		First degree $b_{11} = \sum_{-} W_1 y / \sum_{-} W_1 \epsilon$			
n	β_{02}	β_{04}	W_1	$\sum_{-} W_1 \epsilon$	β_{13}	β_{15}
36	−107·916r	8260·38603	Σ18 − Σ6	288	−179·75	24189·5306
37	−114	9554·95385	Σ18 − Σ6	300	−192	27885·1777
38	−120·25	10396·6582	Σ19 − Σ6	325	−198·25	28898·9048
39	−126·6r	11927·1610	Σ19 − Σ6	338	−211	33115·0057
40	−133·25	13617·3606	Σ20 − Σ7	351	−224·25	37136·3294
41	−140	14739·4967	Σ20 − Σ6	378	−231	41372·4706
42	−146·916r	14734·2725	Σ21 − Σ7	392	−244·75	46135·2764
43	−154	16722·8852	Σ21 − Σ7	406	−259	52035·7419
44	−161·25	18901·3609	Σ22 − Σ7	435	−266·25	57467·5970
45	−168·6r	20294·6356	Σ22 − Σ7	450	−281	60364·875
46	−176·25	22803·7383	Σ23 − Σ8	465	−296·25	67631·1107
47	−184	25533·8680	Σ23 − Σ7	496	−304	73189·0720
48	−191·916r	28497·1290	Σ24 − Σ8	512	−319·75	81530·2345
49	−200	30428·0980	Σ24 − Σ8	528	−336	89405·952
50	−208·25	30449·5231	Σ25 − Σ8	561	−344·25	93132·9528
51	−216·6r	33843·3333	Σ25 − Σ8	578	−361	96082·0856
52	−225·25	37506·4392	Σ26 − Σ9	595	−378·25	104845·616
53	−234	39823·0675	Σ26 − Σ8	630	−387	114079·560
54	−242·916r	43950·4346	Σ27 − Σ9	648	−404·75	124075·662
55	−252	48384	Σ27 − Σ9	666	−423	136438·061
56	−261·25	51235·5313	Σ28 − Σ9	703	−432·25	145780·973
57	−270·6r	51270·2815	Σ28 − Σ9	722	−451	159660
58	−280·25	56263·6052	Σ29 − Σ10	741	−470·25	167854·312
59	−290	61604·3855	Σ29 − Σ9	780	−480	178698·824
60	−299·916r	64951·5129	Σ30 − Σ10	800	−499·75	194952·761
61	−310	70886·1768	Σ30 − Σ10	820	−520	209861·796
62	−320·25	77210·0490	Σ31 − Σ10	861	−530·25	225148·283
63	−330·6r	81235·9534	Σ31 − Σ10	882	−551	241793·301
64	−341·25	88224·3429	Σ32 − Σ11	903	−572·25	252899·062
65	−352	88318·3304	Σ32 − Σ10	946	−583	267774·222
66	−362·916r	95783·6593	Σ33 − Σ11	968	−604·75	274119·002
67	−374	100428·859	Σ33 − Σ11	990	−627	296485·181
68	−385·25	108633·631	Σ34 − Σ11	1035	−638·25	313032·982
69	−396·6r	117320·565	Σ34 − Σ11	1058	−661	337636·583
70	−408·25	122805·531	Σ35 − Σ12	1081	−684·25	360013·223
71	−420	132310·504	Σ35 − Σ11	1128	−696	370623·083
72	−431·916r	132437·083	Σ36 − Σ12	1152	−719·75	394725·353
73	−444	142526·586	Σ36 − Σ12	1176	−744	423895·004
74	−456·25	148768·310	Σ37 − Σ12	1225	−756·25	445713·600
75	−468·6r	159758·753	Σ37 − Σ12	1250	−781	477551·503

TABLE 9.8*c* (cont.)

	Single-step functions				
	Second degree			$b_{22} = \sum_{+} W_2 y \Big/ \sum_{+} W_2 \epsilon^2$	
n	W_2			$\sum_{+} W_2 \epsilon^2$	β_{24}
7	$3(\Sigma 3-\Sigma 2)$	−	$2\Sigma 1$	50	−9·64
8	$2(\Sigma 4-\Sigma 3)$	−	$\Sigma 2$	44	−13·40909091
9	$3(\Sigma 4-\Sigma 3)$	−	$2\Sigma 1$	92	−16·65217391
10	$2(\Sigma 5-\Sigma 4)$	−	$\Sigma 2$	76	−21·44736842
11	$5(\Sigma 5-\Sigma 3)$	−	$4\Sigma 2$	370	−23·44324324
12	$3(\Sigma 6-\Sigma 4)$	−	$2\Sigma 3$	268	−29·00746269
13	$5(\Sigma 6-\Sigma 4)$	−	$4\Sigma 2$	570	−33·46315789
14	$3(\Sigma 7-\Sigma 5)$	−	$2\Sigma 3$	400	−40·06
15	$7(\Sigma 7-\Sigma 5)$	−	$4\Sigma 3$	1078	−47·28571429
16	$2(\Sigma 8-\Sigma 6)$	−	$\Sigma 4$	352	−55·13636364
17	$7(\Sigma 8-\Sigma 6)$	−	$4\Sigma 3$	1470	−61·34285714
18	$2(\Sigma 9-\Sigma 7)$	−	$\Sigma 4$	472	−70·22881356
19	$3(\Sigma 9-\Sigma 6)$	−	$2\Sigma 4$	1044	−73·68965517
20	$4(\Sigma 10-\Sigma 7)$	−	$3\Sigma 4$	1624	−80·70689655
21	$3(\Sigma 10-\Sigma 7)$	−	$2\Sigma 4$	1350	−90·76
22	$5(\Sigma 11-\Sigma 8)$	−	$3\Sigma 5$	2480	−101·4419355
23	$11(\Sigma 11-\Sigma 8)$	−	$6\Sigma 5$	5984	−112·75
24	$5(\Sigma 12-\Sigma 9)$	−	$3\Sigma 5$	3080	−121·5181818
25	$11(\Sigma 12-\Sigma 9)$	−	$6\Sigma 5$	7370	−133·8597015
26	$2(\Sigma 13-\Sigma 10)$	−	$\Sigma 6$	1452	−146·8305785
27	$13(\Sigma 13-\Sigma 9)$	−	$8\Sigma 6$	12428	−151·7531381
28	$3(\Sigma 14-\Sigma 10)$	−	$2\Sigma 6$	3200	−161·74
29	$13(\Sigma 14-\Sigma 10)$	−	$8\Sigma 6$	14924	−175·8780488
30	$7(\Sigma 15-\Sigma 11)$	−	$4\Sigma 7$	8624	−190·6428571
31	$13(\Sigma 15-\Sigma 11)$	−	$8\Sigma 6$	17628	−201·9734513
32	$7(\Sigma 16-\Sigma 12)$	−	$4\Sigma 7$	10136	−217·7707182
33	$15(\Sigma 16-\Sigma 12)$	−	$8\Sigma 7$	23140	−234·1972342
34	$7(\Sigma 17-\Sigma 13)$	−	$4\Sigma 7$	11760	−246·8714286
35	$3(\Sigma 17-\Sigma 12)$	−	$2\Sigma 7$	6250	−253
36	$8(\Sigma 18-\Sigma 13)$	−	$5\Sigma 8$	17680	−270·5941176
37	$17(\Sigma 18-\Sigma 13)$	−	$10\Sigma 8$	39780	−288·8153846
38	$8(\Sigma 19-\Sigma 14)$	−	$5\Sigma 8$	20240	−302·7086957
39	$17(\Sigma 19-\Sigma 14)$	−	$10\Sigma 8$	45390	−321·9617978
40	$9(\Sigma 20-\Sigma 15)$	−	$5\Sigma 9$	25320	−341·8440758

TABLE 9.8c (cont.)

	Single-step functions		
	Second degree	$b_{22} = \sum_{+} W_2 y \Big/ \sum_{+} W_2 \epsilon^2$	
n	W_2	$\sum_{+} W_2 \epsilon^2$	β_{24}
41	17(Σ20 – Σ15) – 10Σ8	51340	–357·0821192
42	3(Σ21 – Σ15) – 2Σ9	10800	–364·54
43	19(Σ21 – Σ15) – 12Σ9	71858	–385·5901639
44	5(Σ22 – Σ16) – 3Σ10	19840	–407·2677419
45	19(Σ22 – Σ16) – 12Σ9	80522	–423·7239264
46	5(Σ23 – Σ17) – 3Σ10	22180	–446·4329125
47	7(Σ23 – Σ17) – 4Σ10	32466	–469·7710220
48	11(Σ24 – Σ18) – 6Σ11	53284	–493·7369942
49	7(Σ24 – Σ18) – 4Σ10	35994	–511·9404901
50	11(Σ25 – Σ18) – 7Σ11	65604	–520·8661972
51	23(Σ25 – Σ18) – 14Σ11	142968	–546
52	12(Σ26 – Σ19) – 7Σ12	77672	–571·7602740
53	23(Σ26 – Σ19) – 14Σ11	157458	–591·1840491
54	12(Σ27 – Σ20) – 7Σ12	85400	–617·9780328
55	25(Σ27 – Σ20) – 14Σ12	184800	–645·4
56	12(Σ28 – Σ21) – 7Σ12	93464	–666·1668664
57	25(Σ28 – Σ20) – 16Σ12	221400	–676·4222222
58	13(Σ29 – Σ21) – 8Σ13	119392	–705·0121951
59	27(Σ29 – Σ21) – 16Σ13	256968	–734·2289157
60	13(Σ30 – Σ22) – 8Σ13	130000	–756·2152
61	27(Σ30 – Σ22) – 16Σ13	279432	–786·4650863
62	7(Σ31 – Σ23) – 4Σ14	74928	–817·3430493
63	27(Σ31 – Σ23) – 16Σ13	302760	–840·6732461
64	7(Σ32 – Σ24) – 4Σ14	81088	–872·5828729
65	29(Σ32 – Σ23) – 18Σ14	376188	–884·3043478
66	5(Σ33 – Σ24) – 3Σ15	66960	–916·9774194
67	29(Σ33 – Σ24) – 18Σ14	405942	–941·5263609
68	5(Σ34 – Σ25) – 3Σ15	72180	–975·2321696
69	31(Σ34 – Σ25) – 18Σ15	461280	–1009·566129
70	5(Σ35 – Σ26) – 3Σ15	77580	–1035·459629
71	31(Σ35 – Σ26) – 18Σ15	495318	–1070·825009
72	8(Σ36 – Σ26) – 5Σ16	141440	–1083·876471
73	33(Σ36 – Σ26) – 20Σ16	600490	–1120·005825
74	8(Σ37 – Σ27) – 5Σ16	151520	–1147·117529
75	33(Σ37 – Σ27) – 20Σ16	642730	–1184·279274

TABLE 9.8c (cont.)

	Single-step functions			
	Third degree		$b_{33} = \sum W_3 y / \sum W_3 \epsilon^3$	
n	W_3		$\sum W_3 \epsilon^3$	β_{35}
7	Σ3	– 2Σ2	36	–11·66666667
8	9Σ4	– 16Σ3	504	–15·83333333
9	3Σ4	– 5Σ3	240	–21
10	5(Σ5–Σ4)	– 3(Σ4–Σ1)	540	–27·16666667
11	6(Σ5–Σ4)	– 5Σ3	1140	–30·47368421
12	15(Σ6–Σ5)	– 11(Σ4–Σ1)	3630	–37·77272727
13	5(Σ6–Σ5)	– 3Σ4	1560	–44·84615385
14	24(Σ7–Σ6)	– 13(Σ5–Σ1)	9204	–53·51694915
15	15(Σ7–Σ6)	– 7Σ5	7140	–61·94117647
16	7(Σ8–Σ7)	– 3(Σ6–Σ1)	3990	–71·97368421
17	7(Σ8–Σ7)	– 4(Σ5–Σ1)	5376	–78·75
18	35(Σ9–Σ8)	– 17(Σ6–Σ1)	32130	–88·72222222
19	20(Σ9–Σ8)	– 9(Σ6–Σ1)	21240	–100·8644068
20	48(Σ10–Σ9)	– 19(Σ7–Σ1)	59736	–112·1946565
21	27(Σ10–Σ8)	– 19(Σ7–Σ1)	63612	–117·6881720
22	63(Σ11–Σ9)	– 40(Σ8–Σ1)	172620	–129·7846715
23	5(Σ11–Σ9)	– 3(Σ8–Σ1)	15540	–144·1351351
24	20(Σ12–Σ10)	– 11(Σ9–Σ1)	71280	–157·5740741
25	35(Σ12–Σ10)	– 23(Σ8–Σ1)	154560	–167·25
26	77(Σ13–Σ11)	– 48(Σ9–Σ2)	378840	–183·7195122
27	44(Σ13–Σ11)	– 25(Σ9–Σ1)	244200	–198·7297297
28	24(Σ14–Σ12)	– 13(Σ10–Σ2)	147264	–216·5677966
29	2(Σ14–Σ12)	(Σ10–Σ1)	13716	–232·9265092
30	117(Σ15–Σ13)	– 56(Σ11–Σ2)	881244	–252·1282528
31	54(Σ15–Σ13)	– 29(Σ10–Σ1)	485460	–262·2043011
32	39(Σ16–Σ14)	– 20(Σ11–Σ2)	382590	–282·5244648
33	65(Σ16–Σ14)	– 31(Σ11–Σ1)	701220	–301·4137931
34	35(Σ17–Σ15)	– 16(Σ12–Σ2)	409920	–323·1065574
35	25(Σ17–Σ15)	– 11(Σ12–Σ2)	316800	–345·7916667
36	5(Σ18–Σ15)	– 3(Σ13–Σ2)	93060	–352·4432624
37	88(Σ18–Σ15)	– 51(Σ13–Σ2)	1,768272	–375·9644670
38	11(Σ19–Σ16)	– 7(Σ13–Σ2)	256410	–388·0855856
39	44(Σ19–Σ16)	– 27(Σ13–Σ2)	1,102464	–412·7298851
40	64(Σ20–Σ17)	– 37(Σ14–Σ2)	1,736928	–435·3016360

TABLE 9.8c (cont.)

	Single-step functions			
	Third degree		$b_{33} = \underline{\sum} W_3 y / \underline{\sum} W_3 \epsilon^3$	
n	W_3		$\underline{\sum} W_3 \epsilon^3$	β_{35}
41	34(Σ20 − Σ17)	− 19(Σ14 − Σ2)	988380	−461·3137255
42	17(Σ21 − Σ18)	− 9(Σ15 − Σ2)	533052	−485·2363184
43	39(Σ21 − Σ18)	− 20(Σ15 − Σ2)	1,305720	−512·6129032
44	247(Σ22 − Σ19)	− 123(Σ16 − Σ3)	8,810490	−540·9827586
45	13(Σ22 − Σ19)	− 7(Σ15 − Σ2)	524160	−555·5416667
46	247(Σ23 − Σ20)	− 129(Σ16 − Σ3)	10,578516	−585·0301205
47	133(Σ23 − Σ20)	− 66(Σ16 − Σ2)	6,091932	−611·8587896
48	56(Σ24 − Σ21)	− 27(Σ17 − Σ3)	2,717820	−642·7178952
49	50(Σ24 − Σ21)	− 23(Σ17 − Σ2)	2,587500	−670·8986667
50	280(Σ25 − Σ22)	− 141(Σ17 − Σ3)	16,009140	−690·6405672
51	147(Σ25 − Σ21)	− 94(Σ17 − Σ3)	10,971492	−703·9420655
52	105(Σ26 − Σ22)	− 64(Σ18 − Σ3)	8,336160	−733·0989520
53	165(Σ26 − Σ22)	− 98(Σ18 − Σ3)	13,809180	−766·9110070
54	44(Σ27 − Σ23)	− 25(Σ19 − Σ3)	3,907200	−797·4189189
55	92(Σ27 − Σ23)	− 51(19 − Σ3)	8,595744	−832·5982533
56	391(Σ28 − Σ24)	− 208(Σ20 − Σ3)	38,671464	−864·4558360
57	102(Σ28 − Σ24)	− 53(Σ20 − Σ3)	10,595760	−901
58	16(Σ29 − Σ25)	− 9(Σ20 − Σ4)	1,814400	−924·1666667
59	102(Σ29 − Σ25)	− 55(Σ20 − Σ3)	12,207360	−957·5808824
60	425(Σ30 − Σ26)	− 224(Σ21 − Σ4)	53,264400	−996·2372654
61	75(Σ30 − Σ26)	− 38(Σ21 − Σ3)	9,900900	−1031·003454
62	117(Σ31 − Σ27)	− 58(Σ22 − Σ4)	16,150680	−1071·029412
63	247(Σ31 − Σ27)	− 118(Σ22 − Σ3)	35,849580	−1107·146341
64	39(Σ32 − Σ28)	− 20(Σ22 − Σ4)	6,121440	−1132·597859
65	247(Σ32 − Σ28)	− 122(Σ22 − Σ3)	40,680900	−1169·888889
66	513(Σ33 − Σ28)	− 305(Σ23 − Σ4)	103,892760	−1186·813253
67	266(Σ33 − Σ28)	− 155(Σ23 − Σ4)	56,155260	−1229·792952
68	16(Σ34 − Σ29)	− 9(Σ24 − Σ4)	3,540600	−1268·236655
69	29(Σ34 − Σ29)	− 16(Σ24 − Σ4)	6,681600	−1312·583333
70	609(Σ35 − Σ30)	− 325(Σ25 − Σ4)	146,860350	−1352·376011
71	58(Σ35 − Σ30)	− 33(Σ24 − Σ4)	15,024900	−1380·515924
72	609(Σ36 − Σ31)	− 335(Σ25 − Σ4)	164,844120	−1421·490099
73	63(Σ36 − Σ31)	− 34(Σ25 − Σ4)	17,714340	−1468·321241
74	44(Σ37 − Σ32)	− 23(Σ26 − Σ4)	12,910590	−1510·648148
75	341(Σ37 − Σ32)	− 175(Σ26 − Σ4)	103,834500	−1558·848276

TABLE 9.8c (cont.)

	Single-step functions					
	Fourth degree			$b_{44} = \sum_{+} W_4 y \big/ \sum_{+} W_4 \epsilon^4$		
n	W_4					$\sum_{+} W_4 \epsilon^4$
7	2(Σ3 − Σ2)	−	5(Σ2 − Σ1)	+	2Σ1	168
8	(Σ4 − Σ3)	−	2(Σ3 − Σ2)	+	Σ1	144
9	25(Σ4 − Σ3)	−	46(Σ3 − Σ2)	+	14Σ1	5376
10	9(Σ5 − Σ4)	−	10(Σ4 − Σ2)	+	11Σ1	3600
11	71(Σ5 − Σ4)	−	73(Σ4 − Σ2)	+	50Σ1	39648
12	16(Σ6 − Σ5)	−	15(Σ5 − Σ3)	+	14Σ1	12480
13	72(Σ6 − Σ5)	−	53(Σ5 − Σ2)	+	58Σ1	84768
14	59(Σ7 − Σ6)	−	41(Σ6 − Σ3)	+	32Σ2	90000
15	45(Σ7 − Σ6)	−	29(Σ6 − Σ3)	+	28Σ1	89880
16	20(Σ8 − Σ7)	−	11(Σ7 − Σ3)	+	12Σ2	54960
17	59(Σ8 − Σ7)	−	31(Σ7 − Σ3)	+	26Σ2	200376
18	144(Σ9 − Σ8)	−	71(Σ8 − Σ4)	+	70Σ2	613152
19	590(Σ9 − Σ8)	−	395(Σ7 − Σ3)	+	396Σ2	4,138824
20	206(Σ10 − Σ9)	−	131(Σ8 − Σ4)	+	106Σ3	1,721280
21	415(Σ10 − Σ9)	−	245(Σ8 − Σ4)	+	226Σ2	4,182864
22	284(Σ11 − Σ10)	−	161(Σ9 − Σ5)	+	120Σ3	3,345552
23	235(Σ11 − Σ10)	−	117(Σ9 − Σ4)	+	100Σ3	3,395784
24	415(Σ12 − Σ11)	−	194(Σ10 − Σ5)	+	185Σ3	7,072128
25	217(Σ12 − Σ11)	−	98(Σ10 − Σ5)	+	78Σ3	4,241328
26	288(Σ13 − Σ12)	−	115(Σ11 − Σ5)	+	134Σ3	6,855936
27	2989(Σ13 − Σ12)	−	1155(Σ11 − Σ5)	+	1126Σ3	80,879904
28	16(Σ14 − Σ13)	−	6(Σ12 − Σ6)	+	5Σ4	489312
29	2989(Σ14 − Σ13)	−	1344(Σ11 − Σ5)	+	1450Σ3	125,116488
30	944(Σ15 − Σ14)	−	410(Σ12 − Σ6)	+	379Σ4	44,279712
31	535(Σ15 − Σ14)	−	221(Σ12 − Σ6)	+	226Σ3	28,384776
32	586(Σ16 − Σ15)	−	235(Σ13 − Σ7)	+	206Σ4	34,551408
33	1533(Σ16 − Σ15)	−	561(Σ13 − Σ6)	+	532Σ4	103,700520
34	763(Σ17 − Σ15)	−	502(Σ14 − Σ7)	+	497Σ4	97,162464
35	1491(Σ17 − Σ15)	−	957(Σ14 − Σ7)	+	826Σ4	209,629728
36	2240(Σ18 − Σ16)	−	1405(Σ15 − Σ8)	+	1071Σ5	346,514448
37	9420(Σ18 − Σ16)	−	5397(Σ15 − Σ7)	+	5408Σ4	1680,885360
38	141(Σ19 − Σ17)	−	79(Σ16 − Σ8)	+	70Σ5	27,560016
39	11148(Σ19 − Σ17)	−	6045(Σ16 − Σ8)	+	5792Σ4	2408,213808
40	2820(Σ20 − Σ18)	−	1765(Σ16 − Σ8)	+	1696Σ5	763,126848

TABLE 9.8c (cont.)

	Single-step functions						
	Fourth degree						$b_{44} = \sum_{+} W_4 y / \sum_{+} W_4 \epsilon^4$
n	W_4						$\sum_{+} W_4 \epsilon^4$
41	13332(Σ20 − Σ18)	−	8151(Σ16 − Σ8)	+	7008Σ5		3922,631856
42	415(Σ21 − Σ19)	−	245(Σ17 − Σ9)	+	226Σ5		133,851648
43	15620(Σ21 − Σ19)	−	9031(Σ17 − Σ9)	+	7456Σ5		5452,591056
44	4779(Σ22 − Σ20)	−	2576(Σ18 − Σ9)	+	2271Σ6		1855,459872
45	19074(Σ22 − Σ20)	−	9955(Σ18 − Σ9)	+	9354Σ5		8071,065288
46	5535(Σ23 − Σ21)	−	2834(Σ19 − Σ10)	+	2406Σ6		2521,801584
47	21945(Σ23 − Σ21)	−	10923(Σ19 − Σ10)	+	9894Σ5		10844,513064
48	415(Σ24 − Σ22)	−	194(Σ20 − Σ10)	+	185Σ6		226,308096
49	8489(Σ24 − Σ22)	−	4667(Σ19 − Σ10)	+	3850Σ6		5386,084704
50	3785(Σ25 − Σ23)	−	1693(Σ21 − Σ11)	+	1560Σ6		2388,695712
51	30485(Σ25 − Σ23)	−	15249(Σ20 − Σ10)	+	14080Σ6		22841,591136
52	8680(Σ26 − Σ24)	−	4263(Σ21 − Σ11)	+	3610Σ7		6940,591392
53	34645(Σ26 − Σ24)	−	16549(Σ21 − Σ11)	+	14800Σ6		29764,941336
54	1967(Σ27 − Σ25)	−	924(Σ22 − Σ12)	+	758Σ7		1798,408080
55	46255(Σ27 − Σ25)	−	20515(Σ22 − Σ11)	+	17754Σ7		46002,560208
56	1650(Σ28 − Σ26)	−	713(Σ23 − Σ12)	+	649Σ7		1756,819680
57	10406(Σ28 − Σ26)	−	4427(Σ23 − Σ12)	+	3718Σ7		11752,633464
58	6713(Σ29 − Σ27)	−	2688(Σ24 − Σ12)	+	2690Σ7		8271,067392
59	3355(Σ29 − Σ26)	−	1915(Σ24 − Σ12)	+	1722Σ7		5932,423224
60	844(Σ30 − Σ27)	−	475(Σ25 − Σ13)	+	396Σ8		1579,756368
61	13897(Σ30 − Σ27)	−	7407(Σ25 − Σ12)	+	7280Σ7		28318,400448
62	4220(Σ31 − Σ28)	−	2549(Σ25 − Σ13)	+	2241Σ8		9725,818032
63	37553(Σ31 − Σ28)	−	22355(Σ25 − Σ13)	+	18306Σ8		91341,873936
64	19396(Σ32 − Σ29)	−	10916(Σ26 − Σ13)	+	10465Σ8		51074,547264
65	86190(Σ32 − Σ29)	−	47821(Σ26 − Σ13)	+	42718Σ8		239180,368752
66	10738(Σ33 − Σ30)	−	5830(Σ27 − Σ14)	+	5447Σ8		31584,929376
67	95251(Σ33 − Σ30)	−	51034(Σ27 − Σ14)	+	44434Σ8		294717,067632
68	26286(Σ34 − Σ31)	−	13905(Σ28 − Σ15)	+	11323Σ9		85474,898496
69	108171(Σ34 − Σ31)	−	54349(Σ28 − Σ14)	+	51338Σ8		378953,415120
70	7455(Σ35 − Σ32)	−	3699(Σ29 − Σ15)	+	3269Σ9		27413,402688
71	118643(Σ35 − Σ32)	−	57766(Σ29 − Σ15)	+	53270Σ8		460374,156648
72	14910(Σ36 − Σ33)	−	7857(Σ29 − Σ15)	+	7252Σ9		65312,203824
73	131005(Σ36 − Σ33)	−	68153(Σ29 − Σ15)	+	59066Σ9		601034,805504
74	32655(Σ37 − Σ34)	−	16659(Σ30 − Σ16)	+	15029Σ9		157710,048048
75	143241(Σ37 − Σ34)	−	72200(Σ30 − Σ16)	+	61166Σ9		723517,105920

TABLE 9.8*c* (cont.)

	Single-step functions	
	Fifth degree $b_{55} = \underline{\sum} W_5\, y \big/ \underline{\sum} W_5\, \epsilon^5$	
n	W_5	$\underline{\sum} W_5\, \epsilon^5$
7	(Σ3 − Σ2) − 4(Σ2 − Σ1) + 5Σ1	240
8	15(Σ4 − Σ3) − 49(Σ3 − Σ2) + 35Σ2	6720
9	9(Σ4 − Σ3) − 26(Σ3 − Σ2) + 14Σ2	6720
10	49(Σ5 − Σ4) − 111(Σ4 − Σ3) + 84Σ2	65520
11	26(Σ5 − Σ4) − 55(Σ4 − Σ3) + 30Σ2	51840
12	72(Σ6 − Σ5) − 143(Σ5 − Σ4) + 55Σ3	208560
13	3(Σ6 − Σ5) − 4(Σ5 − Σ3) + 3Σ3	15120
14	112(Σ7 − Σ6) − 130(Σ6 − Σ4) + 143(Σ3 − Σ1)	840840
15	275(Σ7 − Σ6) − 301(Σ6 − Σ4) + 231Σ3	2,808960
16	19(Σ8 − Σ7) − 20(Σ7 − Σ5) + 13(Σ4 − Σ1)	252720
17	481(Σ8 − Σ7) − 464(Σ7 − Σ5) + 364Σ3	8,910720
18	1755(Σ9 − Σ8) − 1360(Σ8 − Σ5) + 1547(Σ4 − Σ1)	47,895120
19	287(Σ9 − Σ8) − 213(Σ8 − Σ5) + 189Σ4	9,954000
20	819(Σ10 − Σ9) − 589(Σ9 − Σ6) + 456(Σ5 − Σ1)	35,112000
21	112(Σ10 − Σ9) − 75(Σ9 − Σ6) + 68Σ4	6,283200
22	697(Σ11 − Σ10) − 455(Σ10 − Σ7) + 357(Σ5 − Σ1)	47,295360
23	1037(Σ11 − Σ10) − 583(Σ10 − Σ6) + 561Σ5	95,729040
24	5586(Σ12 − Σ11) − 3059(Σ11 − Σ7) + 2622(Σ6 − Σ1)	615,593160
25	510(Σ12 − Σ11) − 430(Σ10 − Σ7) + 366Σ5	92,085120
26	228(Σ13 − Σ12) − 115(Σ12 − Σ8) + 100(Σ6 − Σ1)	37,044000
27	1824(Σ13 − Σ12) − 1274(Σ11 − Σ7) + 1235(Σ6 − Σ1)	485,503200
28	2528(Σ14 − Σ13) − 1701(Σ12 − Σ8) + 1413(Σ7 − Σ1)	788,492880
29	3220(Σ14 − Σ13) − 2030(Σ12 − Σ8) + 2009(Σ6 − Σ1)	1202,742240
30	35200(Σ15 − Σ14) − 21518(Σ13 − Σ9) + 18183(Σ7 − Σ1)	15207,265920
31	22149(Σ15 − Σ14) − 13230(Σ13 − Σ9) + 10235(Σ7 − Σ1)	10916,650080
32	14553(Σ16 − Σ15) − 8463(Σ14 − Σ10) + 5735(Σ8 − Σ1)	8211,661920
33	29475(Σ16 − Σ15) − 16344(Σ14 − Σ10) + 12800(Σ7 − Σ1)	19440,662400
34	34125(Σ17 − Σ16) − 16632(Σ15 − Σ10) + 15125(Σ8 − Σ1)	27440,028000
35	10465(Σ17 − Σ16) − 4998(Σ15 − Σ10) + 4199(Σ8 − Σ1)	9460,956960
36	3540(Σ18 − Σ17) − 1652(Σ16 − Σ11) + 1239(Σ9 − Σ1)	3608,463600
37	1925(Σ18 − Σ17) − 861(Σ16 − Σ11) + 732(Σ8 − Σ1)	2257,995600
38	36736(Σ19 − Σ18) − 14874(Σ17 − Σ11) + 14245(Σ9 − Σ1)	51234,261120
39	110374(Σ19 − Σ18) − 43890(Σ17 − Σ11) + 39121(Σ9 − Σ1)	171065,664000
40	12111(Σ20 − Σ19) − 4719(Σ18 − Σ12) + 3809(Σ10 − Σ1)	20909,168280

TABLE 9.8c (cont.)

	Single-step functions	
	Fifth degree	$b_5{}^{5} = \sum W_5 y / \sum W_5 \epsilon^5$
n	W_5	$\sum W_5 \epsilon^5$
41	69223(Σ20 − Σ19) − 25960(Σ18 − Σ12) + 23405(Σ9 − Σ1)	135787,454880
42	524160(Σ21 − Σ20) − 241736(Σ18 − Σ12) + 229395(Σ10 − Σ2)	1,379383,716480
43	17949(Σ21 − Σ20) − 8085(Σ18 − Σ12) + 6944(Σ10 − Σ1)	52229,469600
44	18954(Σ22 − Σ21) − 8385(Σ19 − Σ13) + 6794(Σ11 − Σ2)	60473,424960
45	6075(Σ22 − Σ21) − 2568(Σ19 − Σ13) + 2233(Σ10 − Σ1)	21840,477120
46	104091(Σ23 − Σ22) − 43290(Σ20 − Σ14) + 35445(Σ11 − Σ2)	408520,153080
47	8060(Σ23 − Σ22) − 3289(Σ20 − Σ14) + 2461(Σ11 − Σ1)	34633,959360
48	128(Σ24 − Σ23) − 47(Σ21 − Σ14) + 47(Σ11 − Σ2)	639,576000
49	73437(Σ24 − Σ23) − 26468(Σ21 − Σ14) + 24192(Σ11 − Σ1)	400419,714240
50	34595(Σ25 − Σ24) − 12285(Σ22 − Σ15) + 10619(Σ12 − Σ2)	204535,553040
51	9709(Σ25 − Σ24) − 3400(Σ22 − Σ15) + 2793(Σ12 − Σ2)	62103,392400
52	57967(Σ26 − Σ24) − 36800(Σ23 − Σ16) + 30355(Σ12 − Σ2)	696524,337600
53	13650(Σ26 − Σ24) − 8075(Σ23 − Σ15) + 7514(Σ12 − Σ2)	185253,868800
54	289960(Σ27 − Σ25) − 168883(Σ24 − Σ16) + 144768(Σ13 − Σ2)	4,267843,419840
55	411312(Σ27 − Σ25) − 236698(Σ24 − Σ16) + 193397(Σ13 − Σ2)	6,521985,180480
56	380380(Σ28 − Σ26) − 212355(Σ25 − Σ17) + 183456(Σ13 − Σ2)	6,654265,778160
57	44032(Σ28 − Σ26) − 24310(Σ25 − Σ17) + 19995(Σ13 − Σ2)	827440,922880
58	236844(Σ29 − Σ27) − 122752(Σ26 − Σ17) + 109263(Σ14 − Σ2)	4,995904,858560
59	216376(Σ29 − Σ27) − 136629(Σ25 − Σ17) + 109478(Σ14 − Σ2)	5,466377,064000
60	35200(Σ30 − Σ28) − 21518(Σ26 − Σ18) + 18183(Σ14 − Σ2)	973265,018880
61	1266840(Σ30 − Σ28) − 765289(Σ26 − Σ18) + 617730(Σ14 − Σ2)	37,441514,658240
62	2917642(Σ31 − Σ29) − 1738165(Σ27 − Σ19) + 1310080(Σ15 − Σ2)	92,410608,850560
63	1470160(Σ31 − Σ29) − 853451(Σ27 − Σ19) + 693814(Σ14 − Σ2)	50,516195,501280
64	7135869(Σ32 − Σ30) − 3870815(Σ28 − Σ19) + 3405009(Σ15 − Σ2)	271,871714,130120
65	24596(Σ32 − Σ30) − 13195(Σ28 − Σ19) + 11116(Σ15 − Σ2)	997715,759040
66	1297863(Σ33 − Σ31) − 687360(Σ29 − Σ20) + 543648(Σ16 − Σ2)	56,181483,308160
67	11745(Σ33 − Σ31) − 6071(Σ29 − Σ20) + 5150(Σ15 − Σ2)	549091,623600
68	10165779(Σ34 − Σ32) − 5203055(Σ30 − Σ21) + 4236111(Σ16 − Σ3)	504,334672,086960
69	108205(Σ34 − Σ32) − 52193(Σ30 − Σ20) + 45560(Σ16 − Σ2)	5,908029,356880
70	10465(Σ35 − Σ33) − 4998(Σ31 − Σ21) + 4199(Σ17 − Σ3)	605501,245440
71	1727005(Σ35 − Σ33) − 804517(Σ31 − Σ21) + 707020(Σ16 − Σ2)	107,839889,483760
72	5445(Σ36 − Σ34) − 2513(Σ32 − Σ22) + 2124(Σ17 − Σ3)	359665,639200
73	4477275(Σ36 − Σ34) − 1961375(Σ32 − Σ21) + 1764279(Σ17 − Σ2)	323,451443,343600
74	123375(Σ37 − Σ35) − 63308(Σ32 − Σ22) + 52128(Σ18 − Σ3)	10,241611,002000
75	5066600(Σ37 − Σ35) − 2148025(Σ33 − Σ22) + 1944866(Σ17 − Σ2)	415,718898,302400

TABLE 9.8*d*

Quantities G_{jk} for the calculation of the bias in a_{kN}

κ_2^2	0					0·5					1·0				
κ_3	−1·0	−0·5	0	0·5	1·0	−1·0	−0·5	0	0·5	1·0	−1·0	−0·5	0	0·5	1·0
G_{21}	0	0	0	0	0	0·035	0·051	0·069	0·091	0·117	0·052	0·074	0·099	0·130	0·166
G_{31}	0·055	0·057	0·072	0·101	0·143	0·075	0·079	0·096	0·126	0·172	0·096	0·102	0·120	0·151	0·197
G_{22}	−0·065	−0·035	0	0·040	0·086	−0·063	−0·033	0	0·035	0·074	−0·061	−0·031	0	0·032	0·064
G_{32}	0	0	0	0	0	0·021	0·036	0·053	0·069	0·083	0·033	0·054	0·076	0·099	0·118
G_{23}	0	0	0	0	0	0·001	0·002	0	−0·005	−0·015	0·002	0·002	0	−0·006	−0·017
G_{33}	−0·051	−0·027	0	0·029	0·059	−0·050	−0·025	0	0·027	0·056	−0·050	−0·025	0	0·025	0·052

TABLE 9.8*e*

Step functions for the unequally-spaced case

n	First degree	Second degree	Third degree
10	$\Sigma3+\Sigma7-\Sigma10$	$\Sigma1+\Sigma2-\Sigma8-\Sigma9+\Sigma10$	$2\Sigma1+\Sigma2-\Sigma4-\Sigma6+\Sigma8+2\Sigma9-2\Sigma10$
11	$\Sigma4+\Sigma7-\Sigma11$	$\Sigma1+\Sigma3-\Sigma8-\Sigma10+\Sigma11$	$2\Sigma1+\Sigma2-\Sigma5-\Sigma6+\Sigma9+2\Sigma10-2\Sigma11$
12	$\Sigma4+\Sigma8-\Sigma12$	$\Sigma1+\Sigma3-\Sigma9-\Sigma11+\Sigma12$	$2\Sigma1+\Sigma2-\Sigma5-\Sigma7+\Sigma10+2\Sigma11-2\Sigma12$
13	$\Sigma4+\Sigma9-\Sigma13$	$\Sigma1+\Sigma3-\Sigma10-\Sigma12+\Sigma13$	$2\Sigma1+\Sigma2-\Sigma6-\Sigma7+\Sigma11+2\Sigma12-2\Sigma13$
14	$\Sigma5+\Sigma9-\Sigma14$	$\Sigma2+\Sigma3-\Sigma11-\Sigma12+\Sigma14$	$2\Sigma1+\Sigma2-\Sigma6-\Sigma8+\Sigma12+2\Sigma13-2\Sigma14$
15	$\Sigma5+\Sigma10-\Sigma15$	$\Sigma2+\Sigma4-\Sigma11-\Sigma13+\Sigma15$	$2\Sigma1+\Sigma2-\Sigma7-\Sigma8+\Sigma13+2\Sigma14-2\Sigma15$
16	$\Sigma5+\Sigma11-\Sigma16$	$\Sigma2+\Sigma4-\Sigma12-\Sigma14+\Sigma16$	$2\Sigma1+\Sigma2-\Sigma7-\Sigma9+\Sigma14+2\Sigma15-2\Sigma16$
17	$\Sigma6+\Sigma11-\Sigma17$	$\Sigma2+\Sigma4-\Sigma13-\Sigma15+\Sigma17$	$2\Sigma1+\Sigma3-\Sigma7-\Sigma10+\Sigma14+2\Sigma16-2\Sigma17$
18	$\Sigma6+\Sigma12-\Sigma18$	$\Sigma2+\Sigma4-\Sigma14-\Sigma16+\Sigma18$	$2\Sigma1+\Sigma3-\Sigma8-\Sigma10+\Sigma15+2\Sigma17-2\Sigma18$
19	$\Sigma6+\Sigma13-\Sigma19$	$\Sigma2+\Sigma5-\Sigma14-\Sigma17+\Sigma19$	$2\Sigma1+\Sigma3-\Sigma8-\Sigma11+\Sigma16+2\Sigma18-2\Sigma19$
20	$\Sigma7+\Sigma13-\Sigma20$	$\Sigma2+\Sigma5-\Sigma15-\Sigma18+\Sigma20$	$2\Sigma1+\Sigma3-\Sigma9-\Sigma11+\Sigma17+2\Sigma19-2\Sigma20$
21	$\Sigma7+\Sigma14-\Sigma21$	$\Sigma2+\Sigma5-\Sigma16-\Sigma19+\Sigma21$	$2\Sigma1+\Sigma3-\Sigma9-\Sigma12+\Sigma18+2\Sigma20-2\Sigma21$
22	$\Sigma7+\Sigma15-\Sigma22$	$\Sigma2+\Sigma5-\Sigma17-\Sigma20+\Sigma22$	$2\Sigma2+\Sigma3-\Sigma10-\Sigma12+\Sigma19+2\Sigma20-2\Sigma22$
23	$\Sigma8+\Sigma15-\Sigma23$	$\Sigma2+\Sigma6-\Sigma17-\Sigma21+\Sigma23$	$2\Sigma2+\Sigma3-\Sigma10-\Sigma13+\Sigma20+2\Sigma21-2\Sigma23$
24	$\Sigma8+\Sigma16-\Sigma24$	$\Sigma3+\Sigma6-\Sigma18-\Sigma21+\Sigma24$	$2\Sigma2+\Sigma4-\Sigma11-\Sigma13+\Sigma20+2\Sigma22-2\Sigma24$
25	$\Sigma8+\Sigma17-\Sigma25$	$\Sigma3+\Sigma6-\Sigma19-\Sigma22+\Sigma25$	$2\Sigma2+\Sigma4-\Sigma11-\Sigma14+\Sigma21+2\Sigma23-2\Sigma25$
26	$\Sigma9+\Sigma17-\Sigma26$	$\Sigma3+\Sigma6-\Sigma20-\Sigma23+\Sigma26$	$2\Sigma2+\Sigma4-\Sigma11-\Sigma15+\Sigma22+2\Sigma24-2\Sigma26$
27	$\Sigma9+\Sigma18-\Sigma27$	$\Sigma3+\Sigma7-\Sigma20-\Sigma24+\Sigma27$	$2\Sigma2+\Sigma4-\Sigma12-\Sigma15+\Sigma23+2\Sigma25-2\Sigma27$
28	$\Sigma9+\Sigma19-\Sigma28$	$\Sigma3+\Sigma7-\Sigma21-\Sigma25+\Sigma28$	$2\Sigma2+\Sigma4-\Sigma12-\Sigma16+\Sigma24+2\Sigma26-2\Sigma28$
29	$\Sigma10+\Sigma19-\Sigma29$	$\Sigma3+\Sigma7-\Sigma22-\Sigma26+\Sigma29$	$2\Sigma2+\Sigma4-\Sigma13-\Sigma16+\Sigma25+2\Sigma27-2\Sigma29$
30	$\Sigma10+\Sigma20-\Sigma30$	$\Sigma3+\Sigma7-\Sigma23-\Sigma27+\Sigma30$	$2\Sigma2+\Sigma5-\Sigma13-\Sigma17+\Sigma25+2\Sigma28-2\Sigma30$
31	$\Sigma10+\Sigma21-\Sigma31$	$\Sigma3+\Sigma8-\Sigma23-\Sigma28+\Sigma31$	$2\Sigma2+\Sigma5-\Sigma14-\Sigma17+\Sigma26+2\Sigma29-2\Sigma31$
32	$\Sigma11+\Sigma21-\Sigma32$	$\Sigma3+\Sigma8-\Sigma24-\Sigma29+\Sigma32$	$2\Sigma2+\Sigma5-\Sigma14-\Sigma18+\Sigma27+2\Sigma30-2\Sigma32$
33	$\Sigma11+\Sigma22-\Sigma33$	$\Sigma4+\Sigma8-\Sigma25-\Sigma29+\Sigma33$	$2\Sigma2+\Sigma5-\Sigma15-\Sigma18+\Sigma28+2\Sigma31-2\Sigma33$
34	$\Sigma11+\Sigma23-\Sigma34$	$\Sigma4+\Sigma8-\Sigma26-\Sigma30+\Sigma34$	$2\Sigma2+\Sigma5-\Sigma15-\Sigma19+\Sigma29+2\Sigma32-2\Sigma34$
35	$\Sigma12+\Sigma23-\Sigma35$	$\Sigma4+\Sigma9-\Sigma26-\Sigma31+\Sigma35$	$2\Sigma2+\Sigma5-\Sigma15-\Sigma20+\Sigma30+2\Sigma33-2\Sigma35$
36	$\Sigma12+\Sigma24-\Sigma36$	$\Sigma4+\Sigma9-\Sigma27-\Sigma32+\Sigma36$	$2\Sigma3+\Sigma5-\Sigma16-\Sigma20+\Sigma31+2\Sigma33-2\Sigma36$
37	$\Sigma12+\Sigma25-\Sigma37$	$\Sigma4+\Sigma9-\Sigma28-\Sigma33+\Sigma37$	$2\Sigma3+\Sigma6-\Sigma16-\Sigma21+\Sigma31+2\Sigma34-2\Sigma37$
38	$\Sigma13+\Sigma25-\Sigma38$	$\Sigma4+\Sigma9-\Sigma29-\Sigma34+\Sigma38$	$2\Sigma3+\Sigma6-\Sigma17-\Sigma21+\Sigma32+2\Sigma35-2\Sigma38$
39	$\Sigma13+\Sigma26-\Sigma39$	$\Sigma4+\Sigma10-\Sigma29-\Sigma35+\Sigma39$	$2\Sigma3+\Sigma6-\Sigma17-\Sigma22+\Sigma33+2\Sigma36-2\Sigma39$
40	$\Sigma13+\Sigma27-\Sigma40$	$\Sigma4+\Sigma10-\Sigma30-\Sigma36+\Sigma40$	$2\Sigma3+\Sigma6-\Sigma18-\Sigma22+\Sigma34+2\Sigma37-2\Sigma40$
41	$\Sigma14+\Sigma27-\Sigma41$	$\Sigma4+\Sigma10-\Sigma31-\Sigma37+\Sigma41$	$2\Sigma3+\Sigma6-\Sigma18-\Sigma23+\Sigma35+2\Sigma38-2\Sigma41$
42	$\Sigma14+\Sigma28-\Sigma42$	$\Sigma5+\Sigma10-\Sigma32-\Sigma37+\Sigma42$	$2\Sigma3+\Sigma6-\Sigma19-\Sigma23+\Sigma36+2\Sigma39-2\Sigma42$
43	$\Sigma14+\Sigma29-\Sigma43$	$\Sigma5+\Sigma11-\Sigma32-\Sigma38+\Sigma43$	$2\Sigma3+\Sigma6-\Sigma19-\Sigma24+\Sigma37+2\Sigma40-2\Sigma43$
44	$\Sigma15+\Sigma29-\Sigma44$	$\Sigma5+\Sigma11-\Sigma33-\Sigma39+\Sigma44$	$2\Sigma3+\Sigma7-\Sigma19-\Sigma25+\Sigma37+2\Sigma41-2\Sigma44$
45	$\Sigma15+\Sigma30-\Sigma45$	$\Sigma5+\Sigma11-\Sigma34-\Sigma40+\Sigma45$	$2\Sigma3+\Sigma7-\Sigma20-\Sigma25+\Sigma38+2\Sigma42-2\Sigma45$
46	$\Sigma15+\Sigma31-\Sigma46$	$\Sigma5+\Sigma11-\Sigma35-\Sigma41+\Sigma46$	$2\Sigma3+\Sigma7-\Sigma20-\Sigma26+\Sigma39+2\Sigma43-2\Sigma46$
47	$\Sigma16+\Sigma31-\Sigma47$	$\Sigma5+\Sigma12-\Sigma35-\Sigma42+\Sigma47$	$2\Sigma3+\Sigma7-\Sigma21-\Sigma26+\Sigma40+2\Sigma44-2\Sigma47$
48	$\Sigma16+\Sigma32-\Sigma48$	$\Sigma5+\Sigma12-\Sigma36-\Sigma43+\Sigma48$	$2\Sigma3+\Sigma7-\Sigma21-\Sigma27+\Sigma41+2\Sigma45-2\Sigma48$
49	$\Sigma16+\Sigma33-\Sigma49$	$\Sigma5+\Sigma12-\Sigma37-\Sigma44+\Sigma49$	$2\Sigma3+\Sigma7-\Sigma22-\Sigma27+\Sigma42+2\Sigma46-2\Sigma49$
50	$\Sigma17+\Sigma33-\Sigma50$	$\Sigma5+\Sigma12-\Sigma38-\Sigma45+\Sigma50$	$2\Sigma4+\Sigma8-\Sigma22-\Sigma28+\Sigma42+2\Sigma46-2\Sigma50$
51	$\Sigma17+\Sigma34-\Sigma51$	$\Sigma6+\Sigma12-\Sigma39-\Sigma45+\Sigma51$	$2\Sigma4+\Sigma8-\Sigma22-\Sigma29+\Sigma43+2\Sigma47-2\Sigma51$
52	$\Sigma17+\Sigma35-\Sigma52$	$\Sigma6+\Sigma13-\Sigma39-\Sigma46+\Sigma52$	$2\Sigma4+\Sigma8-\Sigma23-\Sigma29+\Sigma44+2\Sigma48-2\Sigma52$
53	$\Sigma18+\Sigma35-\Sigma53$	$\Sigma6+\Sigma13-\Sigma40-\Sigma47+\Sigma53$	$2\Sigma4+\Sigma8-\Sigma23-\Sigma30+\Sigma45+2\Sigma49-2\Sigma53$
54	$\Sigma18+\Sigma36-\Sigma54$	$\Sigma6+\Sigma13-\Sigma41-\Sigma48+\Sigma54$	$2\Sigma4+\Sigma8-\Sigma24-\Sigma30+\Sigma46+2\Sigma50-2\Sigma54$
55	$\Sigma18+\Sigma37-\Sigma55$	$\Sigma6+\Sigma13-\Sigma42-\Sigma49+\Sigma55$	$2\Sigma4+\Sigma8-\Sigma24-\Sigma31+\Sigma47+2\Sigma51-2\Sigma55$

TABLE 9.8*e* (cont.)

n	First degree	Second degree	Third degree
56	Σ19+Σ37−Σ56	Σ6+Σ14−Σ42−Σ50+Σ56	2Σ4+Σ8−Σ25−Σ31+Σ48+2Σ52−2Σ56
57	Σ19+Σ38−Σ57	Σ6+Σ14−Σ43−Σ51+Σ57	2Σ4+Σ9−Σ25−Σ32+Σ48+2Σ53−2Σ57
58	Σ19+Σ39−Σ58	Σ6+Σ14−Σ44−Σ52+Σ58	2Σ4+Σ9−Σ26−Σ32+Σ49+2Σ54−2Σ58
59	Σ20+Σ39−Σ59	Σ6+Σ14−Σ45−Σ53+Σ59	2Σ4+Σ9−Σ26−Σ33+Σ50+2Σ55−2Σ59
60	Σ20+Σ40−Σ60	Σ6+Σ15−Σ45−Σ54+Σ60	2Σ4+Σ9−Σ26−Σ34+Σ51+2Σ56−2Σ60
61	Σ20+Σ41−Σ61	Σ7+Σ15−Σ46−Σ54+Σ61	2Σ4+Σ9−Σ27−Σ34+Σ52+2Σ57−2Σ61
62	Σ21+Σ41−Σ62	Σ7+Σ15−Σ47−Σ55+Σ62	2Σ4+Σ9−Σ27−Σ35+Σ53+2Σ58−2Σ62
63	Σ21+Σ42−Σ63	Σ7+Σ15−Σ48−Σ56+Σ63	2Σ4+Σ10−Σ28−Σ35+Σ53+2Σ59−2Σ63
64	Σ21+Σ43−Σ64	Σ7+Σ16−Σ48−Σ57+Σ64	2Σ4+Σ10−Σ28−Σ36+Σ54+2Σ60−2Σ64
65	Σ22+Σ43−Σ65	Σ7+Σ16−Σ49−Σ58+Σ65	2Σ5+Σ10−Σ29−Σ36+Σ55+2Σ60−2Σ65
66	Σ22+Σ44−Σ66	Σ7+Σ16−Σ50−Σ59+Σ66	2Σ5+Σ10−Σ29−Σ37+Σ56+2Σ61−2Σ66
67	Σ22+Σ45−Σ67	Σ7+Σ16−Σ51−Σ60+Σ67	2Σ5+Σ10−Σ30−Σ37+Σ57+2Σ62−2Σ67
68	Σ23+Σ45−Σ68	Σ7+Σ17−Σ51−Σ61+Σ68	2Σ5+Σ10−Σ30−Σ38+Σ58+2Σ63−2Σ68
69	Σ23+Σ46−Σ69	Σ7+Σ17−Σ52−Σ62+Σ69	2Σ5+Σ10−Σ30−Σ39+Σ59+2Σ64−2Σ69
70	Σ23+Σ47−Σ70	Σ8+Σ17−Σ53−Σ62+Σ70	2Σ5+Σ11−Σ31−Σ39+Σ59+2Σ65−2Σ70
71	Σ24+Σ47−Σ71	Σ8+Σ17−Σ54−Σ63+Σ71	2Σ5+Σ11−Σ31−Σ40+Σ60+2Σ66−2Σ71
72	Σ24+Σ48−Σ72	Σ8+Σ18−Σ54−Σ64+Σ72	2Σ5+Σ11−Σ32−Σ40+Σ61+2Σ67−2Σ72
73	Σ24+Σ49−Σ73	Σ8+Σ18−Σ55−Σ65+Σ73	2Σ5+Σ11−Σ32−Σ41+Σ62+2Σ68−2Σ73
74	Σ25+Σ49−Σ74	Σ8+Σ18−Σ56−Σ66+Σ74	2Σ5+Σ11−Σ33−Σ41+Σ63+2Σ69−2Σ74
75	Σ25+Σ50−Σ75	Σ8+Σ18−Σ57−Σ67+Σ75	2Σ5+Σ11−Σ33−Σ42+Σ64+2Σ70−2Σ75
76	Σ25+Σ51−Σ76	Σ8+Σ19−Σ57−Σ68+Σ76	2Σ5+Σ11−Σ34−Σ42+Σ65+2Σ71−2Σ76
77	Σ26+Σ51−Σ77	Σ8+Σ19−Σ58−Σ69+Σ77	2Σ5+Σ12−Σ34−Σ43+Σ65+2Σ72−2Σ77
78	Σ26+Σ52−Σ78	Σ8+Σ19−Σ59−Σ70+Σ78	2Σ5+Σ12−Σ34−Σ44+Σ66+2Σ73−2Σ78
79	Σ26+Σ53−Σ79	Σ9+Σ19−Σ60−Σ70+Σ79	2Σ6+Σ12−Σ35−Σ44+Σ67+2Σ73−2Σ79
80	Σ27+Σ53−Σ80	Σ9+Σ20−Σ60−Σ71+Σ80	2Σ6+Σ12−Σ35−Σ45+Σ68+2Σ74−2Σ80
81	Σ27+Σ54−Σ81	Σ9+Σ20−Σ61−Σ72+Σ81	2Σ6+Σ12−Σ36−Σ45+Σ69+2Σ75−2Σ81
82	Σ27+Σ55−Σ82	Σ9+Σ20−Σ62−Σ73+Σ82	2Σ6+Σ12−Σ36−Σ46+Σ70+2Σ76−2Σ82
83	Σ28+Σ55−Σ83	Σ9+Σ20−Σ63−Σ74+Σ83	2Σ6+Σ13−Σ37−Σ46+Σ70+2Σ77−2Σ83
84	Σ28+Σ56−Σ84	Σ9+Σ21−Σ63−Σ75+Σ84	2Σ6+Σ13−Σ37−Σ47+Σ71+2Σ78−2Σ84
85	Σ28+Σ57−Σ85	Σ9+Σ21−Σ64−Σ76+Σ85	2Σ6+Σ13−Σ37−Σ48+Σ72+2Σ79−2Σ85
86	Σ29+Σ57−Σ86	Σ9+Σ21−Σ65−Σ77+Σ86	2Σ6+Σ13−Σ38−Σ48+Σ73+2Σ80−2Σ86
87	Σ29+Σ58−Σ87	Σ9+Σ21−Σ66−Σ78+Σ87	2Σ6+Σ13−Σ38−Σ49+Σ74+2Σ81−2Σ87
88	Σ29+Σ59−Σ88	Σ10+Σ22−Σ66−Σ78+Σ88	2Σ6+Σ13−Σ39−Σ49+Σ75+2Σ82−2Σ88
89	Σ30+Σ59−Σ89	Σ10+Σ22−Σ67−Σ79+Σ89	2Σ6+Σ13−Σ39−Σ50+Σ76+2Σ83−2Σ89
90	Σ30+Σ60−Σ90	Σ10+Σ22−Σ68−Σ80+Σ90	2Σ6+Σ14−Σ40−Σ50+Σ76+2Σ84−2Σ90
91	Σ30+Σ61−Σ91	Σ10+Σ22−Σ69−Σ81+Σ91	2Σ6+Σ14−Σ40−Σ51+Σ77+2Σ85−2Σ91
92	Σ31+Σ61−Σ92	Σ10+Σ23−Σ69−Σ82+Σ92	2Σ6+Σ14−Σ41−Σ51+Σ78+2Σ86−2Σ92
93	Σ31+Σ62−Σ93	Σ10+Σ23−Σ70−Σ83+Σ93	2Σ7+Σ14−Σ41−Σ52+Σ79+2Σ86−2Σ93
94	Σ31+Σ63−Σ94	Σ10+Σ23−Σ71−Σ84+Σ94	2Σ7+Σ14−Σ41−Σ53+Σ80+2Σ87−2Σ94
95	Σ32+Σ63−Σ95	Σ10+Σ23−Σ72−Σ85+Σ95	2Σ7+Σ14−Σ42−Σ53+Σ81+2Σ88−2Σ95
96	Σ32+Σ64−Σ96	Σ10+Σ24−Σ72−Σ86+Σ96	2Σ7+Σ14−Σ42−Σ54+Σ82+2Σ89−2Σ96
97	Σ32+Σ65−Σ97	Σ10+Σ24−Σ73−Σ87+Σ97	2Σ7+Σ15−Σ43−Σ54+Σ82+2Σ90−2Σ97
98	Σ33+Σ65−Σ98	Σ11+Σ24−Σ74−Σ87+Σ98	2Σ7+Σ15−Σ43−Σ55+Σ83+2Σ91−2Σ98
99	Σ33+Σ66−Σ99	Σ11+Σ24−Σ75−Σ88+Σ99	2Σ7+Σ15−Σ44−Σ55+Σ84+2Σ92−2Σ99
100	Σ33+Σ67−Σ100	Σ11+Σ25−Σ75−Σ89+Σ100	2Σ7+Σ15−Σ44−Σ56+Σ85+2Σ93−2Σ100

CHAPTER 10

FUNCTIONS WHICH ARE NOT POLYNOMIALS

10.1 LINEAR FUNCTIONS

The problem to be discussed in this section is that in which the variable Y is a linear function of $p+1$ variables $X_0, X_1, \ldots, X_p$,

$$Y = \sum_{j=0}^{p} B_{pj} X_j. \tag{1}$$

When n observations y_i, of weight w_i, are made at values x_{ji} of the independent variables X_j (supposed free from error), the least-squares function

$$u_p[x] = \sum_{j=0}^{p} b_{pj} x_j \tag{2}$$

will satisfy the condition that

$$\Sigma w_i \left(y_i - \sum_j b_{pj} x_{ji} \right)^2$$

should be a minimum. Differentiation with respect to b_{pk} leads to the normal equations

$$\Sigma b_{pj} \phi_{jk} = M_k, \quad k = 0 \text{ to } p, \tag{3a}$$

with
$$\phi_{jk} = \sum_i w_i x_{ji} x_{ki} \tag{3b}$$

and
$$M_k = \sum_i w_i y_i x_{ki}. \tag{3c}$$

These equations are formally identical with the polynomial equations if x_j is replaced by x^j. Hence they can be solved by the standard Doolittle technique described in Ch. 7.

Corresponding to the orthogonal polynomials will be the orthogonal functions

$$T_j(x) = \sum_{k=0}^{j} \beta_{kj} x_k, \tag{4a}$$

for which
$$\sum_i w_i T_j(x_i) T_k(x_i) = 0. \tag{4b}$$

x_k can be expanded in terms of these functions in the form

$$x_k = \sum_{j=0}^{k} \alpha_{jk} T_j(x). \tag{4c}$$

Clearly the treatment will be precisely analogous to that of § 7.2. The fitted function in terms of the orthogonal functions will be

$$u_p(x) = \sum_{j=0}^{p} a_j T_j(x), \tag{5}$$

where the coefficients a_j are the quantities occurring in the Doolittle scheme. The standard deviations of the estimated coefficients b_{pj} and of the fitted function $u_p(x)$ will be given by the formulae

$$\operatorname{var} b_{pj} = \sigma^2 \sum_{k=j}^{p} \beta_{jk}^2/S_{kk} = \sigma^2 \chi_{jj} \tag{6a}$$

and

$$\operatorname{var} u_p(x) = \sigma^2 \sum_{j=0}^{p} T_j^2(x)/S_{jj} = \sigma^2 \sum_{j,k} x_j x_k \chi_{jk}, \tag{6b}$$

χ_{jk} being an element of the inverse matrix.

If it is desired to obtain the form of the fitted function when the variables $x_p, x_{p-1}, \ldots$ are omitted, this can be done very simply by calculating b_{pj} in the form

$$b_{pj} = \sum_{k=j}^{p} \beta_{jk} a_k. \tag{7}$$

For example, the value $b_{p-1,j}$ would be obtained by dropping the term $\beta_{jp} a_p$.

The variance of an observation can be estimated from the formula

$$\Sigma w_i v_i^2/(n-p-1), \tag{8}$$

and the usual significance tests can be applied to the various estimates.

The variables denoted by X_j may be complicated functions of other parameters—for example, (1) may be of the form

$$Y = B_0 + B_1 Z + B_2 e^{-Z^2/2} + B_3 \cos \omega t + B_4 \Gamma(Z)$$

where $X_2 = e^{-Z^2/2}$, $X_3 = \cos \omega t$, etc. The only restriction is that the function must be linear in all the parameters B_{pj} to be estimated.

10.1.1 *Example*

In Table 10.1, the quantities y correspond to the difference in seconds between the time as given by a quartz clock and a transit circle. The values are actually averages over a 25-day period. The quantities x represent the corresponding dates, referred to an origin near the centre of the range. It is desired to represent the clock error by an equation of the form

$$u = b_0 + b_1 x + b_2 x^2 + b_3 \cos \theta + b_4 \sin \theta, \tag{1}$$

where $\theta = 2\pi x/365$ is an angular coordinate which corresponds to a periodic variation of error whose period is one year. This coordinate would take account of variations due to seasonal temperature changes and to corresponding changes in the earth's rotation.

TABLE 10.1

Clock error y as a function of the date x

$\theta = 2\pi x/365$

y	x	θ	y	x	θ
1·648	−347·9	16° 54′	29·251	25·8	25° 24′
2·854	−328·5	36° 00′	31·286	51·2	50° 30′
4·664	−300·2	63° 54′	33·072	72·8	71° 48′
6·309	−275·8	88° 00′	35·445	100·9	99° 30′
8·012	−250·7	112° 42′	37·407	124·5	122° 48′
10·020	−222·1	140° 54′	39·623	150·1	148° 00′
11·590	−199·8	162° 54′	41·959	176·6	174° 12′
13·411	−175·0	187° 24′	44·085	200·2	197° 30′
15·234	−150·4	211° 42′	46·206	223·9	220° 48′
17·143	−125·5	236° 12′	48·584	250·5	247° 06′
19·198	−99·8	261° 36′	51·084	277·3	273° 30′
20·916	−77·7	283° 24′	53·023	299·0	294° 54′
23·281	−48·0	312° 42′	55·701	327·5	323° 00′
25·078	−25·4	334° 54′	57·394	345·5	340° 48′
27·244	+0·9	0° 54′			

To obtain the normal equations, columns of the values $x^0, x, x^2, \cos\theta$, $\sin\theta$, and y are formed in Table 10.1.1, suitable powers of ten being removed to bring them to the order of unity so that a check column can be used. The various columns are intermultiplied to give the values at the bottom of the table.

The equations are solved in Table 10.1.2 by the square root method. The quantities from Table 10.1.1 have been divided by an extra factor 10^s to bring them to the order of unity. Since the two curves likely to be of interest are the full curve of (1) and the curve omitting the periodic terms, the inverse matrix for each of these cases has been calculated. The estimated standard deviation of an observation is obtained from (10.1,8). Thus $s_2 = 10^{-4}\sqrt{(1825/26)}$. The standard deviation of b_{pj} is obtained from (10.1,6*a*). The fitted curves are obtained by multiplying the coefficients and standard deviations by $10/10^J$, 10 being the factor removed from y and 10^J the factor removed from x_j. The two curves are

$$u_2 = (27{\cdot}047 \pm 0{\cdot}023) + (80{\cdot}806 \pm 0{\cdot}075) \times 10^{-3}\, x + (20{\cdot}77 \pm 0{\cdot}40) \times 10^{-6}\, x^2$$

and

$$u_4 = (27{\cdot}062 \pm 0{\cdot}018) + (80{\cdot}817 \pm 0{\cdot}061) \times 10^{-3}\, x + (20{\cdot}45 \pm 0{\cdot}31) \times 10^{-6}\, x^2$$
$$+ (79 \pm 17) \times 10^{-3} \cos\theta + (8 \pm 18) \times 10^{-3} \sin\theta.$$

It would appear, then, that an annual variation of significant amplitude is present in the series of observations.

TABLE 10.1.1

Calculations of normal equations

Factors removed:						
–	10^2	10^4	10^{-1}	10^{-1}	10	–
x^0	x	x^2	cos θ	sin θ	y	z
1	−3·479	12·103441	+9·568	+2·907	0·1648	+22·264241
1	−3·285	10·791225	+8·090	+5·878	0·2854	+22·759625
1	−3·002	9·012004	+4·399	+8·980	0·4664	+20·855404
1	−2·758	7·606564	+0·349	+9·994	0·6309	+16·822464
1	−2·507	6·285049	−3·859	+9·225	0·8012	+10·945249
1	−2·221	4·932841	−7·760	+6·307	1·0020	+3·260841
1	−1·998	3·992004	−9·558	+2·940	1·1590	−2·464996
1	−1·750	3·062500	−9·917	−1·288	1·3411	−7·551400
1	−1·504	2·262016	−8·508	−5·255	1·5234	−10·481584
1	−1·255	1·575025	−5·563	−8·310	1·7143	−10·838675
1	−0·998	0·996004	−1·461	−9·893	1·9198	−8·436196
1	−0·777	0·603729	+2·317	−9·728	2·0916	−4·492671
1	−0·480	0·230400	+6·782	−7·349	2·3281	+2·511500
1	−0·254	0·064516	+9·056	−4·242	2·5078	+8·132316
1	+0·009	0·000081	+9·999	+0·157	2·7244	+13·889481
1	+0·258	0·066564	+9·033	+4·289	2·9251	+17·571664
1	+0·512	0·262144	+6·361	+7·716	3·1286	+18·979744
1	+0·728	0·529984	+3·123	+9·500	3·3072	+18·188184
1	+1·009	1·018081	−1·650	+9·863	3·5445	+14·784581
1	+1·245	1·550025	−5·417	+8·406	3·7407	+10·524725
1	+1·501	2·253001	−8·480	+5·299	3·9623	+5·535301
1	+1·766	3·118756	−9·949	+1·011	4·1959	+1·142656
1	+2·002	4·008004	−9·537	−3·007	4·4085	−1·125496
1	+2·239	5·013121	−7·570	−6·534	4·6206	−1·231279
1	+2·505	6·275025	−3·891	−9·212	4·8584	+1·535425
1	+2·773	7·689529	+0·610	−9·981	5·1084	+7·199929
1	+2·990	8·940100	+4·210	−9·070	5·3023	+13·372400
1	+3·275	10·725625	+7·986	−6·018	5·5701	+22·538725
1	+3·455	11·937025	+9·444	−3·289	5·7394	+28·286425
29	−0·001	126·904383	−1·793	−0·704	81·0722	234·478583
	126·904383	−0·832790	−0·273380	−171·793438	102·526498	56·530272
		995·933807	172·270611	−8·973763	363·253474	1648·555721
			1449·332775	5·933602	−0·409027	1625·061581
				1450·645518	−140·802630	1134·305289
					309·672756	715·313271

TABLE 10.1.2

The square root scheme for Example 10.1.1

	Factors removed:	x^0 —	x 10^2	x^2 10^4	$\cos\theta$ 10^{-1}	$\sin\theta$ 10^{-1}	y 10	z —	
	All elements divided by 10^s, $s = 2$								
		0·29000000	−0·00001000	1·26904383	−0·01793000	−0·00704000	0·81072200	2·34478583	
			1·26904383	−0·00832790	−0·00273380	−1·71793438	1·02526498	0·56530273	
				9·95933807	+1·72270611	−0·08973763	3·63253474	16·48555722	
					14·49332775	0·05933602	−0·00409027	16·25061581	
						14·50645518	−1·40802630	11·34305289	
									3·09672756
	1·85695338	0·53851648	−0·00001857	2·35655523	−0·03329517	−0·01307295	1·50547296	4·35415798	0·83027873
	+0·00003061	0·88769074	1·12651846	−0·00735375	−0·00242731	−1·52499465	0·91014304	0·50188577	0·00191837
	−2·08477760	+0·00310994	0·47641029	2·09903107	0·85808647	−0·03341777	0·04359036	2·96729014	0·00001825
	+0·49900231	−0·00013856	−0·11022184	0·26962210	3·70889484	0·02261441	0·00292260	3·73443183	0·00000971
	−0·01622726	+0·38793387	+0·00527621	−0·00174717	0·28654533	3·48984920	0·00028899	3·49013818	0·00000963
p	b_{p0}	b_{p1}	b_{p2}	b_{p3}	b_{p4}				
0	2·79559310								
1	2·79562096	0·80792555							
2	2·70474475	0·80806111	0·02076690						
3	2·70620313	0·80806070	0·02044476	0·00078800					
4	2·70619844	0·80817281	0·02044629	0·00078749	0·00008281				
	Inverse matrix $p = 2$						s_2 0·00084		
	7·79457350						$s(b_0)$ 0·0023		
	−0·00645636	0·78800452					$s(b_1)$ 0·00075		
	−0·99320950	0·00148161	0·22696676				$s(b_2)$ 0·00040		
	Inverse matrix $p = 4$						s_4 0·00063		
	8·04384013						$s(b_0)$ 0·0018		
	−0·01282061	0·93849723					$s(b_1)$ 0·00061		
	−1·04829607	0·00354370	0·23914346				$s(b_2)$ 0·00031		
	0·13457040	−0·00071515	−0·02972746	0·07269913			$s(b_3)$ 0·00017		
	−0·00464985	0·11116064	0·00151187	−0·00050064	0·08210823		$s(b_4)$ 0·00018		

10.1.2 *Elimination of non-significant variables*

When the fitted function is examined it will often be found that the contribution of one (or more) of the variables is negligibly small. This will be evidenced by the value of the corresponding coefficient being considerably less than its standard error. It will usually be necessary to fit a new function from which the non-significant variable has been omitted.

If the variable has been placed last in the fitting scheme the values for the coefficients when it is omitted can be obtained without further calculation. For example, if the variable $\sin\theta$ is omitted from Table 10.1.2, the coefficients b_{pj} are those listed in the line corresponding to $p = 3$. However, the elements of the inverse matrix must be recalculated. From (7.3.4.1,1),

$$(\chi_{jk})_{p-1} = (\chi_{jk})_p - r_{pj}\, r_{pk}.$$

It is clearly very desirable that any variable whose significance may be doubtful should be placed last in the fitting scheme. If a variable occurring elsewhere in the scheme is to be eliminated, it is probably best to draw up a completely new scheme.

10.2 NON-LINEAR FUNCTIONS

If in the equation connecting the variables the parameters B_{pj} which are to be estimated occur in a non-linear form, it is necessary to apply a transformation which will convert the equation into a linear one. This can be done either by a change of variable or by using approximation methods based on a Taylor's series expansion.

10.2.1 *Change of variable*

The variables are to be changed so that the equation is linear in the unknown parameters. A commonly used change of variable is that employing logarithms. For example, if

$$Y = AH^B Z^C$$

where Y, H, and Z are measured and A, B, and C are to be estimated, taking logarithms gives

$$\log Y = \log A + B \log H + C \log Z,$$

or
$$Y' = \Sigma B_j X_j,$$

where $X_0 = 1, \quad X_1 = \log H, \quad X_2 = \log Z, \quad Y' = \log Y,$

$$B_0 = \log A, \quad B_1 = B, \quad B_2 = C.$$

The normal equations have the usual form

$$\Sigma b_j \, \Sigma w_i' x_{ji} x_{ki} = \Sigma w_i' y_i' x_{ki}, \tag{1}$$

but the weights w_i' should be the weights of the transformed variable y_i'. Now, from (1.2,14*a*–*c*),

$$w_i' \propto 1/\text{var}\, y_i' = 1 \bigg/ \left\{ \left(\frac{\partial Y_i'}{\partial Y_i} \right)^2 \text{var}\, y_i \right\},$$

or

$$w_i' \propto w_i \bigg/ \left(\frac{\partial Y_i'}{\partial Y_i} \right)^2 . \tag{2a}$$

For example, if a logarithmic transformation is employed,

$$w_i' = w_i \, Y_i^2. \tag{2b}$$

In practice, the true values Y_i are unknown, and the observed values y_i must be used in their place. Thus for the logarithmic transformation the weights would be

$$w_i' = w_i \, y_i^2.$$

Even if the original observations were all of equal weight, the transformed observations must be weighted, for otherwise the estimates will be inefficient and the standard deviation calculations completely incorrect.

10.2.2 *The simple exponential*

The exponential curve

$$Y = A \, e^{-\lambda X} \tag{1}$$

occurs in many branches of science. On taking logarithms,

$$Y' = B_0 + B_1 X, \tag{2a}$$

where $$Y' = \ln Y, \quad B_0 = \ln A, \quad B_1 = -\lambda, \tag{2b}$$

ln signifying logarithms to the base *e*. The weights w_i' are proportional to $w_i y_i^2$.

10.2.2.1 *Example*

The counts N_0 recorded by a Geiger counter in one-minute intervals with thicknesses t of lead interposed between the counter and a radium source are shown in Table 10.2.2. It is proposed to fit an exponential curve $A \, e^{-\mu t}$ to these values in order to determine the absorption coefficient μ of gamma rays in lead. Taking logarithms,

$$\ln N = \ln A - \mu t.$$

The observed values N_0 are made up of two parts, the part N due to gamma rays and the background count ν. With the present counter the average value of ν is 72. The values $N = N_0 - \nu$ are given in the third column of

Table 10.2.2, and the logarithms in the fourth column. The weights w_i are proportional to $1/N_0$, as the counts follow a Poisson distribution (§ 4.3). Hence the weights w' in the logarithmic form are N^2/N_0. The values $\sqrt{w'}$ are listed in the fifth column, and the products $(\sqrt{w'})\,t$ and $(\sqrt{w'})\ln N$ in the two following columns. The normal equations for $\ln A$ and μ are obtained by intermultiplying these columns, and are solved by the abbreviated Doolittle technique. The estimate of μ is

$$\mu = 0{\cdot}627 \pm 0{\cdot}033.$$

TABLE 10.2.2

Absorption of gamma rays by lead (*Example* 10.2.2.1)

Observations N_0 Background $\nu = 72$ Corrected counts $N = N_0 - \nu$
Lead thickness (cm.) t Weights $w' = N^2/N_0$

(*a*) Observed values

t	N_0	N	$y' = \ln N - 4$	$\sqrt{w'}$	$t\sqrt{w'}$	$y'\sqrt{w'}$	z
0	327	255	1·54126	14	0	21·5776	35·5776
0·2	292	220	1·39363	13	2·6	18·1172	33·7172
0·4	244	172	1·14749	11	4·4	12·6224	28·0224
0·6	236	164	1·09987	11	6·6	12·0986	29·6986
0·8	210	138	0·92725	9	7·2	8·3452	24·5452
1·0	192	120	0·78749	9	9·0	7·0874	25·0874
1·2	185	113	0·72739	8	9·6	5·8191	23·4191
1·4	172	100	0·60517	8	11·2	4·8414	24·0414
1·6	170	98	0·58497	7	11·2	4·0948	22·2948
1·8	160	88	0·47734	7	12·6	3·3414	22·9414
2·0	139	67	0·20469	6	12·0	1·2281	19·2281

(*b*) Normal equations and the abbreviated Doolittle scheme

1031	705·6	1093·140	2829·740	
	848·32	519·154	2073·074	
				1306·142
1031	705·6	1093·140	2829·740	1159·026
1	0·684384	1·060272	2·744656	147·116
	365·419	−228·974	136·445	143·476
	1	−0·626605	0·373394	3·640

$s = \sqrt{(3{\cdot}64/9)} = 0{\cdot}636$ $s(\mu) = s/\sqrt{365} = 0{\cdot}033$

10.2.3 *Linearization*

In this section it will be supposed that Y is of the general form

$$Y = F(X_j; B_k). \tag{1}$$

Estimates b_k of the parameters B_k are to be obtained from the observations y_i, the values x_{ji} being supposed free from error. The residuals are

$$v_i = y_i - F(x_{ji}; b_k). \tag{2}$$

The fundamental idea underlying the linearization process is the expansion of the term $F(x_{ji}; b_k)$ as a linear function. Thus

$$F(x_{ji}; b_k) = F(x_{ji}; b_k^0) + \sum_k \left(\frac{\partial F}{\partial b_k}\right)_i b'_k \tag{3a}$$

where

$$b_k = b_k^0 + b'_k, \tag{3b}$$

b_k^0 being an approximation to b_k. Then if

$$y'_i = y_i - F(x_{ji}; b_k^0), \tag{4a}$$

$$x'_k = \frac{\partial F}{\partial b_k^0}, \tag{4b}$$

(2) becomes

$$v_i = y'_i - \sum_k b'_k x'_{ki}. \tag{5}$$

The least-squares principle states that the parameters are to be chosen to minimize $\Sigma w_i v_i^2$. Thus (5) leads to the normal equations

$$\Sigma w_i \left(y'_i - \sum_k b'_k x'_{ki}\right) x'_{mi} = 0 \tag{6}$$

of the standard form, which can be solved for b'_k. Hence the estimates of the parameters will be

$$b_k = b_k^0 + b'_k.$$

Strictly, the b_k are only approximations, since the expansion ($3a, b$) is valid only if b'_k is small. If, in fact, these values are large, it may be necessary to repeat the calculations, with b_k^0 replaced by the value b_k, to give a further approximation. However, if the original estimates b_k^0 were well chosen, this will not be necessary.

In some problems there are no independent variables X_j, but the Y_i are functions of the parameters alone,

$$Y_i = F_i(B_k), \quad i = 1 \text{ to } n. \tag{7}$$

The treatment given above will apply in this case, with

$$y'_i = y_i - F_i(b_k^0).$$

10.2.3.1 *Implicit functions.* The cases to be considered are those in which Y is related to the X_j by means of the function

$$F(Y, X_j; B_k) = 0. \tag{1}$$

Then, since the fitted value $y_i - v_i$ satisfies the same functional relationship,

$$F(y_i - v_i, x_{ji}; b_k) = 0 = F(y_i, x_{ji}; b_k^0) - \left(\frac{\partial F}{\partial y}\right)_i v_i + \sum_k \left(\frac{\partial F}{\partial b_k}\right)_i b'_k, \tag{2}$$

where

$$b_k = b_k^0 + b'_k,$$

b_k^0 being a reasonable approximation to b_k. Hence

$$\Sigma w_i v_i^2 = \Sigma w_i'(y_i' - \Sigma b_k' x_{ki}')^2, \tag{3a}$$

where
$$y_i' = F(y_i, x_{ji}; b_k^0), \tag{3b}$$

$$x_{ki}' = -\left(\frac{\partial F}{\partial b_k}\right)_i, \tag{3c}$$

$$w_i' = w_i \bigg/ \left(\frac{\partial F}{\partial y}\right)_i^2. \tag{3d}$$

For the new variables defined by these equations the normal equations have the standard forms. However, the treatment will be approximate, no matter how close the values b_k^0 are to b_k, because it is assumed that v_i is so small that higher terms of the expansion (2) may be neglected.

TABLE 10.2.3

Observations of delayed coincidences (*Example* 10.2.3.2)

Counts per 100 seconds y	Sensitive interval (microseconds) t
$2{\cdot}32 \pm 0{\cdot}06$	1·02
$4{\cdot}18 \pm 0{\cdot}10$	3·17
$4{\cdot}97 \pm 0{\cdot}12$	4·49
$7{\cdot}27 \pm 0{\cdot}18$	7·46
$10{\cdot}56 \pm 0{\cdot}26$	12·6
$13{\cdot}56 \pm 0{\cdot}30$	18·3
$15{\cdot}51 \pm 0{\cdot}40$	23·0
$18{\cdot}00 \pm 0{\cdot}45$	29·2
$21{\cdot}02 \pm 0{\cdot}50$	40·6
$22{\cdot}70 \pm 0{\cdot}60$	48·2
$23{\cdot}98 \pm 0{\cdot}60$	58·8
$25{\cdot}50 \pm 0{\cdot}65$	71·5
$25{\cdot}51 \pm 0{\cdot}65$	96·0
$25{\cdot}82 \pm 0{\cdot}65$	119
$25{\cdot}87 \pm 0{\cdot}65$	141

10.2.3.2 *Example*

In an experiment on delayed coincidences (Murdoch, 1953), the number of coincidences per second was measured as a function of the time interval over which the recorder was sensitive after a pulse had appeared in the first counter. The observations are shown in Table 10.2.3. It can be shown that

$$y = A - B\,e^{-\lambda t}, \tag{1}$$

where y is the coincidence rate, t the sensitive time interval, and λ is the decay constant to be determined. For convenience y will be taken as the number of coincidences in 100 seconds.

First, approximate values of the three quantities A, B, and λ must be calculated. Since y is small for $t = 0$, $A \doteqdot B$. For $t = \infty$, $y = A \doteqdot 27$. At $\lambda t = 1$,

$$y = 27(1-e^{-1}) = 17,$$

and so from Table 10.2.3 λt is unity for $t \doteqdot 26$. Hence suitable initial approximations are

$$A_0 = B_0 = 27, \quad \lambda_0 = 0{\cdot}04.$$

TABLE 10.2.3.1

Calculation of values x_j', y' (Example 10.2.3.2)

y	t	$e^{-\lambda_0 t} = -x_1'$	$B_0 t e^{-\lambda_0 t} = x_2'$	y'	z'
2·32	1·02	0·96002	26·439	+1·24054	1·54491
4·18	3·17	0·88092	75·398	+0·96484	1·83790
4·97	4·49	0·83561	101·30	+0·53147	1·70886
7·27	7·46	0·74201	149·46	+0·30427	2·05686
10·56	12·6	0·60412	205·52	−0·12876	2·32232
13·56	18·3	0·48095	237·64	−0·45435	2·44110
15·51	23·0	0·39852	247·48	−0·72996	2·34632
18·00	29·2	0·31099	245·18	−0·60327	2·53754
21·02	40·6	0·19711	216·07	−0·65803	2·30556
22·70	48·2	0·14544	189·28	−0·37312	2·37424
23·98	58·8	0·09518	151·11	−0·45014	1·96578
25·50	71·5	0·05727	110·56	+0·04629	2·09462
25·51	96·0	0·02149	55·70	−0·90977	0·62574
25·82	119	0·00857	27·54	−0·94861	0·31822
25·87	141	0·00356	13·55	−1·03388	0·09806

In the notation of § 10.2.3,

$$x_0' = \partial F/\partial A_0 = 1,$$

$$x_1' = \partial F/\partial B_0 = -e^{-\lambda_0 t},$$

$$x_2' = \partial F/\partial \lambda_0 = tB_0 e^{-\lambda_0 t},$$

$$y' = y - A_0 + B_0 e^{-\lambda_0 t}.$$

The calculation of these quantities is shown in Table 10.2.3.1. The calculation of the normal equations is shown at the top of Table 10.2.3.2. The values $\sqrt{w}$ are taken as the integers nearest $1{\cdot}2/s_i$, s_i being the standard deviation given in Table 10.2.3: that is, σ is taken as 1.2.

The solution of the normal equations is

$$b_0' = -0{\cdot}41, \quad b_1' = -1{\cdot}81, \quad b_2' = -0{\cdot}00388,$$

and so the estimates of the coefficients in (1) are

$$A = 27-0{\cdot}41 = 26{\cdot}59, \quad B = 27-1{\cdot}81 = 25{\cdot}19,$$

$$\lambda = 0{\cdot}040-0{\cdot}00388 = 0{\cdot}03612.$$

The sum of the squares of the residuals is 11·36, and so s is 0·97. This compares well with the value 1·2 of σ – in fact, as in § 2.5.4,

$$\chi^2 = 12(s/\sigma)^2 = 7{\cdot}84$$

with 12 d.f., and $P(\chi^2) = 0{\cdot}80$.

The standard deviation of λ is found by dividing s or σ by $\sqrt{S_{22}} = 10^2\sqrt{147}$, and so is 0·0008 estimated from s and 0·0010 estimated from σ. Thus

$$\lambda = 0{\cdot}0361 \pm 0{\cdot}0010,$$

and the half-life, defined as $0{\cdot}6931/\lambda$, is

$$T = 19{\cdot}2 \pm 0{\cdot}5 \text{ microseconds.}$$

The values A, B, and λ may now be used to recalculate the columns of Table 10.2.3.1, and using these new values a further approximation can be obtained. It is found that the next approximation does not differ appreciably from the approximation calculated above.

The column z' in Table 10.2.3.1 is made up of values

$$z' = 1 + x_1' + 10^{-2}\,x_2' + y',$$

the factor 10^{-2} being removed from x_2' to bring it to the order of unity. This column is useful in checking the multiplications by $\sqrt{w}$ in Table 10.2.3.2. If these are free from mistakes, the sum z_1 of a row of Table 10.2.3.2 should equal $z_i'\sqrt{w_i}$.

TABLE 10.2.3.2

The normal equations for Example 10.2.3.2

Factor removed: x_2', 10^2

$\sqrt{w}$	$\sqrt{w}x_1'$	$\sqrt{w}x_2'$	$\sqrt{w}y'$	z	
20	−19·2004	5·2878	24·8108	30·8982	
12	−10·5710	9·0478	11·5781	22·0549	
10	−8·3561	10·1300	5·3147	17·0886	
7	−5·1941	10·4622	2·1299	14·3980	
5	−3·0206	10·2760	−0·6438	11·6116	
4	−1·9238	9·5056	−1·8174	9·7644	
3	−1·1956	7·4244	−2·1899	7·0389	
3	−0·9330	7·3554	−1·8098	7·6126	
2	−0·3942	4·3214	−1·3161	4·6111	
2	−0·2909	3·7856	−0·7462	4·7485	
2	−0·1904	3·0222	−0·9003	3·9315	
2	−0·1145	2·2112	+0·0926	4·1893	
2	−0·0430	1·1140	−1·8195	1·2515	
2	−0·0171	0·5508	−1·8972	0·6365	
2	−0·0071	0·2710	−2·0678	0·1961	
780	−662·0781	553·1592	663·4128	1334·4939	
	592·6210	−404·9195	−643·4726	−1117·8492	
		675·7205	243·9500	1067·9101	
			808·4930	1072·3831	
					808·4930
780	−662·0781	553·1592	663·4128	1334·4939	564·2520
1	−0·848818	0·709178	0·850529	1·710890	244·2410
	30·6372	64·6120	−80·3559	14·8932	210·7591
	1	2·108939	−2·622820	0·486116	33·4819
		147·1694	−57·0621	90·1075	22·1247
		1	−0·387730	0·612271	11·3572
−0·406721	−1·805121	−0·387730			
0·593275	−0·805126	0·612271			

10.3 HARMONIC ANALYSIS

The general form of a periodic function of fundamental period T is

$$Y = A_0 + \sum_j A_j \cos j\theta + \sum_j B_j \sin j\theta, \tag{1a}$$

where
$$\theta = 2\pi t/T. \tag{1b}$$

If the data are in the form of a continuous curve, there are various mechanical, optical, and electrical instruments which have been devised for the evaluation of the coefficients. These will not be discussed here. If the curve is to be analysed mathematically, it is simplest to replace it by a set of points spaced at equal intervals of θ.

If there are n observations y_i, at points θ_i, the least-squares coefficients are given by the normal equations

$$\Sigma w_i(y_i - a_0 - \Sigma a_j \cos j\theta_i - \Sigma b_j \sin j\theta_i) \begin{Bmatrix} \sin k\theta_i \\ \cos m\theta_i \end{Bmatrix} = 0. \tag{2}$$

10.3.1 *Equally-spaced observations*

When the n observations are uniformly spaced throughout the cycle, and are all of equal weight, the solution of the normal equations is greatly simplified. It will be supposed that the observations are at angles

$$\theta_i = 2\pi i/n = i\phi, \quad i = 0 \text{ to } n-1. \tag{1}$$

Now the standard formula for the sum of a cosine progression gives

$$\sum_{i=0}^{n-1} \cos mi\phi = \sin \tfrac{1}{2}mn\phi \cos \tfrac{1}{2}m(n-1)\,\phi/\sin \tfrac{1}{2}m\phi,$$

and this will vanish for integral values of m since $\frac{1}{2}mn\phi = m\pi$. Thus the sums of the form

$$\sum_i \sin j\theta_i \sin k\theta_i = \tfrac{1}{2}\sum_i \{\cos(k-j)\,i\phi - \cos(k+j)\,i\phi\}$$

will vanish unless $k = j$, the value then being $\frac{1}{2}n$. In fact, the functions $\cos j\theta$, $\sin j\theta$ are orthogonal over the set of angles 0 to $(n-1)2\pi/n$. The normal equations (10.3,2) then have the simple forms

$$a_j = 2\{\Sigma y_i \cos j\theta_i\}/n, \quad a_0 = \Sigma y_i/n, \tag{2a}$$

$$b_j = 2\{\Sigma y_i \sin j\theta_i\}/n. \tag{2b}$$

Some authors include a factor $\frac{1}{2}$ before A_0 in (10.3,1a); their formula for a_0 is then $2\Sigma y_i/n$.

From (2*a*) the variance of a_j is given by

$$(2/n)^2\left(\sum_i \cos^2 j\theta_i\right)\sigma^2,$$

and so

$$\operatorname{var} a_j = 2\sigma^2/n = \operatorname{var} b_j, \tag{3a}$$

$$\operatorname{var} a_0 = \sigma^2/n. \tag{3b}$$

The variances of all the coefficients (except a_0) are the same. The covariance of a_j, b_k is

$$(4\sigma^2/n^2)\sum_i \cos j\theta_i \sin k\theta_i,$$

and so

$$\operatorname{cov}(a_j, b_k) = 0 = \operatorname{cov}(a_j, a_k) = \operatorname{cov}(b_j, b_k). \tag{3c}$$

The variance of the fitted value

$$u_p(\theta) = a_0 + \sum_{j=1}^{p} a_j \cos j\theta + \sum_{j=1}^{p} b_j \sin j\theta \tag{4a}$$

will then be

$$\operatorname{var} a_0 + \sum_{j=1}^{p} (\cos^2 j\theta \operatorname{var} a_j + \sin^2 j\theta \operatorname{var} b_j),$$

or

$$\operatorname{var} u_p(\theta) = (2p+1)\,\sigma^2/n. \tag{4b}$$

The variance of the fitted value is independent of the angle θ.

The standard deviation of an observation may be estimated from Σv_i^2 in the usual way. The value Σv_i^2 is

$$\Sigma v_i^2 = \Sigma y_i^2 - na_0^2 - \sum_j a_j^2 \sum_i \cos^2 j\theta_i - \sum_j b_j^2 \sum_i \sin^2 j\theta_i,$$

or

$$\Sigma v_i^2 = \Sigma y_i^2 - na_0^2 - \tfrac{1}{2}n \sum_j (a_j^2 + b_j^2). \tag{5}$$

The expectation of this is $(n-2p-1)\,\sigma^2$, and so

$$s^2 = \Sigma v_i^2/(n-2p-1) \tag{6}$$

will provide an estimate of the standard deviation of an observation.

10.3.2 *n a multiple of four*

If $n = 4q$, the values of $\cos j\theta_i$ for $i = r,\ 2q-r,\ 2q+r,\ 4q-r$ all have the same magnitude. Hence (10.3.1,2*a*) becomes

$$2qa_j = \sum_{r=0}^{q-1} \{y_r + y_{4q-r} + (-)^j (y_{2q-r} + y_{2q+r})\} \cos j\theta_r. \tag{1a}$$

Similarly, (10.3.1,2*b*) becomes

$$2qb_j = \sum_{r=0}^{q-1} \{y_r - y_{4q-r} - (-)^j (y_{2q-r} - y_{2q+r})\} \sin j\theta_r. \tag{1b}$$

If the observations are grouped to give the quantities

$$\alpha_r = y_r + y_{4q-r} + y_{2q-r} + y_{2q+r}, \tag{2a}$$

$$\alpha'_r = y_r + y_{4q-r} - y_{2q-r} - y_{2q+r}, \tag{2b}$$

$$\beta_r = y_r - y_{4q-r} + y_{2q-r} - y_{2q+r}, \tag{2c}$$

$$\beta'_r = y_r - y_{4q-r} - y_{2q-r} + y_{2q+r}, \tag{2d}$$

then

$$2qa_k = \Sigma\alpha_r \cos r\theta_k, \; k \text{ even}, \tag{3a}$$

$$2qa_k = \Sigma\alpha'_r \cos r\theta_k, \; k \text{ odd}; \tag{3b}$$

$$2qb_k = \Sigma\beta'_r \sin r\theta_k, \; k \text{ even}, \tag{3c}$$

$$2qb_k = \Sigma\beta_r \sin r\theta_k, \; k \text{ odd}. \tag{3d}$$

The values $\cos r\theta_k$ and $\sin r\theta_k$ are listed in Table 10.6*a* for $n = 8$, 12, 16, and 24. From (10.3.1,1), $r\theta_k = k\theta_r$.

The formation of the sums (3*a–d*) is illustrated in Table 10.3.2 for the specific case $n = 12$. The equation

$$\Sigma\alpha_j + \Sigma\alpha'_j + \Sigma\beta_j + \Sigma\beta'_j = 4\sum_0^{q-1} y_i \tag{4}$$

will provide a check on the formation of these quantities.

TABLE 10.3.2

Formation of the sums α, β when n is a multiple of four

List	$+ + + +$ α	$+ - - +$ α'	$+ + - -$ β	$+ - + -$ β'
$y_0 \quad y_6$	$y_0 \quad + y_6$	$y_0 \quad - y_6$	$y_0 \quad - y_6$	$y_0 \quad + y_6$
$y_1\, y_5\, y_7\, y_{11}$	$y_1 + y_5 + y_7 + y_{11}$	$y_1 - y_5 - y_7 + y_{11}$	$y_1 + y_5 - y_7 - y_{11}$	$y_1 - y_5 + y_7 - y_{11}$
$y_2\, y_4\, y_8\, y_{10}$	$y_2 + y_4 + y_8 + y_{10}$	$y_2 - y_4 - y_8 + y_{10}$	$y_2 + y_4 - y_8 - y_{10}$	$y_2 - y_4 + y_8 - y_{10}$
$y_3 \quad y_9$	$y_3 \quad + y_9$	$y_3 \quad - y_9$	$y_3 \quad - y_9$	$y_3 \quad + y_9$

10.3.3 *Harmonic curve through all the points*

If n coefficients are determined, the curve will pass through all the n observed points, and $\Sigma v_i^2 = 0$.

If n is even, only one of the coefficients $A_{\frac{1}{2}n}, B_{\frac{1}{2}n}$ can be estimated. It is usual to estimate $A_{\frac{1}{2}n}$. Since

$$\Sigma \cos^2 \tfrac{1}{2}n\theta_i = \Sigma \cos^2 i\pi = n,$$

$$a_{\frac{1}{2}n} = \{\Sigma y_i \cos \tfrac{1}{2}n\theta_i\}/n. \tag{1}$$

If all the n coefficients are evaluated, the following equations will provide a check on the arithmetical calculations:

$$\Sigma a_k = y_0; \tag{2a}$$

$$2\Sigma b_k \sin k(2\pi/n) = y_1 - y_{n-1}. \tag{2b}$$

These are obtained by equating the fitted values at $\theta = 0$, $2\pi/n$, and $2\pi(n-1)/n$ to the observed values y_0, y_1, and y_{n-1}.

10.3.4 *Example*

Fig. 10.3.4 shows the waveform of the output voltage from an overloaded transformer-coupled amplifier. The amplitudes at intervals of one-twelfth of the cycle are recorded at the left of the top section of Table 10.3.4, and the quantities α, β are calculated at the right.

The coefficients a_j, b_j are then calculated by multiplication of these columns by the columns of Table 10.6a for $n = 12$. The calculations are checked using (10.3.3,2a, b).

TABLE 10.3.4

Calculation of harmonic amplitudes for the waveform of Fig. 10.3.4

	y_i			+ + + + α	+ − − + α'	+ + − − β	+ − + − β'
−6·0		+11·9		+5·9	−17·9	−17·9	+5·9
−0·3	+11·2	+8·4	−10·7	+8·6	−30·6	+13·2	+7·6
+4·4	+9·8	−1·4	−11·4	+1·4	−15·4	+27·0	+4·6
+7·9		−8·0		−0·1	+15·9	+15·9	−0·1
Check $\Sigma\alpha$ = 24·0							

$12a_0$	15·8		
$12a_6$	−1·2		
$6a_2$	9·60	$6b_2$	10·56520
$6a_4$	0·80	$6b_4$	2·59800
$6a_1$	−52·09960	$6b_1$	45·882
$6a_5$	+0·89960	$6b_5$	−0·882
$6a_3$	−2·5	$6b_3$	−2·7
a_0	1·3167		
a_1	−8·6833	b_1	7·6470
a_2	1·6000	b_2	1·7609
a_3	−0·4167	b_3	−0·4500
a_4	0·1333	b_4	0·4330
a_5	0·1499	b_5	−0·1470
a_6	−0·1000		
Check Σa_j =	−6·0001	$2\Sigma b_j \sin j\pi/6$ =	10·3998

10.3.5 *Amplitude and phase*

The fitted curve may be required in the form

$$Y = C_0 + \Sigma C_j \sin(j\theta + D_j), \tag{1}$$

where C_j is the amplitude and D_j the phase of the jth harmonic. On expanding the sine term, and comparing with (10.3,1a),

$$C_j \cos D_j = B_j, \quad C_j \sin D_j = A_j, \tag{2a}$$

or

$$C_j^2 = A_j^2 + B_j^2, \quad \tan D_j = A_j/B_j. \tag{2b}$$

Hence the least-squares estimates of C_j and D_j will be

$$c_j^2 = a_j^2 + b_j^2, \quad d_j = \tan^{-1} a_j/b_j. \tag{2c}$$

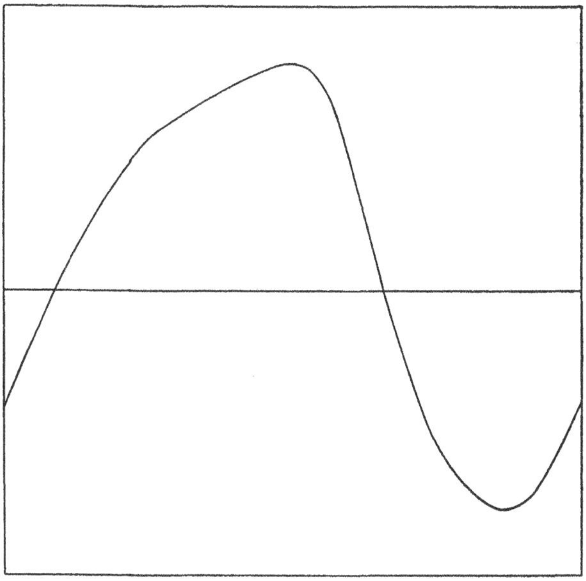

Fig. 10.3.4. Output voltage of an overloaded amplifier.

The variance of c_j will be given by

$$\operatorname{var} c_j = \operatorname{var} \sqrt{(a_j^2 + b_j^2)} = (A_j^2 \operatorname{var} a_j + B_j^2 \operatorname{var} b_j)/(A_j^2 + B_j^2),$$

and so

$$\operatorname{var} c_j = \operatorname{var} a_j = 2\sigma^2/n, \tag{3a}$$

$$\operatorname{var} c_0 = \sigma^2/n. \tag{3b}$$

The variance of d_j can be calculated from

$$\operatorname{var} \tan d_j = \sec^4 D_j \operatorname{var} d_j.$$

Thus

$$\operatorname{var} d_j = B_j^4 (A_j^2 + B_j^2)^{-2} \operatorname{var}(a_j/b_j),$$

or

$$\operatorname{var} d_j = 2\sigma^2/nC_j^2. \tag{4}$$

10.3.5.1 *Example*

The amplitudes and phases of the components of the waveform drawn in Fig. 10.3.4 are calculated in Table 10.3.5. The standard deviations are evaluated using for σ a value 0·2, this representing the error in the measurement of the waveform amplitudes y_i at the points θ_i.

The components of the waveform up to the fifth harmonic are given by

$$u(\theta) = 1{\cdot}32 + 11{\cdot}57 \sin(\theta - 48^\circ\, 38') + 2{\cdot}38 \sin(2\theta + 42^\circ) + 0{\cdot}61 \sin(3\theta - 137^\circ) + 0{\cdot}45 \sin(4\theta + 17^\circ) + 0{\cdot}21 \sin(5\theta + 134^\circ).$$

TABLE 10.3.5

Amplitudes c_j and phases d_j for the waveform of Fig. 10.3.4

$\sigma = 0{\cdot}2$

$\sigma^2(c) = 2\sigma^2/n \quad \sigma(c) = 0{\cdot}082 \quad \sigma(c_0) = \sigma(c)/\sqrt{2} = 0{\cdot}058$

$\sigma(d) = \{\sigma(c)\}/c$ radians $= 4{\cdot}7/c$ degrees

	c	$\sigma(c)$		d	$\sigma(d)$
c_0	1·3167	0·058			
c_1	11·5705	0·082	d_1	−48° 38′	24′
c_2	2·3792	0·082	d_2	42° 16′	2°
c_3	0·6133	0·082	d_3	−137° 12′	8°
c_4	0·4531	0·082	d_4	17° 07′	10°
c_5	0·2100	0·082	d_5	134° 26′	22°

10.3.6 *Correction for grouping*

In certain cases the observed values y_i will be mean values over a certain interval—for example, mean monthly temperatures. The mean value will differ from the value at the centre of the region by an amount which depends on the shape of the curve in that region. Thus if Y° is the central value,

$$Y = Y^\circ + \frac{dY}{d\theta}\Delta\theta + \frac{1}{2}\frac{d^2Y}{d\theta^2}(\Delta\theta)^2,$$

and the average value over the interval $\pm\frac{1}{2}\phi$ is

$$Y_i = Y^\circ + \frac{1}{2}\frac{d^2Y}{d\theta^2}\phi^{-1}\int_{-\frac{1}{2}\phi}^{\frac{1}{2}\phi}(\Delta\theta)^2\,d(\Delta\theta),$$

or

$$Y_i = Y_i^\circ + \frac{1}{24}\phi^2\frac{d^2Y}{d\theta^2}. \tag{1}$$

Hence if

$$Y_i = A_0 + \Sigma A_j \cos j\theta_i + \Sigma B_j \sin j\theta_i, \tag{2a}$$

then

$$Y_i^\circ = A_0 + \Sigma A_j' \cos j\theta_i + \Sigma B_j' \sin j\theta_i, \tag{2b}$$

where

$$A_j' = A_j(1 + \tfrac{1}{24}j^2\phi^2), \quad B_j' = B_j(1 + \tfrac{1}{24}j^2\phi^2).$$

Therefore the coefficients a_j, b_j obtained by fitting a curve to the grouped values y_i should be corrected by multiplying by the factors $(1+\frac{1}{24}j^2\phi^2)$.

10.3.7 *Observations over several periods*

If observations are available over several periods, the values at θ, $\theta+2\pi$, $\theta+4\pi$, etc., can be grouped and the harmonic coefficients determined from the grouped values. Thus if there are rn observations, and

$$y_{ri} = \sum_q y(\theta_i + 2\pi q), \tag{1}$$

then

$$a_j = 2\left\{\sum_{i=0}^{n-1} y_{ri}\cos j\theta_i\right\} \Big/ rn. \tag{2}$$

Also

$$\Sigma v_i^2 = \sum_{i=0}^{nr-1} (y_i - a_0 - \Sigma a_j \cos j\theta_i - \Sigma b_j \sin j\theta_i)^2,$$

or

$$\Sigma v_i^2 = \Sigma y_i^2 - nra_0^2 - \frac{1}{2}\sum_j nr(a_j^2 + b_j^2). \tag{3}$$

Hence

$$s^2 = \Sigma v_i^2/(nr-2p-1)$$

will give an estimate of σ^2 based on $nr-2p-1$ degrees of freedom.

10.3.8 *The search for unknown periods*

It will be supposed that n observations y_i are taken, at unit intervals of time, over a length of time much greater than the periods which are present. Then if an oscillation of period T is present, it will contribute

$$A\cos i\omega + B\sin i\omega \tag{1}$$

to y_i, where $\omega = 2\pi/T$ is the angular frequency.

If the first term in (1) is multiplied by $\cos i\omega'$ and summed over the n values of i, the formula

$$\tfrac{1}{2}A\left\{\frac{\sin\frac{1}{2}n(\omega-\omega')\cos\frac{1}{2}(n-1)(\omega-\omega')}{\sin\frac{1}{2}(\omega-\omega')} + \frac{\sin\frac{1}{2}n(\omega+\omega')\cos\frac{1}{2}(n-1)(\omega+\omega')}{\sin\frac{1}{2}(\omega+\omega')}\right\}$$

is obtained. The first term will be large only if $\omega-\omega'$ is very small, when it has the value $\frac{1}{2}An\cos\frac{1}{2}(n-1)(\omega-\omega')$, while the second term is never large. Hence

$$\sum_0^{n-1} A\cos i\omega\cos i\omega' = \tfrac{1}{2}nA\cos\tfrac{1}{2}(n-1)(\omega-\omega'), \quad \omega\sim\omega'.$$

By using similar expansions of products it follows that, when ω' is nearly equal to ω,

$$E(2/n)\,\Sigma y_i \cos i\omega' = A\cos\tfrac{1}{2}(n-1)(\omega-\omega') + B\sin\tfrac{1}{2}(n-1)(\omega-\omega'), \qquad (2a)$$

$$E(2/n)\,\Sigma y_i \sin i\omega' = -A\sin\tfrac{1}{2}(n-1)(\omega-\omega') + B\cos\tfrac{1}{2}(n-1)(\omega-\omega'). \qquad (2b)$$

Thus if

$$a = (2/n)\,\Sigma y_i \cos i\omega', \quad b = (2/n)\,\Sigma y_i \sin i\omega', \qquad (3a)$$

then

$$c^2 = a^2+b^2 \qquad (3b)$$

will provide an estimate of the amplitude of the frequency ω in the neighbourhood of ω'.

10.3.9 *Significance tests for the amplitude of a period*

If the deviations in y_i are assumed normally distributed, then a and b will also be normally distributed. When the true values of the coefficients A and B are zero, the probability that the values determined will lie in ranges $a \pm \frac{1}{2}da, b\pm\frac{1}{2}db$ is

$$dP(a,b) = (n/4\pi\sigma^2)\{\exp - n(a^2+b^2)/4\sigma^2\}\,da\,db. \qquad (1a)$$

On transforming to angular coordinates c, ψ, and integrating over the angle

$$dP(c) = (n/2\sigma^2)\{\exp - nc^2/4\sigma^2\}\,c\,dc. \qquad (1b)$$

The probability of obtaining a value greater than or equal to c^2 is then

$$Q(c^2) = (n/4\sigma^2)\int_{c^2}^{\infty}\{\exp - nz^2/4\sigma^2\}\,dz^2,$$

or

$$Q(c^2) = \exp - nc^2/4\sigma^2 = e^{-\kappa}, \qquad (2a)$$

where

$$\kappa = nc^2/4\sigma^2 = c^2/E(c^2). \qquad (2b)$$

Hence the significance of a frequency component can be tested. It will be noted that $E(c^2)$ is here equal to $4\sigma^2/n$. The analysis of § 10.3.5 does not hold when A and B are both zero.

Sometimes the amplitudes of a certain number, m say, of frequency components are determined, and it is desired to test the significance of the largest of these. The probability that the square of the largest of m amplitudes is greater than or equal to c^2 is, from (2a),

$$Q_m(c^2) = 1-(1-e^{-\kappa})^m. \qquad (3)$$

No exact test is available for the other periods, but it is suggested that those periods which, when used in (3), give values of Q less than 0·5 can be at least tentatively accepted as real.

In these tests σ is assumed known. In cases where σ has to be estimated from the observations, Fisher (1929) has given a more exact test. Tables for use with this test and with the test based on (3) are given by Davis (1941).

Hartley (1949) discusses a test based on the F-ratio

$$\frac{nc^2}{4\sigma^2}\Big/\frac{\Sigma v^2}{(n-2p-1)\,\sigma^2} = \tfrac{1}{4}n(n-2p-1)\,c^2/\Sigma v^2, \tag{4}$$

where p periods have been determined and Σv^2 is the sum of the squares of the residuals. If the true value of the amplitude is zero, this ratio will be distributed as F with $(2, n-2p-1)$ d.f. To test whether the largest of p amplitudes is significant at a level α, the value F given by (4) is tested at a level α/p. If this period is significant, the next largest amplitude is tested at a level $\alpha/(p-1)$, and the tests continued till an amplitude below the significance level is reached.

The difficulty with periodic analysis is that readings have to be taken over many periods in order to sort out the various frequency components, while in many branches of science the periods do not always persist unchanged for such long intervals of time. It is always advisable to divide the observations into two or more groups and to examine each group separately to see whether the major terms do remain unchanged.

10.4 SMOOTHING

In the process of smoothing, graduation, or trend elimination, the smoothed value is obtained from the observations in the immediate neighbourhood of the point, rather than from the whole set of observations as in the standard least-squares methods. This is an advantage if it is suspected that the form of the curve changes radically from one end of the range to the other. The standard smoothing methods only apply to observations which are equally spaced.

10.4.1 *Least-squares smoothing*

One method of smoothing is to fit a least-squares curve of degree p to the $n = 2m+1$ values centred at each point. Then the smoothed value u_p at any given point is

$$u_p(0) = \Sigma a_j \beta_{0j},$$

or

$$u_p(0) = \sum_{\epsilon} \{(1/n) + (\beta_{02}/S_{22})\,T_2(\epsilon) + (\beta_{04}/S_{44})\,T_4(\epsilon)\}\,y(\epsilon). \tag{1}$$

Now (7.7.3,1a, b) gives when $\epsilon = 0$ the equation

$$\rho_j = -\beta_{0,\, j+1}/\beta_{0,\, j-1},$$

and S_{jj} is $\Pi \rho_j$. Equation (1) can then be written

$$u_p(0) = (1/\rho_0) \sum_{\epsilon} \{1 - (1/\rho_2)\, T_2(\epsilon) + (1/\rho_2\, \rho_4)\, T_4(\epsilon)\}\, y(\epsilon).$$

But from (7.7.3,1a, b),

$$T_2(\epsilon) - \rho_2 = T_3(\epsilon)/\epsilon, \quad T_4(\epsilon) - \rho_4\, T_3(\epsilon)/\epsilon = T_5(\epsilon)/\epsilon,$$

and so

$$u_3(0) = -(1/\rho_0\, \rho_2) \sum_{\epsilon} \{T_3(\epsilon)/\epsilon\}\, y(\epsilon)$$

$$= -(1/\rho_0\, \rho_2) \sum_{\epsilon} (\epsilon^2 + \beta_{13})\, y(\epsilon), \qquad (2a)$$

$$u_5(0) = (1/\rho_0\, \rho_2\, \rho_4) \sum_{\epsilon} \{T_5(\epsilon)/\epsilon\}\, y(\epsilon)$$

$$= (1/\rho_0\, \rho_2\, \rho_4) \sum_{\epsilon} \{\epsilon^4 + \beta_{35}\, \epsilon^2 + \beta_{15}\}\, y(\epsilon). \qquad (2b)$$

Explicit formulae can be written down by using the expressions for ρ_j and β_{kj} listed in Table 7.10a.

It is therefore not necessary to calculate the least-squares curve in its usual form to obtain $u_p(0)$. It is only necessary to multiply the observations at each value of ϵ by the factors given in (2a, b). For tabulation of these factors the form

$$u_p(0) = \sum_{\epsilon} z_{p0}(\epsilon)\, y(\epsilon)/Z_{p0}, \qquad (3)$$

where the $z_{p0}(\epsilon)$ are the least integers, is more useful. Table 10.6b gives the values z_{30} for n from 5 to 13. Values of z_{30} and z_{50} for n up to 21 are given by Kendall (Vol. II, 1948, p. 374) and by Whittaker and Robinson (1944, p. 295). Values for n up to 30 have been given by Kerawala (see § 7.6.4).

10.4.2 *Example*

Table 10.4.2 shows the calculation of the smoothed values, using a 9-point cubic, for the observations of Table 7.6.2.1. The observations are listed in order. The factors z_{30} (ϵ) are written down from Table 10.6b on a separate slip of paper which is placed beside the observations. Table 10.4.2 shows the position of the slip for the calculation of the fitted value corresponding to the observed value 35. The y and z are intermultiplied and the sum of the products is divided by Z_{30} to give the fitted value u_3. The slip of paper is then moved down so that the arrow is opposite the next observation.

10.4.3 *Fitted values at the ends of the range*

The method described above is not applicable to the $m = (n-1)/2$ points at each end of the range. If fitted values are required at these points the polynomial fitting the first n points and the

TABLE 10.4.2

Smoothing by means of a 9-point cubic (Example 10.4.2)

		y	Σyz	u_3
		7		
		3		
		1		
(on a separate		3		
slip of paper)		0	371	1·6
		4	850	3·7
	z_{30}	6	1653	7·2
		10	2414	10·5
	−21	15	2747	11·9
	14	18	3223	14·0
	39	15	4848	21·0
	54	15	5807	25·1
	→59	35	5268	22·8
	54	44	6773	29·3
	39	19	8733	37·8
	14	22	9916	42·9
	−21	74	10267	44·4
		50	10711	46·4
Z_{30}	231	38	10943	47·4
		37	8984	38·9
		29	5306	23·0
		16	3797	16·4
		7	2652	11·5
		3	1827	7·9
		10	1634	7·1
		13	2052	8·9
		10	2378	10·3
		8		
		10		
		6		
		5		

polynomial fitting the last n points may be calculated. Again it is possible to use formulae of the form

$$u_p(j) = \Sigma z_{pj}(\epsilon)\, y(\epsilon)/Z_{pj}. \tag{1}$$

$z_{3j}(\epsilon)$ is tabulated in Table 10.6c for value of n up to 13. The two fitted values at the very ends of the range will be rather inefficient.

10.4.3.1 *Example*

Table 10.4.3 shows the calculation of the fitted values at the extremes of the range for the example of § 10.4.2. The fitted value at $\epsilon = j$ is found by intermultiplying the z_{3j} and y columns and dividing by Z_{3j}. The observations at the end of the range are written down in reverse order.

TABLE 10.4.3

Smoothed values at the extremes of the range

ϵ \ j	$z_{3j}(\epsilon)$ 4	3	2	1	Beginning of range y	u_3	End of range y	u_3
4	1190	392	−28	−168	7	6·7	5	4·5
3	392	455	392	252	3	3·5	6	7·0
2	−28	392	515	432	1	1·6	10	8·8
1	−168	252	432	435	3	1·0	8	9·9
0	−126	84	234	324	0		10	
−1	0	−63	12	162	4		13	
−2	112	−140	−143	12	6		10	
−3	112	−98	−140	−63	10		3	
−4	−98	112	112	0	15		7	
Z_{3j}		1386						

10.4.4 *Summation formulae*

When the number of values n over which smoothing is to take place is large, the calculation of least-squares values becomes rather tedious. In such problems it may be better to use a less efficient method such as the summation method to be described in this section.

In deriving the summation formulae it is assumed that the function can be represented as a cubic over the range of the observations used to calculate a smoothed value. Then, using central difference formulae (cf. Whittaker and Robinson, 1944, Ch. 3),

$$Y_i = Y_0 + \tfrac{1}{2}i^2\delta^2 Y_0 + \text{odd differences}, \qquad (1a)$$

where

$$\delta^2 Y_0 = Y_{-1} - 2Y_0 + Y_{+1}. \qquad (1b)$$

The symbol $n^{-1}[n]Y_0$ will be used to denote the mean value of the n observations centred on Y_0. Then

$$En^{-1}[n]y_0 = Y_0 + \tfrac{1}{24}(n^2-1)\delta^2 Y_0, \qquad (2)$$

since Σi^2 is $n(n^2-1)/12$. Now

$$n^{-1}[n]\delta^2 Y_0 = \delta^2 Y_0 + \tfrac{1}{24}(n^2-1)\delta^4 Y_0 = \delta^2 Y_0$$

if differences beyond the third are neglected, and so

$$u_0 = n^{-1}[n]\{1 - \tfrac{1}{24}(n^2-1)\delta^2\}y_0 \qquad (3)$$

will be an unbiased estimate of Y_0. The means so formed may be averaged in a similar way, leading to general formulae which are written symbolically

$$u_0 = n_1^{-1}n_2^{-1}n_3^{-1}[n_1][n_2][n_3]\{1 - \tfrac{1}{24}\Sigma(n^2-1)\delta^2\}y_0. \qquad (4)$$

Spencer's formulae are the most commonly used. If the values n_i are 4, 4, and 5,

$$u_0 = \tfrac{1}{80}[4][4][5]\{1-\tfrac{9}{4}\delta^2\}y_0,$$

which is written

$$u_0 = \tfrac{1}{320}[4][4][5][-9, 22, -9]y_0, \tag{5a}$$

where
$$[-9, 22, -9]y_0 \equiv -9y_{-1}+22y_0-9y_1. \tag{5b}$$

The factors are still rather large. Since

$$\delta^4 y_0 = y_{-2}-4y_{-1}+6y_0-4y_1+y_2$$

is supposed negligible, $3\delta^4 y_0$ can be subtracted from the right-hand side of (5*b*), the formula

$$u_0 = \tfrac{1}{320}[4][4][5][-3, +3, +4, +3, -3]y_0 \tag{6}$$

being obtained. This is Spencer's 15-point formula, each smoothed value using 15 observed values. Spencer's 21-point formula

$$u_0 = \tfrac{1}{350}[5][5][7][-1, 0, 1, 2, 1, 0, -1]y_0 \tag{7}$$

is obtained in a similar way. In words, (6) states that the sums $-3y_{-2}+3y_{-1}+4y_0+3y_1-3y_2$ are formed for all sets of 5 successive observations in the series. These are then summed in fives, the results summed in fours and then again in fours, and the final smoothed values obtained by dividing by 320.

Values obtained by summation are not as accurate as the least-squares values obtained from the same set of points, but they are more rapidly calculated. The 15-point formula gives an efficiency of 0·78. For a $(2m+1)$-point formula, no fitted values are obtained for the m points at each end of the sequence.

10.4.4.1 *Example*

In Table 10.4.4 the observations of Table 10.4.2 are smoothed by means of Spencer's 15-point formula. The observations are listed in the first column. The sums $-3y_{i-2}+3y_{i-1}+4y_i+3y_{i+1}-3y_{i+2}$, formed on a calculating machine, are given in the second column. These are then summed in fives (column 3), fours (column 4), and fours (column 5). The final smoothed values u are obtained by division by 320.

10.4.5 *Comparison of smoothing methods*

Figure 10.4.5 shows the plots of the smoothed values obtained from a least-squares curve fitted to the whole set (Table 7.6.2.3*a*), from a 9-point least-squares cubic formula (Table 10.4.2), and from a 15-point Spencer's formula (Table 10.4.4).

The higher the value of n the smoother is the fitted curve. Alternatively, the lower the value of n the more closely does the fitted curve follow local variations. Clearly the appropriate value of n will depend on the purpose for which the curve is required. No general rule can be given, and the choice of the amount of smoothing desirable is largely a matter of personal judgment.

TABLE 10.4.4

Smoothing by means of Spencer's 15-point formula

	y	2	3	4	5	u
	7					
	3					
	1	1				
	3	−6				
	0	0	11			
	4	−5	47			
	6	21	134	413		
	10	37	221	637	2988	9·3
	15	81	235	828	4026	12·6
	18	87	238	1110	5027	15·7
	15	9	416	1451	6097	19·1
	15	24	562	1638	7284	22·8
	35	215	422	1898	8527	26·6
	44	227	498	2297	9997	31·2
	19	−53	815	2694	11749	36·7
	22	85	959	3108	13382	41·8
	74	341	836	3650	14465	45·2
	50	359	1040	3930	14713	46·0
	38	104	1095	3777	13666	42·7
	37	151	806	3356	11417	35·7
	29	140	415	2603	8627	27·0
	16	52	287	1681	5946	18·6
	7	−32	173	987	3901	12·2
	3	−24	112	675	2836	8·9
	10	37	103	558		
	13	79	170	616		
	10	43	231			
	8	35				
	10	37				
	6					
	5					
Sum	533	2045	9826	37907	144648	452·1

The sums of the columns can be used as a check on the calculations. Thus

$$37907 = 4 \times 9826 - 3 \times (11 + 231) - 2 \times (47 + 170) - (134 + 103).$$

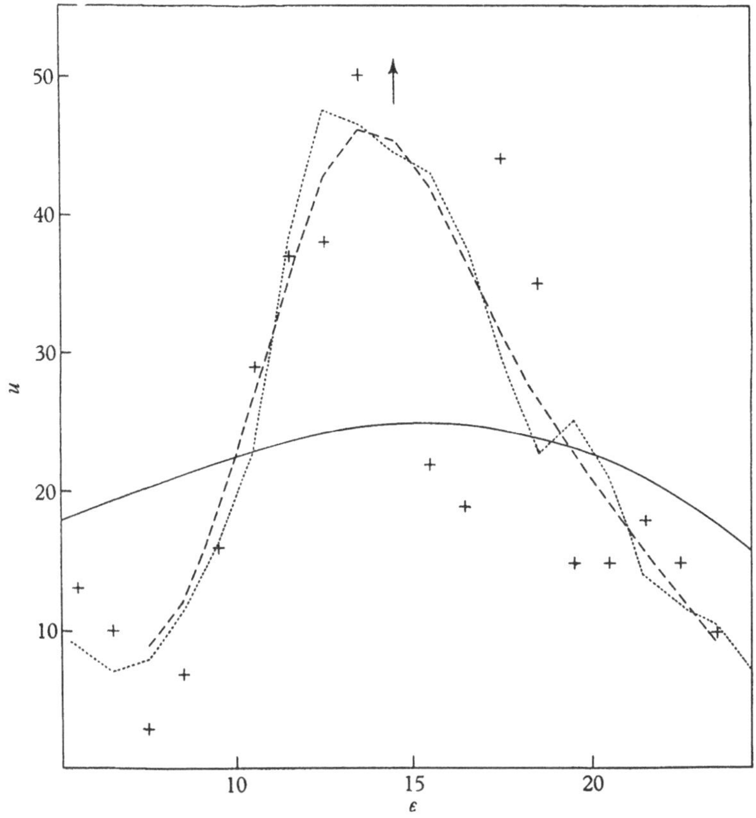

Fig. 10.4.5. Comparison of smoothing methods. The observations are marked with crosses. The solid line is the 62-point least-squares curve, the dotted line is the join of the 9-point least-squares smoothed values, and the dashed line is the join of the 15-point Spencer's formula values.

10.5 NOTES AND REFERENCES

(10.2) The linearization procedure is described by Deming (1943). The following two recent applications of the linearization procedure are of interest in physics. Dumond and Cohen (1953), Cohen *et al.* (1955), use it in the determination of the best values of the fundamental constants, and Breitenberger (1956) has discussed its use in the representation of nuclear scattering data by a series of Legendre polynomials. Price (1954) has considered how, by a suitable choice of angles, these polynomials may be made orthogonal to one another.

Methods of representing data by a series of exponentials $\Sigma A_i \exp -\lambda_i x$ have received some attention. If the values x are spaced at equal intervals, the values λ_i may be derived by Prony's method (Hildebrand, 1956, p. 378; Whittaker and Robinson, 1944, p. 369). The values so obtained are often not very accurate, and the method may give rise to spurious oscillatory terms. Other references are: Householder (1949), Keeping (1951), and Cornell (1956).

(10.3) References on Fourier analysis include: Jackson (1941), Whittaker and Robinson (1944), Danielson and Lanczos (1942), and Worthing and Geffner (1943). Most textbooks on a.c. circuit theory and on optics include a treatment of Fourier series.

For the determination of unknown periods by the periodogram and the correlogram see the books by Brunt (1917), Whittaker and Robinson (1944), and Kendall (1948), and the paper by Kendall (1946*a*). The determination of periods using the Prony method is discussed by Hildebrand (1956).

Fisher's paper on tests of significance in harmonic analysis is reproduced in his book *Contributions to Mathematical Statistics*, John Wiley and Sons, New York, 1950.

(10.4) Summation formulae are described by Whittaker and Robinson (1944) and Kendall (1948). A book by Sasuly (1934) gives details of various methods of smoothing.

10.6 TABLES

TABLE 10.6a

Multiplying factors in harmonic analysis

8 observations					
Upper	$8a_0$	$4a_2$		$4a_1$	
Lower	$8a_4$			$4a_3$	
Sign					
α_0	1	1	α_0'	1	
α_1	± 1	0	α_1'	$\pm 0{\cdot}7071$	
α_2	1	-1	α_2'	0	
Upper	$4b_1$			$4b_2$	
Lower	$4b_3$				
Sign					
β_0	0		β_0'	0	
β_1	$0{\cdot}7071$		β_1'	1	
β_2	± 1		β_2'	0	
12 observations					
Upper	$12a_0$	$6a_2$		$6a_1$	$6a_3$
Lower	$12a_6$	$6a_4$		$6a_5$	
Sign					
α_0	1	1	α_0'	1	1
α_1	± 1	$\pm 0{\cdot}5$	α_1'	$\pm 0{\cdot}8660$	0
α_2	1	$-0{\cdot}5$	α_2'	$0{\cdot}5$	-1
α_3	± 1	∓ 1	α_3'	0	0
Upper	$6b_1$	$6b_3$		$6b_2$	
Lower	$6b_5$			$6b_4$	
Sign					
β_0	0	0	β_0'	0	
β_1	$0{\cdot}5$	1	β_1'	$0{\cdot}8660$	
β_2	$\pm 0{\cdot}8660$	0	β_2'	$\pm 0{\cdot}8660$	
β_3	1	-1	β_3'	0	

TABLE 10.6a (cont.)

16 observations

Upper	$16a_0$	$8a_2$	$8a_4$		$8a_1$	$8a_3$
Lower	$16a_8$	$8a_6$			$8a_7$	$8a_5$
Sign						
α_0	1	1	1	α_0'	1	1
α_1	± 1	$\pm 0{\cdot}7071$	0	α_1'	$\pm 0{\cdot}9239$	$\pm 0{\cdot}3827$
α_2	1	0	-1	α_2'	$0{\cdot}7071$	$-0{\cdot}7071$
α_3	± 1	$\mp 0{\cdot}7071$	0	α_3'	$\pm 0{\cdot}3827$	$\mp 0{\cdot}9239$
α_4	1	-1	1	α_4'	0	0
Upper	$8b_1$	$8b_3$			$8b_2$	$8b_4$
Lower	$8b_7$	$8b_5$			$8b_6$	
Sign						
β_0	0	0		β_0'	0	0
β_1	$0{\cdot}3827$	$0{\cdot}9239$		β_1'	$0{\cdot}7071$	1
β_2	$\pm 0{\cdot}7071$	$\pm 0{\cdot}7071$		β_2'	± 1	0
β_3	$0{\cdot}9239$	$-0{\cdot}3827$		β_3'	$0{\cdot}7071$	-1
β_4	± 1	∓ 1		β_4'	0	0

24 observations

Upper	$24a_0$	$12a_2$	$12a_4$	$12a_6$		$12a_1$	$12a_3$	$12a_5$
Lower	$24a_{12}$	$12a_{10}$	$12a_8$			$12a_{11}$	$12a_9$	$12a_7$
Sign								
α_0	1	1	1	1	α_0'	1	1	1
α_1	± 1	$\pm 0{\cdot}8660$	$\pm 0{\cdot}5$	0	α_1'	$\pm 0{\cdot}9659$	$\pm 0{\cdot}7071$	$\pm 0{\cdot}2588$
α_2	1	$0{\cdot}5$	$-0{\cdot}5$	-1	α_2'	$0{\cdot}8660$	0	$-0{\cdot}8660$
α_3	± 1	0	∓ 1	0	α_3'	$\pm 0{\cdot}7071$	$\mp 0{\cdot}7071$	$\mp 0{\cdot}7071$
α_4	1	$-0{\cdot}5$	$-0{\cdot}5$	1	α_4'	$0{\cdot}5$	-1	$0{\cdot}5$
α_5	± 1	$\mp 0{\cdot}8660$	$\pm 0{\cdot}5$	0	α_5'	$\pm 0{\cdot}2588$	$\mp 0{\cdot}7071$	$\pm 0{\cdot}9659$
α_6	1	-1	1	-1	α_6'	0	0	0
Upper	$12b_1$	$12b_3$	$12b_5$			$12b_2$	$12b_4$	$12b_6$
Lower	$12b_{11}$	$12b_9$	$12b_7$			$12b_{10}$	$12b_8$	
Sign								
β_0	0	0	0		β_0'	0	0	0
β_1	$0{\cdot}2588$	$0{\cdot}7071$	$0{\cdot}9659$		β_1'	$0{\cdot}5$	$0{\cdot}8660$	1
β_2	$\pm 0{\cdot}5$	± 1	$\pm 0{\cdot}5$		β_2'	$\pm 0{\cdot}8660$	$\pm 0{\cdot}8660$	0
β_3	$0{\cdot}7071$	$0{\cdot}7071$	$-0{\cdot}7071$		β_3'	1	0	-1
β_4	$\pm 0{\cdot}8660$	0	$\mp 0{\cdot}8660$		β_4'	$\pm 0{\cdot}8660$	$\mp 0{\cdot}8660$	0
β_5	$0{\cdot}9659$	$-0{\cdot}7071$	$0{\cdot}2588$		β_5'	$0{\cdot}5$	$-0{\cdot}8660$	1
β_6	± 1	∓ 1	± 1		β_6'	0	0	0

TABLE 10.6*b*

Values $z_{30}(\epsilon)$ and Z_{30} for use in smoothing by means of a cubic fitted to n points

n	5	7	9	11	13
ϵ					
6					−11
5				−36	0
4			−21	9	9
3		−2	14	44	16
2	−3	3	39	69	21
1	12	6	54	84	24
0	17	7	59	89	25
−1	12	6	54	84	24
−2	−3	3	39	69	21
−3		−2	14	44	16
−4			−21	9	9
−5				−36	0
−6					−11
Z_{30}	35	21	231	429	143

TABLE 10.6*c*

Values $z_{3j}(\epsilon)$ for calculating the smoothed values at the ends of the range

	$n = 5$		
j	2	1	0
ϵ			
2	69	4	−6
1	4	54	24
0	−6	24	34
−1	4	−16	24
−2	−1	4	−6
Z_{3j}		70	

	$n = 7$			
j	3	2	1	0
ϵ				
3	39	8	−4	−4
2	8	19	16	6
1	−4	16	19	12
0	−4	6	12	14
−1	1	−4	2	12
−2	4	−7	−4	6
−3	−2	4	1	−4
Z_{3j}			42	

	$n = 9$				
j	4	3	2	1	0
ϵ					
4	1190	392	−28	−168	−126
3	392	455	392	252	84
2	−28	392	515	432	234
1	−168	252	432	435	324
0	−126	84	234	324	354
−1	0	−63	12	162	324
−2	112	−140	−143	12	234
−3	112	−98	−140	−63	84
−4	−98	112	112	0	−126
Z_{3j}			1386		

TABLE 10.6*c* (cont.)

j / ϵ	5	4	3	2	1	0
			$n = 11$			
5	678	288	48	−72	−102	−72
4	288	246	192	132	72	18
3	48	192	246	232	172	88
2	−72	132	232	251	212	138
1	−102	72	172	212	206	168
0	−72	18	88	138	168	178
−1	−12	−24	2	52	112	168
−2	48	−48	−64	−23	52	138
−3	78	−48	−88	−64	2	88
−4	48	−18	−48	−48	−24	18
−5	−72	48	78	48	−12	−72
Z_{3j}			858			

j / ϵ	6	5	4	3	2	1	0
				$n = 13$			
6	2915	1452	462	−132	−407	−440	−308
5	1452	1100	792	528	308	132	0
4	462	792	920	888	738	512	252
3	−132	528	888	1004	932	728	448
2	−407	308	738	932	939	808	588
1	−440	132	512	728	808	780	672
0	−308	0	252	448	588	672	700
−1	−88	−88	0	148	328	512	672
−2	143	−132	−202	−116	77	328	588
−3	308	−132	−312	−288	−116	148	448
−4	330	−88	−288	−312	−202	0	252
−5	132	0	−88	−132	−132	−88	0
−6	−363	132	330	308	143	−88	−308
Z_{3j}				4004			

CHAPTER 11

GENERAL REGRESSION AND FUNCTIONAL RELATIONSHIP PROBLEMS IN SEVERAL VARIABLES

11.1 MULTIPLE REGRESSIONS

In multiple regression problems, observations x_{ji} of p variables x_j are made. The regression surface for the variable x_k will be given by the function

$$X_k = E(x_k \mid x_1, \ldots, x_{k-1}, x_{k+1}, \ldots, x_p), \tag{1a}$$

a function of the $p-1$ variables x_j.

It will be assumed for simplicity that the functions are linear, so that

$$X_k = \Sigma' B_{kj} x_j. \tag{1b}$$

The prime superscript on the summation sign indicates that the value k is omitted. If the regression is, in fact, curvilinear, extra terms in x_j^r can be added to (1b). The procedure for the estimation of the regression surface is unchanged if the regression is curvilinear, but the allotment of weights may be much more complicated.

If n sets of observations x_{ji} are made, the least-squares estimates b_{kj} of the regression coefficients B_{kj} will be found from the condition that

$$\Sigma w_i (x_{ki} - \Sigma' b_{kj} x_{ji})^2$$

should be a minimum. This leads to the usual normal equations

$$\Sigma w_i (x_{ki} - \Sigma' b_{kj} x_{ji}) x_{mi} = 0. \tag{2}$$

The weights w_i should be proportional to the variance of x_{ki} for fixed x_{ji}. In many cases it is not possible to ascertain these weights, and it is often assumed that w_i is constant.

The coefficient B_{kj} is a measure of the dependence of x_k on x_j when all the other variables are held constant. This coefficient is often written $B_{kj.12\ldots p}$, the variables held constant being indicated by suffixes after the dot. There will be p regression functions of the form (1a). In particular, there will be a function

$$X_j = \sum_k B_{jk} x_k,$$

B_{jk} being a measure of the dependence of x_j on x_k when all the other variables are held constant. The product $B_{kj} B_{jk}$ will then be a measure of the interdependence of x_j and x_k. This product is written

$$\rho^2_{jk.q} = B_{jk.q} B_{kj.q}, \tag{3}$$

where q stands for the $p-2$ suffixes of the variables held constant. $\rho_{jk.q}$ is called the partial correlation coefficient of the variables x_j and x_k. The estimate of ρ obtained from the n observations will be

$$r_{jk.q} = \sqrt{(b_{jk.q} b_{kj.q})}. \tag{4}$$

11.1.1 *Recurrence relations for partial correlation coefficients*

From (7.1.2,7a), the equation giving $b_{kj.q}$ will be of the form

$$\phi_{jj.q} b_{kj.q} = \phi_{kj.q}, \tag{1}$$

where ϕ_{jk} is $\Sigma x_{ji} x_{ki}$. The theory of the method of single division gives the recurrence relation (7.1.2,2a)

$$\phi_{kj.q} = \phi_{kj.q-1} - \phi_{kq.q-1} \phi_{qj.q-1}/\phi_{qq.q-1}. \tag{2}$$

Hence
$$b_{kj.q} = \frac{\phi_{kj.q-1}\phi_{qq.q-1} - \phi_{kq.q-1}\phi_{qj.q-1}}{\phi_{jj.q-1}\phi_{qq.q-1} - \phi_{jq.q-1}\phi_{qj.q-1}}, \tag{3a}$$

which can be put in the form

$$b_{kj.q} = \frac{b_{kj.q-1} - b_{kq.q-1} b_{qj.q-1}}{1 - b_{jq.q-1} b_{qj.q-1}}. \tag{3b}$$

Since the quantities $\phi_{jk.q-1}$ are symmetrical, it follows from (3a) that

$$\{b_{kj.q} b_{jk.q}\}^{\frac{1}{2}}$$
$$= \frac{\phi_{kj.q-1}\phi_{qq.q-1} - \phi_{kq.q-1}\phi_{jq.q-1}}{\{\phi_{jj.q-1}\phi_{qq.q-1} - \phi_{jq.q-1}\phi_{qj.q-1}\}^{\frac{1}{2}}\{\phi_{kk.q-1}\phi_{qq.q-1} - \phi_{kq.q-1}\phi_{qk.q-1}\}^{\frac{1}{2}}},$$

which simplifies to

$$r_{jk.q} = \frac{r_{jk.q-1} - r_{jq.q-1} r_{kq.q-1}}{(1 - r^2_{jq.q-1})^{\frac{1}{2}}(1 - r^2_{kq.q-1})^{\frac{1}{2}}}. \tag{4a}$$

The correlation coefficients can then be expressed in terms of those of lower order. In particular,

$$r_{12.3} = \frac{r_{12} - r_{13} r_{23}}{(1 - r^2_{13})^{\frac{1}{2}}(1 - r^2_{23})^{\frac{1}{2}}}, \tag{4b}$$

where the r_{jk} are the zero-order coefficients,

$$r_{jk} = \frac{\Sigma(x_{ji} - \bar{x}_j)(x_{ki} - \bar{x}_k)}{\{\Sigma(x_{ji} - \bar{x}_j)^2 \Sigma(x_{ki} - \bar{x}_k)^2\}^{\frac{1}{2}}}. \tag{4c}$$

11.1.2 *Calculation of the partial correlation coefficients*

If there are only three variables, the partial correlation coefficients are calculated most easily from formula (11.1.1,4*b*). With more than three variables, Goulden (1952) recommends a systematic approach using the Doolittle scheme to calculate the $b_{jk.q}$. This requires the solution of p sets of normal equations, one for each value of k.

TABLE 11.1.3

Rainfall and temperature at Toronto, Canada (Example 11.1.3)

x_1	x_2	x_3	x_1	x_2	x_3	x_1	x_2	x_3	x_1	x_2	x_3
1	3·9	31	26	3·9	32	51	5·7	32	76	3·9	38
2	2·9	41	27	6·1	26	52	5·7	31	77	6·5	31
3	3·7	29	28	7·0	48	53	5·9	31	78	6·5	35
4	4·1	33	29	3·8	29	54	2·4	36	79	5·8	37
5	3·4	42	30	5·3	35	55	4·8	31	80	7·3	26
6	1·3	28	31	6·2	27	56	6·3	31	81	9·7	27
7	2·0	41	32	5·6	25	57	4·2	32	82	8·0	37
8	4·7	33	33	2·2	34	58	6·3	30	83	8·1	24
9	4·2	40	34	4·1	29	59	5·8	33	84	6·2	25
10	3·7	28	35	1·7	33	60	5·7	34	85	6·6	27
11	4·2	34	36	3·9	35	61	6·7	29	86	6·9	28
12	4·2	34	37	4·4	26	62	3·5	32	87	7·9	33
13	4·7	33	38	3·0	26	63	7·2	29	88	8·3	26
14	5·0	37	39	5·8	31	64	5·3	27	89	6·7	28
15	5·5	33	40	5·4	38	65	6·2	35	90	5·4	35
16	4·0	39	41	6·0	32	66	5·7	32	91	8·4	26
17	4·4	30	42	4·7	30	67	2·5	34	92	7·2	38
18	3·4	34	43	3·8	40	68	5·0	34	93	5·5	33
19	3·7	40	44	6·9	30	69	7·5	30	94	7·4	31
20	6·1	46	45	4·4	28	70	5·1	30	95	6·3	41
21	4·3	33	46	5·4	29	71	9·1	27	96	8·1	29
22	3·5	25	47	6·0	32	72	7·3	29	97	7·3	33
23	2·2	32	48	7·2	31	73	5·4	34	98	7·8	28
24	3·8	24	49	5·8	29	74	4·3	34	99	9·7	25
25	0·5	30	50	7·1	30	75	5·8	30	100	6·7	34

11.1.3 *Example*

Table 11.1.3 gives the annual mean temperature x_2 (in °F, with zero at 40° F) and the annual rainfall x_3 (in inches) for the year x_1 (zero at 1850), at Toronto, Canada. The calculation of the correlation coefficients is shown in Table 11.1.3*a*, using (11.1.1, 4 *a–c*). The values $\Sigma(x_j-\bar{x}_j)^2$, $\Sigma(x_j-\bar{x}_j)(x_k-\bar{x}_k)$ are obtained from the formulae

$$\Sigma(x_j-\bar{x}_j)^2 = \Sigma x_j^2 - (\Sigma x_j)^2/n,$$

$$\Sigma(x_j-\bar{x}_j)(x_k-\bar{x}_k) = \Sigma x_j x_k - (\Sigma x_j \Sigma x_k)/n.$$

It is clear from the value $r_{12.3}$ that temperature is strongly correlated with time—the mean annual temperature shows a definite rise with time. The other correlations are much less certain.

Table 11.1.3*b* shows the linear regression functions obtained by the standard Doolittle method. It is seen that the dependence of rainfall on temperature and time is not established—the regression coefficients are of the same order as their estimated standard errors. The correlation coefficients obtained from the regression coefficients are of course identical with those obtained in Table 11.1.3*a*.

TABLE 11.1.3*a*

Calculation of correlation coefficients for the observations of Table 11.1.3

$$n = 100$$

$$\Sigma x_1 = 5050 \qquad \Sigma x_1^2 = 338350 \qquad \Sigma(x_1-\bar{x}_1)^2 = 83325$$
$$\Sigma x_2 = 536{\cdot}7 \qquad \Sigma x_2^2 = 3212{\cdot}47 \qquad \Sigma(x_2-\bar{x}_2)^2 = 332{\cdot}0011$$
$$\Sigma x_3 = 3197 \qquad \Sigma x_3^2 = 104425 \qquad \Sigma(x_3-\bar{x}_3)^2 = 2216{\cdot}91$$

$$\Sigma x_1 x_2 = 30651{\cdot}8 \qquad \Sigma(x_1-\bar{x}_1)(x_2-\bar{x}_2) = 3548{\cdot}45$$
$$\Sigma x_1 x_3 = 158283 \qquad \Sigma(x_1-\bar{x}_1)(x_3-\bar{x}_3) = -3165{\cdot}50$$
$$\Sigma x_2 x_3 = 16942{\cdot}1 \qquad \Sigma(x_2-\bar{x}_2)(x_3-\bar{x}_3) = -216{\cdot}199$$

$$r_{12} = 3548{\cdot}45/5259{\cdot}66 = 0{\cdot}674654$$
$$r_{13} = -3165{\cdot}50/13591{\cdot}3 = -0{\cdot}232906$$
$$r_{23} = -216{\cdot}199/857{\cdot}914 = -0{\cdot}252005$$

$$r_{12.3} = 0{\cdot}615961/0{\cdot}972499 \times 0{\cdot}967726 = 0{\cdot}654503$$
$$r_{13.2} = -0{\cdot}062890/0{\cdot}967726 \times 0{\cdot}738134 = -0{\cdot}088043$$
$$r_{23.1} = -0{\cdot}094874/0{\cdot}972499 \times 0{\cdot}738134 = -0{\cdot}132167$$

TABLE 11.1.3*b*

(*a*) *Fitted regression functions for the observations of Table* 11.1.3

$$u_3 = 35{\cdot}336 - (0{\cdot}018828 \pm 0{\cdot}022)\,x_1 - (0{\cdot}44997 \pm 0{\cdot}34)\,x_2$$
$$u_2 = 4{\cdot}532 + (0{\cdot}041111 \pm 0{\cdot}0048)\,x_1 - (0{\cdot}038821 \pm 0{\cdot}030)\,x_3$$
$$u_1 = 7{\cdot}738 + 10{\cdot}4200x_2 - 0{\cdot}41171x_3$$

(*b*) *Correlation coefficients calculated from regression coefficients*

$$r_{12.3} = +\sqrt{(0{\cdot}041111 \times 10{\cdot}4200)} = +0{\cdot}65450$$
$$r_{13.2} = -\sqrt{(0{\cdot}41171 \times 0{\cdot}018828)} = -0{\cdot}08804$$
$$r_{23.1} = -\sqrt{(0{\cdot}44997 \times 0{\cdot}038821)} = -0{\cdot}13217$$

11.1.4 *Orthogonal functions*

The symbol $x_{j.q}$ will be used for the function of x_j and the q variables x_r which is orthogonal to each of the x_r over the n points of observation. Then from the relations between the orthogonal functions and the quantities occurring in the Doolittle scheme (7.2,4*a*–*e*),

$$\phi_{jj.q} = \Sigma x_{j.q}^2, \tag{1a}$$

$$\phi_{jk.q} = \Sigma x_j x_{k.q} = \Sigma x_{j.q} x_{k.q}. \tag{1b}$$

The last result follows because $x_{j.q}$ is x_j plus a linear function of the q variables x_r, and $x_{k.q}$ is orthogonal to each of these q variables.

From (11.1.1,1),

$$b_{kj.q} = \Sigma x_{j.q} x_{k.q} / \Sigma x_{j.q}^2, \tag{2}$$

and so

$$r_{jk.q} = \frac{\Sigma x_{j.q} x_{k.q}}{\{\Sigma x_{j.q}^2 \Sigma x_{k.q}^2\}^{\frac{1}{2}}}. \tag{3}$$

It follows from the last equation that r is less than or equal to unity.

11.1.5 *Significance test for a correlation coefficient*

Significance tests for correlation coefficients can be derived by postulating that the variables follow a multivariate normal distribution—a generalization of the bivariate normal distribution discussed in § 2.3. It can be shown that the significance of a value r can be tested by means of a t-test, using the quantity

$$t = \frac{r}{\sqrt{(1-r^2)}}(n-q-2)^{\frac{1}{2}}, \tag{1}$$

with $n-q-2$ d.f., q being the number of variables held constant. Also if ρ is the true correlation coefficient, and

$$z = \tfrac{1}{2}\ln\frac{1+r}{1-r} = \tanh^{-1} r, \quad Z = \tfrac{1}{2}\ln\frac{1+\rho}{1-\rho} = \tanh^{-1}\rho, \tag{2}$$

then $z-Z$ is distributed approximately normally about a mean value $\rho/2(n-q-1)$ with variance $1/(n-q-3)$. This enables the significance of the departure of an observed value r from the value ρ to be tested.

Proofs of these results are too long to be given here. They will be found in Chs. 14 and 15 of Kendall's book.

11.1.5.1 *Example*

For the correlation coefficients of Table 11.1.3a, ν is 97. Hence (11.1.5,1) gives

(i) for $r_{12.3}$, $t = \dfrac{0{\cdot}655}{0{\cdot}756}\, 9{\cdot}85 = 8{\cdot}5;$

(ii) for $r_{13.2}$, $t = -\dfrac{0{\cdot}088}{0{\cdot}996}\, 9{\cdot}85 = -0{\cdot}87;$

(iii) for $r_{23.1}$, $t = -\dfrac{0{\cdot}132}{0{\cdot}991}\, 9{\cdot}85 = -1{\cdot}31.$

The significance of the last two coefficients is doubtful.

Tables 13 and 15 of the *Biometrika Tables* give the significance levels of r directly.

11.1.6 *Serial correlation of residuals*

The estimate of the serial correlation coefficient for the residuals v_i is given by

$$r = \frac{\sum_{i=1}^{n-1} (v - \bar{v}_i)(v_{i+1} - \bar{v}_{i+1})}{\left\{\sum_{i=1}^{n-1} (v_i - \bar{v}_i)^2 \sum_{i=1}^{n-1} (v_{i+1} - \bar{v}_{i+1})^2\right\}^{\frac{1}{2}}} \tag{1a}$$

Since $\sum_{i=1}^{n} v_i$ is zero, $\bar{v}_i = \sum_{i=1}^{n-1} v_i/(n-1)$ is very small, and

$$r \doteqdot \frac{\sum_{i=1}^{n-1} v_i v_{i+1}}{\sum_{i=1}^{n} v_i^2}. \tag{1b}$$

If the deviations of the observations from the 'true' function are random, $E(v_i v_{i+1})$ is of the order of $-\sigma^2/n$ (§ 2.6.1) and r will be small. Hence the serial correlation coefficient will test the hypothesis that the deviations are random. If they are not random the usual calculations of standard deviation will not be valid, and the values obtained for the standard deviations can only be regarded as at best rough approximations.

Unfortunately, the significance of a value r is not easily tested. Durbin and Watson (1951) give limits for the quantity

$$d = \frac{\Sigma(\Delta v)^2}{\Sigma v^2} = \frac{\Sigma(v_{i+1} - v_i)^2}{\Sigma v^2} \doteqdot 2(1-r) \tag{2}$$

in terms of the number n of observations and the number p of coefficients estimated. Quenouille (1952) has put this test in the form of a test for r considered as an ordinary correlation coefficient based on $n + N_1$ or $n - N_2$ observations. If

$$t_1 = \frac{r}{\sqrt{(1-r^2)}} \sqrt{(n + N_1 - 2)}, \quad t_2 = \frac{r}{\sqrt{(1-r^2)}} \sqrt{(n - N_2 - 2)}, \quad t_1 > t_2, \tag{3}$$

there is no evidence of serial correlation if the larger value t_1 is not significant, while there is evidence of serial correlation if the smaller value t_2 is significant. If the value of t at the particular level of significance lies between t_1 and t_2, the test is inconclusive. Values of N_1 and N_2 for two levels of significance are given in Table 11.1.6.

If the deviations are serially correlated, it is possible that a different form of analysis is desirable. In economics considerable use has been made of regressions of the dependent variable on the independent variables and one or more previous values of the dependent variable. This work is described in monographs edited by Koopmans (1950) and Hood (1953).

TABLE 11.1.6

Values of N_1 and N_2 for use in (11.1.6,3) (*p is the number of coefficients determined*)

	5% level		2% level	
p	N_1	N_2	N_1	N_2
1	20	−1	16	−1
2	35	5	30	5
3	60	10	50	10
4	100	15	75	15

11.1.6.1 *Example*

For the example of § 7.1.1 the value of $\Sigma v_i v_{i+1}$ is found to be $-0{\cdot}632975$, while Σv_i^2 is $3{\cdot}720544$. Hence r is $-0{\cdot}170$, and there is certainly no evidence of positive correlation between successive residuals.

For the example of § 7.6.2.1 the residuals are given in Table 7.6.2.3*a*. From these values, $\Sigma v_i v_{i+1}$ is $3963{\cdot}93$ and Σv_i^2 $7515{\cdot}44$. Hence r is $0{\cdot}527$, and so there is positive correlation between the residuals and the deviations from the fitted curve are not random. The formulae (11.1.6,3) give at the 5% level for $p = 3$

$$t_1 = (0{\cdot}527/0{\cdot}850)\sqrt{(62+60-2)} = 6{\cdot}8,$$
$$t_2 = (0{\cdot}527/0{\cdot}850)\sqrt{(62-10-2)} = 4{\cdot}4.$$

11.2 TWO VARIABLES, FUNCTIONALLY RELATED AND SUBJECT TO ERROR

If the variables Y, X, are connected by the equation

$$F(Y, X;\, B_k) = 0, \tag{1}$$

then, since the adjusted point $y_i - v_{yi}, x_i - v_{xi}$ lies on a curve of the same form,

$$F(y_i - v_{yi}, x_i - v_{xi};\, b_k) = 0 = F(y_i, x_i;\, b_k^0) - \left(\frac{\partial F}{\partial y}\right)_i v_{yi} - \left(\frac{\partial F}{\partial x}\right)_i v_{xi} + \Sigma\left(\frac{\partial F}{\partial b_k}\right)_i b'_k, \tag{2a}$$

where
$$b_k = b_k^0 + b'_k \tag{2b}$$

is an estimate of B_k. Now the least-squares principle requires that

$$\Sigma w_{yi} v_{yi}^2 + \Sigma w_{xi} v_{xi}^2$$

should be a minimum, and so the least-squares conditions are

$$\Sigma w_{yi} v_{yi} \Delta v_{yi} + \Sigma w_{xi} v_{xi} \Delta v_{xi} = 0. \tag{3a}$$

The small variations $\Delta v_{yi}, \Delta v_{xi}$, are not independent, for, from (2a), for a given set of values b_k,

$$\left(\frac{\partial F}{\partial y}\right)_i \Delta v_{yi} + \left(\frac{\partial F}{\partial x}\right)_i \Delta v_{xi} = 0. \qquad (3b)$$

On substituting this expression for Δv_{xi} in (3a) and setting the coefficient of Δv_{yi} equal to zero,

$$v_{yi}/v_{xi} = \frac{1}{w_{yi}}\left(\frac{\partial F}{\partial y}\right)_i \Big/ \frac{1}{w_{xi}}\left(\frac{\partial F}{\partial x}\right)_i. \qquad (3c)$$

Using these results, it is easy to show that

$$w_i\left\{\left(\frac{\partial F}{\partial y}\right)_i v_{yi} + \left(\frac{\partial F}{\partial x}\right)_i v_{xi}\right\}^2 = w_{yi}\, v_{yi}^2 + w_{xi}\, v_{xi}^2, \qquad (4a)$$

where
$$\frac{1}{w_i} = \frac{1}{w_{yi}}\left(\frac{\partial F}{\partial y}\right)_i^2 + \frac{1}{w_{xi}}\left(\frac{\partial F}{\partial x}\right)_i^2. \qquad (4b)$$

Hence, from (2a),

$$\Sigma w_{yi}\, v_{yi}^2 + \Sigma w_{xi}\, v_{xi}^2 = \Sigma w_i (y_i' - \Sigma b_k' x_{ki}')^2, \qquad (5a)$$

where
$$y_i' = F(y_i, x_i;\, b_k^0) \qquad (5b)$$

and
$$x_{ki}' = -\left(\frac{\partial F}{\partial b_k}\right)_i. \qquad (5c)$$

Equation (5a) corresponds to the condition that the sum of the squares of the residuals should be a minimum for a given set of values b_k. To find the values b_k which minimize the sum of the squares of the residuals, (5a) is differentiated with respect to b_k'. This gives rise to the familiar form

$$\Sigma w_i (y_i' - b_k' x_{ki}') x_{mi}' = 0. \qquad (6)$$

It will be observed that, except for the weights, these equations are identical in form with the equations of § 10.2.3.1, where X was free from error. For instance, if Y were a linear function of X, the straight line obtained from (6) would be the regression line and not the line of functional relationship. It would only be possible to differentiate between the regression and functional relationship curves if higher-order differential coefficients were retained in the expansion (2a).

11.2.1 *Example*

In the example of § 10.2.3.2, it was assumed that the time intervals t were free from error. If it is now assumed that the standard error of t is 5% of t, then, from (11.2,4b),

$$\sigma_i^2 = \sigma_{vi}^2 + (B\lambda e^{-\lambda t})^2\, \sigma_{ti}^2 = \sigma_{vi}^2 + (0{\cdot}05\lambda)^2\, x_2'^2.$$

The approximation for λ is 0·04, while the approximations for x_2' are given in Table 10.2.3.1. Taking as in Example 10.2.3.2 a value 1·2 for σ,

$$\sqrt{w_i} = 1{\cdot}2/(\sigma_{yi}^2 + 4\times 10^{-6}\, x_{2i}'^2)^{\frac{1}{2}}.$$

The values $\sqrt{w_i}$ are calculated in Table 11.2.1, and the normal equations are formed and solved in Table 11.2.1*a*. From this table

$$b_2' = (-0{\cdot}304 \pm 0{\cdot}095)\times 10^{-2},$$

and so

$$\lambda = 0{\cdot}0370 \pm 0{\cdot}0010.$$

TABLE 11.2.1

Calculation of weights when both variables are subject to error (Example 11.2.1)

σ_y	$x_2'^2$	$(\sigma_y^2 + 4\times 10^{-6}\, x_2'^2)$	$\sqrt{w}$
0·06	699	0·0064	15
0·10	5685	0·0327	7
0·12	10262	0·0554	5
0·18	22338	0·1218	3
0·26	42238	0·2366	2
0·30	56473	0·3159	2
0·40	61246	0·4050	2
0·45	60113	0·4430	2
0·50	46686	0·4367	2
0·60	35827	0·5033	2
0·60	22834	0·4513	2
0·65	12224	0·4714	2
0·65	3102	0·4349	2
0·65	758	0·4255	2
0·65	184	0·4232	2

11.2.2 *Geometrical interpretation*

The slope of the curve $F(y, x)$ at a particular point can be found from the condition that

$$\Delta F = \frac{\partial F}{\partial x}\Delta x + \frac{\partial F}{\partial y}\Delta y$$

is zero along the curve. The slope is then

$$\frac{dy}{dx} = -\left(\frac{\partial F}{\partial x}\right)\Big/\left(\frac{\partial F}{\partial y}\right). \qquad (1)$$

The slope of the line joining the observed point to the adjusted point on the least-squares curve is given by

$$\tan\theta = v_y/v_x. \qquad (2)$$

Hence, from (11.2,3c),

$$\tan\theta_i = -1\Big/\frac{w_{yi}}{w_{xi}}\left(\frac{dy}{dx}\right)_i. \tag{3}$$

Thus the least-squares principle effectively minimizes the sum of the squares of the deviations in the directions θ_i. If the ratio w_{yi}/w_{xi} is constant for all the points of observation and the observations are plotted on a scale such that w_{yi}/w_{xi} is unity, the deviations are all normal to the curve.

TABLE 11.2.1*a*

The normal equations for Example 11.2.1

Factor removed: x_2', 10^2

$\sqrt{w}$	$\sqrt{w}x_1'$	$\sqrt{w}x_2'$	$\sqrt{w}y'$	z	
15	−14·4003	3·9658	18·6081	23·1736	
7	−6·1664	5·2779	6·7539	12·8654	
5	−4·1780	5·0650	2·6574	8·5444	
3	−2·2260	4·4838	0·9128	6·1706	
2	−1·2082	4·1104	−0·2575	4·6447	
2	−0·9619	4·7528	−0·9087	4·8822	
2	−0·7970	4·9496	−1·4599	4·6927	
2	−0·6220	4·9036	−1·2065	5·0751	
2	−0·3942	4·3214	−1·3161	4·6111	
2	−0·2909	3·7856	−0·7462	4·7485	
2	−0·1904	3·0222	−0·9003	3·9315	
2	−0·1145	2·2112	+0·0926	4·1893	
2	−0·0430	1·1140	−1·8195	1·2515	
2	−0·0171	0·5508	−1·8972	0·6365	
2	−0·0071	0·2710	−2·0678	0·1961	
352	−296·0299	203·1939	317·4500	576·6140	
	271·5026	−141·0222	−318·6228	−484·1723	
		226·0170	93·8149	382·0036	
			418·5443	511·1865	
					418·5443
352	−296·0299	203·1939	317·4500	576·6140	286·2913
1	−0·840994	0·577255	0·901847	1·638108	132·2530
				108	
	22·5432	29·8627	−51·6493	0·7566	118·3349
	1	1·324686	−2·291123	0·033563	13·9181
				563	
		69·1636	−21·0156	48·1480	6·3857
		1	−0·303853	0·696147	7·5324
				147	
−0·511065	−1·888613	−0·303853	$s = \sqrt{(7{\cdot}5324/12)} = 0{\cdot}792$		
0·488935	−0·888613	0·696147	$s(b_2') = s/\sqrt{69{\cdot}16} = 0{\cdot}095$		

11.3 GENERAL LEAST-SQUARES THEORY FOR FUNCTIONALLY RELATED VARIABLES

In earlier discussions the 'true' values were supposed to satisfy equations of the form

$$F(Y_i, X_i;\, B_j) = 0, \quad i = 1 \text{ to } n,$$

each observed coordinate appearing in only one equation. In the more general case, the observed values Y_i may appear in any number of the conditional equations. There will, in general, be no obvious division of observed quantities into dependent variables Y and independent variables X, and so each variable will be represented by the general symbol Y. The conditions which are satisfied by the n values Y_i will be written

$$F_q(Y_i;\, B_j) = 0, \quad q = 1 \text{ to } r. \tag{1}$$

If y_i are the observed values, v_i the residuals, and

$$b_j = b_j^0 + b_j'$$

the estimated values of the p parameters B_j, then

$$F_q(y_i - v_i;\, b_j^0 + b_j') = 0, \tag{2a}$$

and if
$$F'_{qi} = \partial F_q/\partial y_i, \quad F'_{qj} = \partial F_q/\partial b_j, \tag{2b}$$

a Taylor's expansion gives

$$\sum_i F'_{qi} v_i = F_q(y_i;\, b_j^0) + \sum_j F'_{qj} b'_j. \tag{2c}$$

The least-squares principle states that the parameters and adjusted values are to be chosen to minimize $\Sigma w_i v_i^2$. This requires that, for small variations Δv_i,

$$\Sigma w_i v_i \Delta v_i = 0. \tag{3a}$$

From (2c), the variations Δv_i are not independent, but are connected by the r equations

$$\sum_i F'_{qi} \Delta v_i - \sum_j F'_{qj} \Delta b'_j = 0. \tag{3b}$$

Introducing r Lagrangian multipliers λ_q, as in § 8.3, these equations can be combined to give

$$\sum_i \left(w_i v_i - \sum_q \lambda_q F'_{qi} \right) \Delta v_i - \sum_j \sum_q \lambda_q F'_{qj} \Delta b'_j = 0, \tag{4}$$

where the coefficients of Δv_i and $\Delta b'_j$ can now be equated to zero.

Hence

$$w_i v_i = \sum_q \lambda_q F'_{qi} \quad (n \text{ equations}), \tag{5a}$$

$$\sum_q \lambda_q F'_{qj} = 0 \qquad (p \text{ equations}). \tag{5b}$$

Substitution for v_i in (2*c*) gives

$$\sum_i w_i^{-1} \sum_s \lambda_s F'_{si} F'_{qi} - \sum_j F'_{qj} b'_j = F_q(y_i;\, b_j^0). \tag{5c}$$

If the quantity L_{qs} is defined by the equation

$$L_{qs} = \sum_i F'_{qi} w_i^{-1} F'_{si}, \tag{6}$$

then (5*c*) and (5*b*) become

$$\sum_s L_{qs}\lambda_s - \sum_j F'_{qj} b'_j = F_q(y_i;\, b_j^0) \quad (r \text{ equations}), \tag{7a}$$

$$\sum_s F'_{sj} \lambda_s = 0 \qquad (p \text{ equations}). \tag{7b}$$

These $r+p$ equations are the general normal equations. They form a symmetrical set which may be solved to give the p parameters b'_j, and the r multipliers λ_q which determine the residuals and hence the adjusted values.

11.3.1 *Example*

In Table 11.3.1*a* are listed the results of the measurements of the three sides a, b, c, and the three angles A, B, C, of a triangular area. These values are to be adjusted so that they satisfy the three conditional equations

$$F_0 = A+B+C-180^\circ = 0,$$

$$F_1 = (\sin A/a)-(\sin B/b) = 0,$$

$$F_2 = (\sin A/a)-(\sin C/c) = 0,$$

connecting the sides and angles of a triangle. Factors 10^3 have been removed from the length measurements (originally in feet) to bring them to the order of unity. In this example there are no parameters B_j to be estimated, and there will be just $r = 3$ normal equations (11.3,7*a*).

In section (ii) of the table the weights are allotted, σ being chosen so that the weights of the angles are unity. The actual values F_j are calculated in section (iii), and the differential coefficients in section (iv). The normal equations are found by intermultiplying these columns (11.3,6). The resulting equations are given in Table 11.3.1*a*. A factor 10^{-3} is removed from F_j to simplify the calculations, and the equations are solved by the abbreviated Doolittle method.

The solutions λ_j are entered in section (v) of Table 11.3.1. When the λ_j are multiplied by the rows of section (iv) the values $\sqrt{w_i}\, v_i$ are obtained (11.3,5*a*). The adjusted values $u_i = y_i - v_i$ are calculated in section (vi).

TABLE 11.3.1

Adjustment of observations on a triangular area (Example 11.3.1)

(i) Observations

Factors removed: $a, b, c, 10^3$.

A	$90° 17' \pm 2'$	a	$1{\cdot}0000 \pm 0{\cdot}0007$	unit 1000 feet
B	$44° 28' \pm 2'$	b	$0{\cdot}6997 \pm 0{\cdot}0005$	
C	$45° 18' \pm 2'$	c	$0{\cdot}7107 \pm 0{\cdot}0005$	

(ii) Weights

$2' = 0{\cdot}58 \times 10^{-3}$ radians $\qquad \sigma = 0{\cdot}58 \times 10^{-3}$

$\sqrt{w_A} = \sqrt{w_B} = \sqrt{w_C} = 1$

$\sqrt{w_a} = 0{\cdot}83 \qquad \sqrt{w_b} = \sqrt{w_c} = 1{\cdot}16$

(iii) Values F_j

$F_0 = A+B+C-180° = 0{\cdot}8727 \times 10^{-3}$ (radians)

$F_1 = (\sin A/a) - (\sin B/b) = -1{\cdot}1473 \times 10^{-3}$

$F_2 = (\sin A/a) - (\sin C/c) = -0{\cdot}1522 \times 10^{-3}$

(iv) Differential coefficients

$F'_{0A} = 1 = F'_{0B} = F'_{0C}$ $\qquad F'_{0a} = F'_{0b} = F'_{0c} = 0$

$F'_{1A} = \cos A/a \qquad F'_{1B} = -\cos B/b \qquad F'_{1C} = 0$

$F'_{1a} = -\sin A/a^2 \qquad F'_{1b} = \sin B/b^2 \qquad F'_{1c} = 0$

$F'_{2A} = \cos A/a \qquad F'_{2B} = 0 \qquad F'_{2C} = -\cos C/c$

$F'_{2a} = -\sin A/a^2 \qquad F'_{2b} = 0 \qquad F'_{2c} = \sin C/c^2$

i	$F'_{0i}/\sqrt{w_i}$	$F'_{1i}/\sqrt{w_i}$	$F'_{2i}/\sqrt{w_i}$
A	1	$-0{\cdot}0049$	$-0{\cdot}0049$
B	1	$-1{\cdot}0199$	0
C	1	0	$-0{\cdot}9897$
a	0	$-1{\cdot}2048$	$-1{\cdot}2048$
b	0	$1{\cdot}2335$	0
c	0	0	$1{\cdot}2132$

(v) Solution of normal equations

$\lambda_j \qquad +0{\cdot}239517 \times 10^{-3} \qquad -0{\cdot}268853 \times 10^{-3} \qquad +0{\cdot}122032 \times 10^{-3}$

(vi) Fitted values

i	$\sqrt{w_i}\, v_i$	v_i	u_i
A	$+0{\cdot}240 \times 10^{-3}$	$+0{\cdot}240 \times 10^{-3}$	90° 16·2′
B	$+0{\cdot}514 \times 10^{-3}$	$+0{\cdot}514 \times 10^{-3}$	44° 26·2′
C	$+0{\cdot}119 \times 10^{-3}$	$+0{\cdot}119 \times 10^{-3}$	45° 17·6′
a	$+0{\cdot}177 \times 10^{-3}$	$+0{\cdot}213 \times 10^{-3}$	999·79 ft
b	$-0{\cdot}332 \times 10^{-3}$	$-0{\cdot}286 \times 10^{-3}$	699·99 ft
c	$+0{\cdot}148 \times 10^{-3}$	$+0{\cdot}128 \times 10^{-3}$	710·57 ft

TABLE 11.3.1*a*

Solutions of normal equations for Example 11.3.1

Factor removed: F_j, 10^{-3}					
	3	−1·0248	−0·9946	0·8727	1·853300
		4·013285	1·451567	−1·1473	3·292752
			3·902927	−0·1522	4·207694
1	3	−1·024800	−0·994600	0·872700	1·853300
0·333333	1	−0·341600	−0·331533	0·290900	0·617767
0·341600	1	3·663213	1·111812	−0·849186	3·925839
0·093251	0·272984	1	0·303507	−0·231815	1·071692
0·227855	−0·303507	1	3·235742	0·394863	3·630604
0·070418	−0·093798	0·309048	1	0·122032	1·122031
	0·239517	−0·268853	0·122032		
	1·239517	0·731148	1·122031		
Inverse matrix					
0·381233					
0·071879	0·301452				
0·070418	−0·093798	0·309048			

11.3.2 *Summary of standard deviation formulae*

The derivation of the formulae for the standard deviations of the parameters and fitted values is quite involved, and it will be convenient to first list the formulae and defer the proofs till § 11.3.4. The symbol χ_{jk} will be used for an element of the matrix which is the inverse of the matrix

$$\begin{bmatrix} L_{rs} & F'_{rt} \\ F'_{ts} & 0 \end{bmatrix}$$

occurring in the normal equations. Then for the residuals

$$E\Sigma w_i v_i^2 = (r-p)\,\sigma^2, \tag{1}$$

$$\operatorname{var} v_i = Ev_i^2 = w_i^{-2} \sum_{s=1}^{r} \sum_{t=1}^{r} \chi_{st} F'_{si} F'_{ti}\,\sigma^2. \tag{2}$$

For the fitted value $u_i = y_i - v_i$,

$$\operatorname{var} u_i = w_i^{-1}\left\{1 - w_i^{-1} \sum_{s,\,t=1}^{r} \chi_{st} F'_{si} F'_{ti}\right\}\sigma^2, \tag{3}$$

the term in brackets giving the change in variance brought about by the least-squares adjustment of the observations. For the parameters,

$$\operatorname{var} b'_j = -\chi_{r+j,\,r+j}\,\sigma^2. \tag{4}$$

The sum $\Sigma w_i v_i^2$ can be obtained without finding the individual residuals. From (11.3,5*a*),

$$\sum_i w_i v_i^2 = \sum_i w_i^{-1} \sum_q \lambda_q \sum_s \lambda_s F'_{si} F'_{qi},$$

and so, from (11.3,5*c*) and (11.3,5*b*),

$$\sum_i w_i v_i^2 = \sum_q \lambda_q F_q(y_i;\, b_j^0). \quad (5)$$

11.3.2.1 *Example*

The standard deviations of the fitted values obtained in Example 11.3.1 will now be calculated, using (11.3.2,3). The values χ_{jk} from Table 11.3.1*a* are entered in Table 11.3.2, together with the products $w_i^{-1} F'_{ji} F'_{ki}$ calculated from Table 11.3.1(iv). The quantities

$$S_i = 1 - \sum_{j,k} \chi_{jk}\, w_i^{-1} F'_{ji} F'_{ki}$$

are calculated, and checked by the formula

$$\Sigma S_i = n - r\ (+p).$$

This latter formula follows from (11.3.3,7*c*). Here there are no parameters and p is zero. Finally the standard deviations of the fitted values are found from

$$\sigma[u_i] = \sigma \sqrt{S_i}/\sqrt{w_i}.$$

TABLE 11.3.2

Standard deviations of adjusted values

$S_i = 1 - \sum_{j,k} \chi_{jk} F'_{ji} F'_{ki}/w_i$								
χ_{jk}	χ_{00} 0·3812	$2\chi_{01}$ 0·1438	$2\chi_{02}$ 0·1408	χ_{11} 0·3015	$2\chi_{12}$ −0·1876	χ_{22} 0·3090	σ $0{\cdot}58 \times 10^{-3}$	
i	$F'_{ji} F'_{ki}/w_i$						S_i	$\sigma[u]$
A	1	−0·0049	−0·0049	0	0	0	0·6202	1·6 min.
B	1	−1·0199	0	1·0402	0	0	0·4518	1·3 min.
C	1	0	−0·9897	0	0	0·9795	0·4555	1·3 min.
a	0	0	0	1·4515	1·4515	1·4515	0·3862	0·43 ft.
b	0	0	0	1·5215	0	0	0·5413	0·37 ft.
c	0	0	0	0	0	1·4719	0·5452	0·37 ft.
						Check	3·0002	

11.3.2.2 *The residuals.* The value $\Sigma w_i v_i^2$, obtained either by direct summation from Table 11.3.1(vi) or from $\Sigma \lambda_j F_j$, is

$$\Sigma w_i v_i^2 = \Sigma \lambda_j F_j = 0{\cdot}499 \times 10^{-6}.$$

Thus the estimate s of σ, from (11.3.2,1), is $0{\cdot}408 \times 10^{-3}$, based on 3 d.f. This agrees well with the value $0{\cdot}58 \times 10^{-3}$, and so no doubt is cast on the standard deviations of the estimates given in Table 11.3.1(i).

11.3.3 *The general case in matrix notation*

The symbols which will be used for the various matrices are defined in Table 11.3.3.

TABLE 11.3.3

Matrix symbols

Symbol	*Order*	*Element*
$\mathbf{f}$	$r \times 1$	$F_q\ (y_i\,;\ b_j^0)$
$\mathbf{y}$	$n \times 1$	y_i
$\mathbf{v}$	$n \times 1$	v_i
$\mathbf{W}$	$n \times n$	$W_{ii} = w_i$ (diagonal matrix)
$\mathbf{b}$	$p \times 1$	b'_j
$\mathbf{J}$	$p \times r$	$F'_{qj} = \partial F_q/\partial b_j$
$\mathbf{F}$	$n \times r$	$F'_{qi} = \partial F_q/\partial y_i$
$\mathbf{L}$	$r \times r$	L_{qs}
$\boldsymbol{\lambda}$	$r \times 1$	λ_s
$\boldsymbol{\phi}$	$p+r \times p+r$	$\begin{bmatrix} \mathbf{L} & \mathbf{J}^T \\ \mathbf{J} & \mathbf{0} \end{bmatrix}$
$\boldsymbol{\chi}$	$p+r \times p+r$	$[\phi^{-1}]_{jk}$

The matrix $\boldsymbol{\phi}$ can be written in the form

$$\boldsymbol{\phi} = \begin{bmatrix} \boldsymbol{\phi}^{rr} & \boldsymbol{\phi}^{rp} \\ \boldsymbol{\phi}^{pr} & \mathbf{0} \end{bmatrix}, \tag{1a}$$

where
$$\boldsymbol{\phi}^{rr} = \mathbf{L}, \quad \boldsymbol{\phi}^{pr} = \mathbf{J}, \tag{1b}$$

the superscripts indicating the order of the sub-matrices. The inverse can be written similarly as

$$\boldsymbol{\chi} = \begin{bmatrix} \boldsymbol{\chi}^{rr} & \boldsymbol{\chi}^{rp} \\ \boldsymbol{\chi}^{pr} & \boldsymbol{\chi}^{pp} \end{bmatrix}. \tag{2}$$

When two matrices, partitioned in the same way, are multiplied together, the product is obtained by formally treating each sub-matrix as a single element. Thus

$$\boldsymbol{\phi\chi} = \begin{bmatrix} \boldsymbol{\phi}^{rr}\boldsymbol{\chi}^{rr} + \boldsymbol{\phi}^{rp}\boldsymbol{\chi}^{pr} & \boldsymbol{\phi}^{rr}\boldsymbol{\chi}^{rp} + \boldsymbol{\phi}^{rp}\boldsymbol{\chi}^{pp} \\ \boldsymbol{\phi}^{pr}\boldsymbol{\chi}^{rr} & \boldsymbol{\phi}^{pr}\boldsymbol{\chi}^{rp} \end{bmatrix}. \tag{3}$$

But this product is **I**, and so

$$\boldsymbol{\phi}^{rr}\boldsymbol{\chi}^{rr}+\boldsymbol{\phi}^{rp}\boldsymbol{\chi}^{pr}=\mathbf{I}=\boldsymbol{\chi}^{rr}\boldsymbol{\phi}^{rr}+\boldsymbol{\chi}^{rp}\boldsymbol{\phi}^{pr}; \tag{4a}$$

$$\boldsymbol{\phi}^{pr}\boldsymbol{\chi}^{rp}=\mathbf{I}=\boldsymbol{\chi}^{pr}\boldsymbol{\phi}^{rp}; \tag{4b}$$

$$\boldsymbol{\phi}^{rr}\boldsymbol{\chi}^{rp}+\boldsymbol{\phi}^{rp}\boldsymbol{\chi}^{pp}=\mathbf{0}=\boldsymbol{\chi}^{pr}\boldsymbol{\phi}^{rr}+\boldsymbol{\chi}^{pp}\boldsymbol{\phi}^{pr}; \tag{4c}$$

$$\boldsymbol{\phi}^{pr}\boldsymbol{\chi}^{rr}=\mathbf{0}=\boldsymbol{\chi}^{rr}\boldsymbol{\phi}^{rp}. \tag{4d}$$

The following formulae are required in the subsequent discussion:

$$\boldsymbol{\chi}^{rr}\boldsymbol{\phi}^{rr}\boldsymbol{\chi}^{rr}=\boldsymbol{\chi}^{rr}; \tag{5a}$$

$$\boldsymbol{\chi}^{rr}\boldsymbol{\phi}^{rr}\boldsymbol{\chi}^{rp}=\mathbf{0}=\boldsymbol{\chi}^{pr}\boldsymbol{\phi}^{rr}\boldsymbol{\chi}^{rr}; \tag{5b}$$

$$\boldsymbol{\chi}^{pr}\boldsymbol{\phi}^{rr}\boldsymbol{\chi}^{rp}=-\boldsymbol{\chi}^{pp}. \tag{5c}$$

Equation (5*a*) is proved by multiplying (4*a*) by $\boldsymbol{\chi}^{rr}$ and using (4*d*). It can be shown that $\boldsymbol{\chi}^{rr}$ is a singular matrix, and so it cannot be divided out of (5*a*). Equation (5*b*) is established by multiplying (4*c*) by $\boldsymbol{\chi}^{rr}$ and using (4*d*), (5*c*) by multiplying (4*c*) by $\boldsymbol{\chi}^{pr}$ and using (4*b*).

The sum of the diagonal elements of the product $\boldsymbol{\phi}^{rr}\boldsymbol{\chi}^{rr}$ is, from (4*a*),

$$\sum_{t=1}^{r}\left\{I_{tt}-\sum_{u=1}^{p}\phi_{tu}^{rp}\chi_{ut}^{pr}\right\}=r-\sum_{u=1}^{p}\left\{\sum_{t=1}^{r}\chi_{ut}^{pr}\phi_{tu}^{rp}\right\}.$$

From (4*b*), the last term is $\sum_{u=1}^{p} I_{uu}$, and so

$$\sum_{t=1}^{r}(\boldsymbol{\phi}^{rr}\boldsymbol{\chi}^{rr})_{tt}=\sum_{t=1}^{r}\sum_{u=1}^{r}\phi_{tu}^{rr}\chi_{ut}^{rr}=r-p. \tag{6}$$

In the present discussion, from (11.3,6),

$$\boldsymbol{\phi}^{rr}=\mathbf{L}=\mathbf{F}^T\mathbf{W}^{-1}\mathbf{F}, \tag{7a}$$

and

$$\boldsymbol{\phi}^{pr}=\mathbf{J}. \tag{7b}$$

Hence (6) can be written

$$\sum_{i}\sum_{t,u=1}^{r}\chi_{ut}^{rr}w_i^{-1}F'_{ti}F'_{ui}=r-p. \tag{7c}$$

11.3.3.1 *The normal equations.* In the notation of this section the normal equations (11.3,7*a*, *b*) are

$$\boldsymbol{\phi}\{\boldsymbol{\lambda},-\mathbf{b}\}=\{\mathbf{f},\mathbf{0}\}, \tag{1a}$$

or

$$\{\boldsymbol{\lambda},-\mathbf{b}\}=\boldsymbol{\chi}\{\mathbf{f},\mathbf{0}\}. \tag{1b}$$

The matrix $\{\boldsymbol{\lambda},-\mathbf{b}\}$ is the column vector of order $r+p$ whose elements are λ_s, $-b'_j$. Equation (1*b*) can be written in the form

$$\boldsymbol{\lambda}=\boldsymbol{\chi}^{rr}\mathbf{f},\quad \mathbf{b}=-\boldsymbol{\chi}^{pr}\mathbf{f}. \tag{1c}$$

The elements of $\mathbf{f}$ are

$$F_q(y_i;\, b_j^0) = F_q(Y_i;\, B_j) + \Sigma(y_i - Y_i)\, F'_{qi} + \Sigma(b_j^0 - B_j)\, F'_{qj},$$

where Y_i and B_j are the true values, and so

$$\mathbf{f} = \mathbf{F}^T(\mathbf{y} - \mathbf{Y}) + \mathbf{J}^T(\mathbf{b}^0 - \mathbf{B}). \tag{2}$$

Hence from (11.3.3,4*d*),

$$\boldsymbol{\lambda} = \boldsymbol{\chi}^{rr}\,\mathbf{F}^T(\mathbf{y} - \mathbf{Y}), \tag{3a}$$

and from (11.3.3,4*b*),

$$\mathbf{b} = -\boldsymbol{\chi}^{pr}\,\mathbf{F}^T(\mathbf{y} - \mathbf{Y}) - (\mathbf{b}^0 - \mathbf{B}). \tag{3b}$$

If the prefix δ is used to denote deviations from the true value, so that

$$\boldsymbol{\delta}\mathbf{y} = \mathbf{y} - \mathbf{Y}, \quad \boldsymbol{\delta}\mathbf{b} = \mathbf{b}^0 + \mathbf{b} - \mathbf{B},$$

then (3*a*) and (3*b*) become

$$\boldsymbol{\lambda} = \boldsymbol{\chi}^{rr}\,\mathbf{F}^T\,\boldsymbol{\delta}\mathbf{y}, \quad \boldsymbol{\delta}\mathbf{b} = -\boldsymbol{\chi}^{pr}\,\mathbf{F}^T\,\boldsymbol{\delta}\mathbf{y}. \tag{4}$$

The normal equations for the parameters b'_j can be put in the same form as those in the discussion of correlated variables in § 8.3.1. The set of equations (11.3,7*a*) can be written

$$\boldsymbol{\lambda} = \mathbf{L}^{-1}(\mathbf{f} + \mathbf{J}^T\,\mathbf{b}).$$

On multiplying by $\mathbf{J}$ and using (11.3,7*b*), this vanishes, and so

$$\mathbf{J}\mathbf{L}^{-1}\,\mathbf{J}^T\,\mathbf{b} = -\mathbf{J}\mathbf{L}^{-1}\,\mathbf{f}. \tag{5a}$$

This corresponds to (8.3.1,3*a*). The matrix $\mathbf{L}^{-1}$ is the weight matrix for the quantities $\mathbf{f}$, since for a given $\mathbf{b}^0$, from (2),

$$E(\mathbf{f}\mathbf{f}^T) = \mathbf{F}^T\,\mathbf{W}^{-1}\,\mathbf{F}\sigma^2 = \mathbf{L}\sigma^2. \tag{5b}$$

11.3.4 *Covariance matrix for* $\{\boldsymbol{\lambda}, \mathbf{b}\}$

The covariance matrix for $\{\boldsymbol{\lambda}, \mathbf{b}\}$ is

$$E\{\boldsymbol{\lambda}, \boldsymbol{\delta}\mathbf{b}\}\{\boldsymbol{\lambda}, \boldsymbol{\delta}\mathbf{b}\}^T = E\begin{bmatrix} \boldsymbol{\lambda}\boldsymbol{\lambda}^T & \boldsymbol{\lambda}\boldsymbol{\delta}\mathbf{b}^T \\ \boldsymbol{\delta}\mathbf{b}\boldsymbol{\lambda}^T & \boldsymbol{\delta}\mathbf{b}\,\boldsymbol{\delta}\mathbf{b}^T \end{bmatrix}.$$

Now $$E(\boldsymbol{\lambda}\boldsymbol{\lambda}^T) = \boldsymbol{\chi}^{rr}\,\mathbf{F}^T\,E(\boldsymbol{\delta}\mathbf{y}\,\boldsymbol{\delta}\mathbf{y}^T)\,\mathbf{F}\boldsymbol{\chi}^{rr},$$

and $$E(\boldsymbol{\delta}\mathbf{y}\,\boldsymbol{\delta}\mathbf{y}^T) = \mathbf{W}^{-1}\,\sigma^2. \tag{1}$$

Hence $$E(\boldsymbol{\lambda}\boldsymbol{\lambda}^T) = \boldsymbol{\chi}^{rr}\,\boldsymbol{\phi}^{rr}\,\boldsymbol{\chi}^{rr}\,\sigma^2 = \boldsymbol{\chi}^{rr}\,\sigma^2. \tag{2a}$$

Similarly, $$E(\boldsymbol{\lambda}\boldsymbol{\delta}\mathbf{b}^T) = -\boldsymbol{\chi}^{rr}\,\boldsymbol{\phi}^{rr}\,\boldsymbol{\chi}^{rp}\,\sigma^2 = \mathbf{0} \tag{2b}$$

and $$E(\boldsymbol{\delta}\mathbf{b}\,\boldsymbol{\delta}\mathbf{b}^T) = \boldsymbol{\chi}^{pr}\,\boldsymbol{\phi}^{rr}\,\boldsymbol{\chi}^{rp}\,\sigma^2 = -\boldsymbol{\chi}^{pp}\,\sigma^2. \tag{2c}$$

It will be observed that λ_s and b_j are independent, and that (11.3.2,4) has been established.

11.3.4.1 *The residuals.* Since, from (11.3,5*a*),

$$\mathbf{v} = \mathbf{W}^{-1}\mathbf{F}\boldsymbol{\lambda}, \tag{1}$$

the covariance matrix for the v_i is

$$\operatorname{cov}\mathbf{v} = E(\mathbf{W}^{-1}\mathbf{F}\boldsymbol{\lambda})(\mathbf{W}^{-1}\mathbf{F}\boldsymbol{\lambda})^T = \mathbf{W}^{-1}\mathbf{F}\boldsymbol{\chi}^{rr}\mathbf{F}^T\mathbf{W}^{-1}\sigma^2. \tag{2a}$$

Since $\mathbf{W}$ is a diagonal matrix,

$$Ev_i^2 = \sum_{t=1}^{r}\sum_{u=1}^{r} w_i^{-1}F_{it}\chi_{tu}^{rr}F_{ui}^T w_i^{-1}\sigma^2, \tag{2b}$$

which is (11.3.2,2). Also

$$E\Sigma w_i v_i^2 = \sum_{t=1}^{r}\sum_{u=1}^{r}\chi_{tu}^{rr}\sum_i F_{ui}^T w_i^{-1}F_{it}\sigma^2 = \sum_{t=1}^{r}\sum_{u=1}^{r}\chi_{tu}^{rr}\phi_{ut}^{rr}\sigma^2.$$

Hence, using (11.3.3,6),

$$E\Sigma w_i v_i^2 = (r-p)\sigma^2, \tag{3}$$

which is (11.3.2,1).

The residuals v_h are statistically independent of both the adjusted values u_i and the parameters b_j. For

$$E(\mathbf{v}\boldsymbol{\delta}\mathbf{y}^T) = \mathbf{W}^{-1}\mathbf{F}\boldsymbol{\chi}^{rr}\mathbf{F}^T E(\boldsymbol{\delta}\mathbf{y}\,\boldsymbol{\delta}\mathbf{y}^T),$$

and so from (2*a*)
$$E(\mathbf{v}\boldsymbol{\delta}\mathbf{y}^T) = E(\mathbf{v}\mathbf{v}^T),$$

or
$$\operatorname{cov}(v_h, y_i) = \operatorname{cov}(v_h, v_i). \tag{4a}$$

Hence
$$\operatorname{cov}(v_h, u_i) = \operatorname{cov}(v_h, y_i - v_i) = 0, \tag{4b}$$

which establishes the independence of v_h and u_i. The independence of v_h and b_j follows from (11.3.4,2*b*).

11.3.4.2 *Covariance matrix for* $\{\mathbf{u}, \mathbf{b}\}$. The adjusted values $u_i = y_i - v_i$ will form a vector

$$\mathbf{u} = \mathbf{y} - \mathbf{W}^{-1}\mathbf{F}\boldsymbol{\lambda} = \mathbf{y} - \mathbf{W}^{-1}\mathbf{F}\boldsymbol{\chi}^{rr}\mathbf{F}^T\boldsymbol{\delta}\mathbf{y}, \tag{1}$$

and so the covariance matrix for $\{\mathbf{u}, \mathbf{b}\}$ is

$$\begin{aligned}\operatorname{cov}\{\mathbf{u},\mathbf{b}\} = E\{&\boldsymbol{\delta}\mathbf{y} - \mathbf{W}^{-1}\mathbf{F}\boldsymbol{\chi}^{rr}\mathbf{F}^T\boldsymbol{\delta}\mathbf{y},\ -\boldsymbol{\chi}^{pr}\mathbf{F}^T\boldsymbol{\delta}\mathbf{y}\}\\ &\times\{\boldsymbol{\delta}\mathbf{y} - \mathbf{W}^{-1}\mathbf{F}\boldsymbol{\chi}^{rr}\mathbf{F}^T\boldsymbol{\delta}\mathbf{y},\ -\boldsymbol{\chi}^{pr}\mathbf{F}^T\boldsymbol{\delta}\mathbf{y}\}^T.\end{aligned} \tag{2}$$

Now

$$\begin{aligned}\mathbf{W}^{-1}\mathbf{F}\boldsymbol{\chi}^{rr}\mathbf{F}^T\mathbf{W}^{-1}\mathbf{F}\boldsymbol{\chi}^{rr}\mathbf{F}^T\mathbf{W}^{-1} &= \mathbf{W}^{-1}\mathbf{F}\boldsymbol{\chi}^{rr}\boldsymbol{\phi}^{rr}\boldsymbol{\chi}^{rr}\mathbf{F}^T\mathbf{W}^{-1}\\ &= \mathbf{W}^{-1}\mathbf{F}\boldsymbol{\chi}^{rr}\mathbf{F}^T\mathbf{W}^{-1}\end{aligned}$$

and
$$\boldsymbol{\chi}^{pr}\mathbf{F}^T\mathbf{W}^{-1}\mathbf{F}\boldsymbol{\chi}^{rr}\mathbf{F}^T\mathbf{W}^{-1} = \boldsymbol{\chi}^{pr}\boldsymbol{\phi}^{rr}\boldsymbol{\chi}^{rr}\mathbf{F}^T\mathbf{W}^{-1} = \mathbf{0}.$$

Hence
$$\operatorname{cov}\{\mathbf{u},\mathbf{b}\} = \boldsymbol{\omega}\sigma^2 = \begin{bmatrix}\boldsymbol{\omega}^{rr} & \boldsymbol{\omega}^{rp}\\ \boldsymbol{\omega}^{pr} & \boldsymbol{\omega}^{pp}\end{bmatrix}\sigma^2, \tag{3a}$$

where
$$\boldsymbol{\omega}^{rr} = \mathbf{W}^{-1} - \mathbf{W}^{-1}\mathbf{F}\boldsymbol{\chi}^{rr}\mathbf{F}^T\mathbf{W}^{-1}, \tag{3b}$$
$$\boldsymbol{\omega}^{pr} = -\boldsymbol{\chi}^{pr}\mathbf{F}^T\mathbf{W}^{-1}, \tag{3c}$$
$$\boldsymbol{\omega}^{pp} = -\boldsymbol{\chi}^{pp}. \tag{3d}$$

Formula (11.3.2,3) follows from (3b).

11.3.5 *Variance of a function*

If $G(Y_i;\, B_j)$ is some function, the deviation of the value $G(u_i;\, b_j)$ obtained by using the adjusted values and estimated parameters is
$$\delta G = \mathbf{g}^T\{\boldsymbol{\delta}\mathbf{u}, \boldsymbol{\delta}\mathbf{b}\}, \tag{1a}$$
where
$$\mathbf{g} = \{\partial G/\partial Y_i, \partial G/\partial B_j\}. \tag{1b}$$
Hence
$$\operatorname{var} G = E(\delta G)^2 = E\mathbf{g}^T\{\boldsymbol{\delta}\mathbf{u}, \boldsymbol{\delta}\mathbf{b}\}\{\boldsymbol{\delta}\mathbf{u}, \boldsymbol{\delta}\mathbf{b}\}^T\mathbf{g},$$
or
$$\operatorname{var} G = \mathbf{g}^T\boldsymbol{\omega}\mathbf{g}\sigma^2. \tag{2}$$

11.3.5.1 *Special cases.* If both adjusted values and parameters are present in the function G, the evaluation of the standard deviation of G is very complicated. If only one of these types is present, the calculation is not quite so difficult.

If adjusted values alone are present in the function G, $\mathbf{g}$ is a vector of n elements, and
$$\operatorname{var} G = \mathbf{g}^T\{\mathbf{W}^{-1} - \mathbf{W}^{-1}\mathbf{F}\boldsymbol{\chi}^{rr}\mathbf{F}^T\mathbf{W}^{-1}\}\mathbf{g}\sigma^2,$$
which reduces to
$$\operatorname{var} G = \left\{\Sigma w_i^{-1} g_i^2 - \sum_{q=1}^{r}\sum_{s=1}^{r}\chi_{qs}^{rr}\left(\sum_i w_i^{-1} g_i F'_{qi}\right)\left(\sum_i w_i^{-1} g_i F'_{si}\right)\right\}\sigma^2. \tag{1}$$

On the other hand, if parameters alone are present in the function G, $\mathbf{g}$ is a vector of p elements, and
$$\operatorname{var} G = -\mathbf{g}^T\boldsymbol{\chi}^{pp}\mathbf{g}\sigma^2,$$
or
$$\operatorname{var} G = -\sum_{j=1}^{p}\sum_{k=1}^{p}\chi_{jk}^{pp} g_j g_k \sigma^2. \tag{2}$$

In either case, it is possible to compute $\operatorname{var} G$ without inverting the matrix by adding an extra column to the right-hand side of the Doolittle scheme. In the first case, a column of values
$$\sum_i w_i^{-1} g_i F'_{qi}$$
is added. The solution using these values is
$$K_s = \sum_{q=1}^{r}\chi_{qs}\sum_i w_i^{-1} g_i F'_{qi}, \tag{3a}$$
and so
$$\operatorname{var} G = \left\{\Sigma w_i^{-1} g_i^2 - \sum_s K_s \sum_i w_i^{-1} g_i F'_{si}\right\}\sigma^2. \tag{3b}$$

Similarly, in the second case a column of values g_k is added. The solution of the Doolittle scheme using these values is

$$K_j = \sum_{k=1}^{p} \chi_{jk}^{pp}\, g_k, \tag{3c}$$

and

$$\operatorname{var} G = -\sum_j g_j K_j \sigma^2. \tag{3d}$$

However, it is usually simpler to complete the inversion of the matrix and use (1) and (2), especially if the variances of several functions are required.

11.3.5.2 *Example*

The area of the triangle discussed in § 11.3.1 can be evaluated from the formula $\frac{1}{2}bc \sin A$. The standard deviation can be calculated using (11.3.5.1,1), the steps being shown in Table 11.3.5.

TABLE 11.3.5

Evaluation of the standard deviation of a function of the adjusted values

(*a*) Area

$G = \frac{1}{2}bc \sin A = 0{\cdot}248693$

(*b*) Differential coefficients

$g_A = \frac{1}{2}bc \cos A$	$-0{\cdot}0012$	$g_A/\sqrt{w_A}$	$-0{\cdot}0012$
$g_B = g_C = 0 = g_a$			
$g_b = \frac{1}{2}c \sin A$	$0{\cdot}3553$	$g_b/\sqrt{w_b}$	$0{\cdot}3063$
$g_c = \frac{1}{2}b \sin A$	$0{\cdot}3500$	$g_c/\sqrt{w_c}$	$0{\cdot}3017$

(*c*) Variance

	$\Sigma g_i F'_{ji}/w_i$			
j	0	1	2	$\Sigma g_i^2/w_i$
	$-0{\cdot}0012$	$0{\cdot}3778$	$0{\cdot}3660$	$0{\cdot}1848$

	$(\Sigma g_i F'_{ji}/w_i)\,(\Sigma g_i F'_{ki}/w_i)$					
j, k	0, 0	0, 1	0, 2	1, 1	1, 2	2, 2
	0	$-0{\cdot}0005$	$-0{\cdot}0004$	$0{\cdot}1427$	$0{\cdot}1383$	$0{\cdot}1340$

$\Sigma g_i^2/w_i - \Sigma \chi_{jk} (\Sigma g_i F'_{ji}/w_i)(\Sigma g_i F'_{ki}/w_i) = 0{\cdot}1264$

$\sigma[G] = 0{\cdot}356\,\sigma = 0{\cdot}206 \times 10^{-3}$

Area 248,700 ± 200 square feet

The alternative method, using (11.3.5.1,3*a*–*b*), is illustrated in Table 11.3.5.1. The values $\Sigma g_i F'_{ji}/w_i$ are evaluated as in Table 11.3.5, and these values are then treated as an extra column of Table 11.3.1*a*. The solutions using this column are multiplied by the values $\Sigma w_i^{-1} F'_{ji} g_i$, and the sum subtracted from $\Sigma w_i^{-1} g_i^2$ to give $\operatorname{var} G/\sigma^2$.

TABLE 11.3.5.1

Evaluation of standard deviation using (11.3.5.1,3*a*–*b*)

			−0·0012
			0·3778
			0·3660
			−0·001200
			−0·000400
			0·377390
			0·103022
			0·251062
	K_j		0·077590
0·052472	0·079473	0·077590	
$\Sigma K_j \, \Sigma g_i \, F'_{ji}/w_i$	0·058360		
$\Sigma g_i^2/w_i - \Sigma K_j \, \Sigma g_i \, F'_{ji}/w_i$		0·1264	

11.4 NOTES AND REFERENCES

(11.1) Multiple correlation and regression are treated adequately in most books on statistics; see Kendall (1948) and Quenouille (1952). References on serial correlation include: Durbin and Watson (1950), Hannan (1955*a*, 1955*b*), Daniels (1956), and Watson (1955).

(11.2) This treatment is due to Deming (1943).

(11.3) Discussions of this topic are given in many older books on least-squares; Wright (1884), Leland (1921), etc. A good account (without mathematical proofs) is given by Deming (1943). An interesting reference is Benthem (1954).

CHAPTER 12

FURTHER ILLUSTRATIVE EXAMPLES

In this final chapter a number of examples illustrating the types of problem usually encountered in practice will be discussed. There is in the first section a guide to the more commonly used calculating schemes to help the reader identify readily the method best suited to his particular problem. The five examples which follow illustrate the five commonest types of problem—the straight line, the polynomial, the polynomial with equally-spaced observations, the linear function, and the non-linear function.

12.1 GUIDE TO THE MORE IMPORTANT CALCULATING SCHEMES

12.1.1 *The straight line*

Observations y_i, x_i. Fitted line $u(x) = b_0 + b_1 x$.

Estimates of standard deviation: s, standard deviation of an observation; $s(b_1)$, standard deviation of b_1; $s[u(x)]$, standard deviation of fitted value.

(*a*) *Standard calculating scheme*

Section 6.1.4, Tables 6.1.4 and 6.1.4*a*.

$s[u(x)]$, Table 6.1.5, with $\rho_{10}(k)$ from Table 6.7*a*.

(*b*) *Equally-spaced observations*

Variable $\epsilon = (x-\bar{x})/\Delta x$, § 6.3; $\bar{x}$ mean, Δx interval between neighbouring x_i values. $\Sigma\epsilon_i^2$ in Table 6.7*b*.

$u(\epsilon) = a_0 + a_1\epsilon$; Table 6.3.3 ($n$ even), Table 6.3.3*a* (n odd).

$s[u(\epsilon)] = sn^{-\frac{1}{2}}\rho_{10}(k)$, $k = 2|\epsilon|/n$, $\rho_{10}(k)$ from Table 6.7*a*.

$u(x) = b_0 + b_1 x$; $b_1 = a_1/\Delta x$, $b_0 = a_0 - b_1\bar{x}$.

(*c*) *Special methods*

Straight line passing through the origin, § 6.1.6.

Calculation of slope from differences of successive equally-spaced observations, Table 6.3.4 (W_1 from Table 6.7*c*).

Calculation of slope by double summation of equally-spaced observations, § 6.3.6.

(*d*) *Both variables subject to error*

If the regression line only is required (e.g. for estimation or prediction of the value y corresponding to an observed x), the presence of error in the variable x is immaterial.

If the line giving the functional relation between the 'true' variables is required, refer to Table 6.5.3. The estimate of the error-free line lies between the two regression lines, but a more exact estimate can only be made if one of the three quantities σ_γ (standard error of x), σ_δ (standard error of y), or $k = (\sigma_\delta/\sigma_\gamma)^2$, is known.

12.1.2 *Polynomials*

Observations y_i, x_i.

Fitted polynomial of degree p, $u_p(x) = \sum_{j=0}^{p} b_{pj} x^j$.

(*a*) *Standard calculating scheme*

(1) Calculation of normal equations, §§ 7.1.1 and 7.1.1.1

$$\phi_{jk} = \sum_i (w_i)\, x_i^j x_i^k, \quad M_j = \sum_i (w_i)\, x_i^j y_i,$$

$$z_i = \left\{\sum_i x_i^j + y_i\right\}, \quad C_j = \sum_i (w_i)\, x_i^j z_i.$$

Check formula: $C_j = \sum_k \phi_{jk} + M_j$.

In many problems the weights w_i are taken as unity, and the symbol w_i omitted from the equations.

(2) Solution of normal equations.

(i) Gauss–Doolittle scheme (simplest), Table 7.1.8; abbreviated Doolittle scheme, Table 7.1.8.1.

Check formulae:

$$c_j = \sum_{k=j}^{p} \alpha_{jk} + a_j, \quad \sum_{j=0}^{p} b_{pj}\, \phi_{jk} = M_k.$$

(ii) Square root scheme, Table 7.1.8.2, § 7.1.5.

Check formulae:

$$c_j = \sum_{k=j}^{p} s_{kj} + m_j, \quad \sum_{j=0}^{p} b_{pj}\, \phi_{jk} = M_k.$$

(iii) If polynomials of different degrees, standard deviations of fitted values, or elements χ_{jk} of the inverse matrix, are required, the corresponding schemes of Tables 7.2.3, 7.2.3*a*, and 7.2.4 should be used.

Additional check formulae:

$$\Sigma R_{kj}\, \phi_{0j} = 0, \quad \Sigma r_{kj}\, \phi_{0j} = 0; \qquad \Sigma \phi_{jk}\, \chi_{jk} = 1.$$

(3) Residuals $v_{pi} = y_i - u_p(x_i)$.

In the Doolittle scheme,

$$\Sigma(w_i)\,v_{pi}^2 = \Sigma(w_i)\,y_i^2 - \sum_{j=0}^{p} a_j \mathcal{M}_j = \Sigma(w_i)\,v_{p-1,i}^2 - a_p \mathcal{M}_p.$$

Individual residuals, Table 7.2.5.

Check formulae:

$$\sum_i u_p(x_i) = \sum_j b_{pj}\left(\sum_i x_i^j\right), \quad \Sigma v_{pi} = \Sigma y_i - \Sigma u_p(x_i).$$

(4) Estimated standard deviations:

of b_{pj}, § 8.1.2;
of $u_p(x)$, Table 8.1.2 (approximate values, § 8.5.6).

(5) Restoration of powers of ten. If 10^r is removed from y_i and 10^q from x_i to bring them to the order of unity, powers of ten are restored by multiplying b_{pj} by $10^r 10^{-qj}$. See Table 7.1.7.

(*b*) *Equally-spaced observations*

Variable x replaced by $\epsilon = (x - \bar{x})/\Delta x$.

(1) Fitting by power moments.

(i) Calculation of moments: Table 7.6.2.1 (n even) or Table 7.6.2.2 (n odd).

(ii) Calculation of coefficients: Table 7.6.2.1*a* ($p \leqslant 3$) or Table 7.6.2.2*a* ($p \leqslant 5$); S_{jj} and β_{kj} from Table 7.10*b*.

(iii) Residuals: $\Sigma v_{pi}^2 = \Sigma y_i^2 - \Sigma a_j \mathcal{M}_j = \Sigma v_{p-1,i}^2 - a_p \mathcal{M}_p$. Individual residuals, Table 7.6.2.3*a* (n even), Table 7.6.2.3*b* (n odd).

(2) Fitting by orthogonal moments. Most useful when the degree of the polynomial is uncertain.

(i) Calculation of coefficients, § 7.6.3.2, Table 7.6.3.

(ii) Residuals: $\Sigma v_{pi}^2 = \Sigma y_i^2 - \Sigma a_j' \mathcal{M}_j' = \Sigma v_{p-1,i}^2 - a_p' \mathcal{M}_p'$. Individual fitted values and residuals, § 7.6.3.3, Table 7.6.3.1.

(3) Estimated standard deviations:

of an observation, $s_p = \{\Sigma v_{pi}^2/(n-p-1)\}^{\frac{1}{2}}$;
of a fitted value, § 8.4.2.

(4) Return to the original variable x, § 7.4, Table 12.4*f*.

(*c*) *Both variables subject to error*

The only practical procedure is to ignore the error in the independent variable, except perhaps in the allotting of the weights. See § 11.2.

12.1.3 *Other functions*

(*a*) *Linear functions*

Treatment similar to polynomial fitting, with x^j replaced by the variable x_j:

(i) Formation of normal equations, Tables 10.1.1 and 12.5*a*.

(ii) Solution of normal equations, Tables 12.5*b* (Doolittle) and 10.1.2 (square root).

(iii) Restoration of powers of ten. If 10^r is removed from y_i and 10^q from x_j to bring them to the order of unity, powers of ten are restored by multiplying b_{pj} by $10^r\,10^{-q}$.

(*b*) *Changes of variable*

For the simple exponential, see § 10.2.2.

Note that a change to a new variable must usually be accompanied by a change in the weights of the observations.

(*c*) *Linearization*

The fitted value is expressed as a linear function in the form

$$u(b_k) = u(b_k^0) + \Sigma b_j' x_j',$$

where the x_j' are the differential coefficients of the non-linear function with respect to the parameters b_j which are to be estimated. The values x_j' and $u(b_k^0)$ are found from approximate values b_k^0 of b_k obtained by any method appropriate to the particular problem. The corrections b_k' to the b_k^0 are found by the usual least-squares methods.

Examples are given in §§ 10.2.3.2, 12.3.1, and 12.6.

12.2 THE FITTING OF A STRAIGHT LINE—VARIATION OF COSMIC RAY INTENSITY WITH ATMOSPHERIC PRESSURE

The number of cosmic rays reaching the earth's surface depends on the atmospheric pressure, since the larger the pressure the greater the number of air molecules and so the greater the probability of absorption. Usually it is the rate of arrival of particles from outside the earth which is of interest, and so a correction must be made to take account of the variation of absorption with pressure. The corrected number of counts N_0 will be given by the formula

$$N_i - N_0 = k(B_i - B_0),$$

where N_i is the observed number, B_i the atmospheric pressure, and B_0 the standard pressure.

The factor k is clearly the slope of the regression line of N_i on B_i, and it can be determined from a series of observations of these two quantities. Table 12.2 gives the counts in successive hourly intervals and the atmospheric pressures in an experiment quoted by Janossy (*Cosmic Rays*, Oxford University Press, 1948, p. 382). The calculating scheme of § 6.1.4 can be used, the independent variable x_i being identified with $B_i - 700$ and the dependent variable y_i with $N_i - 3000$. The subtraction of the numbers 700 and 3000 simplifies the calculations while leaving the slope of the regression line unaltered.

TABLE 12.2

Cosmic-ray counts N_i (on a scale-of-eight recorder) and atmospheric pressure B_i (mm. of Hg*) in successive hourly intervals*

N_i	B_i	N_i	B_i		N_i	B_i
3454	757·0	3388	752·5		3530	741·7
3420	756·6	3455	751·6		3538	740·5
3412	756·5	3491	750·0		3538	738·7
3407	756·2	3439	748·8		3539	737·7
3387	755·5	3486	746·9		3564	736·8
3414	754·9	3476	745·6		3590	736·8
3388	754·2	3490	744·5		3578	736·7
3438	753·5	3530	743·5		3590	736·5
				SUM	83542	17933·2

The detailed calculations are set out in Table 12.2*a*. The slope k is the quantity b_1 in this Table, and its standard deviation is $s(b_1)$. The value obtained for the variation in counts with barometric pressure is

$$k = -8{\cdot}34 \pm 0{\cdot}65 \text{ counts/mm. Hg}$$

It is very desirable to evaluate the fitted values $u_i = b_0 + b_1 x_i$ and the residuals $v_i = y_i - u_i$. Examination of the residuals may show a systematic departure of the observations from a straight line, while the agreement of Σv_i^2 with the value found in Table 12.2*a* provides a check on the arithmetical calculations. In the present example the fitted values and residuals are evaluated in Table 12.2*b*.

If the only factor causing the readings to vary was the atmospheric pressure, the residual variation should be of the Poisson type (Ch. 4). The average number of counts is $3000 + \Sigma y/n = 3481$, and, since a scale-of-eight recorder was used, the average number of particles counted is $8 \times 3481 = 27848$. Hence the standard deviation should be the square root of this, which is 166·8, and the standard deviation of N_i $166{\cdot}8/8 = 20{\cdot}8$. This is so close to the value s obtained in Table 12.2*a* that it is probable that the residual variation is largely random.

TABLE 12.2*a*

Calculation of slope of regression line; y is $N-3000$, x is $B-700$

(*a*) Summations

Σy	11542	Σxy	533897·0	Σy^2	5655294
n	24	Σx	1133·2	Σx^2	54834·86

(*b*) Coefficients b_0 and b_1

$D = n\Sigma x^2 - (\Sigma x)^2 = 31894{\cdot}40 \qquad D/n = 1328{\cdot}933$

$b_1 = (n\Sigma xy - \Sigma x \Sigma y)/D = -265866{\cdot}4/31894{\cdot}40 = -8{\cdot}335833$

$b_0 = (-\Sigma x \Sigma xy + \Sigma x^2 \Sigma y)/D = 27891873{\cdot}72/31894{\cdot}40 = 874{\cdot}5069$

Check $nb_0 + \Sigma x b_1 = 11541{\cdot}9996 = \Sigma y$

(*c*) Standard deviations

Σy^2	5655294
$-(\Sigma y)^2/n$	5550740
$-b_1^2 D/n$	92342
$= \Sigma v^2$	12212

$s^2 = \Sigma v^2/(n-2) = 555 \qquad s = 23{\cdot}6$

$s(b_1) = s/\sqrt{(D/n)} = s/36{\cdot}5 = 0{\cdot}65$

TABLE 12.2*b*

The fitted values u_i and the residuals v_i

u_i	v_i	u_i	v_i	u_i	v_i
399	+55	437	−49	527	+3
403	+17	444	+11	537	+1
404	+8	458	+33	552	−14
406	+1	468	−29	560	−21
412	−25	484	+2	568	−4
417	−3	494	−18	568	+22
423	−35	504	−14	569	+9
429	+9	512	+18	570	+20
			SUM	11545	−3
			Σv_i^2	12247	

12.3 A POLYNOMIAL CURVE—CALIBRATION OF A PRISM SPECTROMETER

The positions of spectral lines on a photographic plate can be found with the aid of a microscope which is moved across the plate by a micrometer screw. It is necessary to find a formula giving the wavelength λ of a line in terms of the micrometer screw setting d. This will be the formula for the regression curve of λ on d, and it can be found from a series of observations of these quantities.

Table 12.3 gives the readings obtained for seven standard iron lines in the violet region of a spectrum taken with a constant deviation spectrometer. It will be noted that the wavelengths λ are free from experimental error, but the values of d, which corresponds to the independent variable, are subject to error. Since the curve required is a calibration curve, no special procedure is necessary to take account of the errors in the independent variable.

TABLE 12.3

Positions of standard iron lines on a photographic plate

Wavelength λ (A.U.)	*Micrometer Reading d*
4045·81	18·9840
4198·31	18·0049
4307·91	17·4000
4383·55	17·0272
4415·12	16·8793
4736·78	15·6099
4791·25	15·4338
SUM 30878·73	119·3391

It is hoped that a quadratic or cubic curve may prove satisfactory. The first step in determining these least-squares curves is the evaluation of the sums of powers and of the moments, as described in § 7.1.1.1. To simplify the arithmetic a factor 10^4 is removed from λ ($y = 10^{-4}\lambda$), and the origin of d is moved to 17 ($x = d-17$). Table 12.3*a* gives the calculation of the quantities $\phi_{jk} = \Sigma x^j x^k$ and $M_j = \Sigma x^j y$. The z column is a check column, as explained in § 7.1.1.1.

In Table 12.3*b* the Doolittle scheme of Table 7.2.3 is used to solve the normal equations, since this gives the coefficients for both the quadratic and cubic curves. A factor 10 has been removed from all the quantities entered from Table 12.3*a* to bring them nearer to unity. Since the curve is expected to fit the points very accurately, eight decimals have been retained throughout.

The quadratic curve is $u_2(x) = \Sigma b_{2j} x^j$. The fitted values at the points of observation and the residuals v_{2i} are evaluated in Table 12.3*c* (cf. § 7.2.5.1). An examination of the residuals shows them to be a few A.U. in magnitude, with a systematic variation which is of the form of a cubic curve. It appears, therefore, that the observations cannot be represented satisfactorily by a quadratic curve.

Table 12.3*d* gives the fitted values and residuals for the cubic curve. The residuals are now appreciably smaller, but they still seem to show some evidence of a systematic variation. To increase the degree of the polynomial further would make the formula for λ too complicated for practical use. Probably the best procedure is to use the quadratic curve in conjunction with a graph giving an additional small correction term. Such a graph could be obtained by measuring more standard lines of known wavelength in the region, determining their deviations from the quadratic, and drawing a smooth curve through these deviations.

TABLE 12.3*a*

Calculation of moments and sums of powers

Factor removed : y, 10^r, $r = 4$

x^0	x	x^2	x^3	y	z
1	1·9840	3·93625600	7·80953190	0·404581	15·13436890
1	1·0049	1·00982401	1·01477215	0·419831	4·44932716
1	0·4000	0·16000000	0·06400000	0·430791	2·05479100
1	0·0272	0·00073984	0·00002012	0·438355	1·46631496
1	−0·1207	0·01456849	−0·00175842	0·441512	1·33362207
1	−1·3901	1·93237801	−2·68619867	0·473678	−0·67024266
1	−1·5662	2·45298244	−3·84186110	0·479125	−1·47595366

			ϕ_{jk}		M_j	C_j	Check
x^0	7	0·3391	9·50674879	2·35850598	3·087873	22·29222777	777
x		9·50674879	2·35850599	26·29087624	−0·05333933	38·44189169	169
x^2			26·29087624	17·16050342	4·18278301	59·49941745	745
x^3				83·99821011	0·49929920	130·30739495	495
y				Σy^2	1·36654431	9·08316019	019

12.3.1 *Use of a special function*

Any continuous relationship between the two variables may be represented satisfactorily by a polynomial, but the degree of the polynomial required may be very high. It is often possible to find, either theoretically or empirically, a special function with a very much smaller number of parameters to represent the relation between the two variables. Sawyer (*Experimental Spectroscopy*, 2nd ed., 1951, Prentice-Hall Inc., New York, p. 240)

TABLE

The Doolittle scheme

Factors removed : y, 10^r, $r = 4$; elements divided by 10^s, $s = 1$

	β_{00}	1	1. Enter		ϕ_{00}	0·70000000	ϕ_{01}	0·03391000
	R_{00}	1·42857143	2. $\div \phi_{00}$		α_{00}	1	α_{01}	0·04844286
	$R_{00}\,\phi_{00}$		Check					
					3. Enter		ϕ_{11}	0·95067488
	α_{01}	+0·04844286			4. $1 \times \alpha_{01}$			0·00164270
	β_{01}	−0·04844286	β_{11}	1	5. Subtract		S_{11}	0·94903218
	R_{10}	−0·05104449	R_{11}	1·05370505	6. $\div S_{11}$		α_{11}	1
	$\Sigma R_{1j}\,\phi_{0j}$				Check			
							7. Enter	
	α_{02}	1·35810697					8. $1 \times \alpha_{02}$	
		−0·00968810	α_{12}	0·19999026			9. $5 \times \alpha_{12}$	
	β_{02}	−1·34841887	β_{12}	−0·19999026	β_{22}	1	10. Subtract	
	R_{20}	−1·03723583	R_{21}	−0·15383726	R_{22}	0·76922376	11. $\div S_{22}$	
	$\Sigma R_{2j}\,\phi_{0j}$						Check	
	α_{03}	0·33692943						
		−0·13361723	α_{13}	2·75824403				
		−0·90471135		−0·13418194	α_{23}	0·67094237		
	β_{03}	+0·70139915	β_{13}	−2·62406209	β_{23}	−0·67094237	β_{33}	1
	R_{30}	+1·36197612	R_{31}	−5·09540095	R_{32}	−1·30283517	R_{33}	1·94179892
	$\Sigma R_{3j}\,\phi_{0j}$							
$a_0 =$	b_{00}	0·44112471						
$+\beta_{j1}\,a_1$		+0·00103582	a_1	−0·02138228				
	b_{10}	0·44216053	b_{11}	−0·02138228				
Check		$\Sigma b_{1j}\,\phi_{j1}$						
$+\beta_{j2}\,a_2$		−0·00308103		−0·00045696	a_2	+0·00228492		
	b_{20}	0·43907950	b_{21}	−0·02183924	b_{22}	+0·00228492		
Check		$\Sigma b_{2j}\,\phi_{j2}$						
$+\beta_{j3}\,a_3$		−0·00017844		+0·00066756		+0·00017069	a_3	−0·00025440
	b_{30}	0·43890106	b_{31}	−0·02117168	b_{32}	+0·00245561	b_{33}	−0·00025440
Check		$\Sigma b_{3j}\,\phi_{j3}$						

12.3*b*

for Example 12.3

								Σy^2	0·13665443
ϕ_{02}	0·95067488	ϕ_{03}	0·23585060	M_0	0·30878730	C_0	2·22922278	$M_0\,a_0$	0·13621371
α_{02}	1·35810697	α_{03}	0·33692943	a_0	0·44112471	c_0	3·18460397	Σv_0^2	0·00044072
							397		
ϕ_{12}	0·23585060	ϕ_{13}	2·62908762	M_1	−0·00533393	C_1	3·84418917	$\mathscr{M}_1\,a_1$	0·00043390
	0·04605341		0·01142528		0·01495854		0·10798993	Σv_1^2	0·00000682
S_{21}	0·18979719	S_{31}	2·61766234	$\mathscr{M}_1$	−0·02029247	$\mathscr{C}_1$	3·73619924		
α_{12}	0·19999026	α_{13}	2·75824403	a_1	−0·02138228	c_1	3·93685201		
							201		
ϕ_{22}	2·62908762	ϕ_{23}	1·71605034	M_2	0·41827830	C_2	5·94994174	$\mathscr{M}_2\,a_2$	0·00000679
	1·29111818		0·32031034		0·41936618		3·02752300	Σv_2^2	0·00000003
	0·03795759		0·52350697		−0·00405830		0·74720346		
S_{22}	1·30001185	S_{32}	0·87223303	$\mathscr{M}_2$	+0·00297042	$\mathscr{C}_2$	2·17521528		
α_{22}	1	α_{23}	0·67094237	a_2	+0·00228492	c_2	1·67322727		
							729		
12. Enter		ϕ_{33}	8·39982101	M_3	0·04992992	C_3	13·03073950	$\mathscr{M}_3\,a_3$	0·00000003
13. $1\times\alpha_{03}$			0·07946501		0·10403953		0·75109076	Σv_3^2	0·00000000
14. $5\times\alpha_{13}$			7·22015152		−0·05597158		10·30534925		
15. $10\times\alpha_{23}$			0·58521810		0·00199298		1·45944410		
16. Subtract		S_{33}	0·51498638	$\mathscr{M}_3$	−0·00013101	$\mathscr{C}_3$	0·51485539		
17. $\div S_{33}$		α_{33}	1	a_3	−0·00025440	c_3	0·99974564		
Check							560		

suggests for the present example the formula

$$\lambda = \lambda_0 + \frac{C}{d_0 - d},$$

where λ_0, C, and d_0 are constants. In terms of y and x,

$$u(x) = b_0 + b_1/(b_2 - x),$$

or

$$\{u(x) - b_0\}\{b_2 - x\} = b_1.$$

Approximate values b_j^0 of the three constants b_j can be found from the readings (x_1, y_1), (x_2, y_2), (x_3, y_3), for three lines in the

TABLE 12.3*c*

Fitted values and residuals for the quadratic curve

b_{2j}	0·43907950	−0·02183924	+0·00228492		
	b_{20}	$b_{21}x_i$	$b_{22}x_i^2$	$u_2(x_i)$	v_{2i}
	0·43907950	−0·04332905	+0·00899403	0·40474448	−0·00016348
		−0·02194625	+0·00230737	0·41944062	+0·00039038
		−0·00873570	+0·00036559	0·43070939	+0·00008161
		−0·00059403	+0·00000169	0·43848716	−0·00013216
		+0·00263600	+0·00003329	0·44174879	−0·00023679
		+0·03035873	+0·00441533	0·47385356	−0·00017556
		+0·03420462	+0·00560487	0·47888899	+0·00023601
SUM	3·07355650	−0·00740568	+0·02172217	3·08787299	+0·00000001
Check				Σv_{2i}^2	$34{\cdot}584 \times 10^{-8}$

TABLE 12.3*d*

Fitted values and residuals for the cubic curve

b_{3j}	0·43890106	−0·02117168	+0·00245561	−0·00025440		
	b_{30}	$b_{31}x_i$	$b_{32}x_i^2$	$b_{33}x_i^3$	$u_3(x_i)$	v_{3i}
	0·43890106	−0·04200461	0·00966591	−0·00198674	0·40457562	+0·00000538
		−0·02127542	0·00247973	−0·00025816	0·41984721	−0·00001621
		−0·00846867	0·00039290	−0·00001628	0·43080901	−0·00001801
		−0·00057587	0·00000182	−0·00000001	0·43832700	+0·00002800
		+0·00255542	0·00003577	+0·00000045	0·44149270	+0·00001930
		+0·02943075	0·00474517	+0·00068337	0·47376035	−0·00008235
		+0·03315909	0·00602357	+0·00097737	0·47906109	+0·00006391
SUM	3·07230742	−0·00717931	0·02334487	−0·00060000	3·08787298	+0·00000002
Check					Σv_{3i}^2	$1{\cdot}264 \times 10^{-8}$

spectrum. It can be shown that

$$b_0^0 = \{x_1 y_1(y_2 - y_3) + x_2 y_2(y_3 - y_1) + x_3 y_3(y_1 - y_2)\}/D$$

and $$b_2^0 = -\{x_1 y_1(x_2 - x_3) + x_2 y_2(x_3 - x_1) + x_3 y_3(x_1 - x_2)\}/D,$$

where $$D = x_1(y_2 - y_3) + x_2(y_3 - y_1) + x_3(y_1 - y_2).$$

Once b_0^0 and b_2^0 have been evaluated,

$$b_1^0 = (y_1 - b_0^0)(b_2^0 - x_1).$$

From the calculations in Table 12.3.1, the following three approximate values are obtained:

$$b_0^0 = 0{\cdot}250069, \quad b_1^0 = -1{\cdot}6856, \quad b_2^0 = -8{\cdot}9249.$$

TABLE 12.3.1

Calculation of the approximate values b_j^0

x_3	$-1{\cdot}5662$	$y_1 - y_2$	$-0{\cdot}033774$	$x_1 - x_2$	$1{\cdot}9568$	$x_3 y_3$	$-0{\cdot}75040558$
x_1	$1{\cdot}9840$	$y_2 - y_3$	$-0{\cdot}040770$	$x_2 - x_3$	$1{\cdot}5934$	$x_1 y_1$	$0{\cdot}80268870$
x_2	$0{\cdot}0272$	$y_3 - y_1$	$+0{\cdot}074544$	$x_3 - x_1$	$-3{\cdot}5502$	$x_2 y_2$	$0{\cdot}01192326$

$D = \Sigma x_i (y_j - y_k) = -0{\cdot}0259632444$

$b_0^0 = +\Sigma x_i y_i (y_j - y_k)/D = -0{\cdot}0064926127/D = +0{\cdot}25006939$

$b_2^0 = -\Sigma x_i y_i (x_j - y_k)/D = +0{\cdot}2317194/D = -8{\cdot}924902$

$b_1^0 = (y_1 - b_0^0)(b_2^0 - x_1) = -0{\cdot}15451161 \times 10{\cdot}908902 = -1{\cdot}6855520$

As in § 10.2.3, the linear variables

$$x_0' = \partial u(x)/\partial b_0 = 1,$$
$$x_1' = \partial u(x)/\partial b_1 = 1/(b_2^0 - x),$$
$$x_2' = \partial u(x)/\partial b_2 = -b_1^0/(b_2^0 - x)^2,$$
$$y' = y - u^0(x) = y - b_0^0 - b_1^0/(b_2^0 - x),$$

are introduced. The least-squares corrections b_j' to the values b_j^0 are found by solving the normal equations for y' as a linear function of the x_j'. The normal equations are calculated in Table 12.3.1*a*, and the solution using the scheme of Table 7.1.8 is found in Table 12.3.1*b*.

It is apparent that the value b_2' is not at all well determined, since S_{22} and $\mathscr{M}_2$ are both small. The normal equations are said to be ill-conditioned. The variations of x_1' and x_2' with x are very similar, and so a change in the value b_2' may be compensated by an opposing change in the value b_1', and a large range of pairs of values b'_1, b'_2 can be found which satisfy the normal equations almost equally as well as the exact solution. The value b'_2 actually obtained is very susceptible to round-off errors—errors due to retaining only a fixed number of decimals throughout the calculation. Small changes in the observations also cause very large changes in the value b'_2.

Since it is only a calibration curve which is required, this indeterminancy in the value of b'_2 is of no importance. Adopting the values b'_j found in Table 12.3.1*b*, and restoring the powers of ten by multiplying by $10^r 10^{-q}$, the least-squares values of the constants are

$$b_0 = 0{\cdot}250069 + 0{\cdot}000109 = 0{\cdot}250178,$$

$$b_1 = -1{\cdot}6856 + 0{\cdot}001125 = -1{\cdot}684475,$$

$$b_2 = -8{\cdot}9249 - 0{\cdot}0002 = -8{\cdot}9251,$$

and the least-squares formula for λ is

$$\lambda = 2501{\cdot}78 + \frac{1684{\cdot}475}{d - 8{\cdot}0749}.$$

The fitted values for the observed lines are listed in Table 12.3.1*c*. It is seen that the sum Σv^2 is about half that for the cubic curve (Table 12.3*d*), while there are only three constants as against four with the cubic curve. Hence the special function gives a considerably better fit.

TABLE 12.3.1*a*

The normal equations for the least-squares corrections to the values b_j^0

Factor removed: 10^q =			10^{-1}	10^{-2}	$10^r = 10^{-4}$		
x	$b_2^0 - x$	x'_0	x'_1	x'_2	y'	z	
1·9840	−10·9089	1	−0·916683	+1·416423	−0·04	1·459740	
1·0049	−9·9298	1	−1·007070	+1·709519	+0·10	1·802449	
0·4000	−9·3249	1	−1·072398	+1·938502	−0·41	1·456104	
0·0272	−8·9521	1	−1·117056	+2·103315	−0·05	1·936259	
−0·1207	−8·8042	1	−1·135822	+2·174579	−0·11	1·928757	
−1·3901	−7·5348	1	−1·327175	+2·969004	−1·00	1·641829	
−1·5662	−7·3587	1	−1·358936	+3·112809	−0·06	2·693873	
SUM			ϕ_{jk}		M_j	C_j	Check
+0·3391	−62·8134	7	−7·935140	+15·424151	−1·570000	12·919011	011
			9·150541	−18·088813	+1·965148	−14·908264	264
				36·343792	−4·180633	29·498497	497
				Σy^2	1·197900	−2·587585	585

TABLE 12.3.1*b*

Solution of normal equations

Factors removed : 10^q = –	x_0'	x_1' 10^{-1}	x_2' 10^{-2}	y' $10^r = 10^{-4}$		
All elements divided by 10^s, $s = 1$						Σy^2 0·119790
1. Enter	ϕ_{00} 0·700000	ϕ_{01} −0·793514	ϕ_{02} +1·542415	M_0 −0·157000	C_0 1·291901	$a_0 M_0$ 0·035213
2. $\div\phi_{00}$	α_{00} 1	α_{01} −1·133591	α_{02} +2·203450	a_0 −0·224286	c_0 1·845573	Σv_0^2 0·084577
Check					573	
	3. Enter	ϕ_{11} 0·915054	ϕ_{12} −1·808881	M_1 +0·196515	C_1 −1·490826	$a_1 \mathscr{M}_1$ 0·022130
	4. $1\times\alpha_{01}$	0·899520	−1·748468	+0·177974	−1·464487	Σv_1^2 0·062447
	5. Subtract	S_{11} 0·015534	S_{21} −0·060413	$\mathscr{M}_1$ +0·018541	$\mathscr{C}_1$ −0·026339	
	6. $\div S_{11}$	α_{11} 1	α_{12} −3·889082	a_1 +1·193575	c_1 −1·695571	
	Check				5507	
		7. Enter	ϕ_{22} 3·634379	M_2 −0·418063	C_2 2·949850	$a_2 \mathscr{M}_2$ 0·000000
		8. $1\times\alpha_{02}$	3·398634	−0·345942	2·846639	Σv_2^2 0·062447
		9. $5\times\alpha_{12}$	0·234951	−0·072107	0·102435	
		10. Subtract	S_{22} 0·000794	$\mathscr{M}_2$ −0·000014	$\mathscr{C}_2$ 0·000776	
		11. $\div S_{22}$	α_{22} 1	a_2 −0·017632	c_2 0·977329	
		Check			982368	
$-\alpha_{j2} b_2'$	+0·038851	−0·068572	$a_2 = b_2'$ −0·017632			
$-\alpha_{01} b_1'$	+1·275293	$+a_1$ +1·193575 $= b_1'$ +1·125003				
Check	$+a_0$ −0·224286 $= b_0'$ +1·089858 $\Sigma b_j' \phi_{j2}$	= −0·418065				

TABLE 12.3.1c

Fitted values for the observed lines

d	λ *fitted*	*Residual* v
18·9840	4045·89	−0·08
18·0049	4198·13	+0·18
17·4000	4308·17	−0·26
17·0272	4383·39	+0·16
16·8793	4415·00	+0·12
15·6099	4737·30	−0·52
15·4338	4790·80	+0·45
	SUM 30878·68	+0·05
	Σv^2	0·6193

12.4 POLYNOMIAL WITH EQUALLY-SPACED OBSERVATIONS—VARIATION OF VISCOSITY OF WATER WITH TEMPERATURE

Table 12.4 gives the viscosity of water y at temperatures t in the range 0(1)20° C. It is desired to fit a polynomial to these values so that the viscosity at any temperature in this range can be accurately determined.

TABLE 12.4

Viscosity of water y (centipoises) at temperature t° C

t	y	t	y	t	y
0	1·7921	7	1·4284	14	1·1709
1	1·7313	8	1·3860	15	1·1404
2	1·6728	9	1·3462	16	1·1111
3	1·6191	10	1·3077	17	1·0828
4	1·5674	11	1·2713	18	1·0559
5	1·5188	12	1·2363	19	1·0299
6	1·4728	13	1·2028	20	1·0050
				SUM	28·1490

A cubic is first fitted, using the scheme of Tables 7.6.2.2 and 7.6.2.1*a*, in which the variable t is replaced by the special variable $\epsilon = t - \bar{t}$. The calculations are given in Tables 12.4*a* and 12.4*b*. The sums and differences are formed in Table 12.4*a*, and the moments M_j are calculated by multiplying these by ϵ^j and summing. The moments are entered in Table 12.4*b*, together with the values S_{jj} and β_{kj} from Table 7.10*b*, and the orthogonal coefficients a_j and the power-series coefficients b_{3j} are calculated.

TABLE 12.4*a*

Calculation of moments

$M_j = \Sigma\epsilon^j [y_+ + (-)^j y_-]$

ϵ^3	ϵ	Diff.	y_+	y_-	Sum	ϵ^2
0	0		1·3077			0
1	1	−0·0749	1·2713	1·3462	2·6175	1
8	2	−0·1497	1·2363	1·3860	2·6223	4
27	3	−0·2256	1·2028	1·4284	2·6312	9
64	4	−0·3019	1·1709	1·4728	2·6437	16
125	5	−0·3784	1·1404	1·5188	2·6592	25
216	6	−0·4563	1·1111	1·5674	2·6785	36
343	7	−0·5363	1·0828	1·6191	2·7019	49
512	8	−0·6169	1·0559	1·6728	2·7287	64
729	9	−0·7014	1·0299	1·7313	2·7612	81
1000	10	−0·7871	1·0050	1·7921	2·7971	100
	SUM	−4·2285	11·3064	15·5349	26·8413	
			y_0 1·3077			

TABLE 12.4*b*

Evaluation of coefficients

S_{00} 21
S_{11} 770
S_{22} 22432·6r = 67298/3
S_{33} 622987·2
β_{02} −36·6r = −110/3
β_{13} −65·8

$a_j = \mathscr{M}_j/S_{jj}$		Σy^2	38·90059014	$a_3 = b_{33}$	−0·000020016912
$M_0 = \mathscr{M}_0$	28·1490	$a_0\mathscr{M}_0$	37·73172386	$\beta_{13}a_3$	+0·001317113
a_0	1·34042857	Σv_0^2	1·16886628	$+a_1$	−0·038651169
				$= b_{31}$	−0·037334056
$M_1 = \mathscr{M}_1$	−29·7614	$a_1\mathscr{M}_1$	1·15031290		
a_1	−0·038651169	Σv_1^2	0·01855338	$a_2 = b_{32}$	+0·00090313828
M_2	1052·3898	$a_2\mathscr{M}_2$	0·01829740		
$+\beta_{02}M_0$	−1032·13	Σv_2^2	0·00025598	$\beta_{02}a_2$	−0·03311507
$= \mathscr{M}_2$	20·2598			$+a_0$	1·34042857
a_2	0·00090313828			$= b_{30}$	1·30731350
M_3	−1970·7704	$a_3\mathscr{M}_3$	0·00024962		
$+\beta_{13}M_1$	+1958·30012	Σv_3^2	0·00000636		
$= \mathscr{M}_3$	−12·47028				
a_3	−0·000020016912				

Table 12.4c shows the calculation of the fitted values u and the residuals v for the cubic curve. The scheme is explained in § 7.6.2.3, the basic formulae being

$$u'' = b_{31}\epsilon + b_{33}\epsilon^3, \quad u' = b_{30} + b_{32}\epsilon^2,$$

$$u_+ = u' + u'', \qquad u_- = u' - u''.$$

An examination of the residuals shows a pronounced systematic variation, and it appears that a polynomial of higher degree is required.

TABLE 12.4c

Evaluation of residuals for the cubic curve

Odd powers
b_{31} −0·037334056 $\qquad$ b_{30} 1·3073135
b_{33} −0·0000200169 $\qquad$ b_{32} 0·000903138

ϵ	u''	u_{3+}	v_{3+}	v_{3-}	u_{3-}	u'
0		1·307314	+0·000386			
1	−0·037354	1·270863	+0·000437	+0·000629	1·345571	1·308217
2	−0·074828	1·236098	+0·000202	+0·000246	1·385754	1·310926
3	−0·112543	1·202899	−0·000099	+0·000415	1·427985	1·315442
4	−0·150617	1·171147	−0·000247	+0·000419	1·472381	1·321764
5	−0·189173	1·140719	−0·000319	−0·000265	1·519065	1·329892
6	−0·228328	1·111498	−0·000398	−0·000754	1·568154	1·339826
7	−0·268204	1·083363	−0·000563	−0·000671	1·619771	1·351567
8	−0·308921	1·056193	−0·000293	−0·001235	1·674035	1·365114
9	−0·350599	1·029869	+0·000031	+0·000233	1·731067	1·380468
10	−0·393357	1·004270	+0·000730	+0·001116	1·790984	1·397627
SUM	−2·113924	11·306919	−0·000519	+0·000133	15·534767	13·420843

It is simplest to use the tables of the orthogonal polynomials $T'_j(\epsilon)$ to investigate the decrease in the residuals for polynomials of higher degree. The values of $T'_4(\epsilon)$ and $T'_5(\epsilon)$ (for positive ϵ) are entered from the tables of Fisher and Yates (1948) in Table 12.4d. The moments $\mathscr{M}'_4$ and $\mathscr{M}'_5$ are calculated by multiplying these columns by the sum and the difference columns of Table 12.4a respectively. The orthogonal coefficients a'_j and the sums $\Sigma v_p^2 = \Sigma v_{p-1}^2 - a'_p \mathscr{M}'_p$ are calculated. The individual residuals are found from the formulae

$$v_4 = v_3 - a'_4 T'_4(\epsilon), \quad v_{5+} = v_{4+} - a'_5 T'_5(\epsilon), \quad v_{5-} = v_{4-} + a'_5 T'_5(\epsilon).$$

It is clear that the addition of a quartic term $a'_4 T'_4(\epsilon)$ considerably reduces the residuals. The residuals v_4 still show evidence

TABLE 12.4*d*

Residuals for fourth- and fifth-degree polynomials

ϵ	$T'_4(\epsilon)$	Factor 10^6 removed $a'_4 T'_4(\epsilon)$	v_{4+}	v_{4-}	$T'_5(\epsilon)$	Factor 10^6 removed $a'_5 T'_5(\epsilon)$	v_{5+}	v_{5-}
0	+594	+556	−170		0	0	−170	
1	+540	+506	−69	+123	+1404	−97	+28	+26
2	+385	+361	−159	−115	+2444	−168	+9	−283
3	+150	+140	−239	+275	+2819	−194	−45	+81
4	−130	−122	−125	+541	+2354	−162	+37	+379
5	−406	−380	+61	+115	+1063	−73	+134	+42
6	−615	−576	+178	−178	−788	+54	+124	−124
7	−680	−637	+74	−34	−2618	+180	−106	+146
8	−510	−478	+185	−757	−3468	+239	−54	−518
9	0	0	+31	+233	−1938	+134	−103	+367
10	+969	+907	−177	+209	+3876	−267	+90	−58
SUM (1–10)	−297	−279	−240	+412	+5148	−354	+114	+58
		Σv_4^2 1352828				Σv_5^2 775152		

$\mathscr{M}'_4 = \Sigma y T'_4(\epsilon) = +5{\cdot}3565$	$\mathscr{M}'_5 = \Sigma y T'_5(\epsilon) = -8{\cdot}3840$
$S'_{44} = 5{,}720{,}330$	$S'_{55} = 121{,}687{,}020$
$a'_4 = \mathscr{M}'_4/S'_{44} = 0{\cdot}93639703 \times 10^{-6}$	$a'_5 = \mathscr{M}'_5/S'_{55} = -6{\cdot}8898063 \times 10^{-8}$
$a'_4 \mathscr{M}'_4 = 502 \times 10^{-8}$	$a'_5 \mathscr{M}'_5 = 58 \times 10^{-8}$
$\Sigma v_4^2 = 134 \times 10^{-8}$	$\Sigma v_5^2 = 76 \times 10^{-8}$

of a systematic trend, and so it is probably advisable to use the fifth-degree curve, where the residuals are more random.

The power-series coefficients b_{5j} for the fifth-degree curve

$$u_5(\epsilon) = \sum_{j=0}^{5} b_{5j}\,\epsilon^j$$

may be obtained by adding $\beta'_{j4} a'_4 + \beta'_{j5} a'_5$ to the cubic coefficients b_{3j}, where the β'_{jk} are tabulated in Table 7.10*c*. The values obtained are listed in Table 12.4*e*.

TABLE 12.4*e*

Evaluation of power-series coefficients for the fifth-degree polynomial

β'_{55}	$0{\cdot}525$	$\beta'_{55} a'_5$	$-3{\cdot}617 \times 10^{-8}$	b_{55}	$-0{\cdot}0000000362$
β'_{44}	$0{\cdot}583r$	$\beta'_{44} a'_4$	$+0{\cdot}546232 \times 10^{-6}$	b_{54}	$+0{\cdot}000000546$
β'_{35}	$-63{\cdot}2916r$	$\beta'_{35} a'_5$	$+4{\cdot}360673 \times 10^{-6}$	b_{53}	$-0{\cdot}00001566$
β'_{24}	$-54{\cdot}583r$	$\beta'_{24} a'_4$	$-5{\cdot}111167 \times 10^{-5}$	b_{52}	$+0{\cdot}0008520$
β'_{15}	$+1466{\cdot}76r$	$\beta'_{15} a'_5$	$-1{\cdot}010574 \times 10^{-4}$	b_{51}	$-0{\cdot}037435$
β'_{04}	$+594$	$\beta'_{04} a'_4$	$+5{\cdot}562198 \times 10^{-4}$	b_{50}	$1{\cdot}30787$

Check $\Sigma b_{5j}\,10^j = 1{\cdot}00490,\ v_5(+10) = +0{\cdot}00010$

$\Sigma b_{5j}\,(-10)^j = 1{\cdot}79216,\ v_5(-10) = -0{\cdot}00006$

If it is assumed that the residuals in Table 12.4*d* are due to random errors of observation, the quantity

$$s_5 = \{\Sigma v_5^2/(n-6)\}^{\frac{1}{2}} = (77{\cdot}5 \times 10^{-8}/15)^{\frac{1}{2}} = 2{\cdot}27 \times 10^{-4}$$

will provide an estimate of the standard deviation of an observation. The standard deviation of a fitted value will be given by the formula (§ 8.4.2)

$$s[u_5(\epsilon)] = (s_5/\sqrt{n})\,\rho_{50}(k,n) = (0{\cdot}50 \times 10^{-4})\,\rho_{50}(k,n),$$

where k is $2|\epsilon|/n$ and $\rho_{50}(k,n)$ is tabulated in Table 8.8*a*. For example, when ϵ is ± 10, k is $0{\cdot}95$ and $\rho_{50}(0{\cdot}95, 21)$ is $4{\cdot}1$, and so the standard deviation of the fitted value is 0·00020. For most of the temperature range the rough approximation (§ 8.4.3) $2{\cdot}2\, s_5/\sqrt{n} = 0{\cdot}00011$ will be adequate.

The polynomial has been obtained in terms of the variable $\epsilon = t - \bar{t} = t - 10$. If the polynomial

$$u_5(t) = \sum_{j=0}^{5} c_{5j}\, t^j$$

in terms of the variable t is required, a change of origin is necessary. This may be accomplished by the scheme of Table 7.4.1. The value g is $\epsilon - t$, here equal to -10. It is convenient to reduce the scale of the variables ϵ and t by a factor 10, so that the coefficients b_{5j} are increased by a factor 10^j and so are all of the same magnitude. This changes the value of g to -1. The calculations are shown in Table 12.4*f*. The fitted polynomial is

$$u_5(t) = 1{\cdot}7922 - 0{\cdot}6317 \times 10^{-1}\,t + 0{\cdot}2011 \times 10^{-2}\,t^2$$
$$- 0{\cdot}0737 \times 10^{-3}\,t^3 + 0{\cdot}0236 \times 10^{-4}\,t^4 - 0{\cdot}0036 \times 10^{-5}\,t^5.$$

TABLE 12.4*f*

Polynomial expressed in terms of the variable t, $u_5(t) = \Sigma c_{5j}\, t^j$

Factor 10^j removed from b_{5j} ; $g = -1$: multiply b_{5j} by the power kg^m shown						
b_{50} 1·30787	b_{51} −0·37435	b_{52} +0·08520	b_{53} −0·01566	b_{54} +0·00546	b_{55} −0·00362	Sum = c_{5j}
1 1·30787	g +0·37435	g^2 +0·08520	g^3 +0·01566	g^4 +0·00546	g^5 +0·00362	1·79216
	1 −0·37435	$2g$ −0·17040	$3g^2$ −0·04698	$4g^3$ −0·02184	$5g^4$ −0·01810	−0·63167
		1 +0·08520	$3g$ +0·04698	$6g^2$ +0·03276	$10g^3$ +0·03620	+0·20114
			1 −0·01566	$4g$ −0·02184	$10g^2$ −0·03620	−0·07370
				1 +0·00546	$5g$ +0·01810	+0·02356
					1 −0·00362	−0·00362
Check $\Sigma c_{5j}(-g)^j = 1{\cdot}30787 = b_{50}$						

12.5 A LINEAR FUNCTION—VARIATION OF VAPOUR PRESSURE OF ETHYL ALCOHOL WITH TEMPERATURE

Table 12.5 lists the vapour pressure p of ethyl alcohol at temperatures t in the range 0(5)50° C. Kirchhoff derived a formula, discussed in textbooks on heat, for $\log p$ in terms of the absolute temperature $T = t + 273{\cdot}16$, of the form

$$\log p = A - B/T - C \log T.$$

TABLE 12.5

The vapour pressure p of ethyl alcohol (mm. of Hg) *at temperatures t° C*

t	p	t	p
0	12·24	30	78·06
5	17·31	35	102·60
10	23·78	40	133·70
15	32·44	45	172·20
20	44·00	50	220·00
25	58·86		
		SUM	895·19

TABLE 12.5*a*

Calculation of quantities occurring in the normal equations

t	x_0	x_1	x_2	y	z	
0	1	0·660858	−0·38583	0·087781	1·362809	
5	1	0·595053	−0·30705	0·238297	1·526300	
10	1	0·531572	−0·22968	0·376212	1·678104	
15	1	0·470294	−0·15366	0·511081	1·827715	
20	1	0·411107	−0·07895	0·643453	1·975610	
25	1	0·353904	−0·00551	0·769820	2·118214	
30	1	0·298588	+0·06672	0·892428	2·257736	
35	1	0·245068	+0·13776	1·011147	2·393975	
40	1	0·193256	+0·20766	1·126131	2·527047	
45	1	0·143073	+0·27646	1·236033	2·655566	
50	1	0·094442	+0·34418	1·342423	2·781045	
		ϕ_{jk}		M_j	C_j	Check
	11	3·997215	−0·127900	8·234806	23·104121	121
		1·804774	−0·500581	2·212646	7·514054	054
			0·587214	0·909419	0·868152	152
			Σy^2	7·890779	19·247650	650

TABLE 12.5*b*

Solution of the normal equations using the Doolittle scheme

											Σy^2	7·890779
1. Enter	ϕ_{00}	11·000000	ϕ_{01}	3·997215	ϕ_{02}	−0·127900	M_0	8·234806	C_0	23·104121	$a_0 M_0$	6·164730
2. $\div \phi_{00}$	α_{00}	1	α_{01}	0·3633832	α_{02}	−0·0116273	a_0	0·7486187	c_0	2·1003746	Σv_0^2	1·726049
Check										746		
3. Enter			ϕ_{11}	1·804774	ϕ_{12}	−0·500581	M_1	2·212646	C_1	7·514054	$a_1 \mathscr{M}_1$	1·726034
4. $1 \times \alpha_{01}$				1·452521		−0·046477		2·992390		8·395649	Σv_1^2	0·000015
5. Subtract			S_{11}	0·352253	S_{21}	−0·454104	$\mathscr{M}_1$	−0·779744	$\mathscr{C}_1$	−0·881595		
6. $\div S_{11}$			α_{11}	1	α_{12}	−1·289142	a_1	−2·213591	c_1	−2·502732		
Check										733		
7. Enter					ϕ_{22}	0·587214	M_2	0·909419	C_2	0·868152	$a_2 \mathscr{M}_2$	0·000003
8. $1 \times \alpha_{02}$						0·001487		−0·095749		−0·268639	Σv_2^2	0·000012
9. $5 \times \alpha_{12}$						0·585405		+1·005201		+1·136501		
10. Subtract					S_{22}	0·000322	$\mathscr{M}_2$	−0·000033	$\mathscr{C}_2$	+0·000290		
11. $\div S_{22}$					α_{22}	1	a_2	−0·102484	c_2	0·900621		
Check										897516		

		$a_1 =$	
$-\alpha_{01} b_1$	0·8043818	b_1	−2·213591
$+a_0$	0·7486187		
$= b_0$	1·5530005		
Check	$\Sigma b_j \phi_{j1} =$		2·212645

To find the values of the constants A, B, and C in the present example, this equation is first converted into the form

$$u(x_j) = b_0 x_0 + b_1 x_1 + b_2 x_2,$$

where

$$x_0 = 1, \quad x_1 = (T^{-1} - 0{\cdot}003) \times 10^3, \quad x_2 = (\log_{10} T - 2{\cdot}475) \times 10,$$

$$y = \log_{10} p - 1,$$

the new variables being chosen so that their values at the points of observation are all of the order of unity.

The quantities occurring in the normal equations are calculated in Table 12.5*a*, z being as usual the check column. The solution of the normal equations is carried out in Table 12.5*b*. It will be seen that S_{22} and $\mathscr{M}_2$ are both very small. The equations are ill-conditioned; there is a considerable range of pairs b_1, b_2 which satisfy the equations almost as well as the exact solution, and the values obtained are very susceptible to round-off and observational errors. That this would be so is apparent from Table 12.5*a*, since, over the range of values of t, x_1 and x_2 both vary roughly linearly with t. Only over a much larger range of t would it be possible to distinguish between the two variables. It is simplest to omit the variable x_2, and use the form

$$y = b_0 x_0 + b_1 x_1.$$

The values b_0 and b_1 are calculated in the 'backward' section at the bottom of Table 12.5*b*, and the fitted values and residuals in Table 12.5*c*. The fitted pressures are given by

$$\begin{aligned}\log_{10} P &= 1 + u(x_j)\\ &= 2{\cdot}553000 - 2{\cdot}213591(T^{-1} - 0{\cdot}003) \times 10^3\\ &= 9{\cdot}193773 - \frac{2213{\cdot}591}{T}.\end{aligned}$$

This formula will be suitable for the prediction of the vapour pressure at a given temperature, but the coefficient of $1/T$ will not be an estimate of the constant B in Kirchhoff's formula. For small variations t about T_0 such that powers of t/T_0 may be neglected,

$$\frac{T_0}{T_0 + t} = (1 + t/T_0)^{-1} = 1 - t/T_0,$$

$$\begin{aligned}\log_e (T_0 + t) &= \log_e T_0 + \log_e (1 + t/T_0) = \log_e T_0 + t/T_0\\ &= \log_e T_0 + 1 - T_0/(T_0 + t),\end{aligned}$$

and the Kirchhoff formula becomes

$$\log p = A' - B'/T,$$

where $A' = A - C(\log T_0 + \log e), \quad B' = B - CT_0 \log e.$

It is the quantities A' and B' that are estimated when the $\log T$ term is omitted from the Kirchhoff formula.

TABLE 12.5c

Fitted values and residuals

b_0 1·5530005,	b_1 −2·213591			
t	$u(x)$	v	P	p–P
0	0·090131	−0·002350	12·306	−0·066
5	0·235797	+0·002500	17·211	+0·099
10	0·376318	−0·000106	23·786	−0·006
15	0·511962	−0·000881	32·506	−0·066
20	0·642978	+0·000475	43·952	+0·048
25	0·769602	+0·000218	58·831	+0·029
30	0·892049	+0·000379	77·992	+0·068
35	1·010520	+0·000627	102·452	+0·148
40	1·125211	+0·000920	133·417	+0·283
45	1·236295	−0·000262	172·304	−0·104
50	1·343945	−0·001522	220·772	−0·772
SUM	8·234808	−0·000002	895·529	−0·339
	Σv^2	$16·60 \times 10^{-6}$		

The formula which has been fitted is the simple least-squares formula for the values y, each value y being assumed to have the same weight. This would correspond to the assumption that the standard deviations in the values p are proportional to p, the percentage standard deviation being the same at all temperatures. If it were assumed that the standard deviations were the same at all temperatures, then

$$\operatorname{var} y = \left(\frac{\partial}{\partial p}\log p\right)^2 \operatorname{var} p = \frac{1}{p^2}\operatorname{var} p,$$

and the weight attached to each observation would be proportional to p^2. The values in the columns x_j, y, z, of Table 12.5a would then be multiplied by some factor proportional to p (say $10^{-2}p$) and the resulting columns intermultiplied to give the normal equations.

It is found that the straight line obtained by this method is determined almost entirely by the observations at the higher temperatures. The residuals at all temperatures from 0° to 30° are negative. This would seem to indicate that the weighting is much too severe, and that equal weights for the values $\log p$ are more appropriate in fitting Kirchhoff's formula.

12.6 A NON-LINEAR FUNCTION—THE COUNTING RATE OF A TYPE I COUNTER

When the average rate of occurrence of events is unity, the average number $N(t)$ of counts in time t, given that the counter is free at time $t = 0$, may be calculated from the asymptotic formula

$$N(t) = \frac{t}{1+\tau} + \frac{\tau^2}{2(1+\tau)^2},$$

where τ is the resolving time of the counter. Table 12.6 shows the divergences $V(t)$ of the asymptotic formula from the true values when τ has the value 0·2. It is desired to find a formula which will represent this variation adequately. One possible form which suggests itself is

$$u(t) = \mathrm{e}^{-\lambda t}(\alpha t^4 - \beta t^2 + \gamma).$$

The least-squares values for the constants in this formula will now be determined.

TABLE 12.6

Divergences $V(t)$ of the asymptotic formula for $N(t)$ from the true values for a type I counter with $\tau = 0{\cdot}2$. A factor 10^{-2} has been removed from $V(t)$ and a factor 10 from t

t	$V(t)$	t	$V(t)$
0·0	+1·3889	1·6	−0·0634
0·2	+1·0755	1·8	−0·0841
0·4	+0·8011	2·0	−0·0713
0·6	+0·5654	2·2	−0·0456
0·8	+0·3672	2·4	−0·0263
1·0	+0·2059	2·6	−0·0121
1·2	+0·0809	2·8	−0·0029
1·4	−0·0086	3·0	+0·0028
		SUM	+4·1734

It is first necessary to obtain approximate values for the constants. If the zeros of $V(t)$ are assumed to be at $t_1 = 1{\cdot}4$ and $t_2 = 2{\cdot}9$, then from the theory of quadratic equations

$$\beta/\alpha = t_1^2 + t_2^2 = 10{\cdot}37, \quad \gamma/\alpha = t_1^2 t_2^2 = 16{\cdot}4836.$$

At $t = 0$, $u(t)$ is just γ, and so γ may be taken as 1·3889. Using this value of γ, estimates of α and β follow. At $t = 2$,

$$u(2) = \mathrm{e}^{-2\lambda}(16\alpha - 4\beta + \gamma),$$

which gives, on equating this expression to $V(2)$, the value 1·182 for λ. The first approximations which will be adopted are

$$\beta^\circ = b_0^0 = +0{\cdot}874; \quad \alpha^\circ = b_1^0 = +0{\cdot}0843;$$

$$\lambda^\circ = b_2^0 = +1{\cdot}18; \quad \gamma^\circ = b_3^0 = +1{\cdot}389.$$

From § 10.2.3 the least-squares adjustments b'_j to these values are found by solving the normal equations

$$\Sigma\left(y' - \sum_j b'_j x'_j\right) x'_k = 0,$$

where $x'_0 = \partial u/\partial\beta = -t^2\,\mathrm{e}^{-\lambda t}, \quad x'_1 = \partial u/\partial\alpha = t^4\,\mathrm{e}^{-\lambda t},$

$$x'_2 = \partial u/\partial\lambda = -t\,\mathrm{e}^{-\lambda t}(\alpha t^4 - \beta t^2 + \gamma) = -tu^\circ(t),$$

$$x'_3 = \partial u/\partial\gamma = \mathrm{e}^{-\lambda t}, \quad y' = V(t) - u^\circ(t),$$

the first approximations being used to calculate the values x'_j, y'. In Table 12.6*a* the values $x'_3, x'_0, x'_1, u^\circ(t), x'_2$, and y' are calculated in that order, and the columns are intermultiplied to give the sums for the normal equations. It will be noted that the correction to γ is labelled b'_3. This is because it is hoped that this correction will prove negligible. If this variable appears last in the Doolittle scheme it can be dropped without affecting the previous calculations.

Table 12.6*b* gives the solution of the normal equations. It is apparent that the inclusion of the variable x'_3, causes only a very slight reduction in Σv^2, and so this variable may be omitted and the equations solved for b'_0, b'_1, and b'_2. Restoring the 10^{-2} factor removed from y', the least-squares values are

$$\alpha = b_1^0 + b'_1 = 0{\cdot}0843 + 0{\cdot}00637 = 0{\cdot}09067;$$

$$\beta = b_0^0 + b'_0 = 0{\cdot}874 + 0{\cdot}0038 = 0{\cdot}8778;$$

$$\lambda = b_2^0 + b'_2 = 1{\cdot}18 - 0{\cdot}0728 = 1{\cdot}1072; \quad \gamma = 1{\cdot}3889.$$

Table 12.6*c* shows the fitted values $u(t)$ and the residuals $v(t)$ obtained with these values for the constants. The sum $\Sigma v^2(t)$ is in reasonable agreement with the value obtained in Table 12.6*b*. Some divergence is to be expected when a linearization procedure is used. The residuals

$$v(x'_j) = y' - \Sigma b'_j x'_j$$

are also shown in Table 12.6*c*, and it is found that $\Sigma v^2(x'_j)$ agrees almost exactly with the value in Table 12.6*b*. From the normal equations,

$$\Sigma v(x'_j)\, x'_k = \Sigma(y' - \Sigma b'_j x'_j)\, x'_k = 0,$$

and the sum of the residuals will only vanish if one of the variables x'_k is constant at all points of observation. Since this is not so here, $\Sigma v(x'_j)$ is different from zero.

TABLE 12.6a

Formation of the normal equations

Factor removed : y', 10^{-2}

t	x_3'	x_0'	x_1'	x_2'	y'	z	$u^\circ(t)$
0·0	1·000000	0·000000	0·000000	0·000000	−0·0100	0·990000	1·389000
0·2	0·789781	−0·031591	0·001264	−0·213900	+0·5998	1·145354	1·069502
0·4	0·623754	−0·099801	0·015968	−0·312206	+2·0586	2·286315	0·780514
0·6	0·492628	−0·177346	0·063845	−0·320785	+3·0758	3·134142	0·534642
0·8	0·389068	−0·249004	0·159362	−0·268976	+3·0980	3·128450	0·336220
1·0	0·307279	−0·307279	0·307279	−0·184152	+2·1748	2·297927	0·184152
1·2	0·242683	−0·349464	0·503227	−0·088892	+0·6823	0·989854	0·074077
1·4	0·191666	−0·375665	0·736304	+0·000052	−0·8563	−0·303943	−0·000037
1·6	0·151374	−0·387517	0·992045	+0·071683	−1·8598	−1·032215	−0·044802
1·8	0·119552	−0·387348	1·255009	+0·120037	−1·7413	−0·634050	−0·066687
2·0	0·094420	−0·377680	1·510720	+0·143178	+0·0289	1·399538	−0·071589
2·2	0·074571	−0·360924	1·746870	+0·142135	+1·9007	3·503352	−0·064607
2·4	0·058895	−0·339235	1·953995	+0·119914	+2·3664	4·159969	−0·049964
2·6	0·046514	−0·314435	2·125578	+0·080657	+1·8922	3·830514	−0·031022
2·8	0·036736	−0·288010	2·258000	+0·028966	+0·7445	2·780192	−0·010345
3·0	0·029013	−0·261117	2·350053	−0·030576	−0·7392	1·348173	+0·010192
SUM	4·647934	−4·306416	15·979519	−0·712865	13·4154	29·023572	4·039246
		ϕ_{jk}			M_j	C_j	Check
x_3'	2·656437	−0·742982	1·400281	−0·644190	5·026109	7·695655	655
x_0'		1·400280	−5·266322	+0·003364	−2·847197	−7·452857	857
x_1'			27·770052	+0·915776	+9·028290	33·848077	077
x_2'				0·443199	−2·639644	−1·921494	495
y'				$\Sigma y'^2$	49·969826	58·537384	384

TABLE 12.6*b*

Solution of the normal equations

Factor 10^{-2} removed from y'													$\Sigma y'^2$	49·969826
1. Enter	ϕ_{00}	1·400280	ϕ_{01}	−5·266322	ϕ_{02}	+0·003364	ϕ_{03}	−0·742982	M_0	−2·847197	C_0	−7·452857	$M_0 a_0$	5·789220
2. $\div\phi_{00}$	α_{00}	1	α_{01}	−3·760906	α_{02}	+0·002402	α_{03}	−0·530595	a_0	−2·033305	c_0	−5·322405	Σv_0^2	44·180606
Check												404		
	3. Enter		ϕ_{11}	27·770052	ϕ_{12}	+0·915776	ϕ_{13}	+1·400281	M_1	+9·028290	C_1	33·848077	$\mathscr{M}_1 a_1$	0·354293
	4. $1\times\alpha_{01}$			19·806142		−0·012652		+2·794285		+10·708040		28·029495	Σv_1^2	43·826313
	5. Subtract		S_{11}	7·963910	S_{21}	0·928428	S_{31}	−1·394004	$\mathscr{M}_1$	−1·679750	$\mathscr{C}_1$	5·818582		
	6. $\div S_{11}$		α_{11}	1	α_{12}	0·116579	α_{13}	−0·175040	a_1	−0·210920	c_1	0·730619		
	Check											619		
			7. Enter		ϕ_{22}	0·443199	ϕ_{23}	−0·644190	M_2	−2·639644	C_2	−1·921494	$\mathscr{M}_2 a_2$	17·730314
			8. $1\times\alpha_{02}$			0·000008		−0·001785		−0·006839		−0·017902	Σv_2^2	26·095999
			9. $5\times\alpha_{12}$			0·108235		−0·162512		−0·195824		+0·678324		
			10. Subtract		S_{22}	0·334956	S_{32}	−0·479893	$\mathscr{M}_2$	−2·436981	$\mathscr{C}_2$	−2·581916		
			11. $\div S_{22}$		α_{22}	1	α_{23}	−1·432705	a_2	−7·275524	c_2	−7·708224		
			Check									229		
					12. Enter		ϕ_{23}	2·656437	M_3	5·026109	C_3	7·695655	$\mathscr{M}_3 a_3$	0·054824
					13. $1\times\alpha_{03}$			0·394223		1·510708		3·954449	Σv_3^2	26·041175
					14. $5\times\alpha_{13}$			0·244006		0·294023		−1·018485		
					15. $10\times\alpha_{23}$			0·687545		3·491475		3·699124		
					16. Subtract		S_{33}	1·330663	$\mathscr{M}_3$	−0·270097	$\mathscr{C}_3$	1·060567		
					17. $\div S_{33}$		α_{33}	1	a_3	−0·202979	c_3	0·797021		
					Check							021		
					$a_2 =$									
$-\alpha_{j2} b_2'$		+0·017476		+0·848173	b_2'	−7·275524								
			$+a_1$	−0·210920										
$-\alpha_{01} b_1'$		+2·396649	$= b_1'$	+0·637253										
	$+a_0$	−2·033305												
	$= b_0'$	+0·380820												
Check	$\Sigma b_j'\phi_{j2} =$			−2·639643										
	$\Sigma b_j'\phi_{j1} =$			9·028278										

TABLE 12.6c

The fitted values and the residuals

	Non-linear function		Linear representation	
Factor removed:	–	10^{-2}	10^{-2}	10^{-2}
t	$u(t)$	$v(t)$	$u(x'_j)$	$v(x'_j)$
0·0	1·388900	0	0	−0·0100
0·2	1·084993	−0·9493	+1·5450	−0·9452
0·4	0·803226	−0·2126	+2·2436	−0·1850
0·6	0·558182	+0·7218	+2·3070	+0·7688
0·8	0·356415	+1·0785	+1·9637	+1·1343
1·0	0·198875	+0·7025	+1·4186	+0·7562
1·2	0·082862	−0·1962	+0·8344	−0·1521
1·4	0·003551	−1·2151	+0·3258	−1·1821
1·6	−0·044909	−1·8491	−0·0369	−1·8229
1·8	−0·068603	−1·5497	−0·2210	−1·5203
2·0	−0·073349	+0·2049	−0·2227	+0·2516
2·2	−0·064387	+1·8787	−0·0583	+1·9590
2·4	−0·046222	+1·9922	+0·2437	+2·1227
2·6	−0·022574	+1·0474	+0·6481	+1·2441
2·8	+0·003605	−0·6505	+1·1186	−0·3741
3·0	+0·030066	−2·7266	+1·6207	−2·3599
SUM	+4·190631	−1·7231	+13·7303	−0·3149
	$\Sigma v^2(t)$	26·9546	$\Sigma v^2(x'_j)$	26·0957

BIBLIOGRAPHY

BOOKS

BARLOW'S *Tables of Squares, Cubes, etc.* (ed. L. J. Comrie), 4th ed., 1947, E. and F. N. Spon Ltd, London.

Biometrika Tables for Statisticians, ed. E. S. Pearson and H. O. Hartley, Cambridge University Press, 1954.

BLEULER, E., and GOLDSMITH, G. J. (1952): *Experimental Nucleonics*, Reinhart and Co., New York.

BODEWIG, E. (1956): *Matrix Calculus*, North-Holland Publishing Co., Amsterdam.

BOOTH, A. D. (1955): *Numerical Methods*, Butterworths Scientific Publications, London.

BRUNT, D. (1917): *The Combination of Observations*, Cambridge University Press.

CLARK, D., and CLENDINNING, J. (1951): *Plane and Geodetic Surveying*, 4th ed., Constable and Co., London.

CROXTON, F. E., and COWDEN, D. J. (1955): *Applied General Statistics*, 2nd ed., Prentice-Hall, New York.

DAVIS, H. T. (1935): *Tables of the Higher Mathematical Functions*, 2 vols., Principia Press, Bloomington, Indiana.

—— (1941): *The Analysis of Economic Time Series*, Principia Press, Bloomington, Indiana.

DEMING, W. E. (1943): *Statistical Adjustment of Data*, John Wiley and Sons, New York.

DWYER, P. S. (1951): *Linear Computations*, John Wiley and Sons, New York.

ELDERTON, W. P. (1938): *Frequency Curves and Correlation*, 3rd ed., Cambridge University Press.

FISHER, R. A. (1948): *Statistical Methods for Research Workers*, 10th ed., Oliver and Boyd, Edinburgh.

—— and YATES, F. (1948): *Statistical Tables for Biological, Agricultural, and Medical Research*, 3rd ed., Oliver and Boyd, Edinburgh.

FLETCHER, A., MILLER, J. C. P., and ROSENHEAD, L. (1946): *An Index of Mathematical Tables*, Sci. Computing Service, London.

GOULDEN, C. H. (1952): *Methods of Statistical Analysis*, 2nd ed., John Wiley and Sons, New York.

HALD, A. (1952*a*): *Statistical Theory with Engineering Applications*, John Wiley and Sons, New York.

—— (1952*b*): *Statistical Tables and Formulae*, John Wiley and Sons, New York.

Handbook of Chemistry and Physics, 38th ed., 1956, Chemical Rubber Publishing Co., Cleveland, Ohio.

HARTREE, D. R. (1952): *Numerical Analysis*, Clarendon Press, Oxford.

HILDEBRAND, F. B. (1956): *Introduction to Numerical Analysis*, McGraw-Hill Book Co., New York.

HOOD, W. C. (ed.) (1953): *Studies in Econometric Method*, John Wiley and Sons, New York.

HOUSEHOLDER, A. S. (1953): *Principles of Numerical Analysis*, McGraw-Hill Book Co., New York.

JACKSON, D. (1941): *Fourier Series and Orthogonal Polynomials*, Carus Math. Monographs, Math. Assoc. Amer., Oberlin, Ohio.

JEFFREYS, H. (1948): *Theory of Probability*, 2nd ed., Clarendon Press, Oxford.

KENDALL, M. G. (1948): *The Advanced Theory of Statistics*, 2 vols., 2nd ed., Charles Griffin and Company, London.

KOOPMANS, T. (ed.) (1950): *Statistical Inference in Dynamic Economic Models*, John Wiley and Sons, New York.

KORFF, S. A. (1955): *Electron and Nuclear Counters*, 2nd ed., D. van Nostrand Company, New York.

LELAND, O. M. (1921): *Practical Least Squares*, McGraw-Hill Book Company, New York.

LEWIS, W. B. (1942): *Electrical Counting*, Cambridge University Press.

MILNE, W. E. (1949): *Numerical Analysis*, Princeton University Press.

MOLINA, E. C. (1942): *Poisson's Exponential Binomial Limit*, D. van Nostrand Company, New York.

PAIGE, L. J., and TAUSSKY, O. (ed.) (1953): *Simultaneous Linear Equations and the Determination of Eigenvalues*, Applied Math. Series, Vol. 29, U.S. Govt. Printing Office, Washington.

PEARSON, K. (1934): *Tables of the Incomplete B-function*, Cambridge University Press.

QUENOUILLE, M. H. (1952): *Associated Measurements*, Butterworths Scientific Publications, London.

SASULY, M. (1934): *Trend Analysis of Statistics*, The Brookings Institution, Washington.

TAUSSKY, O. (ed.) (1954): *Contributions to the Solution of Linear Systems and the Determination of Eigenvalues*, Applied Math. Series, Vol. 39, U.S. Govt. Printing Office, Washington.

TRUMPLER, R. J., and WEAVER, H. F. (1953): *Statistical Astronomy*, Univ. Calif. Press, Berkeley and Los Angeles.

WHITTAKER, E. T., and ROBINSON, G. (1944): *The Calculus of Observations*, 4th ed., Blackie and Sons, Glasgow.

—— and WATSON, G. N. (1940): *Modern Analysis*, 4th ed., Cambridge University Press.

WILSON, E. B. (1952): *An Introduction to Scientific Research*, McGraw-Hill Book Company, New York.

WORTHING, A. G., and GEFFNER, J. (1943): *Treatment of Experimental Data*, John Wiley and Sons, New York.

WRIGHT, T. W. (1884): *Treatise of the Adjustment of Observations*, D. van Nostrand Company, New York.

PAPERS

AITKEN, A. C. (1933*a*): On the graduation of data by the orthogonal polynomials of least squares, *Proc. Roy. Soc. Edin.*, **53**, 54.

—— (1933*b*): On fitting polynomials to weighted data by least squares, *Proc. Roy. Soc. Edin.*, **54**, 1.

—— (1933*c*): On fitting polynomials to data with weighted and correlated errors, *Proc. Roy. Soc. Edin.*, **54**, 12.

—— (1935): On least squares and linear combination of observations, *Proc. Roy. Soc. Edin.*, **55**, 42.

—— (1950): Studies in practical mathematics. V. On the iterative solution of a system of linear equations, *Proc. Roy. Soc. Edin.*, **A**, **63**, 52.

ALLAN, F. E. (1930): The general form of the orthogonal polynomials for simple series with proofs of their simple properties, *Proc. Roy. Soc. Edin.*, **50**, 310.

ANDERSON, R. L. (1954): The problem of autocorrelation in regression analysis, *J. Am. Statist. Assoc.*, **49**, 113.

—— and HOUSEMAN, E. E. (1942): Tables of orthogonal polynomial values extended to N = 104, *Iowa State Coll. Agric. Expt. Station, Res. Bull.* 297.

Aspin, A. A. (1949): Tables for use in comparisons whose accuracy involves two variances, separately estimated, *Biometrika*, **36**, 290.

Barnard, G. A. (1950): On the Fisher–Behrens test, *Biometrika*, **37**, 203.

Bartlett, M. S. (1937): Properties of sufficiency and statistical tests, *Proc. Roy. Soc., London*, **A**, **160**, 268.

—— (1949): Fitting a straight line when both variables are subject to error, *Biometrics*, **5**, 207.

—— *et al.* (1946): Symposium on auto-correlation in time series, *J. Roy. Statist. Soc. Suppl.*, **8**, 27.

Benthem, J. P. (1954): Note on minimizing a quadratic function with additional linear conditions by matrix methods, with application to stress analysis, *Nationaal Luchtvaart-Laboratorium, Amsterdam*, Report S 437.

Berkson, J. (1950): Are there two regressions?, *J. Am. Statist. Assoc.*, **45**, 164.

Birge, R. T. (1947): Least-squares' fitting of data by means of polynomials, *Rev. Mod. Physics*, **19**, 298.

—— (1949): The exact representation of a series of points by a polynomial in power series form, *Am. J. Physics*, **17**, 196.

—— and Shea, J. D. (1927): A rapid method for calculating the least-squares solution of a polynomial of any degree, *Univ. Calif. Publ. Math.*, **2**, 67.

Blackman, M., and Michiels, J. L. (1948): Efficiency of counting systems, *Proc. Phys. Soc., London*, **60**, 549.

Blom, G. (1954): Transformations of the binomial, negative binomial, Poisson, and χ^2 distributions, *Biometrika*, **41**, 302.

Bodewig, E. (1947): Bericht über die verschiedenen Methoden zur Lösung eines Systems linearer Gleichungen mit reellen Koeffizienten, *Nederl. Akad. Wetensc. Proc.*, **50**, 930, 1104, 1285; (1948) **51**, 53, 211.

Breitenberger, E. (1956): Remarks on the least-squares reduction of angular correlation data, *Proc. Phys. Soc., London*, **A**, **69**, 489.

Cadwell, J. H. (1954): The statistical treatment of mean deviation, *Biometrika*, **41**, 12.

Cochran, W. G. (1952): The χ^2 test of goodness of fit, *Ann. Math. Statist.*, **23**, 315.

—— (1954*a*): The combination of estimates from different experiments, *Biometrics*, **10**, 101.

—— (1954*b*): Some methods for strengthening the common χ^2 tests, *Biometrics*, **10**, 417.

Cohen, E. R. (1953): Basis for the criterion of least squares, *Rev. Mod. Physics*, **25**, 709.

—— (1956): Standard errors of the residues in a least-squares analysis, *Phys. Rev.*, **101**, 1641.

—— Dumond, J. W. M. *et al.* (1955): Analysis of variance of the 1952 data on the atomic constants and a new adjustment, 1955, *Rev. Mod. Physics*, **27**, 363.

Cornell, R. G. (1956): A new estimation procedure for a linear combination of exponentials (abstract only), *Ann. Math. Statist.*, **27**, 207.

Cornish, E. A., and Fisher, R. A. (1937): Moments and cumulants in the specification of distributions, *Rev. de l'Inst. Int. Statist.*, **5**, 307.

Cox, D. R. (1953): Some simple approximate tests for Poisson variates, *Biometrika*, **40**, 354.

—— (1954): The mean and coefficient of variation of range in small samples from non-normal populations, *Biometrika*, **41**, 469.

Cox, G. J., and Matuschak, M. C. (1941): An abbreviation of the method of least squares, *J. Phys. Chem.*, **45**, 362.

Creasey, M. A. (1956): Confidence limits for the gradient in the linear functional relationship, *J. Roy. Statist. Soc.*, **B**, **18**, 65.

Crow, J. F. (1945): A chart of the χ^2 and t-distributions, *J. Am. Statist. Assoc.*, **40**, 376.

Daniel, C., and Heerema, N. (1950): Design of experiments for most precise slope estimation or linear extrapolation, *J. Am. Statist. Assoc.*, **45**, 546.

Daniels, H. E. (1956): The approximate distribution of serial correlation coefficients, *Biometrika*, **43**, 169.

Danielson, G. C., and Lanczos, C. (1942): Some improvements in practical Fourier analysis, *J. Franklin Inst.*, **233**, 365, 435.

David, F. N., and Neyman, J. (1938): Extension of the Markoff theorem on least squares, *Statist. Res. Mem. Uni. Coll., London*, **2**, 105.

Davis, P., and Rabinowitz, P. (1954): A multiple purpose orthonormalizing code and its uses, *J. Assoc. Comp. Mach.*, **1**, 183.

de la Garza, A. (1954): Spacing of information in polynomial regression, *Ann. Math. Statist.*, **25**, 123.

—— *et al.* (1955): Some minimum cost experimental procedures in quadratic regression, *J. Am. Statist. Assoc.*, **50**, 178.

Deming, W. E. (1937): On the significant figures of least squares and correlations, *Science*, **85**, 451.

—— (1938): Some thoughts on curve fitting and the chi test, *J. Am. Statist. Assoc.*, **33**, 543.

—— and Birge, R. T. (1934): On the statistical theory of errors, *Rev. Mod. Physics*, **6**, 122.

Dent, B. M. (1935): On observations of points connected by a linear relation, *Proc. Phys. Soc., London*, **47**, 92.

Dixon, W. J. (1950): Analysis of extreme values, *Ann. Math. Statist.*, **21**, 488.

—— (1951): Ratios involving extreme values, *Ann. Math. Statist.*, **22**, 68.

—— (1953): Processing data for outliers, *Biometrics*, **9**, 74.

Dumond, J. W. M., and Cohen, E. R. (1953): Least-squares adjustment of the atomic constants, 1952, *Rev. Mod. Physics*, **25**, 691.

Durbin, J., and Watson, G. S. (1950): Testing for serial correlation in least-squares regression I, *Biometrika*, **37**, 409; (1951) II, **38**, 159.

Elfving, G. (1952): Optimal allocation in linear regression theory, *Ann. Math. Statist.*, **23**, 255.

Elmore, W. C. (1950): Statistics of counting, *Nucleonics*, **6**, 26.

Feller, W. (1948): On probability problems in the theory of counters, Courant Anniversary Volume, Interscience Publishers, New York.

Fisher, R. A. (1920): A mathematical examination of the methods of determining the accuracy of an observation by the mean error and by the mean square error, *Month. Not. R. Astron. Soc.*, **80**, 758.

—— (1929): Tests of significance in harmonic analysis, *Proc. Roy. Soc., London*, **A**, **125**, 54.

Forsythe, G. E. (1953*a*): Solving linear algebraic equations can be interesting, *Bull. Am. Math. Soc.*, **59**, 299.

—— (1953*b*): A numerical analyst's 15 foot shelf, *Math. Tables Aids Comp.*, **7**, 221.

—— (1957): Generation and use of orthogonal polynomials for data fitting with a digital computer, *J. Soc. Indust. Appl. Math.*, **5**, 74.

Galton, F. (1886): Regression towards mediocrity in hereditary stature, *J. Anthrop. Inst.*, **15**, 246.

Geary, R. C. (1949): Determination of linear relations between systematic parts of variables with errors of observation the variances of which are unknown, *Econometrica*, **17**, 30.

—— (1953): Non-linear functional relationship between two variables when one is controlled, *J. Am. Statist. Assoc.*, **48**, 94.

GINI, C. (1921): Sull' interpolazione di una retta quando i valori della variabile indipendente sono affetti da errori accidentali, *Metron*, **1**, No. 3, 63.

GRUBBS, F. E. (1950): Sample criteria for testing outlying observations, *Ann. Math. Statist.*, **21**, 27.

GUEST, P. G. (1950*a*): Orthogonal polynomials in the least squares fitting of observations, *Phil. Mag.*, **41**, 124.

—— (1950*b*): Estimation of the error at a point on a least-squares curve, *Austral. J. Sci. Res.*, **A**, **3**, 173.

—— (1950*c*): Estimation of the errors of the least-squares polynomial coefficients, *Austral. J. Sci. Res.*, **A**, **3**, 364.

—— (1951): The estimation of standard error from successive finite differences, *J. Roy. Statist. Soc.*, **B**, **13**, 233.

—— (1952): Tables of certain functions occurring in the fitting of polynomials to equally-spaced observations, *Math. Tables Aids Comp.*, **6**, 40.

—— (1953*a*): On the standard errors in the fitting of polynomials to unequally-spaced observations, *Austral. J. Physics*, **6**, 131.

—— (1953*b*): The Doolittle method and the fitting of polynomials to weighted data, *Biometrika*, **40**, 229.

—— (1954): Grouping methods in the fitting of polynomials to equally-spaced observations, *Biometrika*, **41**, 62.

—— (1956): Grouping methods in the fitting of polynomials to unequally-spaced observations, *Biometrika*, **43**, 149.

—— and SIMMONS, W. M. (1953): An experiment on cosmic rays, *Am. J. Physics*, **21**, 357.

HALMOS, P. R. (1946): The theory of unbiased estimation, *Ann. Math. Statist.*, **17**, 34.

HANNAN, E. J. (1955*a*): Exact tests for serial correlation, *Biometrika*, **42**, 133.

—— (1955*b*): An exact test for correlation between time series, *Biometrika*, **42**, 316.

HARTLEY, H. O. (1948): The estimation of non-linear parameters by 'internal least squares', *Biometrika*, **35**, 32.

—— (1949): Tests of significance in harmonic analysis, *Biometrika*, **36**, 194.

—— (1950): The maximum F-ratio as a short-cut test for heterogeneity of variance, *Biometrika*, **37**, 308.

—— (1951): The fitting of polynomials to equidistant data with missing values, *Biometrika*, **38**, 410.

HAYES, G. J., and VICKERS, T. (1951): The fitting of polynomials to unequally-spaced data, *Phil. Mag.*, **42**, 1387.

HEALY, M. J. R., and DYKE, G. V. (1953): A Hollerith technique for the solution of normal equations, *J. Am. Statist. Assoc.*, **48**, 809.

HELMERT, F. R. (1876): Die Genauigkeit der Formel von Peters zur Berechnung des wahrscheinlichen Beobachtungsfehlers direkter Beobachtungen gleicher Genauigkeit, *Astron. Nachrichten*, **88**, No. 2096.

HORSNELL, G. (1953): The effect of unequal group variances on the F-test for the homogeneity of group means, *Biometrika*, **40**, 128.

HOUSEHOLDER, A. S. (1949): Analyzing exponential decay curves, Seminar on Sci. Comput., Int. Bus. Machines Corp.

—— (1956): Bibliography on numerical analysis, *J. Assoc. Comp. Mach.*, **3**, 85.

HSU, P. L. (1938): On the best unbiassed quadratic estimate of the variance, *Statist. Res. Mem. Uni. Coll., London*, **2**, 91.

JAEGER, W., and VON STEINWEHR, H. (1921): Wärmekapazität des Wassers zwischen 5° und 50° in internationalen Wattsekunden, *Ann. der Physik*, **64**, 305.

JAMES, G. S. (1951): The comparison of several groups of observations when the ratios of population variances are unknown, *Biometrika*, **38**, 324.

—— (1954): Linear hypotheses when the ratios of the population variances are unknown, *Biometrika*, **41**, 19.

—— (1956): On the accuracy of weighted means and ratios, *Biometrika*, **43**, 304.

JESSOP, W. N. (1952): One line or two?, *Appl. Statist.*, **1**, 131.

KAVANAGH, A. J. (1941): Note on the adjustment of observations, *Ann. Math. Statist.*, **12**, 111.

KEEPING, E. S. (1951): A significance test for exponential regression, *Ann. Math. Statist.*, **22**, 180.

KENDALL, M. G. (1946*a*): Contributions to the study of oscillatory time-series, *Nat. Inst. Econ. Soc. Res. Occas. Pap.*, IX, Cambridge University Press.

—— (1946*b*): On autoregressive time series, *Biometrika*, **33**, 105.

—— (1949): On the reconciliation of theories of probability, *Biometrika*, **36**, 101.

—— (1951): Regression, structure, and functional relationship, Part I, *Biometrika*, **38**, 11; (1952): Part II, **39**, 96.

KERAWALA, S. M. (1941): A rapid method for calculating the least-squares solution of a polynomial of degree not exceeding the fifth, *Indian J. Physics*, **15**, 241.

KIMBALL, B. F. (1953): Note on computation of orthogonal predictors, *Ann. Math. Statist.*, **24**, 299.

LADERMAN, J. (1948): The square root method for solving simultaneous linear equations, *Math. Tables Aids Comp.*, **3**, 13.

LINDLEY, D. V. (1947): Regression lines and the linear functional relationship, *Suppl. J. Roy. Statist. Soc.*, **B**, **9**, 218.

—— (1953): Estimation of a functional relationship, *Biometrika*, **40**, 47.

LORD, E. (1947): The use of range in place of standard deviation in the t-test, *Biometrika*, **34**, 41.

—— (1950): Power of the modified t-test (u-test) based on range, *Biometrika*, **37**, 64.

LOTKIN, M. (1956): Note on the sensitivity of least-squares solutions, *J. Maths. and Physics*, **35**, 309.

MEIER, P. (1953): Variance of a weighted mean, *Biometrics*, **9**, 59.

MURDOCH, H. S. (1953): The half-life of $^{181}Ta^{m}$ and the delayed coincidence method, *Proc. Phys. Soc., London*, **A**, **66**, 944.

NAGLER, H. (1950): On the best unbiased quadratic estimate of variance, *Biometrika*, **37**, 444.

NAIR, K. R. (1948): The distribution of the extreme deviate from the sample mean and its studentized form, *Biometrika*, **35**, 118.

—— (1952): Tables of percentage points of the 'Studentized' extreme deviate from the sample mean, *Biometrika*, **39**, 189.

—— and SHRIVASTAVA, M. P. (1942): On a simple method of curve-fitting, *Sankhya*, **6**, 121.

NEILSEN, K. L., and GOLDSTEIN, L. (1947): An algorithm for least squares, *J. Maths. and Physics*, **26**, 120.

NEKRASSOFF, V. A. (1930): Nomogram for t test, *Metron*, **8**, 95.

NEYMAN, J., and SCOTT, E. L. (1951): On certain methods of estimating the linear structural relation, *Ann. Math. Statist.*, **22**, 352; corr., **23**, 135.

PEARSON, E. S. (1950): Some notes on the use of range, *Biometrika*, **37**, 88.

—— and CHANDRA SEKAR, C. (1936): The efficiency of statistical tools and a criterion for the rejection of outlying observations, *Biometrika*, **28**, 308.

PEARSON, K. (1901): On lines and planes of closest fit to systems of points in space, *Phil. Mag.* (6), **2**, 559.

PLACKETT, R. L. (1949): A historical note on the method of least squares, *Biometrika*, **36**, 458.

—— (1950): Some theorems in least squares, *Biometrika*, **37**, 149.

PRICE, P. C. (1954): On simplifying the reduction by least squares of angular distribution experiments in nuclear physics, *Phil. Mag.*, **45**, 237.

PROSCHAN, F. (1953): Rejection of outlying observations, *Am. J. Physics*, **21**, 520.

REIERSÖL, O. (1950): Identifiability of a linear relation between variables which are subject to error, *Econometrica*, **18**, 375.

RILEY, J. D. (1955): Solving systems of linear equations with a positive definite, symmetric, but possibly ill-conditioned matrix, *Math. Tables Aids Comp.*, **9**, 96.

SHERMAN, J. (1951): Mathematical tables—errata, *Math. Tables Aids Comp.*, **5**, 81.

—— and MORRISON, W. J. (1950): Adjustment of an inverse matrix corresponding to a change in one element of a given matrix, *Ann. Math. Statist.*, **21**, 124.

SMITH, H. F. (1956): Estimating a linear functional relationship (abstract only), *Ann. Math. Statist.*, **27**, 210.

TOCHER, K. D. (1952): On the concurrence of a set of regression lines, *Biometrika*, **39**, 109.

TRICKETT, W. H., and WELCH, B. L. (1954): On the comparison of two means: further discussion of iterative methods for calculating tables. *Biometrika*, **41**, 361.

UTTAM CHAND (1950): Distributions related to comparison of two means and two regression coefficients, *Ann. Math. Statist.*, **21**, 507.

VAN DER REYDEN, D. (1943): Curve fitting by the orthogonal polynomials of least squares, *Onderstepoort J. Vet. Sci. Animal Industry*, **18**, 355.

VAN IJZEREN, J. (1952): Elementary proof of independence of mean and variance of samples from a normal distribution, *Statistica Rijswijk*, **6**, 113.

VAN KLINKEN, J., and PRINS, H. J. (1954): Survey of testing and estimation methods with respect to Poisson distributions, *Math. Centrum Amsterdam Statist. Afdeling*, Rep. S 133.

WALD, A. (1940): The fitting of straight lines if both variables are subject to error, *Ann. Math. Statist.*, **11**, 284.

WALSH, J. E. (1954): Analytic tests and confidence intervals for the mean value, probabilities, and percentage points of a Poisson distribution, *Sankhya*, **14**, 25.

WATSON, G. S. (1955): Serial correlation in regression analysis I, *Biometrika*, **42**, 327; II (with E. J. Hannan) (1956), **43**, 436.

WELCH, B. L. (1937): The significance of the difference between two means when the population variances are unequal, *Biometrika*, **29**, 350.

—— (1947): The generalization of 'Student's' problem when several different population variances are involved, *Biometrika*, **34**, 28.

—— (1951): On the comparison of several means: an alternative approach, *Biometrika*, **38**, 330.

WILLIAMS, E. J. (1953): Test of significance for concurrent regression lines, *Biometrika*, **40**, 297.

WISHART, J., and METAKIDES, T. (1953): Orthogonal polynomial fitting, *Biometrika*, **40**, 361.

SUPPLEMENTARY REFERENCES

2.6 NOETHER, G. E. (1955): Use of the range instead of the standard deviation, *J. Am. Statist. Assoc.*, **50**, 1040.

3.2 COX, D. R. (1958): Some problems concerned with statistical inference, *Ann. Math. Statist.*, **29**, 106. FISHER, R. A. (1955): Statistical methods and scientific induction, *J. Roy. Statist. Soc.*, **B**, **17**, 69. PEARSON, E. S. (1955): Statistical concepts and their relation to reality, *ibid.*, 204.

3.5 FISHER, R. A., and HEALY, M. J. R. (1956): New tables of Behrens' test of significance, *J. Roy. Statist. Soc.*, **B**, **18**, 212. WELCH, B. L. (1956): Note on some criticisms made by Sir Ronald Fisher, *ibid.*, 297.

3.7 GUMBEL, E. J. (1958): *Statistics of Extremes*, Columbia University Press.

4.6 GUEST, P. G. (1958): Methods for numerical calculations with the Type I counter, *Austral. J. Physics*, **11**, 143.

6.5 ACTON, F. S. (1959): *Analysis of Straight-line Data*, John Wiley and Sons, New York. BARTON, D. E., and MALLOWS, C. L. (1959): Estimation of linear and non-linear structural relations, *Nature*, **184**, 1086. SCHEFFÉ, H. (1958): Fitting straight lines when one variable is controlled, *J. Am. Statist. Assoc.*, **53**, 106.

7.2 ASCHER, M., and FORSYTHE, G. E. (1958): SWAC experiments on the use of orthogonal polynomials for data fitting, *J. Assoc. Comp. Mach.*, **5**, 9. DAVIS, P., and RABINOWITZ, P. (1956): Numerical experiments in potential theory using orthonormal functions, *J. Washington Acad. Sci.*, **46**, 12.

7.6 DELURY, D. B. (1950): *Values and Integrals of the Orthogonal Polynomials up to* $n = 26$, University of Toronto Press.

7.9 HEAD, J. W., and OULTEN, G. M. (1958): The solution of 'ill-conditioned' linear simultaneous equations, *Aircraft Eng.*, **30**, 309.

8.3 DWYER, P. S. (1958): Generalizations of a Gaussian theorem, *Ann. Math. Statist.*, **29**, 106.

8.6 HOEL, P. G. (1958): Efficiency problems in polynomial estimation, *Ann. Math. Statist.*, **29**, 1134. KIEFER, J., and WOLFOWITZ, J. (1959): Optimum design in regression problems, *ibid.*, **30**, 271.

10.2 FINNEY, D. J. (1958): The efficiencies of alternative estimators for an asymptotic regression equation, *Biometrika*, **45**, 370. PATTERSON, H. D. (1958): The use of autoregression in fitting an exponential curve, *ibid.*, 389.

10.3 Symposium on spectral approach to time series, *J. Roy. Statist. Soc.*, **B**, **19**, 1 (1957). SALZER, H. E. (1957): Formulas for calculating Fourier coefficients, *J. Maths. and Physics*, **36**, 96.

11.1 FOOTE, R. J. (1958): A modified Doolittle approach for multiple and partial correlation and regression, *J. Am. Statist. Assoc.*, **53**, 133. DURBIN, J. (1957): Testing for serial correlation in systems of simultaneous regression equations, *Biometrika*, **44**, Parts 3 and 4. HANNAN, E. J. (1957): Testing for serial correlation in least-squares regression, *Biometrika*, **44**, 57. QUENOUILLE, M. H. (1949): Approximate tests of correlation in time series, *Proc. Camb. Phil. Soc.*, **45**, Part 3. WHITE, J. S. (1957): A t-test for the serial correlation coefficient, *Ann. Math. Statist.*, **28**, Part 4. WILLIAMS, E. J. (1959): *Regression Analysis*, John Wiley and Sons, New York.

11.3 RAINSFORD, H. F. (1957): *Survey Adjustments and Least Squares*, Constable, London.

INDEX